Botrytis – the Fungus, the Pathogen and its Management in Agricultural Systems

Sabine Fillinger • Yigal Elad
Editors

Botrytis – the Fungus, the Pathogen and its Management in Agricultural Systems

Springer

Editors
Sabine Fillinger
UMR 1290 BIOGER, INRA
AgroParisTech
Thiverval-Grignon, France

Yigal Elad
Department of Plant Pathology
and Weed Research
Agricultural Research Organization,
The Volcani Center
Bet Dagan, Israel

ISBN 978-3-319-23370-3 ISBN 978-3-319-23371-0 (eBook)
DOI 10.1007/978-3-319-23371-0

Library of Congress Control Number: 2015959018

Springer Cham Heidelberg New York Dordrecht London

Printed on acid-free paper

Springer International Publishing AG Switzerland is part of Springer Science+Business Media (www.springer.com)

Preface

Botrytis, a fungal genus, is the focus of intensive scientific research worldwide. The complex interactions between this pathogen and the plants it infects (referred to collectively as the pathosystem) and the economic importance of the diseases caused by *Botrytis* on more than 1400 species of cultivated plants, many of which are important agricultural crops, render this pathogen of particular interest to farmers, agriculture experts, advisers, extension staff, students and researchers in many fields worldwide.

The idea for this book arose from the *Botrytis* symposium that took place in June 2013 near Bari, Italy. A book published on *Botrytis* in 2004 (*Botrytis*: *Biology, Pathology and Control*, Elad Y, Williamson B, Tudzynski P. Delen N (eds), Kluwer Academic Publishers (Springer)) has become the reference book on this subject. However, new aspects of the biology of *Botrytis* have since come to light, making it essential to publish a new edition.

This book is the product of intensive work by 41 authors, all of whom are leading scientists from various scientific disciplines studying *Botrytis* as a fungus and as a pathogen. The authors of this book have amassed state-of-the-art knowledge on diverse topics, including *Botrytis* epidemiology, disease management, biological and chemical control and aspects of the plant-pathogen interactome, including virulence factors and defence processes, signalling cascades, the oxidative burst and general biological aspects, such as vegetative incompatibility, mycoviruses and the revised *Botrytis* species concept. This book also provides reviews of the genetic and postgenomic analyses, such as transcriptomics and proteomics, used to study *Botrytis* biology and pathogenicity.

This 20-chapter book is a comprehensive treatise covering the rapidly developing science of *Botrytis* and reflecting the major developments in studies of this fungus.

It will serve as a source of general information for specialists in agriculture and horticulture, but also for students and scientists interested in the biology of this fascinating, multifaceted phytopathogenic fungal species.

Thiverval-Grignon, France Sabine Fillinger
Bet Dagan, Israel Yigal Elad

Acknowledgements

The editors thank Professor Melané Vivier from Stellenbosch University for her critical reading and constructive correction of Chapters 17–19. We also thank all the recognised and anonymous authors and reviewers for their contributions to this new *Botrytis* book, the production of which was an exciting and rich experience.

Contents

1 ***Botrytis*, the Good, the Bad and the Ugly** 1
Yigal Elad, Melané Vivier, and Sabine Fillinger

2 ***Botrytis*-Biology, Detection and Quantification** 17
Frances M. Dewey (Molly) and Robert Grant-Downton

3 **Genetics of *Botrytis cinerea*** 35
Rita Milvia De Miccolis Angelini, Stefania Pollastro, and Franco Faretra

4 **Vegetative Incompatibility in *Botrytis*** 55
Nadia Korolev and Yigal Elad

5 **RNA Mycoviruses and Their Role in *Botrytis* Biology** 71
Mingde Wu, Jing Zhang, Long Yang, and Guoqing Li

6 **Diversity Within and Between Species of *Botrytis*** 91
Anne-Sophie Walker

7 **Epidemiology and Aerobiology of *Botrytis* spp.** 127
Odile Carisse

8 **Cultural and Integrated Control of *Botrytis* spp.** 149
Yigal Elad

9 **Biological Control and Biopesticide Suppression of *Botrytis*-Incited Diseases** 165
Philippe C. Nicot, Alison Stewart, Marc Bardin, and Yigal Elad

10 **Chemical Control and Resistance Management of *Botrytis* Diseases** 189
Sabine Fillinger and Anne-Sophie Walker

11 **Control Strategies for Postharvest Grey Mould on Fruit Crops** ... 217
Gianfranco Romanazzi and Samir Droby

12 **Infection Process and Fungal Virulence Factors** ... 229
Celedonio González, Nélida Brito, and Amir Sharon

13 **Signal Transduction Cascades Regulating Differentiation and Virulence in *Botrytis cinerea*** ... 247
Julia Schumacher

14 **Reactive Oxygen Species in the *Botrytis* – Host Interaction** ... 269
Ulrike Siegmund and Anne Viefhues

15 **Secondary Metabolism in *Botrytis cinerea*: Combining Genomic and Metabolomic Approaches** ... 291
Isidro G. Collado and Muriel Viaud

16 **Contribution of Proteomics Research to Understanding *Botrytis* Biology and Pathogenicity** ... 315
Eva Liñeiro, Jesús Manuel Cantoral, and Francisco Javier Fernández-Acero

17 **Overview of Plant Defence Systems: Lessons from Arabidopsis-*Botrytis cinerea* Systems Biology** ... 335
Oliver Windram, Claire Stoker, and Katherine Denby

18 **Hitting the Wall: Plant Cell Walls During *Botrytis cinerea* Infections** ... 361
Barbara Blanco-Ulate, John M. Labavitch, Estefania Vincenti, Ann L.T. Powell, and Dario Cantu

19 **Ripening of Tomato Fruit and Susceptibility to *Botrytis cinerea*** ... 387
Barbara Blanco-Ulate, Estefania Vincenti, Dario Cantu, and Ann L.T. Powell

20 **Plant Hosts of *Botrytis* spp.** ... 413
Yigal Elad, Ilaria Pertot, Alba Marina Cotes Prado, and Alison Stewart

Chapter 1
Botrytis, the Good, the Bad and the Ugly

Yigal Elad, Melané Vivier, and Sabine Fillinger

Abstract *Botrytis* spp. are efficient pathogens, causing devastating diseases and significant crop losses in a wide variety of plant species. Here we outline our review of these pathogens, as well as highlight the major advances of the past 10 years in studying *Botrytis* in interaction with its hosts. Progress in molecular genetics and the development of relevant phylogenetic markers in particular, has resulted in the characterisation of approximately 30 species. The host range of *Botrytis* spp. includes plant species that are members of 170 families of cultivated plants.

Keywords Host range • Control strategies • Genomics • Fungus-plant interaction

1.1 What Makes *Botrytis* Species Such Threats?

The genus *Botrytis* is highly diverse, with numerous species identified that differ in terms of their biology, ecology, morphological features and host range. Intensive research has been carried out since the publication of the previous Botrytis book (Elad et al. 2004). Progress in molecular genetics and the development of relevant phylogenetic markers in particular, has resulted in the characterisation of approximately 30 species, of which 7 new species, a hybrid and a species complex have been identified in the last decade (Chap. 6).

Y. Elad (✉)
Department of Plant Pathology and Weed Research, Agricultural Research Organization, The Volcani Center, Bet Dagan 50250, Israel
e-mail: elady@volcani.agri.gov.il

M. Vivier
Department of Viticulture and Oenology, Institute for Wine Biotechnology, Stellenbosch University, Stellenbosch, South Africa

S. Fillinger (✉)
UMR 1290 BIOGER, INRA, AgroParisTech,
BP01, Avenue Lucien Brétignières, F-78850 Thiverval-Grignon, France
e-mail: sabine.fillinger@versailles.inra.fr

S. Fillinger, Y. Elad (eds.), *Botrytis – the Fungus, the Pathogen and its Management in Agricultural Systems*, DOI 10.1007/978-3-319-23371-0_1

Fig. 1.1 *Botrytis cinerea* on a lisianthus cotyledon (Photo: Yigal Elad)

Species of *Botrytis* are responsible for heavy losses in a number of economically important horticultural and floral crops. Most species have a limited host range attacking either monocotyledonous or dicotyledonous plants. A notable exception is *B. cinerea* with a very wide host range (Chap. 20), but *B. pseudocinerea* was found on diverse plant hosts as well and may be suspected to have also a broad host range as well (Walker et al. 2011; Plesken et al. 2015). The plant hosts of *Botrytis* spp. include species of 170 families of cultivated plants (Chap. 20).

The pathogen displays an extraordinary variability in phenotypic traits, making it a model for studying sources of variation in filamentous fungi. Moreover, evidence that some species can be present as (non-symptomatic) endophytes is increasing (Chap. 2). Epidemics of *Botrytis*-incited grey mould are common in open fields, orchards and greenhouses (Fig. 1.1). These infections are promoted by high humidity. A *Botrytis* epidemic comprises a sequence of processes, each of which is influenced by the host and the surrounding environment (Chap. 7). For most *Botrytis*-induced diseases, the pathogen disperses by large amounts of air-dispersed conidia. Hence aerobiology plays a key role in understanding the epidemiology of *Botrytis* spp., since they have developed a variety of strategies to infect and colonise their host (Chap. 7).

With regard to control, different cultural methods aiming to reduce humidity can be combined for improved disease suppression (Chap. 8), in addition to application of chemical botryticides or biocontrol treatments. Recent years have seen the development of many biological control agents and other biopesticides, such as plant extracts, minerals and organic compounds, against *Botrytis*-incited diseases (Chap. 9). Chemical control based on the application of synthetic fungicides, however still constitutes the principal means of efficient and reliable protection of many crops (Chap. 10). In addition, integrated management programs adopting a holistic approach are the key for minimizing postharvest losses caused by *B. cinerea* (Chap. 11).

1.2 Sex, Drugs and Viruses

Botrytis cinerea is an interesting model system for necrotrophic pathogens, which is not simple to deal with. Among natural strains, variations in karyotypes are frequent and some isolates of this heterothallic fungus behave as dual-maters without changes in the mating-type genes (Chap. 3). Genome plasticity and evolution may be due to mobile genetic elements, such as transposons (Biémont 2010) or inteins, these latter elements representing local hotspots for meiotic recombination in fungi (Liu and Yang 2004). The presence of the active PRP8 intein varies among *Botrytis* spp. and *B. cinerea* can display high rates of meiotic gene conversion in crosses between an isolate that possesses an intein element and an isolate that doesn't (Bokor et al. 2010, 2012). Compared to other fungal species, the *B. cinerea* genome seems to be rather poor in transposable elements (0.7–4 %). Repeat induced point mutation (RIP) of duplicated sequences constitutes another fungal meiotic gene modification mechanism involved in genome evolution (Galagan and Selker 2004) and might be responsible for reduced expansion of transposable elements in this species (Amselem et al. 2011, 2015). Crosses, gene knockouts, site-directed mutagenesis and gene-fusions are current genetic tools, which can be used in addition to functional genomics to address important biological questions (Chaps. 3, 12, 13, 14, 15, 16, and 17).

The genetic basis of vegetative incompatibility in *B. cinerea* is not known, but is presumed to conform to the system found in other ascomycetes. The large number of vegetative compatibility groups (VCGs) and limited occurrence of isolates displaying the same VCG suggest that sexual recombination occurs in field populations of *B. cinerea* (Chap. 4). Evidence of recombination and gene flow, between and within populations, has also been obtained by population genetics as indirect evidence for sexual recombination (Chap. 6). Interestingly, RNA mycoviruses have been found to be widespread in *Botrytis* and some can even attenuate pathogenicity of their fungal host (Chap. 5). The identification of mycoviruses capable of overcoming vegetative incompatibility in horizontal transmission could be useful to establish mycoviruses as a control mechanism against plant diseases caused by *Botrytis* (Chaps. 4 and 5).

1.3 Communicating with the Plant

Molecular research, including the complete genome sequencing and analyses of two *B. cinerea* strains, has led to a working model outlining the molecular strategy used by *B. cinerea* in the infection process. In Chap. 12, the significant advances made in understanding the molecular mechanisms underlying *B. cinerea* attack on susceptible hosts are highlighted. In the last decade, much new knowledge has been gained about the signalling cascades that mediate the communication between environment and the cellular machinery regulating developmental programs in *B. cinerea* (Chap. 13). Genomic analyses have been complemented by proteomics and led to the identification of new virulence factors (Chap. 16).

During host-pathogen interactions reactive oxygen species (ROS) are of key importance for plant defence, but also for fungal attack. Since *Botrytis* exploits this plant defence reaction and even contributes to the oxidative burst, the fungus needs a robust oxidative stress responsive system to cope with ROS. Chapter 14 deals with the role of ROS in *Botrytis* – host interaction and both ROS generating and detoxifying systems and their importance for pathogenicity. *B. cinerea* has the potential to produce many secondary metabolites (SMs) including botrydial and botcinic acid, two unspecific phytotoxins contributing to the necrotrophic and polyphagous lifestyle of the fungus (Chap. 15).

A key challenge to understand the pathology of *Botrytis* spp. involves unravelling host responses. Chapter 17 updates recent knowledge of the complex regulatory mechanisms and multiple downstream defence processes achieved by high-throughput 'omics technologies'. Chapter 18 focuses on the cell walls of host plant tissues during infections by *Botrytis* and on fungal enzymes that could modify plant cell walls as a consequence of infection. Chapter 19 describes how fruit ripening gradually reduces host tissue resistance, and how *Botrytis* modifies its infection strategy accordingly.

1.4 The Good Side of *Botrytis*

Botrytis cinerea not only provokes damaging plant diseases but it is also at the origin of noble rot used for sweet dessert wines (e.g. Sauternes). The incidence of noble rot on the grape berries can be extremely variable according to weather conditions. The grapes are dehydrated, concentrating the juice, and the original composition of botrytized grapes is not only due to the *B. cinerea* metabolism, but also to desiccation, known as *passerillage* (Magyar 2011).

To increase the ability to control the noble rot development, artificial induction of *B. cinerea* strains in harvested grapes has been tested since the middle of the twentieth century and incited researchers to select *B. cinerea* strains with higher noble rot efficacy (Azzolini et al. 2013). However, it was shown recently by population genetics that *B. cinerea* populations isolated from grey mould and noble rot are genetically not different (Fournier et al. 2013). Noble rot symptoms therefore do not seem to be caused by a specific *B. cinerea* population, but instead seem to depend essentially on microclimatic conditions, which infer that continuing research on climate conditions that promote the production of sweet wines is still necessary.

1.5 What Has Changed Since the Last *Botrytis* Book?

1.5.1 One Fungus: One Name

Pleomorphism is encountered in many of the most important ascomycete plant pathogens such as the different morphological forms and different sexual stages. A dual system of fungal nomenclature was proposed in 1905, recommending both asexual and sexual names for fungi. In that sense, *Botrytis cinerea* is the name of the asexual stage (anamorph) and *Botryotinia fuckeliana* that of the sexual stage (teleomorph) (Faretra et al. 1988). DNA sequence comparisons have made it possible to reliably connect asexual states of fungi to their sexual states. A community effort of the mycologists was launched in 2011 to simplify the taxonomy and nomenclature of fungi, known as the 'one fungus – one name' initiative (reviewed in Wingfield et al. 2012). The *Botrytis* community agreed in 2013 at the Botrytis-Symposium in Bari to use *Botrytis cinerea* as generic name. For all other species of the same genus, the species prefix *Botrytis* is adopted.

1.5.2 Agreement on a Common Gene Nomenclature

The completion of the *B. cinerea* genome sequences and the automatic annotation of more than 10.000 genes have increased the frequency of newly identified genes, albeit still different formats of *Botrytis* gene and protein names exist in literature. The *Botrytis* community involved in the genome project agreed at the 'Botrytis-Sclerotinia genome workshop' in Versailles in 2006 on a common gene nomenclature that is summarized in Table 1.1. Gene names should be written in a three-letters-one-digit code. A prefix "*Bc*" can be used for *B. cinerea* genes if the species is ambiguous. Genes and alleles are to be written in lower case italics (*abc1*), while the unified format of protein names is standard characters with the first letter in upper case (Abc1). The mutant alleles are in superscript (numbers, known mutation or phenotype abbreviation) following the gene name, ex: $abc1^{T324S}$, knockouts: Δ*abc1*. Genotype symbols are italicised, phenotype symbols are not.

Table 1.1 Proposed nomenclature for *Botrytis* genes and proteins

Gene	Protein	Allele	Comment
Bcpde1	BcPde1	Δ*Bcpde1*	Gene name preceded by *B. cinerea*'s initials; the mutant allele is the gene deletion
bos1	Bos1	$bos1^{I365S}$	The amino-acid replacement of the mutant allele is indicated in superscript
cyp684	Cyp684	$cyp684^{HydR2}$	The phenotype associated to the mutant allele is indicated in superscript

The following exceptions may apply: (i) for gene families which have ubiquitous nomenclature (e.g. ABC-transporters; extracellular polysaccharide degrading enzymes, cytochrome P540s, etc.) this general nomenclature should be adapted accordingly; (ii) gene names published before 2015 shall not be renamed. This convention may be used for publication purposes.

1.5.3 **Botrytis** *Infection: Which Way to Choose?*

Botrytis spp. have long been considered exclusively as necrotrophic pathogens, but evidence is accumulating that some *Botrytis* species can be present inside plant tissues without triggering disease symptoms and should therefore be considered as endophytes. These infections may in some cases induce symptoms at a later stage or be transferred by seed propagation. Numerous dicotyledonous plants have been found to harbour endophytic *B. cinerea* infestation, although the full implications of this type of interaction remain unknown (reviewed in Chap. 2 and Van Kan et al. 2014). The newly identified species *B. deweyae* is suspected to have emerged as pathogen from endophytic origin (Grant-Downton et al. 2014). The capacity of endophytic development before switching to disease (necrotrophy) makes the infection cycle more complex and renders *Botrytis* disease management even more challenging. Identifying the factors that determine the choices in lifestyles of the fungus and understanding the underlying regulatory patterns constitute interesting future research *foci*.

1.5.4 The Host Range of **Botrytis**

This book comprises the most complete inventory of *Botrytis* diseases ever reported, increasing the reported number of hosts for *B. cinerea* from over 200 (Jarvis 1977) to more than 1,400 (Chap. 20). Fungi belonging to the *Botrytis* genus infect all plant parts including seeds, and other planting material, seedlings, stems, leaves, flowers and fruits at pre-harvest and postharvest stages. It is now clear that 596 plant genera are known to be affected by *Botrytis* spp. and 586 genera are affected by *B. cinerea* (Chap. 20). Of the 596 genera, the majority belongs to the division of seed plants, few (15) belong to the division of flowerless plants, and only one to the division of spore-bearing vascular plants. *Botrytis* host plants include species that grow in a variety of climate regions spanning from the tropics to cold regions, in humid as well as in dry places, in open fields, in greenhouses, in closed environments and even during cold storage. Host plants affected by *Botrytis* spp. are native to most continents, i.e., the Americas, Africa, Europe, Asia, Australia and various islands (Chap. 20).

1.5.5 Light Regulation

In the past it was demonstrated that *B. cinerea* responds to near-UV, blue, red and far-red light. A "two-receptor-model" was postulated in which near-UV/ blue- and red/ far-red-reversible photoreceptors are closely interacting to regulate asexual reproduction (Epton and Richmond 1980). Later it was demonstrated that *B. cinerea* possesses near-UV-sensing cryptochromes (BcCry1, BcCry2), potential blue light sensors (BcWcl1, BcVvd1, BcLov3, BcRgs1), opsins (Bop1, Bop2) as well as red/ far-red light-sensing phytochromes (BcPhy1-3) (Schumacher and Tudzynski 2012). Exposure of a *B. cinerea* isolate to white light resulted in induction of expression of 249 light-induced genes, among which genes involved in photoperception, oxidative stress response and transcription (Schumacher et al. 2014). Chapter 13 gives an overview about light-regulation in *B. cinerea*.

1.5.6 Secondary Metabolites

Sequencing and annotation of the complete *B. cinerea* genome (Amselem et al. 2011) revealed more than 40 clusters of genes dedicated to the synthesis of polyketides, terpenes, non-ribosomal peptides and alkaloids which indicates that *B. cinerea* has the potential to produce many metabolites that have not been described so far (Chap. 15). It is suggested that toxins and secondary metabolites (SMs) play a significant role in plant tissue colonization and some of these metabolites seem to be crucial for necrotrophic behavior (Chap. 13). Interestingly, the *Velvet* gene complex was shown to link light-dependent development and secondary metabolism in *B. cinerea* (Schumacher et al. 2012). Many *B. cinerea* SMs are detailed in Chap. 13. However, besides botcinic acid and botrydial that were shown to be required for necrotrophy, the biological role of several other SMs needs to be investigated. The role of many SMs was not previously linked to disease development in *Botrytis* and still awaits elucidation (Chap. 13). Findings so far are just a tip of the iceberg of necrotrophic fungal effectors.

1.5.7 *The Plant-*Botrytis *Interactome*

The paradigm of *B. cinerea* infection has been revisited during the past years. It is now clear that the pathogenic development of *B. cinerea* is far more complex than previously appreciated. Disease progression is tightly regulated during the entire infection process, and the fungus makes developmental adaptations that match the different infection stages. *Botrytis* infections are "dynamic events" that involve significant crosstalk and even modulation of the reaction of the host by the pathogen. An exciting example of such modulation is the uncovering of the role of

B. cinerea small interfering RNAs (Bc-siRNAs) as effector molecules. Protein effectors are known to be delivered into host cells by plant and animal pathogens, but the delivery of Bc-siRNAs into plant hosts provides an example of active trans-kingdom movement of siRNAs (Knip et al. 2014). siRNAs could be exchanged between cells of an organism, typically leading to gene silencing events in cells and tissues. Moreover, pathogen genes can be silenced by plant-produced siRNAs through RNA interference (RNAi) when these specific genes are targeted through a process known as Host-Induced Genes Silencing (HIGS). Interestingly, Bc-siRNAs have been shown to utilise the RNAi machinery of the host cell to target and silence defense genes in the hosts. The pathogen thus hijacks an inherent host defence component to enhance its own pathogenicity (Weiberg et al. 2013; Knip et al. 2014). Both plant (host) machinery and pathogen factors are essential to this sophisticated mechanism of modulating the host response (Weiberg et al. 2013).

Other equally subtle examples of the crosstalk between pathogens and hosts include the very interesting observation that *Botrytis* infection can enhance ripening in unripe tomatoes (Chap. 19). On the other hand, ripeness level in tomatoes also influences the success of infections: green tomatoes are resistant to *Botrytis*, whereas ripening tomatoes become susceptible (Fig. 1.2). Therefore, the inherent developmental processes and ripening-specific changes in organs encountered by *Botrytis* cause subtle changes to these "substrates" that either make them easy to infect or perhaps more resistant. The host cell walls are particularly important in this scenario, since these are the barriers that *Botrytis* needs to breach to gain access to the

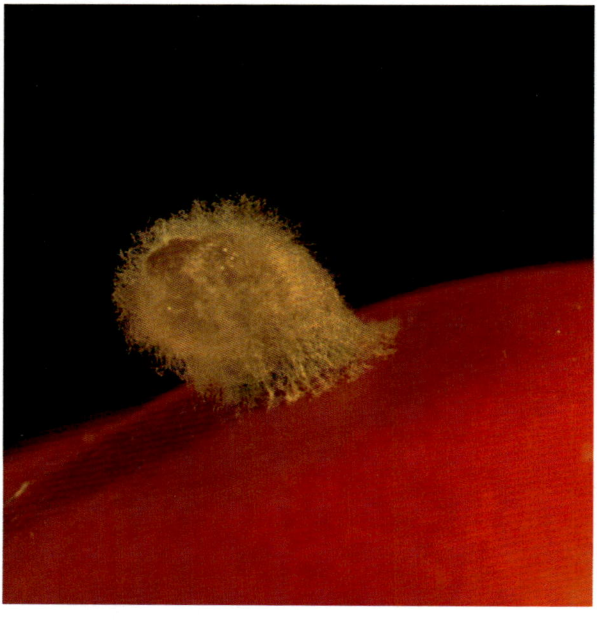

Fig. 1.2 ***Botrytis cinerea* infection of ripe tomato fruit**. Close-up photo of *B. cinerea* growing on a droplet of juice released by a wounded ripe tomato (Photo by Dario Cantu; reproduced with permission from PNAS (2008) 105: 827; Copyright National Academy of Sciences, U.S.A.)

plant's sub-tissue to utilise its resources. Genome-wide transcriptional profiling of *B. cinerea* infecting different hosts has led to important insights regarding genes targeting plant cell walls to facilitate infection (Blanco-Ulate et al. 2014). A large number (275) of putatively secreted Carbohydrate-Active enZymes (CAZymes) were annotated in the *B. cinerea* genome and it is tempting to speculate that the diversity of *Botrytis* CAZymes could contribute to its wide host range (Blanco-Ulate et al. 2014). This study also shows that *Botrytis* probably modulates its enzymatic arsenal depending on the composition of the specific host cell walls, clearly showing the ability to adapt its infection strategy as per host encountered.

These examples highlight the importance of studying the infection process as a dynamic interaction. Indeed, this is not a new idea; it has been acknowledged and appreciated for a while, but implementing strategies to achieve this was particularly challenging from an experimental design and technology point of view. The availability of sequenced genomes of *Botrytis* species, as well as an increasing availability of sequenced plant genomes provide essential baseline information to support experiments where the interactome of pathogens and hosts can be studied. The importance to "follow" an infection cycle with high-resolution using systems-biology was recently shown in a study of the *Botrytis* transcriptome in interaction with *Arabidopsis* and the evaluation of the data using network-modelling approaches (Windram et al. 2012).

1.5.8 Control

Although chemical control of *Botrytis* still largely relies on the use of chemical pesticides, suppression of diseases can significantly benefit from cultural management of crops mainly by restricting the humidity around the plant and by limiting plant surface wetness. Plant and microbial extracts, organic compounds, and biocontrol agents have become available in some countries for *Botrytis* suppression either for use at pre- or postharvest. In the last few years, the definition of biocontrol has been broadened to include a range of alternative means of control (Chap. 9). Biocontrol relies on various modes of action including induced resistance (IR) i.e., application of natural compounds or their derivatives that are able to upregulate plant genes contributing to systemic resistance. A major advancement in this field is the understanding of the plant responses to the resistance inducers. For instance, *Trichoderma* that induces systemic resistance against *B. cinerea* primes salicylic acid (SA) and ethylene (ET)-related genes in strawberry (Meller Harel et al. 2014). Biocontrol induced by biochar primes defence genes in ET and jasmonic acid (JA)-related pathways in tomato, but biochar-mediated IR in the *B. cinerea*-tomato pathosystem exclusively involves the JA pathway (Mehari et al. 2015). More examples regarding the advancement in IR research are given in Chap. 9.

New developments based on the combination of biocontrol agents, physical means, natural antimicrobials, and decontaminating agents have also been achieved with post harvest *Botrytis* management (Chap. 11). These alternative methods are effective in small-scale experiments, as standalone treatments, whereas commercial applications are less effective. It is recommended to combine alternative methods at postharvest (Romanazzi et al. 2012; Mlikota Gabler et al. 2010) along with pre-harvest disease and inoculum suppression.

During the last decades, restriction in fungicide application have become necessary to reduce their impact on the environment (Fenner et al. 2013) and to limit fungicide residues on harvest (Verger and Boobis 2013), requiring optimized protection strategies. At the same time, acquired resistance to most botryticides arose in many agronomical situations, sometimes impeding fungicide efficacy and leading to additional sprays. Not only are the frequently found target-site modifications responsible for *Botrytis*' strong capacity to cope with most fungicides, but active fungicide efflux leading to multiple drug resistance (MDR) is increasingly observed in *B. cinerea* (Kretschmer et al. 2009; Leroch et al. 2013). Chapter 10 gives a detailed overview of the current state of knowledge of fungicide resistance to end up with strategy proposals combining cropping practices for optimal resistance management.

1.6 New Knowledge, New Tools

1.6.1 Botrytis *Genomic Resources*

One of the prominent changes since the previous *Botrytis* book (Elad et al. 2004) is the determination of the genome sequences of three *B. cinerea* strains, as alluded to in many of the book chapters. The first sequencing initiative was launched already in the late 1990s by Syngenta, but the sequence of strain B05.10 was only published at the same time as the sequence of strain T4, resulting from an international community effort (Amselem et al. 2011). Thanks to the improvement of sequencing technologies and the extreme reduction of sequencing costs, improved genome sequences and gene annotations of both strains were published only 1 year later (Staats and Van Kan 2012). The genomes and annotations can be accessed and searched at the website of URGI.[1]

A new, near-complete assembly of the B05.10 genome, based on PacBio sequence data and confirmed by a linkage map and optical map can be accessed through the EBI webserver[2] (J. van Kan, *pers. comm.*). The assembly consists of 18 chromosomes (including two mini-chromosomes of 208 and 247 kb, respectively) without any internal gaps, of which 10 are full length from telomere to telomere. In addition, the genome sequence of the noble rot isolate BcWD1 (Blanco-Ulate et al. 2013) is accessible in public databases.[3] The three strains differ by roughly 162,000 SNPs (≈4 SNPs/kb).

[1] http://urgi.versailles.inra.fr/Species/Botrytis

[2] http://fungi.ensembl.org/Botrytis_cinerea/Info/Index

[3] DBJ/EMBL/GenBank accession no.AORW00000000.

RNA sequencing considerably improved the quality of genome annotation including gene prediction for *B. cinerea*, but also constitutes the current technology for global expression studies (see below). 11,701 genes have been predicted with substantial cDNA support for the last version of the B05.10 genome, while there were 16,448 predicted genes in the first genome version. The noble rot strain does not differ significantly either in genome size or in predicted gene number (11,073). Whether differences in secreted proteins or CAZymes (Blanco-Ulate et al. 2013) constitute a general feature of noble rot strains, remains to be investigated. A community effort has been organized with support from EBI in the framework of the Ensembl Fungi platform for manual annotation of all genes of the *B. cinerea* B05.10 strain.

1.6.2 Functional Genomics

The establishment of the *B. cinerea* genome sequences paved the way to "post-genomic" analyses, such as transcriptomics[4] (Chap. 17), proteomics and phosphoproteomics (Chap. 16), which allowed the identification of new virulence factors (e.g., Frias et al. 2011). In addition, high-throughput functional genomics, such as random mutagenesis (insertion-library) or site-directed mutagenesis with optimized tools (Schumacher 2012, Fig. 1.3) have become possible, but also the identification of transcription-factor binding sites through a yeast-one hybrid approach (Simon et al. 2013). All published gene knockout mutants for *B. cinerea* can be retrieved on

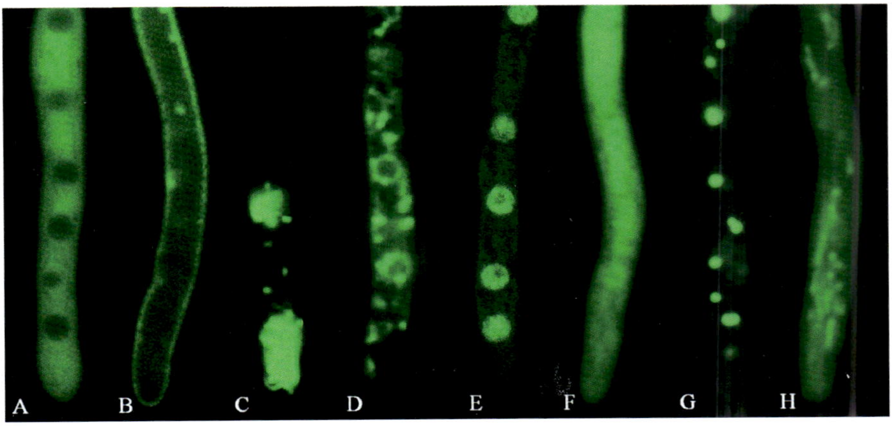

Fig. 1.3 GFP-tagged *Botrytis cinerea* hyphae. The observed green fluorescence is due to the following localisations: (**a**) cytosolic (not nuclear); (**b**) membrane; (**c**) hyphal tip and septum (F-actin); (**d**) endoplasmic reticulum; (**e**) nuclei; (**f**) cytosolic and nuclear; (**g**) peroxisomes; (**h**) mitochondria (Microscopic images kindly provided by R. Marschall and J. Schumacher)

[4] http://www.ncbi.nlm.nih.gov/gds/?term=Botrytis+cinerea[Organism]

a dedicated web site.[5] Genetic and genomic tools are important resources to get deeper insights into the fungal biology, to identify pathogenicity factors and new fungicide targets. They also accelerate the identification of naturally occurring mutations such as those involved in fungicide resistance.

1.6.3 Genome Perspectives

The genomes of six additional *B. cinerea* strains and of 11 different *Botrytis* species (*B. calthae*, *B. convoluta*, *B. elliptica*, *B. galanthina*, *B. hyacinthi*, *B. paeoniae*, *B. porri*, *B. narcissicola*, *B. tulipae*, *B. fabae*, *B. pseudocinerea*) have been sequenced (Staats and Van Kan, unpublished data; M. Hahn, *pers. comm.*). Large scale intra- and inter-species comparison will certainly help to understand some aspects of genome plasticity in *Botrytis*, such as variation in karyotypes (Chap. 3) or infection development (endophytic versus necrotrophic behaviour; grey mould versus noble rot), but also factors involved in the adaptation to the biotic and abiotic environment (plant hosts, fungicides, microclimate conditions). Interspecies comparison may allow to trace the evolutionary history of dispensable genomic elements, e.g., the bikaverin gene cluster (Campbell et al. 2013, Chap. 15) and facilitate our understanding of host adaptation.

1.7 Conclusions

Fungi of the genus *Botrytis* are important pathogens of many economically important crops, such as grapevine and many other crops, at pre- and post-harvest stages (Chap. 20). Intensive research has led to accumulation of important information on *Botrytis* spp. and their establishment in host tissues. The recent advances in *Botrytis* research surely will all have a major impact on disease prevention and control. However, further research efforts are needed on these fungi and their interactions with plants. For instance, it has been clear for a very long time that *Botrytis* spp., particularly *B. cinerea*, are necrotrophs, until recent studies have demonstrated that some species can act as endophytes in plants. The study of the endophtyic processes, the host reactions to the endophytic phase, as well as the switch from an endophytic to an aggressive stage needs to be prioritised to improve *Botrytis* management.

There is a significant importance in genetic analysis of *Botrytis* (Chap. 3). The large number of vegetative compatibility groups (VCGs), the limited occurrence of isolates displaying the same VCG (Chap. 4) and population genetics (Chap. 6) indicate that sexual recombination occurs in *B. cinerea* populations. Despite this, only limited reports of actual apothecia present in natural field settings are available. This calls for further research into the role of the sexual stage in shaping the variability of *Botrytis* spp. There is evidence for a high proportion of *B. cinerea* isolates

[5] http://botbioger.versailles.inra.fr/botmut/

containing mycoviruses (Chap. 5). In view of the vegetative incompatibility abundance in *Botrytis* forming barriers for movement of mycoviruses in fungal populations, it is crucial to discover and to characterize novel mycoviral species in *Botrytis*, but also to test their potential in biocontrol of the pathogen. Furthermore, genetical and phenotypical variation in *Botrytis* populations is immense (Chaps. 3 and 6).

Improved knowledge of the components of variability and of epidemiological (Chap. 7) behaviour like spatial distribution of disease and inoculum has been achieved and needs to be pursued. This may also help to reduce *Botrytis*-incited losses to a level that is acceptable at pre- or post-harvest stages, which still pose a great challenge for producers and handlers. Different cultural (Chap. 8), chemical (Chap. 10) and biological (Chap. 9) methods in the field and various compounds and methods after harvest (Chap. 11) can be used for improved disease suppression. Yet, the combination of all types of disease management in holistic systems needs further attention for effective *Botrytis* suppression along with strategies to minimize the development of resistance to chemical botryticides and of chemical fungicides residues in the produce. Such combination of control methods needs to take advantage of the right fungicide and of epidemiological considerations for the achievement of long lasting efficient disease suppression systems.

Studying the infection process still holds potential for further understanding of the plant pathogen interaction (Chap. 12). Future discoveries of the fungal factors, the identification of their plant targets and response processes are promising for improved plant defense tools. Furthermore, the identification of signal transduction pathways in the pathogen, those recognizing the plant host or the fungal specific light signaling pathways can be a future gateway to understand processes that govern the pathogen (Chap. 13). Oxidative burst is part of the successful infection of *B. cinerea* (Chap. 14). However, ROS scavenging and detoxification during infection is not clear and needs thorough clarification in order to understand how the fungus produces and copes with large amounts of ROS. Part of the infection complex involves production of SMs by the pathogen (Chap. 15). There is much awaiting elucidation as for the mode of action and the actual role of SM in the *Botrytis* infection process; nevertheless, their activities toward plants could have interesting toxic activities and requires further research.

Acknowledgements We are grateful to Adeline Simon, Matthias Hahn, and Jan van Kan for critical reading of the manuscript and for unpublished data. We thank Dario Cantu, Robert Marschall, and Julia Schumacher for providing us with macroscopic and microscopic illustrations.

References

Amselem J, Cuomo CA, Van Kan JAL et al (2011) Genomic analysis of the necrotrophic fungal pathogens *Sclerotinia sclerotiorum* and *Botrytis cinerea*. PLoS Genet 7(8):e1002230

Amselem J, Lebrun MH, Quesneville H (2015) Whole genome comparative analysis of transposable elements provides new insight into mechanisms of their inactivation in fungal genomes. BMC Genomics 16:141

Azzolini M, Tosi E, Faccio S et al (2013) Selection of *Botrytis cinerea* and *Saccharomyces cerevisiae* strains for the improvement and valorization of Italian passito style wines. FEMS Yeast Res 13:540–552

Biémont C (2010) A brief history of the status of transposable elements: from junk DNA to major players in evolution. Genetics 186:1085–1093

Blanco-Ulate B, Allen G, Powell AL et al (2013) Draft genome sequence of *Botrytis cinerea* BcDW1, inoculum for noble rot of grape berries. Genome Announc 1:e00252-13

Blanco-Ulate B, Morales-Cruz A, Amrine KC, et al (2014) Genome-wide transcriptional profiling of Botrytis cinerea genes targeting plant cell walls during infections of different hosts. Front Plant Sci 5:435. doi:10.3389/fpls.2014.00435

Bokor AAM, Van Kan JAL, Poulter RTM (2010) Sexual mating of *Botrytis cinerea* illustrates PRP8 intein HEG activity. Fungal Genet Biol 47:392–398

Bokor AAM, Kohn LM, Poulter RTM et al (2012) PRP8 inteins in species of the genus *Botrytis* and other ascomycetes. Fungal Genet Biol 49:250–261

Campbell MA, Staats M, Van Kan JA et al (2013) Repeated loss of an anciently horizontally transferred gene cluster in *Botrytis*. Mycologia 105:1126–1134

Elad Y, Williamson B, Tudzynski P, Delen N (eds) (2004) *Botrytis*: biology, pathology and control. Kluwer Academic Publishers (Springer), Dordrecht, 416 pp

Epton HAS, Richmond DV (1980) Formation, structure and germination of conidia. In: Coley-Smith JR, Verhoeff K, Jarvis WR (eds) The biology of *Botrytis*. Academic, London, pp 41–83

Faretra F, Antonacci E, Pollastro S (1988) Sexual behaviour and mating system of *Botryotinia fuckeliana*, teleomorph of *Botrytis cinerea*. J Gen Microbiol 134:2543–2550

Fenner K, Canonica S, Wackett LP et al (2013) Evaluating pesticide degradation in the environment: blind spots and emerging opportunities. Science 341:752–758

Fournier E, Gladieux P, Giraud T (2013) The 'Dr Jekyll and Mr Hyde fungus': noble rot versus gray mold symptoms of *Botrytis cinerea* on grapes. Evol Appl 6:960–969

Frias M, Gonzalez C, Brito N (2011) BcSpl1, a cerato-platanin family protein, contributes to *Botrytis cinerea* virulence and elicits the hypersensitive response in the host. New Phytol 192:483–495

Galagan JE, Selker EU (2004) RIP: the evolutionary cost of genome defense. Trends Genet 20:417–423

Grant-Downton RT, Terhem RB, Kapralov MV et al (2014) A novel *Botrytis* species is associated with a newly emergent foliar disease in cultivated hemerocallis. PLoS ONE 9(6):e89272

Jarvis WR (1977) *Botrytinia* and *Botrytis* species: taxonomy, physiology, and pathogenicity, a guide to the literature, Monograph No. 15. Canada Department of Agriculture, Ottawa

Knip M, Constantin ME, Thordal-Christensen H (2014) Trans-kingdom cross-talk: small RNAs on the move. PLoS Genet 10(9):e1004602

Kretschmer M, Leroch M, Mosbach A et al (2009) Fungicide-driven evolution and molecular basis of multidrug resistance in field populations of the grey mould fungus *Botrytis cinerea*. PLoS Pathog 5(12):e1000696

Leroch M, Plesken C, Weber RW et al (2013) Gray mold populations in german strawberry fields are resistant to multiple fungicides and dominated by a novel clade closely related to Botrytis cinerea. Appl Environ Microbiol 79:159–167

Liu X-Q, Yang J (2004) Prp8 intein in fungal pathogens: target for potential antifungal drugs. FEBS Lett 572:46–50

Magyar I (2011) Botrytized wines. In: Ronald SJ (ed) Advances in food and nutrition research, vol 63. Academic, San Diego, pp 147–206

Mehari ZH, Elad Y, Rav-David D et al (2015) Induced systemic resistance in tomato (*Solanum lycopersicum*) against *Botrytis cinerea* by biochar amendment involves jasmonic acid signaling. Plant Soil. doi:10.1007/s11104-015-2445-1

Meller Harel Y, Haile Mehari Z, Rav-David D et al (2014) Induced systemic resistance against gray mold in tomato (*Solanum lycopersicum*) by benzothiadiazole and *Trichoderma harzianum* T39. Phytopathology 104:150–157

Mlikota Gabler F, Mercier J, Jiménez JI et al (2010) Integration of continuous biofumigation with *Muscodor albus* with pre-cooling fumigation with ozone or sulfur dioxide to control postharvest gray mold of table grapes. Postharvest Biol Technol 55:78–84

Plesken C, Weber RW, Rupp S, et al. (2015) Botrytis pseudocinerea is a significant pathogen of several crop plants but susceptible to displacement by fungicide-resistant B. cinerea strains. Appl Environ Microbiol. doi:10.1128/AEM.01719-15

Romanazzi G, Lichter A, Mlikota Gabler F et al (2012) Natural and safe alternatives to conventional methods to control postharvest gray mold of table grapes. Postharvest Biol Technol 63:141–147

Schumacher J (2012) Tools for *Botrytis cinerea*: new expression vectors make the gray mold fungus more accessible to cell biology approaches. Fungal Genet Biol 49:483–497

Schumacher J, Tudzynski P (2012) Morphogenesis and infection in *Botrytis cinerea*. In: Pérez-Martín J, Di Pietro A (eds) Topics in current genetics, vol. morphogenesis and pathogenicity in fungi. Springer, Berlin, pp 225–241

Schumacher J, Pradier JM, Simon A, et al. (2012) Natural variation in the VELVET gene bcvel1 affects virulence and light-dependent differentiation in Botrytis cinerea. PLoS One 7(10):e47840. doi:10.1371/journal.pone.0047840

Schumacher J, Simon A, Cohrs KC et al (2014) The transcription factor BcLTF1 regulates virulence and light responses in the necrotrophic plant pathogen *Botrytis cinerea*. PLoS Genet 10:e1004040

Simon A, Dalmais B, Morgant G et al (2013) Screening of a *Botrytis cinerea* one-hybrid library reveals a Cys2His2 transcription factor involved in the regulation of secondary metabolism gene clusters. Fungal Genet Biol 52:9–19

Staats M, Van Kan JAL (2012) Genome update of *Botrytis cinerea* strains B05.10 and T4. Eukaryot Cell 11:1413–1414

Van Kan JA, Shaw MW, Grant-Downton RT (2014) *Botrytis* species: relentless necrotrophic thugs or endophytes gone rogue? Mol Plant Pathol 15:957–961

Verger PJ, Boobis AR (2013) Global food supply. Reevaluate pesticides for food security and safety. Science 341:717–718

Walker A-S, Gautier A, Confais J et al (2011) *Botrytis pseudocinerea*, a new cryptic species causing gray mold in French vineyards in sympatry with *Botrytis cinerea*. Phytopathology 101:1433–1445

Weiberg A, Wang M, Lin FM et al (2013) Fungal small RNAs suppress plant immunity by hijacking host RNA interference pathways. Science 342:118–123

Wingfield MJ, De Beer ZW, Slippers B et al (2012) One fungus, one name promotes progressive plant pathology. Mol Plant Pathol 13:604–613

Windram O, Madhou P, McHattie S, et al (2012) Arabidopsis Defense against Botrytis cinerea: chronology and regulation deciphered by high-resolution temporal transcriptomic analysis. Plant Cell 24:3530–3557

Chapter 2
Botrytis-Biology, Detection and Quantification

Frances M. Dewey (Molly) and Robert Grant-Downton

Abstract Species of *Botrytis* are responsible for heavy losses in a number of economically important horticultural and floral crops, the most important being *Botrytis cinerea*. Most species, except *B. cinerea*, have a limited host range attacking either monocotyledonous or dicotyledonous plants. Within the genus there are about 28 well-described species but new species continue to be isolated. Most species are opportunist growing as saprophytes on dead and decaying matter but have the ability to become aggressive pathogens under environmental conditions adverse to their hosts. Evidence that some species can be present as endophtyes (non-symptomatic) is increasing, as is evidence that such infections may either become aggressive at a later stage, notably at flowering time, or be transferred non-symptomatically by clonal or seed propagation of the host. Both the sexual and asexual stages are known for *B. cinerea*. The common method of dispersal of nearly all species is the production of asexual spores (macroconidia, common name conidia) dispersed by wind or water. Survival from one season to the next is generally by the production of sclerotia. Infections are most easily recognized by the appearance of characteristic grey conidial clusters on the surfaces of infected material but early detection, pre-conidiation or in non-symptomatically infected material is difficult; commonly surface sterilization or freezing of material followed by plating out on selective media is used. Detection at the species level requires molecular methods with species-specific probes. Detection and quantification, at the genus level, in extracts from infected tissues, juice and wines, is relatively easy using commercially produced rapid Lateral Flow immunological devices.

Keywords Botrytis species • Conidia • PCR • Monoclonal antibodies • ELISA

F.M.D. (Molly) (✉)
Department of Plant Sciences, University of Oxford, South Parks Rd, Oxford OX1 3RB, UK
e-mail: molly@fmdewey.com

R. Grant-Downton
St Peters College, University of Oxford, New Inn Hall Street, Oxford OX1 2DL, UK

S. Fillinger, Y. Elad (eds.), *Botrytis – the Fungus, the Pathogen and its Management in Agricultural Systems*, DOI 10.1007/978-3-319-23371-0_2

2.1 Introduction

The genus *Botrytis* comprises about 28 well-described species (Beever and Weeds 2004). Each year new species continue to be identified such as *B. caroliniana*, pathogenic on blackberries (Li et al. 2012), *B. fabiopsis* on broad beans in Central China (Zhang et al. 2010a), *B. deweyae* on daylilies (*Hemerocallis*) (Grant-Downton et al. 2014) and *B. sinoallii* on *Allium* crops in China (Zhang et al. 2010b; Chap. 6). Most species, with the exception of *B. cinerea*, have a limited host range attacking only a few species of either monocotyledonous plants or dicotyledonous plants while some, such as the snowdrop fungus, *Botrytis galanthina*, appear to be host-specific (Beever and Weeds 2004). Undoubtedly, the most common and the most important species of *Botrytis* is *B. cinerea*; it is responsible for considerable losses in crops, notably grape vines, and post harvest spoilage of many fruits (Droby and Lichter 2004). *B. cinerea* can infect over 200 species of plants both monocots and dicots, gymnosperms, pteridophytes, bryophytes and macroalgae (Anderson 1924; Beever and Weeds 2004; Capieau et al. 2004; Choquer et al. 2007; Mirzaei et al. 2007; Mittal et al. 1987; Chap. 20). Other species of *Botrytis* that are known to cause significant losses in horticulture, agriculture and floriculture are listed in Table 2.1.

In the wine industry *Botrytis*-infections in grape berries used to make table wines are undesirable because the fungus is thought to be responsible for 'off' odours and

Table 2.1 A list of *Botrytis* species with importance in agriculture and horticulture

Name	Date of description	Major plant host genus/genera in agriculture and horticulture
Botrytis aclada	1850	*Allium*
Botrytis allii	1917	*Allium*
Botrytis byssoidea	1925	*Allium*
Botrytis caroliniana	2012	*Rubus*, *Fragaria*
Botrytis cinerea	1794	Multiple host genera
Botrytis convoluta	1932	*Iris*
Botrytis deweyae	2014	*Hemerocallis*
Botrytis elliptica	1881	*Lilium*
Botrytis fabae	1929	*Vicia*
Botrytis fabiopsis	2010	*Vicia*
Botrytis gladiolorum	1941	*Gladiolus*
Botrytis globosa	1938	*Allium*
Botrytis hyacinthi	1928	*Hyacinthus*
Botrytis narcissicola	1906	*Narcissus*
Botrytis paeoniae	1897	*Paeonia*
Botrytis pelargonii	1949	*Pelargonium*
Botrytis polyblastis	1926	*Narcissus*
Botrytis porri	1949	*Allium*

(continued)

Table 2.1 (continued)

Name	Date of description	Major plant host genus/genera in agriculture and horticulture
Botrytis pseudocinerea	2011	*Brassica, Fragaria, Vitis,* likely multiple others?
Botrytis sinoallii	2010	*Allium*
Botrytis sinoviticola	2014	*Vitis*
Botrytis sphaerosperma	1949	*Allium*
Botrytis squamosa	1925	*Allium*
Botrytis tulipae	1913	*Tulipa*
Botrytis sp. Group S	2013	*Fragaria, Vitis,* likely multiple others?

poor keeping qualities of the resulting wines but this may be in part due to the presence of other fungi commonly associated with *B. cinerea* infections, notably species of *Aspergillus* and *Penicillium* (Dewey and Meyer 2004; Rousseaux et al. 2014). In contrast *Botrytis* infections of grape berries late in the season are highly desirable, they are used for the production of much-prized dessert wines (Dewey et al. 2008, 2013; Sivertsen et al. 2005).

2.2 Life Cycle

2.2.1 Asexual Stage

Most species and isolates of *Botrytis* produce numerous conidia (macroconidia) that are asexual conidia borne at the tips of the branching conidiophores (Fig. 2.1). The size of the conidia is species related (Beever and Weeds 2004). In general, conidia are short-lived their survival time depending on temperature, moisture, microbial activity and exposure to sunlight (Kerssies et al. 1995; Blanco et al. 2006; Carisse et al. 2012; Nassr and Bakarat 2013).

Overwintering of *Botrytis* species is generally brought about by the production of melanized resting bodies, sclerotia, which are resistant to adverse environmental conditions (Holtz et al. 2004). Some species, notably *B. cinerea*, produce quite large sclerotia (~4 mm), whereas others for example *B. tulipae* are quite small (~0.8 mm). Under favourable conditions, such as interrupted wet periods in Spring, sclerotia will 'germinate' to produce mycelia and conidia. Sclerotia are believed to be the source of early infections at the start of the growing season (Hsiang and Chastagner 1992).

Botrytis spp. can also produce temporary resting structures known as chlamydospores that have thickened, hyaline, walls. These structures can vary considerably in size and shape; they can survive periods of drought of up to 3 months (Urbasch 1983). They are often found in ageing cultures either within hyphae or at the ends of hyphae. In favourable conditions they 'germinate' to produce hyphae or microconidia.

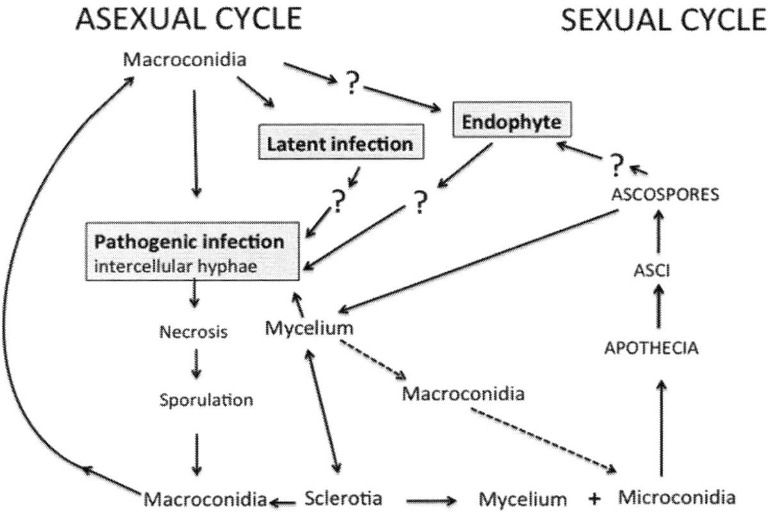

Fig. 2.1 Diagram of life cycle of *Botrytis cinerea* showing known and possible interactions between different phases of growth. Note: Macroconidia common name is conidia

2.2.2 *Sexual Stage*

The sexual stage or teleomorph of *Botrytis* spp., formerly known as *Botryotinia* spp., is not commonly seen in nature. This may in part be due to lack of recognition of the fruiting structures. Microconidia, which can be produced both from post-germination development of macroconidia and from endogenous formation within old hyphae (Brierley 1918; Fukumori et al. 2004), are uninucleate and have now been confirmed to act as spermatia (Fukumori et al. 2004). The 'fertilisation' of receptive sclerotial structures by spermatia induce the production of apothecia, the sexual structures (Urbasch 1983), where asci and ascospores are generated. Faretra and Antonacci (1987) were the first to induce the sexual stage in *B. cinerea* under laboratory conditions. The asci, containing ascospores, are borne on the upper surface of apothecia (Fig. 2.1). Sexual compatibility is controlled by a single mating type locus with two alleles, MAT-1 and MAT-2. Both mating types are widespread in nature and most isolates are heterothallic i.e. can only produce ascospores when crossed with the opposite mating type (Faretra et al. 1988).

2.3 Diversity of the Genus

At present, it remains difficult to estimate the number of species within the genus *Botrytis*. Simply, this is because numerous undescribed species are likely to exist. However, a selected list of the pathogenic taxa with importance in horticulture and agriculture are listed in Table 2.1, along with their major host genus. Morphological circumscription, primarily using features such as macroconidia size, shape and ornamentation, as well as host preference, agrees very well with molecular data (see Table 2.1).

A significant number of species might appear to be specific to certain genera (Table 2.1). However, it is now evident that such rigid host specificity may not be always the true case, especially in man-made environments. For instance, *B. elliptica* has been reported from several genera of distantly related monocots in addition to its classical host *Lilium* – for example daylily, *Hemerocallis* (Chang et al. 2001), tuberose, *Polianthes* (Horst 2013), toad lily, *Tricyrtis* (Furukawa et al. 2005) – as well as the dicot *Stephanotis* (Tompkins and Hansen 1950). Hence, the distinction between polyphagous species – typified by *B. cinerea* – and highly specialised pathogens may not be quite as clear-cut as previously thought.

New pathogenic taxa are still being discovered at a regular frequency. For example, the most recent discoveries are *B. sinoviticola* (Zhou et al. 2014) and *B. deweyae* (Grant-Downton et al. 2014). The former is a cryptic species associated with *Vitis vinifera* in China that, unlike its close relatives, *B. cinerea* and *B. pseudocinerea*, requires injury of the host to permit infection. The latter species has been identified only from cultivated *Hemerocallis* (daylily) and so far appears to be pathogenic only on this genus.

2.4 Plasticity in *Botrytis* Interactions with Plants

Most species of *Botrytis*, including *B. cinerea* only become pathogenic when growing conditions for their hosts are limiting and microclimate is suitable, for example on tomato plants in poorly ventilated, humid greenhouses, or plants suffering from physical damage such as pruning wounds and wind or through insect damage (Fermaud and Le Menn 1992). However, some species such as *B. squamosa* the cause of *Botrytis*-leaf blight of onions, (Carisse et al. 2012) and *B. elliptica* on *Lilium* can rapidly destroy whole crops (Hsieh and Huang 2001). Germination of the conidia on plant surfaces is greatly encouraged by the presence of nutrients such as simple sugars and pollen (Chou and Preece 2008).

2.4.1 *Quiescent/Latent Phases*

Various workers have shown that, although conidia of *Botrytis* spp. commonly germinate on plant tissues particularly, petals and sepals of their host, invasion of inner tissues does not always follow (Holtz et al. 2004). Such infections are often described as quiescent or latent. In such cases spread of the fungus within host tissue is arrested until later in the season when either host tissues become senescent or the sugar levels change as in ripening of grape berries or in storage (Meyer et al. 2000; Sanzani et al. 2012).

2.4.2 *Non-symptomatic Infections and Endophytic Interactions*

Evidence is increasing, that some *Botrytis* species can be present as endophtyes i.e. grow within host plant tissues without generating disease symptoms and also eliciting little or no response from the host. This was first reported from *B. cinerea* where it was evident that even systemic, endophytic colonization of host plants was possible (Barnes and Shaw 2003; Sowley et al. 2010). Such infections may either become aggressive at a later stage, notably at flowering time, or in storage or, be transferred non-symptomatically by clonal or seed propagation of the host (Barnes and Shaw 2003). Endophytic colonization of plant tissues by *B. cinerea* is now known from a range of unrelated dicot plants (M. Shaw *pers. comm.*). Remarkably, there is evidence that *B. cinerea* may also be able to interact with tissues of the seaweed *Fucus serratus* in this manner (Zuccaro et al. 2008). The extent of endophytic interactions between *Botrytis* and various taxonomic groups may well have been underestimated. A number of undescribed endophytic *Botrytis* have been isolated from pteridophyte, gymnosperm and angiosperm hosts (Shipunov et al. 2008; Zhao et al. 2010; Zhou et al. 2010). The full extent of *Botrytis* diversity as endophytic fungi remains unknown. Indeed, the recently identified *B. deweyae* (Grant-Downton et al. 2014), which causes a disease of emerging spring foliage of cultivated *Hemerocallis* (daylily), is suspected to have endophytic origins. Its sudden emergence as a pathogen may be in part due to physiological changes in the host as a result of their recent intense selection and in-breeding in cultivation.

2.5 Dispersal

The common method of dispersal of nearly all species of *Botrytis* is by transmission of conidia in air currents. Water splash (Blanco et al. 2006), insects (Fermaud and le Menn 1992), agricultural machinery and pruning shears are also important means of transmitting conidia. Survival times and distances travelled by air-borne conidia are debatable; within greenhouses conidia are known to travel 1.8 m (Kerssies et al. 1995) but outdoors it is highly likely that they are carried much further in air currents (Chap. 7).

2.6 Detection and Quantification

2.6.1 *Classical Methods*

Late stage infections are easily recognized in symptomatic material by the appearance of grey/brown conidial clusters on the surfaces of infected material (Khazeli et al. 2010). Where proof of identity at the species level is required, such as infections of onions, microscopic examination of conidia is needed (Chilvers and du Toit 2006). The size of the conidia, their shape and method of attachment to the conidiophores varies from species to species (Beever and Weeds 2004).

Detection of infections in non-symptomatic material, as in the early stages of infection or latent and endophytic infections, is difficult. Surface sterilization followed by freezing of tissues or berries for 2 h at −12 °C or dipping in 300 μg/ml paraquat and then by air drying and incubation in a humid atmosphere on damp filter paper or selective media for 7–10 days has been commonly used (Edwards and Seddon 2001; Kerssies 1990; Sanzani et al. 2012). A more comprehensive discussion of plating methods can be found in Dewey and Yohalem (2004).

2.6.2 *Imaging Methods*

2.6.2.1 Visual

Visual inspection in vineyards is the most common method of estimating levels of rot. This method is certainly the least expensive; it requires little training and no equipment.

2.6.2.2 Near-Infra-Red Spectroscopy (NRI) and Mid-Infra-Red Spectroscopy (MIR)

Hill et al. (2013) found scanning homogenized samples of grape berries by either NRI or MRI was relatively quick, highly sensitive and more reliable than visual estimates but expensive.

2.6.2.3 Autofluorescence

Belanger et al. (2011) measured autofluorescence of infected grape berries using various wave lengths. They found using the fluorescence ratio of two different wave lengths, F_{440} /F_{740}, that they could detect infection as early as 4 days after inoculation.

2.6.3 *Assays for Biochemical Markers*

Glycerol, gluconic acid and laccase have been used as indicators and quantifiers of levels of *Botrytis*-rot in grape juice and wines but recent studies have shown that they do not always reflect the levels of rot determined by other methods (Dewey et al. 2008; Urruty et al. 2010).

2.6.4 Analysis of Volatiles

Van den Driessche et al. (2012) using GC-MS identified a number of volatile biomarkers of strawberries infected with *B. cinerea*. They showed that these markers can be used to track infections but could not detect pre-symptomatic infections. If sensitivity could be improved such non-destructive methods could be most useful in the food industry.

2.6.5 Immunological Methods

A review of the early immunological methods and problems with lack of specificity of antisera raised to *B. cinerea* is given by Dewey and Yohalem (2004). Using hybridoma technology, Bossi and Dewey (1992), Salinas and Schots (1994) and Meyer and Dewey (2000) raised monoclonal antibodies to *Botrytis*. These antibodies have been used to immunolabel *Botrytis* spp. *in planta* (Cole et al. 1998; Salinas and Verhoff 1995) and differentiate the biocontrol agent *Ulocladium atrum* from *Botrytis* within tissues of infected cyclamen (Kessel et al. 1999).

2.6.5.1 ELISA-Based Methods

Meyer and Dewey (2000) raised a monoclonal antibody, BC-12.CA4, that recognizes all species of *Botrytis* and members of the related genus *Sclerotinia* but not other fungi commonly involved in pathogenesis of grape berries such as species of *Aspergillus* and *Penicillium*. The antigen recognized by this antibody is a water-soluble, thermostable, glycoprotein that is produced both *in vivo* and *in vitro*. It is present in the walls and extracellular matrix of the germ tubes and hyphae but not on the surfaces of conidia (Dewey and Yohalem 2004). The use of this antibody has led to the development of a number of immunological assays for the detection and quantification of *Botrytis*-antigens in extracts from plants, fruit juices and wines. The first type of ELISA test to be developed for *Botrytis* was a Plate-trapped antigen ELISA (Meyer and Dewey 2000). These tests, which are laboratory-based, originally involved an overnight antigen coating step with extracts from infected plant tissues followed by sequnetial incubations with the monoclonal antibody, a secondary enzyme conjugated anti-mouse antibody and the substrate with washings between each step. By increasing the concentration of the secondary reporter antibody the total time involved can be reduced to 20 min (SAPS 2000).

Fernández-Baldo et al. (2011) used the BC-12.CA4 antibody to develop a competitive ELISA for the quantification of *B. cinerea* in apple, table grape and pear tissues. In their assay *Botrytis*-antigens, present in extracts from infected tissues, compete with 'purified' *B. cinerea* antigens immobilized on the surface of microtitre plates using a cross linking agent. These assays take about 40 min. They calculated that the detection limit of this assay was 0.97 μg ml^{-1} and that the antigen could be detected in non-symptomatic infections. Using the same competitive ELISA system Fernández-Baldo et al. (2009) also developed a faster screen-printed immunosensor assay modified with carbon nanotubes in a continuous-flow system. The limit of detection in this assay system was apparently 0.02 μg ml^{-1}. A further modification involving the use of micromagnetic beads coupled to the carbon-based screen-printed electrodes apparently increased in sensitivity to 0.008 μg ml^{-1} 'pure' *Botrytis* antigen (Fernández-Baldo et al. 2010).

Binder (2014) used the BC.12-CA4 antibody to develop an electrochemical biosensor for species of *Botrytis* involved in neck rot of onions, namely: *B. allii*, *B. aclada*, *B. byssoidea* and *B. cinerea*. They used the antibody in a sandwich ELISA format in which the antibody was immobilized on a screen-printed gold electrode with an on board carbon counter and a Ag/AgCl reference electrode.

2.6.5.2 *Botrytis* Immunochromatographic or Lateral Flow Assays (B-LFDs)

Lateral Flow devices based on the technology developed by Unilever for their 'Clear Blue' pregnancy test have been developed by two commercial companies: Forsite Diagnostics, York, UK and EnviroLogix, Portland, Maine, USA. The format used by both devices is a sandwich assay in which the antigens are captured by the *Botrytis*-antibody, BC-12.CA4. When used in conjunction with a reader that measures the intensity of the positive test band (Dewey et al. 2008, 2013; Urruty et al. 2010) these devices are semi-quantitative. In all cases, as in ELISA tests, it is necessary to dilute extracts from infected plants, juice or wines to eliminate the non-specific competitive binding effect of sugars and to optimize the sensitivity. In general extracts from plants are diluted 1:20 (w/v) in buffer; table wines and juice or must from grape berries 1:40 and juice from late harvest grape berries and *Botrytis*-dessert wines 1:500 (Dewey et al. 2013). The advantages of the *Botrytis*-Lateral Flow devices are that they are simple to use, do not require laboratory facilities and are rapid (10 min or less).

2.6.6 *Molecular Methods*

2.6.6.1 Nucleic Acid Analysis and Detection

For a comprehensive review and discussion of older literature on nucleic acid detection and analysis see Dewey and Yohalem (2004). Here, we focus on new discoveries, technical refinements and novel technologies. Improved DNA sequencing technology has permitted the production and continued refinement of a *B. cinerea* genome sequence (Amselem et al. 2011; Staats and Van Kan 2012), and the genome sequencing and assembly of multiple pathogenic *Botrytis* species is currently in progress (Jan van Kan, *pers. comm.*).

2.6.6.2 Nucleic Acid Sequence as a Phylogenetic Analysis Tool in *Botrytis*

Staats et al. (2005) performed the first molecular phylogenetic analysis using isolates of all the recognized species at the time. Although nuclear ribosomal internal transcribed spacer (*ITS*) sequence had been previously used by Holst-Jensen et al. (1998), its utility was restricted to demonstrating that the genus *Botrytis* was monophyletic and it could not resolve the relationships between species due to limited phylogenetically informative sequence characters. To overcome this, Staats et al. (2005) used PCR-amplified products from three single-copy nuclear genes that encode proteins involved in basic cellular processes and evolve at moderate evolutionary rates – *G3PDH*, *RPB2* and *HSP60*. The sequences from each gene were used to build trees and congruence between the different trees was assessed. There was no significant incongruence between the trees and combined analysis of all three genes revealed two main clades separated by a deep split. Clade 1 contained *B. cinerea*, *B. pelargonii*, *B. calthae* and *B. fabae*, all infecting dicotyledonous plants. Clade 2 could be sub-divided into five smaller clades and consisted largely of *Botrytis* species infecting monocots but with several species that are specialist pathogens of dicots.

Whilst the molecular analysis highly supported the previous taxonomic analysis based on the recognition of species by their morphological features, the host and pathogen phylogenies are not congruent. This suggests that tight co-evolution between *Botrytis* and host plants does not result in *Botrytis* speciation, and that host-shifts may be frequent. Interestingly, this analysis also revealed that sexuality was plastic in *Botrytis*, with evidence for loss of sexual reproduction on at least three occasions. Staats et al. (2005) consider that this may be adaptive. In drier climates with intense solar radiation, thin-walled and easily dehydrated ascospores are unlikely to be an efficient manner of dispersal. The asexual macroconidia have thicker walls and are more tolerant to UV and dehydration stresses. Likewise, the asexual species retain the highly melanised sclerotia that are particularly resistant to environmental stress and can be long-lived in a dormant state (Holtz et al. 2004). Further support for this classification has come from analysis of two paralogous gene sequences, *NEP1* and *NEP2*, which encode known and putative *Botrytis* phytotoxic proteins, homologues of a necrosis and ethylene-inducing protein from *Fusarium* (Staats et al. 2007). Trees based on *NEP1* and *NEP2* sequences are congruent with trees generated using combined data from the housekeeping genes and hence they represent yet another pair of phylogenetically informative sequences for the genus *Botrytis* (for a review see Chap. 6).

There is evidence that *Botrytis* may have acquired, through horizontal gene transfer, genetic material from other species. This type of analysis to screen for candidates for horizontal gene transfer (HGT) has been greatly assisted by the ever-increasing amount of genome sequence data in the fungi as well as other organisms. Richards et al. (2009) show that a plant to ascomycete gene transfer may have taken place, as both *Botrytis* and *Sclerotinia* genomes encode a gene family found elsewhere only in plants. Furthermore, there is evidence of horizontal transfer of genes from *Fusarium* to the *Botrytis* lineage (Campbell et al. 2013).

Transfer of a bikaverin secondary metabolite gene cluster preceded the radiation of the genus *Botrytis*, and subsequent decay of transferred sequences has occurred over evolutionary time. Indeed, this degeneration of gene clusters that have been horizontally transferred suggests that such events may have been underestimated due to their repeated loss. Interestingly, loss of the bikaverin gene cluster may be correlated with the acquisition of increased virulence and transition from endophytic to pathogenic lifestyles. Conversely, there is data to suggest that other HGT events in *B. cinerea* may have promoted pathogenicity. Zhu et al. (2012) identified four candidates for genes that have been horizontally transferred to this species. Three of these may have been acquired from bacteria whilst a fourth may be derived from a plant. Of these genes, three are predicted enzymes (a UDP-glucosyltransferase, two alpha/beta hydrolases) and one is a lipoprotein. Zhu et al. (2012) speculate that they may all be related to phytopathogenicity and their acquisition may have acted to promote fungal virulence.

2.6.6.3 Identification of *Botrytis* Species and Analysis of Their Populations Using Nucleic Acid Sequencing

The comparative ease of PCR fragment sequencing and the robust nature of DNA sequence data that can be readily fitted into existing phylogenetic frameworks have together aided species identification and especially assisted the confirmation of newly discovered species in the genus. Previously, DNA-based molecular techniques assisted the identification of *B. allii* as an allopolyploid hybrid of two other species, *B. aclada* and *B. byssoidea* (Nielsen and Yohohalem 2001; Yohalem et al. 2003) and this has been reconfirmed by analysis by Staats et al. (2005) where data from *G3PDH*, *RPB2* and *HSP60* all agreed with its hybrid status and parental genome contributions. Furthermore, these three phylogenetically informative genes as well as *NEP1* and *NEP2* have had much value in defining new species and their evolutionary relationships, such as *B. sinoallii*, *B. caroliniana*, *B. fabiopsis*, *B. sinoviticola* and *B. deweyae*. The *HSP60* sequence has also been used to identify endophytic isolates as probable novel, undescribed species in *Centaurea* (Shipunov et al. 2008).

DNA sequence analysis has also been instrumental in reviving *B. mali* as a valid species (O'Gorman et al. 2008) whilst *B. pseudocinerea* is a morphologically cryptic species where major support for its recognition as a distinct species was gained by DNA sequence analysis (Walker et al. 2011). Leroch et al. (2013) have also identified a novel clade of *Botrytis*, closely related to *B. cinerea* and *B. fabae*, that is widespread in German strawberry-growing areas but is rare in vineyards. These isolates, named group S, have strong, multi-drug resistance. Combined analysis of sequence data from multiple genes, *HSP60* and *NEP2* but also the zinc finger transcription factor *mrr1*, *ms547* (encoding an RNA helicase-like protein) and *fg1020* (encoding a putative ubiquitin conjugation factor), confirmed the distinction of these isolates.

In phytopathology studies in the field, sequence-based identification can be a valuable tool. For instance, in onion crops several different *Botrytis* species can be problematic; Nielsen et al. (2002) and Chilvers and Du Toit (2006) have employed PCR amplification followed by restriction endonuclease digestion of the products (PCR-RFLP) to generate species-specific banding patterns to assist in diagnosis. Primers for PCR that are highly specific for the detection of *B. cinerea* have also been published (Rigotti et al. 2006) as has a technique to directly PCR, without DNA extraction, *B. cinerea* from environmental (immature grape berries) samples to generate a highly sensitive detection assay (Gindro et al. 2005). Walcott et al. (2004) developed a magnetic capture hybridization and PCR (MCH-PCR) assay for the detection of *B. aclada* in onion seed that was more sensitive and efficient than conventional PCR methods and enabled detection time to be reduced from 10 to 14 days by agar plating to less than 24 h.

A particularly attractive use for nucleic acid sequencing involves the analysis of intraspecific variation in *Botrytis* genotypes within natural populations. In contrast to interspecific analysis of DNA differences, such studies require focus on genomic sequences that evolve rapidly. For example, in *B. cinerea* a recent population analysis surveyed transposon presence (*Boty* and *Flipper*) within isolates, which can be variable, and was combined with microsatellite DNA analysis (Kecskemeti et al. 2014). The use of sequences where there is known to be a huge man-made selective pressure for nucleotide change has also been employed, with nucleotide polymorphisms that are responsible for agricultural fungicide resistance being the focus for population studies in *B. cinerea* and *B. squamosa* (Van der Heyden et al. 2014).

Detection of fungicide-resistant *B. cinerea* in the early stages of grape growth has also been performed by Saito et al. (2009) using a PCR-RFLP method. This is claimed to allow rapid detection of isolates that are resistant to three major fungicides, and thus to permit the early diagnosis and improve integrated pest management strategies. Such molecular studies will prove immensely important for analysing the population genetic diversity of pathogenic *Botrytis*, especially tracking the emergence, maintenance and spread of fungicide resistance from the local to the international level. At a practical level, these studies will permit far more responsive protective measures for crops in the future, and increase the effectiveness of *Botrytis* containment and control.

2.6.6.4 Quantitative Analysis of *Botrytis* Infections Using Nucleic Acids

Whilst useful in systematics, diagnostics and tracking the genetic diversity of *Botrytis* populations, all studies that are reliant primarily on qualitative sequence data, the use of nucleic acid analysis in quantitative analysis of *Botrytis* infections remains challenging outside of highly controlled experiments. Whilst RNA-based methods such as Northern blots and quantitative reverse-transcription PCR are ideal for laboratory investigations to detect the level of *Botrytis* gene expression, translating these rather expensive and technically demanding approaches to field analysis and practical applications is challenging. Naturally, DNA-based quantification methods have been the primary focus due to the relative ease of isolating DNA from environmental samples. However, these studies still present a major challenge as not only does the sampling have to be carefully controlled but also the sensitivity of such techniques means that standardisation of high-quality DNA extraction and care to avoid sample contamination are paramount for reproducible results. qPCR methods also remain expensive in terms of reagents and apparatus. Nevertheless, a number of studies have developed the technique for quantifying *Botrytis* infections in different crops Alexandre (2010).

Suarez et al. (2005) developed a TaqMan real-time PCR system that was highly sensitive and highly specific for *B. cinerea*, which allows detection and quantification prior to symptom development. In addition to *B. cinerea*, qPCR has also been developed for *Allium* crop pathogens *B. aclada*, *B. allii* and *B. byssoidea* (Chilvers et al. 2007) whilst Carisse et al. (2012) have demonstrated that qPCR can quantify the airborne conidia inoculum of *B. squamosa*. Further technical refinements have been made to improve qPCR protocols. Saito et al. (2013) tested different DNA extraction methods to find the optimum purity from grapevine material. Further technical refinements have been made to improve qPCR protocols. Saito et al. (2013) tested different DNA extraction methods to find the optimum purity from grapevine material. Tomlinson et al. (2010) developed a loop-mediated isothermal amplification system for the detection of *B. cinerea* that combined the sensitivity and specificity of previous nucleic acid-based methods with simplified equipment that allowed detection of pre-symptomatic infections in less than 15 min. Their assay has the advantage of being less susceptible to inhibitory substances from plant material than the TaqMan real-time PCR assay described by Suarez et al. (2005).

The latest technological advances will enable detection of specific fungicide-resistant genotypes and/or different species simultaneously with their quantification. Use of pyrosequencing, such as Qiagen's PyroMark system – already widely used in medical applications such as detection of drug-resistant microbes – generates sensitive, quantitative sequence data outputs that would permit discrimination of specific genotypes within multiple samples.

2.6.7 Sampling Issues in Quantification and Detection

Determining the total amount of *Botrytis*-rot in any tissue, organ or crop by any method is problematic because the results are highly dependent on the method sampling. Quantification is most difficult where estimates are made of whole standing crops as in vineyards. In these cases it is imperative that the sampling is done at random and that enough samples are taken to make sure that variations between different areas are taken into account. It is impossible to compare levels of rot unless they are expressed in the same way for example results from determining levels of rot in vineyards will be very different if expressed as the percentage rotten grape berries on the basis of incidence compared with those based on percentage rot by weight (Dewey et al. 2008). Problems are also encountered in quantifying levels of rot in plant tissues in absolute terms in a particular tissue because there is no source of purified antigen or standard. Claims made on sensitivity based on extracts from freeze-dried mycelium grown in culture in different labs inevitably mean that the levels of antigen extracted vary. More recently, in our own studies we have chosen to use as our stock or reference antigen dilutions of a specific dessert wine that is highly stable (1998 Dolce from Far Niente, CA, USA, Dewey et al. 2008).

Similarly preparations of fungal and plant standards for DNA analysis vary from lab to lab. DNA extracts are generally prepared from cultures of *Botrytis* species grown in the lab. In some cases known concentrations of conidia have been used as standards but again direct comparisons between labs is not practical. Of course, with DNA-based quantification measures the inherent problem with variably multinucleate hyphae and conidia is that the final result will provide a quantitative measure reflecting the number of *nuclei* present rather than the number of conidia or hyphae.

Phillips et al. (1987) have shown that the number of nuclei in macroconidia is variable and dependent on glucose concentration in the media. Nuclear number in conidia was correlated with the aggressiveness of infection of the host. This has potential implications for the accuracy of disease prediction from qPCR data. For example, the presence of fewer conidia with more nuclei may have a different ultimate disease outcome than more conidia with fewer nuclei, yet in terms of quantification they could appear equivalent. A further confounding factor is variation in ploidy – in natural *B. cinerea* strains the DNA content per nucleus varies considerably, indicating that aneuploidy and polyploidy may be widespread in wild populations (Büttner et al. 1994). Nevertheless, qPCR may have an advantage over antibody-based methods by allowing earlier detection of *Botrytis* infection (Celik et al. 2009). Indeed, the qPCR detection assay may be able to detect as few as three conidia, and probably even fewer if the high-copy ITS rDNA gene is targeted (Mehli et al. 2005).

2.7 Conclusions

The major technical advances that have been made in the past decade in immunological and molecular techniques have helped considerably in the classification of the genus, identification of new species, expanded our knowledge of host ranges and furthered our understanding of the biology of *Botrytis* species. In particular, quantification of inoculum and infections in the field has improved due to these new technologies. These advances will all have a major impact on disease prevention and control. Whilst *Botrytis* species, particularly *B. cinerea*, have generally been regarded as necrotic, facultative pathogens, new studies have demonstrated that some species can be present in plants as endophytes. Our understanding of the endophtyic stages and their failure to induce host resistance is still in its infancy, nor do we know what the triggers are that effect the switch from endophyte to pathogen, and vice-versa. The possibility that *Botrytis* species can be present, undetected, as endophytes has important implications in their control.

References

Alexandre H (2010) Development of a qPCR assay for specific quantification of *Botrytis cinerea* on grapes. FEMS Microbiol Lett 313:81–87

Amselem JI, Cuomo CA, Van Kan JA et al (2011) Genomic analysis of the necrotrophic fungal pathogens *Sclerotinia sclerotiorum* and *Botrytis cinerea*. PLoS Genet 7, e1002230

Anderson JP (1924) *Botrytis cinerea* in Alaska. Phytopathology 14:152–155

Barnes SE, Shaw ME (2003) Infection of commercial hybrid *Primula* seed by *Botrytis cinerea* and latent disease spread through the plants. Phytopathology 93:573–578

Beever RE, Weeds PL (2004) Taxonomy and genetic variation of *Botrytis* and *Botryotinia*. In: Elad Y, Williamson B, Tudzynski P, Delen N (eds) *Botrytis*: biology, pathology and control. Kluwer Academic Publishers, Dordrecht, pp 30–31

Belanger MC, Roger JM, Cartolaro P et al (2011) Autofluorescence of grape berries following *Botrytis cinerea* infection. Int J Remote Sens 32:3835–3849

Binder M (2014) Development of a *Botrytis* specific immunosensor; towards using PCR species identification. Ph.D. thesis University of Cranfield, UK

Blanco C, de los Santos B, Romero F (2006) Relationship between concentrations of *Botrytis cinerea* conidia in air, environmental conditions, and the incidence of grey Mould in strawberry flowers and fruits. Eur J Plant Pathol 114:415–425

Bossi R, Dewey FM (1992) Development of a monoclonal antibody-immunodetection assay for *Botrytis cinerea* (Pers). Plant Pathol 41:472–482

Brierley WB (1918) Microconidia of *Botrytis cinerea*. Bull Mis Inf R Bot Gardens Kew 4:129–146

Büttner P, Koch F, Voigt K et al (1994) Variations in ploidy among isolates of *Botrytis cinerea*: implications for genetic and molecular analyses. Curr Gen 25:445–450

Campbell MA, Staats M, van Kan AL et al (2013) Repeated loss of an horizontally transferred gene cluster. Mycologia 105:1126–1134

Capieau K, Stenlid J, Stenstrom E (2004) Potential for biological control of *Bortytis cinerea* in *Pinus sylvestris* seedlings. Scand J Forest Res 19:312–319

Carisse O, Levasseur A, Van der Heyden H (2012) A new risk indicator for Botrytis leaf blight of onion caused by Botr*ytis squamosa* based infection efficiency of airborne inoculum. Plant Pathol 61:154–1164

Celik M, Kalpulov T, Zutahy Y et al (2009) Quantitative and qualitative analysis of *Botrytis* inoculated table grapes by qPCR and antibodies. Postharvest BiolTech 52:235–239

Chang SW, Kim SK, Hwang BK (2001) Grey mould of daylily (*Hemerocallis fulva* L.) caused by *Botrytis elliptica* in Korea. Plant Path J 17:305–307

Chilvers MI, du Toit LJ (2006) Detection and identification of *Botrytis* species associated with neck rot, scape blight, and umbel blight of onion. Online Plant Health Prog. doi:10.1094/PHP-2006-1127-01-DG

Chilvers MI, du Toit LJ, Peever TL (2007) A real-time, quantitative PCR seed assay for *Botrytis* spp. that cause neck rot of onion. Plant Dis 91:599–608

Choquer M, Fournier E, Kunz C et al (2007) *Botrytis cinerea* virulence factors: new insights into a necrotrophic and polyphageous pathogen. FEMS Microbiol Lett 277:1–10

Chou MC, Preece TF (2008) The effect of pollen grains on infection caused by *Botrytis cinerea* FR. Ann Appl Biol 62:11–22

Cole L, Dewey FM, Hawes CR (1998) Immunocytochemical studies of the infection mechanisms of *Botrytis fabae*: penetration and post-penetration processes. New Phytol 139:587–609

Dewey FM, Meyer U (2004) Rapid, quantitative tube-immunoassay for on site detection of *Botrytis*, *Aspergillus* and *Penicillium* antigens in grape juice. Anal Chem Acta 513:11–19

Dewey FM, Yohalem D (2004) Detection, quantification and immunolocalisation of *Botrytis* species. In: Elad Y, Williamson B, Tudzynski P, Delen N (eds) Botrytis: biology, pathology and control. Kluwer Academic Publishers, Dordrecht, pp 181–194

Dewey FM, Ebeler SE, Adams DO et al (2000) Quantification of *Botrytis* in grape juice determined by a monoclonal antibody-based immunoassay. Am J Enol Vitic 51:276–282

Dewey FM, Hill M, De Scenzo R (2008) Quantification of *Botrytis* and laccase in wine grapes. Am J Enol Vitic 59:47–54

Dewey FM, Steel CC, Gurr SJ (2013) Lateral flow devices to rapidly determine levels of stable *Botrytis* antigens in table and dessert wines. Am J Enol Vitic 64:291–295

Droby S, Lichter A (2004) Post-harvest *Botrytis* infection: etiology, development and management. In: Elad Y, Williamson B, Tudzynski P, Delen N (eds) *Botrytis*: biology, pathology and control. Kluwer Academic Publishers, Dordrecht, pp 349–362

Edwards SG, Seddon B (2001) Selective media for the specific isolation and enumeration of *Botrytis cinerea* conidia. Lett Appl Microbiol 32:63–66

Faretra F, Antonacci E (1987) Production of apothecia of *Botryotinia fuckeliana* (de bary) Whetz. Under controlled environmental conditions. Phytopathol Mediterr 26:29–35

Faretra F, Antonacci E, Pollastro S (1988) Sexual behaviour and mating system of *Botryotinia fuckeliana*, teleomorph of *Botrytis cinerea*. Microbiology 134:2543–2550

Fermaud M, le Menn R (1992) Transmission of *Botrytis cinerea* to grapes by grape berry moth larvae. Phytopathology 82:1393–1398

Fernández-Baldo MA, Messina GA, Sanz MI et al (2009) Screen-printed immunosensor modified with carbon nanotubes in a continuous-flow system for the *Botrytis cinerea* determination in apple tissues. Talanta 79:681–686

Fernández-Baldo MA, Messina GA, Sanz MI et al (2010) Microfluidic immunosensor with micromagnetic beads coupled to carbon-based screen-printed electrodes (SPCEs) for determination of *Botrytis cinerea* in tissue of fruits. J Agric Food Chem 10:11201–11206

Fernández-Baldo MA, Fernández JG, Pereira SV et al (2011) Development of an indirect competitive enzyme-linked immunosorbent assay applied to the *Botrytis cinerea* quantification in tissues of postharvest fruits. BMC Microbiol 11:220. doi:10.1186/1471-2180-11-220

Fukumori Y, Nakajima M, Akuts K (2004) Microconidia act the role as spermatia in the sexual reproduction of *Botrytis cinerea*. J Gen Plant Pathol 70:256–260

Furukawa T, Ushimaya K, Kishi K (2005) Botrytis blight of Taiwanese toad lily caused by *Botrytis elliptica* (Berkley) Cooke. J Gen Plant Pathol 71:95–97

Gindro K, Pezet R, Viret O et al (2005) Development of a rapid and highly sensitive direct-PCR assay to detect a single conidium of *Botrytis cinerea* Pers:Fr *in vitro* and quiescent forms *in planta*. Vitis 44:139–142

Grant-Downton RT, Razak B, Terhem MV et al (2014) A novel *Botrytis* species is associated with a newly emergent foliar disease in cultivated *Hemerocallis*. Plos One 9, e89272

Hill GN, Evans KJ, Beresford RM et al (2013) Near and mid-infrared spectroscopy for the quantification of botrytis bunch rot in white wine grapes. J Near Infrared Spec 21:467–475

Holst-Jensen A, Vaage M, Schumacher T (1998) An approximation to the phylogeny of *Sclerotinia* and related genera. Nordic J Bot 18:705–719

Holtz G, Coertze S, Williamson B (2004) The ecology of *Botrytis* on plant surfaces. In: Elad Y, Williamson B, Tudzynski P, Delen N (eds) *Botrytis*: biology, pathology and control. Kluwer Academic Publishers, Dordrecht, pp 9–24

Horst RK (2013) Westcott's plant disease handbook. 5th edn. Chapman and Hall, New York

Hsiang T, Chastagner GA (1992) Production and viability of sclerotia from fungicide resistant and sensitive isolates of *B. cinerea*, *B. elliptica* and *B. tulipae*. Plant Pathol 41:600–605

Hsieh T-F, Huang J-W (2001) Leaf disk method for assessment of disease severity of lily leaf blight caused by *Bortytis elliptica*. Plant Pathol Bull 10:37–44

Kecskeméti E, Brathuhn A, Kogel K-H et al (2014) Presence of transposons and mycoviruses in *Botrytis cinerea* isolates collected from a German grapevine growing region. J Phytopathol 1:582–595

Kerssies A (1990) A selective medium for *Botrytis cinerea* to be used in a spore trap. Neth J Plant Pathol 96:247–250

Kerssies A, Van Bosker Z, Frinking HD (1995) Influence of environmental conditions in a glasshouse on conidia of *Botrytis cinerea* and on post-harvest infection of rose flowers. Eur J Plant Pathol 101:201–216

Kessel GT, de Hass BH, Lombaers-Van der Plas CH et al (1999) Quantification of mycelium of *Botrytis* spp. and the antagonist *Ulocladium atrum* in necrotic leaf tissue of cyclamen and lily by fluorescence microscopy and image analysis. Phytopathology 89:868–876

Khazeli P, Zamanizadeh H, Morid B et al (2010) Morphological and molecular identification of *Botrytis cinerea* causal agent of grey mould in rose greenhouses in central regions of Iran. Int J Agric Sci Res 1:19–24

Leroch M, Plesken C, Weber RW et al (2013) Gray mold populations in German strawberry fields are resistant to multiple fungicides and dominated by a novel clade closely related to *Botrytis cinerea*. Appl Environ Microbiol 79:159–167

Li X, Kerrigan J, Chai W et al (2012) *Botrytis caroliniana*, a new species isolated from blackberry in South Carolina. Mycologia 104:650–658

Mehli L, Kjellsen TD, Dewey FM et al (2005) A case study from the interaction of strawberry and *Botrytis cinerea* highlights the benefits of co-monitoring both partners at genomic and mRNA level. New Phytol 168:465–474

Meyer U, Dewey FM (2000) Efficacy of different immunogens for raising monoclonal antibodies to *Botrytis cinerea*. Mycol Res 104:979–987

Meyer U, Spotts RA, Dewey FM (2000) Detection and quantification of *Botrytis cinerea* by ELISA in Pear Stems During cold storage. Plant Dis 84:1099–1103

Mirzaei S, Mohammadi-Goltapeh E, Shams-bakhsh M (2007) Taxonomical studies on the genus *Botrytis* in Iran. J Agric Tech 3:65–76

Mittal RK, Singh P, Wang BSP (1987) *Botrytis*: a hazard to reforestation. Eur J For Pathol 17:369–384

Nassr S, Bakarat R (2013) Effect of factors on conidium germination of *Botrytis cinerea in vivo*. Int J Plant Soil Sci 2:41–54

Nielsen K, Yohohalem DS (2001) Universally primed polymerase chain reaction alleles and internal transcribed spacer restriction fragment length polymorphisms distinguish two subgroups in *Botrytis aclada* distinct from *B. bysoidea*. Phytopathology 91:527–533

Nielsen K, Yohalem DS, Jensen DF (2002) PCR Detection and RFLP differentiation of *Botrytis* species associated with neck rot of onion. Plant Dis 86:682–686

O'Gorman DT, Sholberg PL, Stokes SC et al (2008) DNA sequence analysis of herbarium specimens facilitates the revival of *Botrytis mali*, a postharvest pathogen of apple. Mycologia 100:227–235

Phillips DJ, Margosan DA, Mackey BE (1987) Size, nuclear number, and aggressiveness of *Botrytis cinerea* spores produced on media of varied glucose concentrations. Phytopathology 77:1606–1608

Richards TA, Soanes DM, Foster PG et al (2009) Phylogenomic analysis demonstrates a pattern of rare and ancient horizontal gene transfer between plants and fungi. Plant Cell 21:1897–1911

Rigotti S, Viret O, Gindro K (2006) Two new primers highly specific for the detection of *Botrytis cinerea* Pers.: Fr. Phytopathol Mediterr 45:253–260

Rousseaux S, Diguta CF, Radoï-Matei F et al (2014) Non-Botrytis grape-rotting fungi responsible for earthy and moldy off-flavors and mycotoxins. Food Microbiol 38:104–121

Saito S, Suzuki S, Takavanagi T (2009) Nested PCR-RFLP is a high-speed method to detect fungicide-resistant *Botrytis cinerea* at an early growth stage on grapes. Pest Manag Sci 65:197–204

Saito S, Dunne KJ, Evans KJ et al (2013) Optimisation of techniques for quantification of *Botrytis cinerea* in grape berries and receptacles by quantitative polymerase chain reaction. Aust J Grape Wine Res 19:68–73

Salinas J, Schots A (1994) Monoclonal antibodies-based immunofluorescence test for the detection of conidia of *Botrytis cinerea* on cut flowers. Phytopathology 84:351–356

Salinas J, Verhoff K (1995) Microscopical studies of the infection of gerbera flowers by *Botrytis cinerea*. Eur J Plant Pathol 101:377–386

Sanzani SM, Schena L, De Cicco V et al (2012) Early detection of *Botrytis cinerea* latent infections as a tool to improve postharvest quality of table grapes. Postharvest Biol Tech 68:64–71

SAPS ELISA kit for *Botrytis* (2000) Science and plants for schools. Homerton College, Cambridge

Shipunov A, Newcombe G, Raghavendra AKH et al (2008) Hidden diversity of endophytic fungi in an invasive plant. Am J Bot 95:1096–1108
Sivertsen HK, Dewey FM, Heymann H (2005) Relationship between sensory descriptive analysis and levels of *Botrytis* antigens in dessert wines. Am J Enol Vitic 56:330–335
Sowley ENK, Dewey FM, Shaw MW (2010) Persistent, symptomless, systemic and seed-borne infection of lettuce by *Botrytis cinerea*. Eur J Plant Pathol 126:61–71
Staats M, Van Kan JA (2012) Genome update of *Botrytis cinerea* strains B05.10 and T4. Euk Cell 11:1413–1414
Staats M, Van Baarlen P, Van Kan JA (2005) Molecular phylogeny of the plant pathogenic genus *Botrytis* and the evolution of host specificity. Mol Biol Evol 22:333–346
Staats M, Van Baarlen P, Schouten A et al (2007) Positive selection in phytotoxic protein encoding genes of *Botrytis* species. Fungal Genet Biol 44:52–63
Suarez MB, Walsh K, Boonham N et al (2005) Development of real-time (TaqMan) assays for the detection and quantification of *Botrytis cinerea in planta*. Plant Physiol Biochem 43:890–899
Tomlinson JA, Dickinson MJ, Boonham N (2010) Detection of *Botrytis cinerea* by loop-mediated isothermal amplification. Lett Appl Microbiol 51:650–657
Tompkins CM, Hansen HN (1950) Flower blight of *Stephanotis floribunda*, caused by *Botrytis elliptica*, and its control. Phytopathology 40:780–781
Urbasch I (1983) On the genesis and germination of chlamydospores of *Botrytis cinerea*. J Phytopathol 108:54–60
Urruty L, Audebert J, Snakkers G et al (2010) Relevance of an immunoassay test for rapid detection of *Botrytis cinerea* in 'Ugni Blanc' musts and wines. Qud Vitic Rnol Univ Torino 31:2009–2010
Van den Driessche T, Keulemans J, Geeraerd A et al (2012) Evaluation of fast volatile analysis for detection of *Botrytis cinerea* infections in strawberry. Food Microbiol 32:406–414
Van der Heyden H, Dutilleul P, Brodeur L et al (2014) Spatial distribution of single nucleotide polymorphisms related to fungicide resistance and implications for sampling. Phytopathology 104:604–613
Walcott RR, Gitaitis RD, Langton DB (2004) Detection of *Botrytis aclada* in onion seed using magnetic capture hybridization and polymerase chain reaction. Seed Sci Tech 32:425–438
Walker AS, Gautier AL, Confais J et al (2011) *Botrytis pseudocinerea*, a new cryptic species causing gray mold in French vineyards in sympatry with *Botrytis cinerea*. Phytopathology 101:1433–1445
Yohalem DS, Nielsen K, Nicolaisen M (2003) Taxonomic and nomenclatural clarification of the onion neck rotting *Botrytis* species. Mycotaxon 85:175–182
Zhang J, Wu M-D, Li G-Q et al (2010a) *Botrytis fabiopsis*, a new species causing chocolate spot of broad bean in central China. Mycologia 102:1114–1126
Zhang J, Zhang L, Li G-Q et al (2010b) *Botrytis sinoallii*: a new species of the grey mold pathogen of *Allium* crops in China. Mycoscience 51:421–431
Zhao J, Zhou L, Wang J, et al (2010) Endophytic fungi for producing bioactive compounds originally from their host plants. In: A Mendez-Vilas (ed) Current research, technology and evolution topics in microbiology and microbial biotechnology, vol I Formatex, pp 567–576
Zhou X, Zhu F, Liu L et al (2010) A review: recent advances and future prospects of taxol-producing endophytic fungi. Appl Microbiol Biotechnol 86:1707–1717
Zhou Y, Zhang J, Wang X et al (2014) Morphological and phylogenetic identification of *Botrytis sinoviticola*, a novel cryptic species causing gray mold disease on table grapes (*Vitis vinifera*) in China. Mycologia 106:43–56
Zhu B, Zhou Q, Xie G et al (2012) Interkingdom gene transfer may contribute to the evolution of phytopathogenicity in *Botrytis cinerea*. Evol Bioinform Online 8:105–107
Zuccaro A, Schoch CL, Spatafora JW et al (2008) Detection and identification of fungi intimately associated with the brown seaweed *Fucus serratus*. Appl Environ Microbiol 74:931–941

Chapter 3
Genetics of *Botrytis cinerea*

Rita Milvia De Miccolis Angelini, Stefania Pollastro, and Franco Faretra

Abstract *Botrytis cinerea* displays an extraordinary variability in phenotypic traits, making it a model for studying sources of variation in filamentous fungi and in particular in plant pathogens. The whole genome sequence was recently made available and is sustaining an impressive progress of knowledge. The present review aims at giving an updated picture on the genetic features of this fungal pathogen and, in particular, on mechanisms underlying its broad variation and adaptation capability, *i.e.* mating system and sexual behavior and other sources of variation (chromosome number, mycoviruses, transposons, vegetative compatibility, etc.), as well as on tools available for its genetic manipulation.

Keywords Genetics • Variation • Karyotype • Heterokaryosis • Sexual behavior

3.1 Introduction

The genus *Botrytis* Pers. and its teleomorphic stage *Botryotinia* Whetz. are classified within the family *Sclerotiniaceae* Whetz. of Inoperculate Discomycetes (Hennebert 1973; Yohalem et al. 2003). *Botrytis* species are necrotrophic pathogens decaying infected plant tissues, on which produce sporulating gray mycelium, and also have excellent saprotrophic capabilities. These fungi produce macroconidia, dry spores dispersed by wind over long distance, and microconidia that are not infectious and play the role of spermatia. *Botrytis* overwinters as mycelium in decaying plant debris and as sclerotia. Some species produce a teleomorphic stage, apothecia arising from fertilized sclerotia and releasing ascospores (Fig. 3.1). Most species have a worldwide distribution and occur wherever their host crops are grown. Fungi of the genus *Botrytis* are important pathogens of many economically important crops, such as grapevine and protected crops (strawberry, tomato and other horticultural, flower and ornamental crops). *Botrytis* diseases cause mostly

R.M. De Miccolis Angelini • S. Pollastro • F. Faretra (✉)
Department of Soil, Plant and Food Sciences – Plant Pathology Section, University of Bari, via Amendola 165/A, 70126 Bari, Italy
e-mail: francesco.faretra@uniba.it

S. Fillinger, Y. Elad (eds.), *Botrytis – the Fungus, the Pathogen and its Management in Agricultural Systems*, DOI 10.1007/978-3-319-23371-0_3

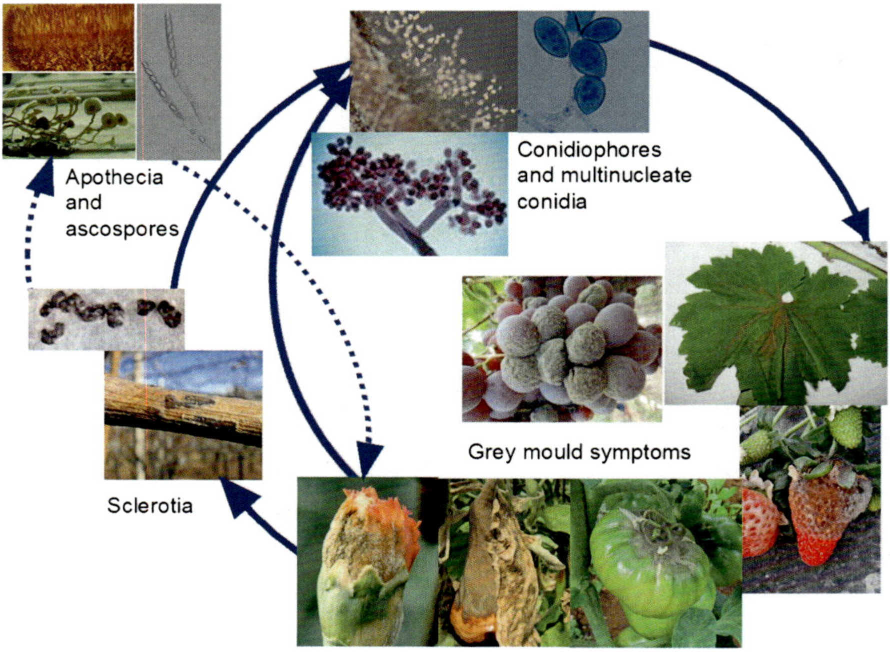

Fig. 3.1 Life cycle of *B. cinerea* and associated symptoms. Apothecia can be produced by sclerotia under laboratory conditions but they have been rarely detected in the field and their role in grey mould epidemiology is still uncertain

blossom blights and fruit rots but also leaf spots and bulb rots, in the field as well as in post-harvest (Jarvis 1977). Most of *Botrytis* species are specialized pathogens with a narrow host range, generally infecting only one or a few closely related species within the same plant genus (Mansfield 1980). However, *B. cinerea* is a polyphagous species infecting over 200 host plants (Jarvis 1977) and since recently suspected as an endophyte (Chap. 2).

A phylogenetic analysis discriminated *Botrytis* species into two clades. Clade 1 contains four species that are pathogens exclusively on Eudicotyledon hosts, whereas clade 2 contains 15 species pathogens on Monocotyledon and only 3 species that are pathogens on Eudicotyledon hosts (Chap. 6). Co-speciation of *Botrytis* species and host plants seems have not occurred in their evolution, but it has been proposed that host shifts have occurred during *Botrytis* speciation, possibly by the acquisition of novel pathogenicity factors (Staats et al. 2005). Aguileta et al. (2012) compared genes among the specialized pathogens *B. tulipae* and *B. ficariarum* and the polyphagous pathogens *B. cinerea* and *Sclerotinia sclerotiorum* and found 21 genes showing footprints of positive selection, some of which with functions related to plant–fungus interactions.

Grey mould, caused by *B. cinerea*, has a disastrous economic impact on numerous crops all over the world, although it is not easy to estimate due to the broad range of host crops and growing areas. The pathogen causes both qualitative (taste, aroma, oxydasic casse in wine) and quantitative (reduced yields of fruit, vegetable crops and ornamentals) losses; its economic impact is mainly on wine- and table-grape, solanaceae (tomato, pepper, eggplant), cucurbits, strawberry, lettuce and other salad vegetables, beans, bulbs and ornamentals. For instance, gray mould is one of the major challenges for strawberry growers in Florida (USA). Losses from fruit rot can exceed 50 %, fungicides are commonly applied once a week and their cost was approximately 7 % of pre-harvest variable costs (about $ 690 per acre) (IFAS 2010). The global cost of crop protection against the pathogen has recently been estimated around $ 310 million, mainly in France and Italy for grapevine and China and Japan for vegetables, but also in Spain, Chile, USA, Germany, Australia and The Netherlands. For its scientific and economic impacts, *B. cinerea* is considered the second most significant fungal pathogen in molecular plant pathology and is an excellent model for the study of infection process in necrotrophic pathogens (Dean et al. 2012).

The aim of the present chapter is to give a picture of current knowledge on essential genetic features of this important phytopathogenic fungus updating previous reviews on the same issue (Faretra and Grindle 1992; Beever and Weeds 2004).

3.2 Numbers of Nuclei and Chromosomes

The life cycle of *B. cinerea* is reported in Fig. 3.1. Hyphal cells and conidia are multinucleate. Conidia contain generally 3–6 nuclei (Grindle 1979; Lorenz and Eichorn 1983; Shirane et al. 1988, 1989) whilst microconidia are uninucleate, usually not able to germinate on culture media, and have the function of male gametes (spermatia) in sexual process. Young asci contain a single diploid nucleus that undergoes meiosis generating a tetrad of haploid nuclei which through a mitotic division produce the eight nuclei around which eight ascospores in each ascus are formed. Mature ascospores usually contain four nuclei (Lorenz and Eichorn 1983; Faretra and Antonacci 1987) (Fig. 3.2). Cytological observation of fungal chromosomes is difficult due to their small size. However, Shirane et al. (1988, 1989) set up a method known as "germ tube burst" and observed 16 chromosomes in mitotic methaphase in germling tips. The same number of chromosomes has been counted in developing asci (Faretra and Grindle 1992) (Fig. 3.3).

B. cinerea karyotype has been investigated by pulsed-field gel electrophoresis (PFGE) which separated up to 13 major bands (1.8–4.6 Mbp), corresponding to 1 or more large chromosomes, and up to 3 minor bands (200–580 Kbp), corresponding to small chromosomes (Fig. 3.4). A broad variability was found in electrophoretic patterns among different isolates (Van Kan et al. 1993; Pollastro 1996; Vallejo et al. 1996, 2002). Genome size estimates varied from 13.2–22.6 Mbp (Vallejo et al. 2002) to 33.9–39.7 Mbp (Van Kan et al. 1993), respectively on the assumption of only one chromosome per band or two for more intense bands.

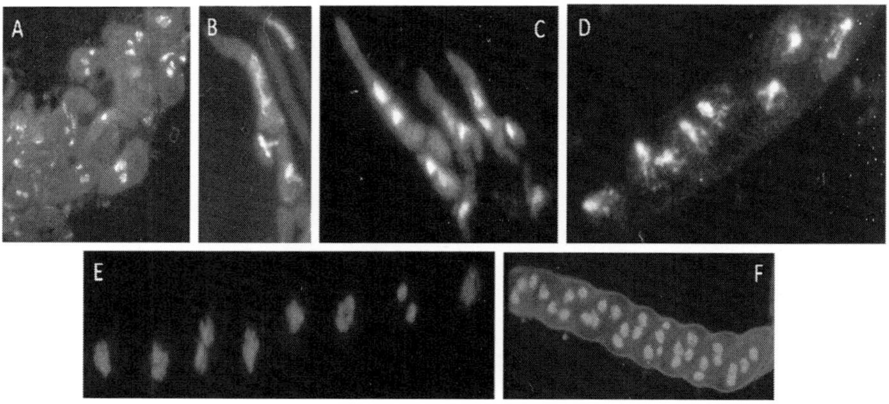

Fig. 3.2 Distribution of nuclei during *B. cinerea* development. (**a**) Diffuse dikaryotic state in sub-hymenium cells; (**b**, **c**) proasci with evident clamp connection (no crozier formation like in *S. sclerotiorum*; Uhm and Fujii 1986) to re-establish dikaryotic status in the mother cell of ascus; (**d**) the tetrad of meiotic nuclei undergoes to mitotic division to form the eight nuclei around which ascospores will be differentiated; (**e**) the eight nuclei divide mitotically; and (**f**) ripened ascospores contain usually four nuclei. Nuclei were DAPI stained and pictures are not at the same magnification

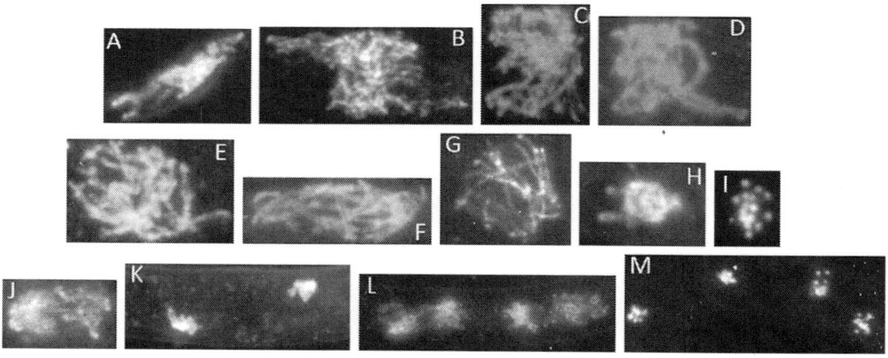

Fig. 3.3 Evolution of DAPI-stained chromosomes in developing asci of *B. cinerea*: (**a**) zygotene; (**b–f**) various stages of transition to pachytene; notice the double structure nature of bivalents in **b**; (**g**) late pachytene; (**h**) diplotene; (**i**) diakinesis; (**j**) anaphase I; (**k**) telophase I; (**l**) anaphase II; (**m**) the meiotic tetrad. Up to 16 chromosomes were counted in some preparations in late pachytene and diachinesis stages

The latter estimates correspond to a number of 12–14 chromosomes that, according to the cytological evidence of 16 chromosomes, indicates some bands should correspond to more than 2 chromosomes. Indeed, the estimated genome size after genome sequencing corroborates this assumption (see Sect. 3.5 and Chap. 1). In some instances, ascospore progenies show the parental electrophoretic patterns, but often they show novel profiles. Vallejo et al. (2002) using an rDNA probe showed

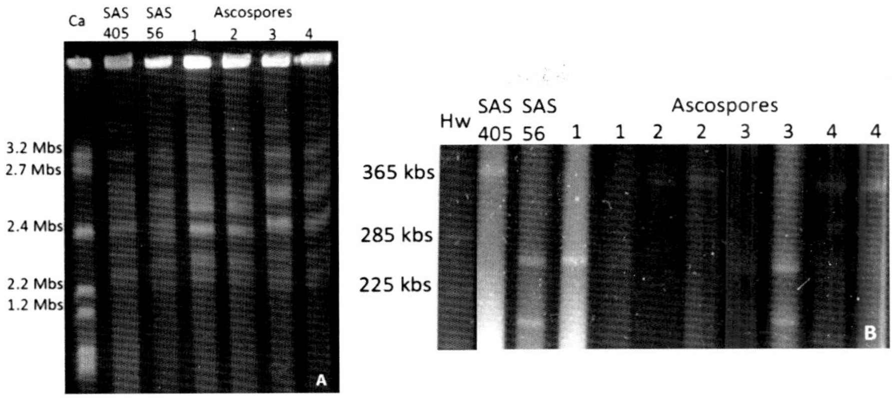

Fig. 3.4 Electrophoretic karyotype of the reference monosacosporic strains SAS56 and SAS405 and an ordered tetrad of the cross SAS405×SAS56 (**a**); minichromosomes of the reference monosacosporic isolates SAS56 and SAS405 and their segregation in an ordered ascus (**b**). Notice the parental combination of minichromosomes in the ascospore pairs n. 2, 3 and 4 and their anomalous segregation in the pair n. 1. *Ca Candida albicans*, *Hw Hansenula wingeii* used as weight markers

that rDNA gene cluster was located in a single high-molecular-weight band, which varied in size depending on the strain. Consequently, they hypothesized that the appearance of new chromosomal bands and loss of others is a result of meiotic crossing-over between pairs of homologous chromosomes having heterologous regions that generate variation in homologous chromosomes length. Small chromosomes show anomalous segregation following meiotic recombination, indicating they are supernumerary (dispensable or "B" chromosomes) (Fig. 3.4). The role of minichromosomes is still unknown, but De Guido et al. (unpublished) found that these hold genes coding for polyketide synthases (PKS), enzymes involved in the production of secondary metabolites that might play a role in pathogenicity (Chap. 15).

It is known that various forms of heteroploidy may occur in fungi. The ploidy level of various strains was examined by Büttner et al. (1994) and Büttner and Tudzynski (1996). They used fluorescence microscopy to estimate DNA content of individual nuclei and reported differences among strains and even among nuclei in the mycelium of a single isolate. These findings must be considered in light of knowledge that most strains are fertile in sexual crosses and yield viable ascospores. It may be that sexual crossing process restricts participation to haploid nuclei, whereas somatic cell function permits heteroploidy. If this is so, ascospores may initially be strictly haploid, but their growth may produce heteroploid colonies (Beever and Weeds 2004).

3.3 Sexual Behavior and Mating System

B. cinerea apothecia have been rarely found in nature, but apothecia might be overlooked in the field. The only reports are by Polach and Abawi (1975) in New York, Blank (1988) in Switzerland, and Vanev (1988) in Bulgaria. This suggests that the sexual process does not play an important role in the population biology of the fungus. As opposite, the broad variability among fungal isolates and the very high number of individual haplotypes (Van der Vlugt-Bergmans et al. 1993) and vegetative compatibility groups (VCGs) (Pollastro et al. unpublished; Beever and Weeds 2004, Chap. 4) would suggest that meiotic recombination occurs in the populations of the fungus. Additional indirect evidence for sexual recombination among natural *Botrytis* populations was provided by population genetics (see Chap. 6).

The mating system in *B. cinerea* was first investigated by Groves and Loveland (1953), who concluded that the fungus is heterothallic. This was confirmed by the obtainment of apothecia under laboratory conditions that showed that mating is controlled by a single locus (*MAT1*) with two idiomorphs. The mating type *MAT1-1* was arbitrarily assigned to the strain SAS56, and the *MAT1-2* to the strain SAS405 (Faretra et al. 1988a, b). Both mating types have been commonly found on different host plants and in different countries, although MAT1-1 isolates were lightly prevailing (Beever and Parkes 1993; Faretra and Pollastro 1993a); mating type indeed does not influence appreciably fitness (Pollastro et al. 1996).

Few *B. cinerea* isolates fertile with both *MAT1-1* and *MAT1-2* strains and often auto-fertile, referred to as pseudo-homothallic (MAT1-1/2) (Faretra and Grindle 1992) or 'dual-mater' (Amselem et al. 2011), were first detected by Lorenz and Eichhorn (1983) in Germany, and then reported in Europe and other countries (Beever and Parkes 1993; Faretra and Pollastro 1993a). Heterokaryosis is the main cause of secondary homothallism of field and monoconidial isolates of *B. cinerea*, since single multinucleate conidia may contain nuclei carrying opposite idiomorphs (Faretra et al. 1988b). This also indicates that the mating type idiomorphs do not cause vegetative incompatibility. Despite genetic and molecular investigations, genetic mechanisms causing secondary homothallism in monoascosporic isolates of the fungus have not yet been clarified. Indeed, all the nuclei in a multinucleate ascospore derive from mitotic divisions of a single meiotic nucleus as showed by cytological observations (Lorenz and Eichhorn 1983; Faretra and Antonacci 1987) as well as by tetrad analysis in complete asci, which also evidenced an apparent unidirectional 'switching' of mating behaviour since homothallic ascospores in complete asci were always expected to be *MAT1-2* (Faretra and Pollastro 1996).

Amselem et al. (2011) first investigated the structure of the *MAT1* locus in the sequenced genomes of *B. cinerea* strains B05.10 and T4, and identified two open reading frames (ORFs) in each idiomorph. These results were confirmed in strains SAS56 and SAS405 (De Miccolis Angelini et al. unpublished) (Fig. 3.5). The *MAT1-1* and *MAT1-2* idiomorphs display unique sequences of 2,513 bp and 2,776 bp, respectively with flanking sequences of the locus almost identical in the opposite mating types. Conventionally, the *MAT1-1* idiomorph contains a gene with the

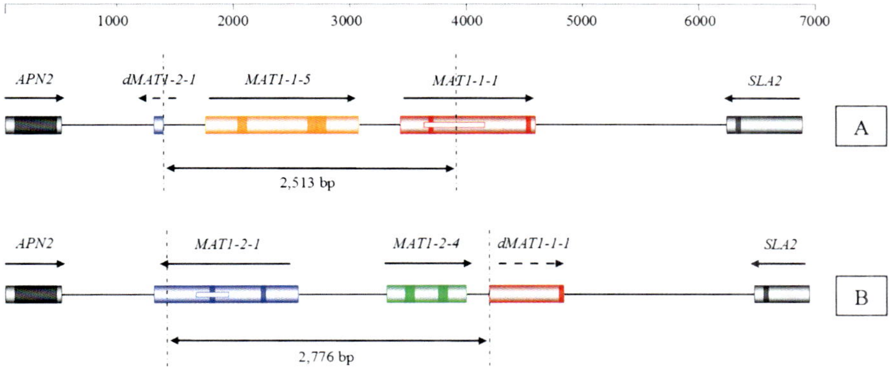

Fig. 3.5 Structures of the mating-type (*MAT1*) locus in *MAT1-1* (**a**) and *MAT1-2* (**b**) idiomorphs of *B. fuckeliana* and adjacent *APN2* (DNA lyase) and *SLA2* (cytoskeletal protein) genes in the near-identical flanking sequences. *Solid boxes* represent the predicted genes interrupted by introns (*full coloured boxes*). Internal *white box* in the *MAT1-1-1* (α-box) and *MAT1-2-1* (*HMG*-box) genes indicate the sequences encoding putative conserved domains. *Arrows* indicate orientation of coding sequences

alpha domain, while the *MAT1-2* a gene with the HMG domain. Hence, the previous casual attribution of the *MAT1-1* and *MAT1-2* mating type to SAS56 and SAS505 strains (Faretra et al. 1988b), respectively, was coincidentally right.

The two genes containing the mating-type specific domains (*MAT1-1-1* and *MAT1-2-1*) show high homology (82–86 %) with the corresponding genes of the closely-related homothallic species *S. sclerotiorum*. Both the alpha and the HMG domains are interrupted by an intron which is at the same position as that of homologous *MAT* genes identified in other ascomycetes (Debuchy and Turgeon 2006). Each idiomorph include an additional ORF (*MAT1-1-5* and *MAT1-2-4*) encoding a putative protein of unknown function which has a homologous sequence in the *MAT1* locus of *S. sclerotiorum*, but no in other fungal species (Amselem et al. 2011). The structural organization of the *MAT1* locus in *B. cinerea* monoascoporic strains showing homothallic behaviour did not display differences with the locus of heterothallic *MAT1-2* strains. Moreover, the *MAT1-1-1* gene sequence was never detected in the genome of homothallic strains (Amselem et al. 2011; De Miccolis Angelini et al. unpublished). Hence, the mating-type switching occurring in yeasts, with the transposition of a silent mating-type gene cassette present in genomic location other than the mating-type locus (Strathern et al. 1980), does not represent the mechanism underlying pseudo-homothallism in monoascosporic *B. cinerea* strains.

In *S. sclerotiorum* an inversion of a 3.6-kb region of the *MAT1* locus regularly occur during the sexual cycle. The rearrangement of this region originates two genetic variants (Inv+ and Inv−) of the *MAT1* locus both conferring a self-fertile homothallic behaviour. The inversion region includes the *MAT1-2-1* and the *MAT1-2-4* genes and a large portion of the *MAT1-1-1* gene and is flanked on either side by a 250-bp inverted repeat, probably involved in inversion loop formation and crossing-over events (Chitrampalam et al. 2013). In *B. cinerea*, the homologous

250-bp motif is uniquely found within the complete or deleted *MAT1-1-1* gene respectively in the *MAT1-1* and in the *MAT1-2* idiomorphs close to the near identical 3′-flanking region. During meiotic chromosome pairing the idiomorphic sequences at *MAT1* locus, highly dissimilar in strains of opposite mating type but flanked by homologous sequences, might originate a large region of heteroduplex DNA resulting in single strand loops (unpaired regions). It might be responsible for unequal recombination events and gene conversion at *MAT1* locus with structural changes, probably unstable or transient.

Expression analysis by RT-PCR revealed that all the mating-type genes of *B. cinerea* are transcriptionally active at low level in different development stages (mycelium, mature sclerotia, carpogenesis-induced sclerotia and fully developed apothecia) (De Miccolis Angelini et al. unpublished), like it occurs in other fungal species, suggesting their possible role in processes other than mating (Ferreira et al. 1996; Coppin and Debuchy 2000).

Terhem and Van Kan (unpublished) generated mutants in the four genes of *B. cinerea MAT1* locus in both SAS56 and SAS405 strains and crossed them with appropriate wild-type strain or knockout mutant. Knockout mutants of the *MAT1-1-1* gene and the *MAT1-2-1* gene proved sterile, thus confirming the essential role of the alpha-domain protein and the HMG-box protein in the mating process. The mutants of the *MAT1-1-5* gene and the *MAT1-2-4* gene were able to produce stipes but failed to develop apothecial disks, evidencing that these two genes jointly play a key role in disk development. Crosses performed by Terhem and Van Kan (2014), with different knockout mutants in the genes *Bhp1* (class I hydrophobin), *Bhl1* (hydrophobin-like), *Bhp2* and *Bhp3* (class II hydrophobins), showed that hydrophobins *Bhp1* and *Bhp3* are important for normal development of sclerotia, and hydrophobins *Bhp1* and *Bhp2* are important in sclerotia for normal development of apothecia.

B. cinerea has been used as model to explore a gene conversion event occurring during meiosis, known as 'homing'. The homing process is mediated by site-specific endonuclease (HEG), typically encoded within self-splicing genetic elements, such as group I introns and inteins, and results in unidirectional inheritance of genomic regions including their host gene and flanking sequences. Analysis of meiotic products from single heterozygous nuclei generated by crossing strains carrying an HEG-containing intein in the *Prp8* gene with strains carrying no intein in the same gene, permitted to estimate the extension of gene conversion tracts, including at least the 25-bp sequence immediately flanking the intein and often affecting a larger region (up to 451 bp from the insertion site), and to demonstrated that intein homing was highly efficient, affecting all ascospore progenies. Hence, inteins represent potential hotspots for meiotic recombination in fungi (Bokor et al. 2010).

Examination of transposable element (TE) sequences in *B. cinerea* and *S. sclerotiorum* provided an evidence for RIP (Repeat-Induced Point mutation), a mechanism of genomic modifications occurring during meiosis and resulting in the irreversible inactivation of duplicated sequences (Galagan and Selker 2004; Martinez et al. 2008). RIP intensity is variable among TEs and seems to be less efficient in *Sclerotinia* than in *Botrytis* (Lebrun and Dickman, unpublished). This mechanism, firstly identified in *Neurospora crassa*, is common in fungi and is considered as a protective strategy against deleterious effects of transposon mobility or recombination between duplicated sequences.

3.4 Other Sources of Variation

Other sources of variation in *B. cinerea* are represented by extrachromosomal genetic elements, such as mitochondrial DNA (De Miccolis Angelini et al. 2012a; Yin et al. 2012), plasmids (Hiratsuka et al. 1987) and mycoviruses.

Since their first observation by Howitt et al. (1995), mycoviruses infecting *B. cinerea* isolates have been reported by several authors (reviewed in Chap. 5; Pearson and Bailey 2013) and have been occasionally characterized (De Guido et al. 2005; Wu et al. 2007, 2010). The genome of most mycoviruses consists of dsRNA and is often associated with isometric, bacilliform or filamentous virus-like particles (VLPs) (Howitt et al. 1995; Santomauro et al. unpublished). Two flexuous rod-shaped mycoviruses, with single-strand RNA (ssRNA) genomes, have also been isolated and characterized (Howitt et al. 2001, 2006); because of their sequence similarity to plant viruses (i.e., *Garlic virus A*), it has been suggested that transmission of genetic materials can occur between plants and *B. cinerea* (Howitt et al. 2006). Analysis of dsRNA electropherotypes of infected isolates indicates a broad variation in number, molecular size and abundance of viral genomic dsRNA segments, thus suggesting a frequent occurrence of mixed infections of multiple viruses in single isolates of the fungus. The majority of mycoviruses in *B. cinerea* seems to be in a cryptic state (Howitt et al. 1995; Santomauro et al. unpublished), but in few cases dsRNAs were associated to hypovirulence (Wu et al. 2007, 2010; Zhang et al. 2010; Potgieter et al. 2013). Mycoviruses are efficiently transmitted to conidia, while transmission to ascospore offspring does not occur (Santomauro et al. unpublished) or is incomplete (Pearson et al. 2009).

Vegetative incompatibility, frequently observed in *B. cinerea*, constitutes a strong barrier to mycovirus spreading in fungal populations. It prevents hyphal anastomosis between isolates belonging to different VCGs, as demonstrated by the use of auxotrophic mutants (Pollastro 1996; Beever and Parkes 2003; Korolev et al. 2008), while the exchange of both nuclear and cytoplasmic elements between compatible isolates can be responsible for establishment of a heterokaryotic or heteroplasmic state. A large number of VCGs (more than 80) has been identified in *B. cinerea*, consistent with the existence of a very high genetic diversity in natural populations, with several VCGs represented by only one or few isolates (Beever and Weeds 2004; Chap. 4).

3.5 Genomics and Transcriptomics

The first *B. cinerea* expressed sequence tags (EST) dataset was generated by Viaud et al. (2005) providing a useful resource for gene discovery (3,026 unigenes) and identification of putative virulence factors (cell wall degrading enzymes, signal transduction components and secondary metabolism enzymes). Silva et al. (2006) sequenced another 11,482 ESTs from an isolate of *B. cinerea*. Much larger datasets have become available through next-generation-RNA-sequencing (http://www.ncbi.nlm.nih.gov/gds/?term=(Botrytis+cinerea[Organism])).

The first genome assemblies of strains B05.10 (Syngenta AG) and T4 (Genoscope) were obtained by using Sanger sequencing technology at low coverage (4.5 and 10 fold sequence depth, respectively). The draft assembly of B05.10 amounts to 42.3 Mb and consists of 588 scaffolds (*N*50 = 257 kb). The draft assembly of T4 amounts to 39.5 Mb and consists of 118 scaffolds (*N*50 = 562 kb) (Amselem et al. 2011). The comparative analysis of the genomes of the *B. cinerea* strains B05.10, T4 and one strain of *S. sclerotiorum* showed that they include 11,860–14,270 predicted genes, which share 83 % amino acid identity on average between the two species. The *S. sclerotiorum* genome was assembled into 16 chromosomes and found large-scale co-linearity with the *B. cinerea* genomes. The *S. sclerotiorum* genome consists of transposable elements for 7 % compared to ~1 % of *B. cinerea*. The arsenal of genes associated with necrotrophic processes, including genes involved in plant cell wall degradation and oxalic acid production, is similar between the species. Analysis of secondary metabolism gene clusters revealed a higher number and broader diversity of *B. cinerea* secondary metabolites as compared to *S. sclerotiorum*, which might be involved in adaptation to specific ecological niches (Chap. 15). No genomic features of the two species distinguish them from other pathogenic and non-pathogenic fungi, reinforcing the hypothesis of a quantitative, multigenic nature of necrotrophic pathogenesis (Amselem et al. 2011).

Staats and Van Kan (2012) improved genome assemblies of B05.10 (GenBank accession number AAID00000000) and T4 (ALOC00000000) by using NGS Illumina technology. The quality-trimmed data consisted of 81.1 million sequencing reads (3.94 Gb) for B05.10 and 114.6 million sequencing reads (6.38 Gb) for T4. The adopted gene finding strategy resulted in 10,427 and 10,467 predicted protein-coding gene structures in B05.10 and in T4, respectively.

More recently, it has been released the draft genome sequence of the BcDW1 *B. cinerea* strain, used as inoculum to induce noble rot in the production of botrytized wine (GenBank accession number AORW00000000), obtained with NGS technology. A number of 11,073 complete gene models have been identified that also include secreted proteins, such as laccases and carboxylesterases, which might be relevant for noble rot (Blanco-Ulate et al. 2013).

Transcriptomic analyses performed during plant-infection (Gioti et al. 2006; Mulema and Denby 2012), conidia germination (Leroch et al. 2013) or after pectate induction (Zhang et al. 2014) permitted to identify a number of fungal genes significantly altered in their expression profiles during pathogenesis-related development and host plant colonization, as well as genes involved in plant defence responses to *B. cinerea* infection (see also Chap. 17). Also proteomic approaches were developed in *B. cinerea* to identify potential virulence factors (see Chap. 16).

3.6 Transposons and Population Biology

About 1.3 % of the *B. cinerea* genome consists of repetitive sequences (Staats and Van Kan 2012). Transposable elements (TEs) belonging to class I retroelements (LTRs) or class II DNA transposons (MITEs, TIRs) have been identified (Amselem et al. 2011). These mobile repetitive DNA elements represent dynamic components of genomes. By transposing themselves from one genome location to another one, they are recognized as a potential cause of structural rearrangements of DNA sequences and chromosomes, gene duplication, inactivation and changes in gene expression. Their presence in *B. cinerea* might explain its plasticity and genetic variability and can be considered as an addition source of diversity in genome evolution. While some TEs within the fungal genome are presumably inactive and could be considered as a genomic remnants of ancient transposition events (Deng et al. 2013), several studies suggested that TEs can still contribute to the genetic diversity within *B. cinerea* populations. Two types of strains, named *transposa* and *vacuma*, have been distinguished on the basis of two transposons, *Boty* and *Flipper* (Giraud et al. 1999). Differences in biological and pathogenic behavior of *transposa* and *vacuma* strains have been associated with a 'partial specialization' for plant organs and season timing along with differences in their response to fungicides (Martinez et al. 2005; Pollastro et al. 2007; Samuel et al. 2012).

3.7 Improvements of Genetic Tools

Great efforts have been made during recent years to establish methods and protocols for studying gene function and pathogenesis determinants at molecular level in *B. cinerea*. DNA extraction protocols for *B. cinerea* generally start with mycelium reduced to powder with liquid nitrogen. Gonzáles et al. (2008) adapted a method based on the homogenization of the fungus with sand and the aid of a common household drill for processing of large number of samples in short times and generating DNA of sufficient quality for PCR and Southern blotting. Genetic transformation was first achieved with genes conferring resistance to hygromycin B (Hamada et al. 1994) or phleomycin (Santos et al. 1996), and the nitrate reductase gene (*niaD*)

(Levis et al. 1997). Ish-Shalom et al. (2011) demonstrated successful transformation with three different DNA constructs using both air-pressure- and sclerotium-mediated transformation, methods which proved fast, simple and reproducible as compared to protocols based on protoplasts. *Agrobacterium tumefaciens*–mediated transformation (ATMT) was first applied to *B. cinerea* by Rolland et al. (2003) and has been shown to be a useful tool for random mutagenesis and the identification of new virulence-associated genes (Giesbert et al. 2012).

Gene replacement is crucial for demonstrating gene function. Constructs without homologous sequences are rarely integrated into the genome and result in low transformation efficiencies. Despite the relatively high frequency of homologous recombination observed in *B. cinerea* constructs containing *B. cinerea* sequences may integrate at undesired gene loci. The targeted integration of constructs at defined loci in the genome is more advantageous as it allows for generation of transformants with identical genomic backgrounds. Noda et al. (2007) developed a system for site-directed integration of expression cassettes by exploiting homologous recombination allowed by the use of a recipient strain and a vector both carrying a truncated copy of the hygromycin resistance gene. Schumacher (2012) constructed a set of basic cloning vectors by yeast recombinational cloning permitting a variety of protein fusions at defined genomic *loci*. These cloning vectors facilitate promoter studies by the insertion of promoters of interest upstream of fluorescent reporter genes, the localization of *B. cinerea* proteins in living hyphae and finally the identification of protein–protein interactions in situ by bimolecular fluorescence complementation (BiFC) assays.

RNAi mediated gene silencing was also developed in *B. cinerea*, but due to the variable silencing levels, interpretations are difficult and gene knockouts for non-essential genes proved to be easier to handle (Patel et al. 2008a; Schumacher et al. 2008).

Molecular tools based on the expression of reporter gene constructs in *B. cinerea*, which is crucial for sub-cellular localization of proteins and identification of protein complexes, have been hampered for long time by their weak expression levels and insufficient fluorescence signal when expression vectors were derived from other fungal systems. Patel et al. (2008b) developed and validated the pOT and pLOB vector systems with a range of exogenous genes, including enhanced green fluorescent protein (eGFP), monomeric red fluorescent protein (mRFP), luciferase (LUC) and ß-glucoronidase (GUS). eGFP was successfully used by Gourgues et al. (2004) and Shlezinger et al. (2011). All these reports showed that the use of regulatory sequences of *B. cinerea* enhances the expression of reporter genes. Moreover, Leroch et al. (2011) obtained improved transcript levels and increased fluorescence by using eGFP and mCherry with codons optimized for *B. cinerea* in addition to an intron in the *bcgfp* gene that further enhances gene expression.

Obligate biotrophic pathogens, such as *Blumeria graminis* f.sp. *tritici* (*Bgt*), are not easily handled under laboratory conditions and their genetic transformation is difficult. *B. cinerea* was successfully used as a heterologous expression system for the characterization of the *Bgt CYP51* gene, encoding 14α-demethylase and involved in resistance to sterol demethylation inhibitors. The gene was transcribed in *B. cinerea* and transformed strains showed reduced sensitivity to the fungicides. These results indicated that this heterologous expression system can be used for functional analysis of other genes of *Bgt* and possibly of other biotrophic pathogens (Yan et al. 2012).

3.8 Genetics of Fungicide Resistance

Acquired resistance to fungicides in phytopathogenic fungi is a relevant challenge in modern crop protection. Due to its broad genetic variability and adaptation capabilities, *B. cinerea* is recognized as a 'high-risk' pathogen, and acquired resistance has been experienced with all the fungicides used against grey mould (www.frac.info; Chap. 10). Genetic bases of resistance to fungicides in *B. cinerea* were reviewed by Leroux (2004) and De Miccolis Angelini et al. (2012b) and in Chap. 10 of this book. Field resistant mutants can be easily derived from naturally infected plants exposed to intensive spray schedules. The diversity of the genetic background of field *B. cinerea* isolates represents a major obstacle to the understanding of genetic bases of resistant phenotypes. Laboratory resistant mutants can be generated from wild-type sensitive strains by direct exposure to appropriate fungicide concentrations to select either spontaneous mutations or mutations induced by physical or chemical mutagens (e.g., De Guido et al. 2007; De Miccolis Angelini et al. 2010). However, the genetic and physiological mechanisms underlying resistance of laboratory mutants might not reflect those occurring in the field (Grindle 1987; Grindle and Faretra 1993). Both field and laboratory resistant mutants often are unstable due to their heterokaryotic or heteroplasmic state. Repeated monoconidial subcultures proved resistant under selective conditions rather than mass-transferred mycelial or conidial cultures are preferentially used to reduce instability in resistant mutants.

The use of well characterized reference strains to obtain isogenic resistant mutants, having virtually identical genomes but differing for the mutated genes, makes it possible to identify the role of specific allelic variants on fungicide resistance and their pleiotropic effects on fitness. This can be accomplished with time- and labour-expensive classic backcrosses of resistant mutants with a recurrent reference strains and selection of suitable offspring (Pollastro et al. 1996) or in a much more quick and easy way by in vitro site-directed mutagenesis (e.g., Billard et al. 2012; Lalève et al. 2013).

As in other fungal pathogens, the most common mechanism of *B. cinerea* resistance to site-specific fungicides is the occurrence of point mutations in major genes underlying the alteration of the target site protein which reduces fungicide binding. A large number of single nucleotide polymorphisms (SNPs) conferring resistance to several groups of fungicides have been identified in field and laboratory resistant mutants of *B. cinerea* (Chap. 10) and can be used as molecular markers for quickly detection and quantification of resistant genotypes in field samples of the pathogen.

Mutations conferring fungicide resistance frequently have negative effects on the fitness of resistant mutants, compromising their chances to compete with wild-type sensitive strains and to survive in nature. For example, mutations in the *Daf1* gene conferring a high level of resistance (HR) to dicarboximides and resistance to phenylpyrroles in laboratory mutants of *B. cinerea* cause, as a pleiotropic effect, a strong reduction of tolerance to high osmotic pressure compared with wild-type sensitive (S) isolates or low resistant (LR) mutants occurring in the field (Faretra and Pollastro 1993b; Pollastro et al. 1996). Moreover, heterokaryon can overcome negative pleiotropic effects of the mutations responsible for fungicide resistance by maintaining wild-type sensitive nuclei and mutate nuclei in a balanced state. For instance, the coexistence of *Daf1HR* nuclei with *Daf1S* or *Daf1LR* nuclei can generate a phenotype highly resistant to dicarboximides showing an almost normal osmotolerance. This is possible because the high resistance is dominant (or partially dominant) on sensitivity, and hypersensitivity to high osmolarity is recessive (or partially recessive) on normal osmotolerance (Faretra and Pollastro 1993a).

So far, there are few examples of genes conferring resistance to antifungal compounds located on extrachromosomal determinants. In several fungal pathogens resistance to QoI fungicides is caused by the G143A mutation in the mitochondrial *cytb* gene (Gisi et al. 2002). In *B. cinerea*, the same mutation has been detected in both field and laboratory QoI-resistant mutants that are often in a heteroplasmic state in which resistant and sensitive mitochondria co-exist in the same hyphal compartment (Banno et al. 2009; Ishii et al. 2009; Jiang et al. 2009). Classical genetic analysis of progeny from sexual crosses between resistant and sensitive isolates confirmed that resistance is due to cytoplasmic determinants showing maternal (i.e. sclerotial) inheritance (De Miccolis Angelini et al. 2012a). Sequence analysis of the *cytb* gene in *B. cinerea* has evidenced a highly polymorphic genetic structure leading to discriminate up to six different types of *cytb* gene on the bases of the presence or absence of four intronic sequences (Yin et al. 2012). The presence of a group I self-splicing intron, adjacent to the codon 143 in sensitive isolates of the fungus, prevents the occurrence of the G143A mutation since it would be lethal impeding the correct excision of the intronic sequence from the gene transcript (Grasso et al. 2006).

References

Aguileta G, Lengelle J, Chiapello H et al (2012) Genes under positive selection in a model plant pathogenic fungus, *Botrytis*. Infect Genet Evol 12:987–996

Amselem J, Cuomo CA, Van Kan JA et al (2011) Genomic analysis of the necrotrophic fungal pathogens *Sclerotinia sclerotiorum* and *Botrytis cinerea*. PLoS Genet 7:e1002230

Banno S, Yamashita K, Fukumori F et al (2009) Characterization of QoI resistance in *Botrytis cinerea* and identification of two types of mitochondrial cytochrome b gene. Plant Pathol 58:120–129

Beever RE, Parkes SL (1993) Mating behaviour and genetics of fungicide resistance of *Botrytis cinerea* in New Zealand. N Z J Crop Hortic Sci 21:303–310

Beever RE, Parkes SL (2003) Use of nitrate non-utilising (Nit) mutants to determine vegetative compatibility in *Botryotinia fuckeliana* (*Botrytis cinerea*). Eur J Plant Pathol 109:607–613

Beever RE, Weeds PL (2004) Taxonomy and genetic variation of *Botrytis* and *Botryotinia*. In: Elad Y, Williamson B, Tudzynski P, Delen N (eds) *Botrytis*: biology, pathology and control. Kluwer Academic Publishers, Dordrecht, pp 29–52

Billard A, Fillinger S, Leroux P et al (2012) Strong resistance to the fungicide fenhexamid entails a fitness cost in *Botrytis cinerea*, as shown by comparisons of isogenic strains. Pest Manag Sci 68:684–691

Blanco-Ulate B, Allen G, Powell AL et al (2013) Draft genome sequence of *Botrytis cinerea* BcDW1, inoculum for noble rot of grape berries. Genome Announc 1:e00252–13

Blank P (1988) Two inoperculate discomycetes on grapevine. Z Mycol 55:115–118

Bokor AA, Van Kan JA, Poulter RT (2010) Sexual mating of *Botrytis cinerea* illustrates PRP8 intein HEG activity. Fungal Genet Biol 47:392–398

Büttner P, Tudzynski P (1996) Variation in DNA content and chromosome numbers of *Botrytis cinerea*. In: Abstract of the XI international *Botrytis* symposium, Wageningen, 23–27 June 1996

Büttner P, Koch F, Voigt K et al (1994) Variations in ploidy among isolates of *Botrytis cinerea*: implications for genetic and molecular analyses. Curr Genet 25:445–450

Chitrampalam P, Inderbitzin P, Maruthachalam K et al (2013) The *Sclerotinia sclerotiorum* mating type locus (*MAT*) contains a 3.6-kb region that is inverted in every meiotic generation. PLoS One 8:e56895

Coppin E, Debuchy R (2000) Co-expression of the mating-type genes involved in internuclear recognition is lethal in *Podospora anserina*. Genetics 155:657–669

De Guido MA, Minafra A, Santomauro A, Pollastro S, De Miccolis Angelini RM, Faretra F (2005) Molecular characterization of mycoviruses from *Botryotinia fuckeliana*. J Plant Pathol 87:293

De Guido MA, De Miccolis Angelini RM et al (2007) Selection and genetic analysis of laboratory mutants of *Botryotinia fuckeliana* resistant to fenhexamid. J Plant Pathol 89:203–210

De Miccolis Angelini RM, Habib W, Rotolo C et al (2010) Selection, characterization and genetic analysis of laboratory mutants of *Botryotinia fuckeliana* (*Botrytis cinerea*) resistant to the fungicide boscalid. Eur J Plant Pathol 128:185–199

De Miccolis Angelini RM, Rotolo C, Masiello M et al (2012a) Genetic analysis and molecular characterisation of laboratory and field mutants of *Botryotinia fuckeliana* (*Botrytis cinerea*) resistant to QoI fungicides. Pest Manag Sci 68:1231–1240

De Miccolis Angelini RM, Pollastro S, Faretra F (2012b) Genetics of fungicide resistance in *Botryotinia fuckeliana* (*Botrytis cinerea*). In: Thind TS (ed) Fungicide resistance in crop protection: risk and management. CAB International, Wallingford, pp 237–250

Dean R, Van Kan JA, Pretorius ZA et al (2012) The top 10 fungal pathogens in molecular plant pathology. Mol Plant Pathol 13:414–430

Debuchy R, Turgeon BG (2006) Mating-type structure, evolution, and function in *Euascomycetes*. In: Kües U, Fischer R (eds) The mycota I. Growth, differentiation, and sexuality. Springer, Berlin, pp 293–323

Deng H, Shu D, Luo D et al (2013) *Scatter*: a novel family of miniature inverted-repeat transposable elements in the fungus *Botrytis cinerea*. J Basic Microbiol 53:815–822

Faretra F, Antonacci E (1987) Production of apothecia of *Botryotinia fuckeliana* (de Bary) Whetz. under controlled environmental conditions. Phytopathol Mediterr 26:29–35

Faretra F, Grindle M (1992) Genetic studies of *Botryotinia fuckeliana* (*Botrytis cinerea*). In: Verhoeff K, Malathrakis NE, Williamson B (eds) Recent advances in *Botrytis* research. PUDOC, Wageningen, pp 7–17

Faretra F, Pollastro S (1993a) Genetics of sexual compatibility and resistance to benzimidazole and dicarboximide fungicides in isolates of *Botryotinia fuckeliana* (*Botrytis cinerea*) from nine countries. Plant Pathol 42:48–57

Faretra F, Pollastro S (1993b) Isolation, characterization and genetic analysis of laboratory mutants of *Botryotinia fuckeliana* resistant to the phenylpyrrole fungicide CGA 173506. Mycol Res 97:620–624

Faretra F, Pollastro S (1996) Genetic studies of the phytopathogenic fungus *Botryotinia fuckeliana* (*Botrytis cinerea*) by analysis of ordered tetrads. Mycol Res 100:620–624

Faretra F, Antonacci E, Pollastro S (1988a) Improvement of the technique used for obtaining apothecia of *Botryotinia fuckeliana* (*Botrytis cinerea*) under controlled conditions. Ann Microbiol 38:29–40

Faretra F, Antonacci E, Pollastro S (1988b) Sexual behaviour and mating system of *Botryotinia fuckeliana*, teleomorph of *Botrytis cinerea*. J Gen Microbiol 134:2543–2550

Ferreira AVB, Saupe S, Glass NL (1996) Transcriptional analysis of the *mtA* idiomorph of *Neurospora crassa* identifies two genes in addition to *mtA-1*. Mol Gen Genet 250:767–774

Galagan JE, Selker EU (2004) RIP: the evolutionary cost of genome defense. Trends Genet 20:417–423

Giesbert S, Schumacher J, Kupas V et al (2012) Identification of pathogenesis-associated genes by T-DNA-mediated insertional mutagenesis in *Botrytis cinerea*: a type 2A phosphoprotein phosphatase and an SPT3 transcription factor have significant impact on virulence. Mol Plant Microbe Interact 25:481–495

Gioti A, Simon A, Le Pêcheur P et al (2006) Expression profiling of *Botrytis cinerea* genes identifies three patterns of up-regulation in planta and an FKBP12 protein affecting pathogenicity. J Mol Biol 358:372–386

Giraud T, Fortini D, Levis C et al (1999) Two sibling species of the *Botrytis cinerea* complex, *transposa* and *vacuma*, are found in sympatry on numerous host plants. Phytopathology 89:967–973

Gisi U, Sierotzki H, Cook A et al (2002) Mechanisms influencing the evolution of resistance to Qo inhibitor fungicides. Pest Manag Sci 58:859–867

González C, Noda J, Espino JJ et al (2008) Drill-assisted genomic DNA extraction from *Botrytis cinerea*. Biotechnol Lett 30:1989–1992

Gourgues M, Brunet-Simon A, Lebrun MH et al (2004) The tetraspanin BcPls1 is required for appressorium-mediated penetration of *Botrytis cinerea* into host plant leaves. Mol Microbiol 51:619–629

Grasso V, Palermo S, Sierotzki H et al (2006) Cytochrome b gene structure and consequences for resistance to Qo inhibitor fungicides in plant pathogens. Pest Manag Sci 62:465–472

Grindle M (1979) Phenotypic differences between natural and induced variants of *Botrytis cinerea*. J Gen Microbiol 111:109–120

Grindle M (1987) Genetic basis of fungicide resistance. In: Ford MG, Hollomon DH, Khambay BPA, Sawicki RM (eds) Combating resistance to xenobiotics. Ellis Horwood Ltd, Chichester, pp 75–93

Grindle M, Faretra F (1993) Genetic aspects of fungicide resistance. In: Lyr H, Polter C (eds) Modern fungicides and antifungal compounds, Proceedings of the 10th International Symposium, Reinhardsbrunn (Germany), May 1992. Ulmer, Stuttgart, pp 33–43

Groves JW, Loveland CA (1953) The connection between *Botryotinia fuckeliana* and *Botrytis cinerea*. Mycologia 45:415–425

Hamada W, Reignault P, Bompeix G et al (1994) Transformation of *Botrytis cinerea* with the hygromycin B resistance gene, *hph*. Curr Genet 26:251–255

Hennebert GL (1973) *Botrytis* and *Botrytis*-like genera. Persoonia 7:183–204

Hiratsuka K, Namba S, Yamashita S et al (1987) Linear plasmid-like DNA's in the fungus *Botrytis cinerea*. Nippon Shokubutsu Byori Gakkaiho 53:638–642

Howitt RLJ, Beever RE, Pearson MN et al (1995) Presence of double-stranded RNA and virus-like particles in *Botrytis cinerea*. Mycol Res 99:1472–1478

Howitt RLJ, Beever RE, Pearson MN et al (2001) Genome characterization of Botrytis virus F, a flexuous rod-shaped mycovirus resembling plant 'potex-like' viruses. J Gen Virol 82:67–78

Howitt RLJ, Beever RE, Pearson MN et al (2006) Genome characterization of a flexuous rod-shaped mycovirus, *Botrytis* virus X, reveals high amino acid identity to genes from plant 'potex-like' viruses. Arch Virol 151:563–579

IFAS (2010) Interactive strawberry budget. Plant city, 2008–2009 season. http://www.fred.ifas.ufl.edu/iatpc/files/strawberries09I.xls

Ishii H, Fountaine J, Chung WH et al (2009) Characterisation of QoI-resistant field isolates of *Botrytis cinerea* from citrus and strawberry. Pest Manag Sci 65:916–922

Ish-Shalom S, Gafni A, Lichter A et al (2011) Transformation of *Botrytis cinerea* by direct hyphal blasting or by wound-mediated transformation of sclerotia. BMC Microbiol 11:266

Jarvis WR (1977) *Botryotinia* and *Botrytis* species; taxonomy, physiology and pathogenicity, Monograph No. 15. Canadian Department of Agriculture, Ottawa

Jiang J, Ding L, Michailides TJ et al (2009) Molecular characterization of field azoxystrobin-resistant isolates of *Botrytis cinerea*. Pestic Biochem Physiol 93:72–76

Korolev N, Elad Y, Katan T (2008) Vegetative compatibility grouping in *Botrytis cinerea* using sulphate non-utilizing mutants. Eur J Plant Pathol 122:369–383

Lalève A, Gamet S, Walker AS et al (2013) Site-directed mutagenesis of the P225, N230 and H272 residues of succinate dehydrogenase subunit B from *Botrytis cinerea* highlights different roles in enzyme activity and inhibitor binding. Environ Microbiol. doi:10.1111/1462-2920.12282

Leroch M, Mernke D, Koppenhoefer D et al (2011) Living colors in the gray mold pathogen *Botrytis cinerea*: codon-optimized genes encoding green fluorescent protein and mCherry, which exhibit bright fluorescence. Appl Environ Microbiol 77:2887–2897

Leroch M, Kleber A, Silva E et al (2013) Transcriptome profiling of *Botrytis cinerea* conidial germination reveals upregulation of infection-related genes during the prepenetration stage. Eukaryot Cell 12:614–626

Leroux P (2004) Chemical control of *Botrytis* and its resistance to chemical fungicides. In: Elad Y, Williamson B, Tudzynski P, Delen N (eds) *Botrytis*: biology, pathology and control. Kluwer Academic Publishers, Dordrecht, pp 195–222

Levis C, Fortini D, Brygoo Y (1997) Transformation of *Botrytis cinerea* with the nitrate reductase gene (*niaD*) shows a high frequency of homologous recombination. Curr Genet 32:157–162

Lorenz DH, Eichhorn KW (1983) Untersuchungen an Botryotinia fuckeliana Whetz., dem Perfektstadium von *Botrytis cinerea* Pers. Z Pflanzenkranh Pflanzenschutz 90:1–11

Mansfield JW (1980) Mechanisms of resistance to *Botrytis*. In: Coley- Smith JR, Verhoeff K, Jarvis WR (eds) The biology of *Botrytis*. Academic, London, pp 181–218

Martinez F, Dubos B, Fermaud M (2005) The role of saprotrophy and virulence in the population dynamics of *Botrytis cinerea* in vineyards. Phytopathology 95:692–700

Martinez F, Corio-Costet MF, Levis C et al (2008) New PCR primers applied to characterize distribution of *Botrytis cinerea* populations in French vineyards. Vitis 47:217–226

Mulema JM, Denby KJ (2012) Spatial and temporal transcriptomic analysis of the *Arabidopsis thaliana-Botrytis cinerea* interaction. Mol Biol Rep 39:4039–4049

Noda J, Brito N, Espino JJ et al (2007) Methodological improvements in the expression of foreign genes and in gene replacement in the phytopathogenic fungus *Botrytis cinerea*. Mol Plant Pathol 8:811–816

Patel RM, Van Kan JA, Bailey AM et al (2008a) RNA-mediated gene silencing of superoxide dismutase (*bcsod1*) in *Botrytis cinerea*. Phytopathology 98:1334–1339

Patel RM, Heneghan MN, Van Kan JA et al (2008b) The pOT and pLOB vector systems: improving ease of transgene expression in *Botrytis cinerea*. J Gen Appl Microbiol 54:367–376

Pearson MN, Bailey AM (2013) Viruses of *Botrytis*. Adv Virus Res 86:249–272

Pearson MN, Beever RE, Boine B, et al (2009) Can mycoviruses be used for the biocontrol of the plant pathogenic fungus *Botrytis cinerea*? In: Elad Y, Maurhofer M, Keel C, Gessler C, Duffy B (eds), Integrated control of plant pathogens. Proceedings of the meeting "Molecular Tools for Understanding and Improving Biocontrol", IOBC/WPRS Interlaken (Switzerland), 43, pp 7–10, 9–12 September 2008

Polach FJ, Abawi GS (1975) The occurrence and biology of *Botryotinia fuckeliana* on beans in New York. Phytopathology 65:657–660

Pollastro S (1996) Study of phenotypic variability of *Botryotinia fuckeliana* with particular regard to fungicide resistance and pathogenicity mechanisms. PhD thesis, University of Bari, Italy [in Italian]

Pollastro S, Faretra F, Santomauro A et al (1996) Studies on pleiotropic effects of mating type, benzimidazole-resistance and dicarboximide-resistance genes in near-isogenic strains of *Botryotinia fuckeliana* (*Botrytis cinerea*). Phytopathol Mediterr 35:48–57

Pollastro S, De Miccolis Angelini RM, Rotolo C, et al (2007) Characterisation of *vacuma* and *transposa* biotypes of *Botryotinia fuckeliana*. In: Abstracts of the XIV international Botrytis symposium, Cape Town, 21–26 October 2005

Potgieter CA, Castillo A, Castro M et al (2013) A wild-type *Botrytis cinerea* strain co-infected by double-stranded RNA mycoviruses presents hypovirulence-associated traits. Virol J 10:220

Rolland S, Jobic C, Fèvre M, Bruel C (2003) *Agrobacterium*-mediated transformation of *Botrytis cinerea*, simple purification of monokaryotic transformants and rapid conidia-based identification of the transfer-DNA host genomic DNA flanking sequences. Curr Genet 44:164–171

Samuel S, Veloukas T, Papavasileiou A et al (2012) Differences in frequency of transposable elements presence in *Botrytis cinerea* populations from several hosts in Greece. Plant Dis 96:1286–1290

Santos M, Vallejo I, Rebordinos L et al (1996) An autonomously replicating plasmid transforms *Botrytis cinerea* to phleomycin resistance. FEMS Microbiol Lett 137:153–158

Schumacher J (2012) Tools for *Botrytis cinerea*: New expression vectors make the gray mold fungus more accessible to cell biology approaches. Fungal Genet Biol 49:483–497

Schumacher J, Viaud M, Simon A et al (2008) The Gα subunit BCG1, the phospholipase C (BcPLC1) and the calcineurin phosphatase co-ordinately regulate gene expression in the grey mould fungus *Botrytis cinerea*. Mol Microbiol 67:1027–1050

Shirane N, Masuko M, Hayashi Y (1988) Nuclear behavior and division in germinating conidia of *Botrytis cinerea*. Phytopathology 78:1627–1630

Shirane N, Masuko M, Hayashi Y (1989) Light microscopic observation of nuclei and mitotic chromosomes of *Botrytis* species. Phytopathology 79:728–730

Shlezinger N, Minz A, Gur Y et al (2011) Anti-apoptotic machinery protects the necrotrophic fungus *Botrytis cinerea* from host-induced apoptotic-like cell death during plant infection. PLoS Pathog 7:e1002185

Silva E, Valdés J, Holmes D et al (2006) Generation and analysis of expressed sequence tags from *Botrytis cinerea*. Biol Res 39:367–376

Staats M, Van Kan JA (2012) Genome update of *Botrytis cinerea* strains B05.10 and T4. Eukaryot Cell 11:1413–1414

Staats M, van Baarlen P, Van Kan JA (2005) Molecular phylogeny of the plant pathogenic genus *Botrytis* and the evolution of host specificity. Mol Biol Evol 22:333–346

Strathern JN, Spatola E, McGill C et al (1980) Structure and organization of transposable mating type cassettes in *Saccharomyces* yeasts. Proc Natl Acad Sci U S A 77:2839–2843

Terhem RB, Van Kan JA (2014) Functional analysis of hydrophobin genes in sexual development of *Botrytis cinerea*. Fungal Genet Biol 71:42–51

Uhm JY, Fujii H (1986) Course of meiosis in *Sclerotinia sclerotiorum* and related species. Trans Mycol Soc Jpn 27:129–141

Vallejo I, Santos M, Cantoral JM et al (1996) Chromosomal polymorphism in *Botrytis cinerea* strains. Hereditas 124:31–38

Vallejo I, Carbú M, Muñoz F et al (2002) Inheritance of chromosome-length polymorphisms in the phytopathogenic ascomycete *Botryotinia fuckeliana* (anam. *Botrytis cinerea*). Mycol Res 106:1075–1085

Van der Vlugt-Bergmans CJB, Brandwagt BF, Vant't Klooster JW et al (1993) Genetic variation and segregation of DNA polymorphisms in *Botrytis cinerea*. Mycol Res 97:1193–1200

Van Kan JA, Goverse A, Van der Vlugt-Bergmans CJB (1993) Electrophoretic karyotype analysis of *Botrytis cinerea*. Neth J Plant Pathol 99:119–128

Vanev S (1988) Anamorph and teleomorph of *Botryotinia fuckeliana* (de Bary) Whetzel in Bulgaria. Fitologiya 34:60–66

Viaud M, Legeai F, Pradier JM et al (2005) Expressed sequence tags from the phytopathogenic fungus *Botrytis cinerea*. Eur J Plant Pathol 111:139–146

Wu MD, Zhang L, Li GQ, Jiang DH, Hou MS, Huang HC (2007) Hypovirulence and double-stranded RNA in *Botrytis cinerea*. Phytopathology 97:1590–1599

Wu M, Zhang L, Li G et al (2010) Genome characterization of a debilitation-associated mitovirus infecting the phytopathogenic fungus *Botrytis cinerea*. Virology 406:117–126

Yan LY, Chen YF, Yang QQ et al (2012) Heterologous expression of the *CYP51* gene of the obligate fungus *Blumeria graminis* in the necrotrophic fungus *Botrytis cinerea*. J Eukaryot Microbiol 59:88–92

Yin YN, Kim YK, Xiao CL (2012) Molecular characterization of pyraclostrobin resistance and structural diversity of the cytochrome b gene in *Botrytis cinerea* from apple. Phytopathology 102:315–322

Yohalem DS, Nielsen K, Nicolaisen M (2003) Taxonomic and nomenclatural clarification of the onion neck rotting *Botrytis* species. Mycotaxon 85:175–182

Zhang L, Wu MD, Li GQ et al (2010) Effect of Mitovirus infection on formation of infection cushions and virulence of *Botrytis cinerea*. Physiol Mol Plant Pathol 75:71–80

Zhang L, Hua C, Stassen JH et al (2014) Genome-wide analysis of pectate-induced gene expression in *Botrytis cinerea*: identification and functional analysis of putative d-galacturonate transporters. Fungal Genet Biol 72:182–191

Chapter 4
Vegetative Incompatibility in *Botrytis*

Nadia Korolev and Yigal Elad

Abstract In ascomycetes, vegetative compatibility is determined by a series of *het* or *vic* genes that exist in two or more allelic states. Strains that carry identical alleles at all *loci* are compatible; those that differ at one or more *loci* are incompatible. The genetic basis of vegetative incompatibility in *Botrytis cinerea* is not known, but is presumed conform to the system found in other ascomycetes. Using nitrate non-utilizing mutants, 66 distinct vegetative compatibility groups (VCGs) were identified among field and single ascospore isolates of *B. cinerea*, an observation that is consistent with the presence of at least six *het* genes. Sulphate non-utilizing mutants have also been used to identify multiple VCGs in this species. The large number of VCGs and the limited occurrence of isolates displaying the same VCG suggest that sexual recombination plays an important role in field populations of *B. cinerea*. Mycelial compatibility groups (MCGs) recognized by the formation of interaction lines between isolates paired on agar media (barrages), indicate incompatibility in *B. cinerea*, however, there is no direct correlation with VCGs recognized by complementation. Sexual crosses have been shown to generate new *B. cinerea* VCGs. Unlike the situation in *Neurospora crassa*, in *B. cinerea*, the *MAT1* locus does not act as a *het* gene. The homologue of the *N. crassa het-c* gene has been cloned and sequenced in *B. cinerea*, but it does not act as a *vic* gene in *B. cinerea*.

Keywords Barrage • Mycelial compatibility group (MCG) • Nitrate non-utilising mutants • Sulphate non-utilising mutants • Vegetative compatibility group (VCG)

N. Korolev (✉) • Y. Elad
Department of Plant Pathology and Weed Research, Agricultural Research Organization, The Volcani Center, Bet Dagan 50250, Israel
e-mail: vpptlg@agri.gov.il

S. Fillinger, Y. Elad (eds.), *Botrytis – the Fungus, the Pathogen and its Management in Agricultural Systems*, DOI 10.1007/978-3-319-23371-0_4

4.1 Heterokaryosis

In filamentous ascomycetes, continual fusion between hyphae (anastomosis) yields networks of interconnected hyphae (mycelia) that make up fungal colonies. Hyphal networks can be formed by intercellular interactions within and between fungal colonies. In an individual colony, networked hyphae are presumably important for intra-hyphal communication and homeostasis during growth and reproduction. Different fungal colonies are also capable of hyphal fusion with each other to form a heterokaryon, in which genetically different nuclei may coexist in a common cytoplasm. Heterokaryon formation has the potential benefits of functional diploidy and mitotic genetic exchange (parasexual cycle) and may increase biomass for cooperation in physiological efforts such as resource exploitation or asexual/sexual reproduction (Glass et al. 2000). Several researchers have observed the transfer of nuclei between *B. cinerea* isolates through hyphal anastomosis maintaining the presumed heterokaryotic condition (Hansen 1938; Lorbeer 1980; Jarvis 1977).

Although there are obvious benefits to the formation of heterokaryons, these interactions can be restricted by a self-/nonself-recognition mechanism, so that networks can only be formed between colonies that are identical at all self-/nonself-recognition *loci* labelled as heterokaryon incompatibility (*het*) or vegetative incompatibility (*vic*) depending on fungal species. The hyphal fusion between colonies that are not identical at these *loci* results in the rapid induction of programmed cell death in the fusion compartment and subtending cells. In this case, the strains are vegetative incompatible and heterokaryon formation is not possible (Hutchinson and Glass 2012). Fungal species differ in the extent to which the resulting heterokaryons proliferate: two main patterns have been described. In fungi that exhibit the first pattern, including *Verticillium dahliae* and *Gibberella fujikuroi*, heterokaryons are limited to actual fusion cells, nuclei do not migrate between cells and heterokaryons are continually reformed by repeated fusion events within mycelia. Among the fungi that exhibit the second pattern, including *Neurospora crassa* and *Podospora anserina*, heterokaryotic cells proliferate and almost all cells within the mycelium are heterokaryotic (Glass et al. 2000; Glass and Kuldau 1992). *B. cinerea* is thought to exhibit the second pattern. The patterns of hyphal growth, branching and anastomosis in *B. cinerea* are typical of *Ascomycetes*. The vegetative cells and conidia are multinucleate containing generally 3-6 nuclei, and if the nuclei are genetically different, conidia are heterokaryotic (Chap. 3). Morphological diversity among monoconidial isolates derived from a single parent has often been regarded as evidence of the existence of heterokaryons among field isolates. Mycelial (M), conidial (C) and intermediate types (M/C) of *B. cinerea* have been recognized (Hansen 1938). The mycelial and conidial types (presumed homotypes) generate only M- and C-type progeny, respectively, whereas inconsistent, intermediate types (presumed heterotypes) can produce progeny of all three types. Hansen (1938) concluded that these intermediate types are natural heterokaryons that contain both C and M nuclei. Similar results were obtained by Menziger (1966) and Gringle (1979). However, Jinks (1959) and Grindle (1979) noted that although heterokaryosis might account for variations among monoconidial progeny, the possible role of cytoplasmic elements should also be considered. In the imperfect fungus *V. dahliae*, known for its morphological instability, the presence or absence of microsclerotia in progeny derived from a single conidia (exhibiting a white, mycelial phenotype or grey-to-black, microsclerotial phenotypes) is at least partially controlled by mitochondria, which are self-replicating and transmitted through the cytoplasm of the conidia (Heale 1988).

Other evidence of heterokaryosis in *B. cinerea* has come from studies of fungicide resistance. Heterokaryosis enables fungi to adapt to changing environmental conditions by adjusting the proportion of the two kinds of nuclei (sensitive and resistant to the fungicide). Summers et al. (1984) demonstrated that a dicarboximide-resistant field isolate was a presumed heterokaryon and, using monoconidial isolates, they were able to resolve it into homokaryotic sensitive and resistant types. They also showed that the relative balance of fungicide-resistant and fungicide-sensitive nuclei in the heterokaryon responded to the presence of fungicide in the medium. Other studies have found that isolates resistant to dicarboximide or dichlofluanid may not always transmit their resistance to their sexual progeny and some asexual progeny might also be non-resistant, indicating that the parent isolates were heterokaryons that possessed both resistant and sensitive nucleotypes (Farerta and Pollastro 1993; Mamiev et al. 2013; Pollastro et al. 1996; Yourman and Jeffers 1999).

The most direct evidence for heterokaryosis in field strains comes from studies of mating-types (Beever and Weeds 2004). Most field strains of *B. cinerea* are heterothallic, carrying one or another allele of the mating-type gene, *MAT1-1* or *MAT1-2*. In addition, some field strains that behave in a homothallic fashion (*MAT1-1/2* strains, or dual-maters) are presumed to be heterokaryotic. Monoconidial isolates of homothallic field isolates either remained homothallic (i.e., were heterokaryotic *MAT-1-1* + *MAT1-2*) or behaved in a heterothallic manner (i.e., were presumably either *MAT1-1* or *MAT1-2* homokaryons). These findings indicate that segregation occurs during conidiogenesis, so that either or both alleles of the mating-type gene can be present in a single conidium (Faretra et al. 1988). Heterokaryosis is the main cause of such secondary homothallism in the field and in the monoconidial isolates, if single multinucleate conidia contains nuclei carrying opposite idiomorphs. This also indicates that the mating type idiomorphs do not cause vegetative incompatibility in *B. cinerea*. Secondary homothallism was also observed in monoascosporic isolates. As the ascospore nuclei are derived from one haploid nucleus, such dual-mater strains cannot be explained by the possession of both *MAT1-1* and *MAT1-2* sequence in the same genome (Beever and Weeds 2004; Chap. 3). Instead, this phenomenon arises by changes elsewhere in the genome – it is unknown whether this form of mating-type switching has a monogenic or polygenic basis (Amsalem et al. 2011).

Sexual crossing has been shown to generate new VCGs in *B. cinerea* (Beever and Parkes 2003). The existence of homothallic strains that are heterokaryotic for mating type (Faretra et al. 1988) indicates that *MAT1* does not act as a *het* gene in this species, unlike the situation in *N. crassa*. Formation of "forced" heterokaryons between auxotrophic mutants of *B. cinerea* under laboratory conditions have been clearly established (Beever and Parkes 2003; Delcan and Melgarejo 2002; Korolev et al. 2008; Weeds et al. 1998), and it is discussed in paragraph 4: Vegetative (heterokaryon) compatibility groups.

Fusion of vegetative hyphae to form heterokaryotic cells is a common feature of the growth and development of filamentous fungi. The other case of hyphal fusion may occur between conidial germlings. Conidia can form specialized hyphae, called conidial anastomosis tubes (CATs), which are morphologically distinct from germ tubes and allow anastomosis with other nearby conidia (Glass et al. 2004). CATs have been recognized in 73 species of filamentous fungi, including *B. cinerea* (reviewed in Roca et al. 2005). Fusion between conidial anastomosis tubes often contributes to the earliest phases of colony establishment, while anastomosis between mature hyphae increases interconnectedness within the mycelium. CATs develop in culture and in host–pathogen systems; they grow towards each other and CAT formation and fusion are increased on nutrient-poor media. Nuclei pass through fused CATs, allowing gene exchange between individuals, although the movement of organelles between two fused conidial germlings is much slower than that between fused hyphae. Heterokaryons resulting from fusion of the mature hyphae are incompatible if the hyphae contain nuclei with different *het* genes. However, such heterokaryons can form via CAT fusions without an incompatibility response. This type of heterokaryon has been obtained in *Colletotrichum* spp., *Neurospora* spp., *Podospora* spp., and other fungi (Fisher-Harman et al. 2012; Read et al. 2012; Tong et al. 2014; Tsukasaki et al. 2014; Vittal et al. 2012). CAT fusion seems to be very common among fungi, but very little is known about the biology or significance of this process (Roca et al. 2005). Researchers have been aware of conidial anastomosis in *B. cinerea* since the early 1930s (Hansen and Smith 1932) and, in the last decade, this process has been the subject of intense study (Roca et al. 2012; Tudzynski et al. 2012).

4.2 Genetic and Molecular Systems Controlling the Formation of Heterokaryons

Genetic studies in several ascomycetes have shown that vegetative incompatibility is determined by a number of *het* loci (at least 11 in *N. crassa*, 9 in *Podospora anserina*, 8 in *Aspergillus nidulans* and 7 in *Cryphonectria parasitica*) that exist in two or three allelic states (reviewed by Glass et al. 2000). *het* loci have been cloned and molecularly characterized in *N. crassa* and *P. anserine*. With the exception of *het-S*, all *het* loci have a non-allelic partner and, in almost all cases, that non-allelic partner encodes a protein containing a HET domain. HET domains are approximately 150 amino acids long and are specific for filamentous fungi. Genes encoding HET domain proteins have been identified as important players for both nonself recognition and programmed cell death, and almost always form genetic interactions with other loci. These other loci encode a diverse set of proteins with diverse cellular functions, including but not limited to membrane proteins, glycolipid transfer proteins, enzymes involved in DNA synthesis, transcriptional regulators, and even prions. Strong balancing selection maintains multiple alleles for many *het* loci, yet the degree of sequence variability and the number of alleles can vary (reviewed by Hutchinson and

Glass 2012). One of the best characterized heterokaryon-incompatibility loci is the *het*-c/*pin*-c system in *N. crassa*. In that system, the *het*-c locus encodes a glycine-rich plasma membrane protein and interacts with its partner for incompatibility, *pin*-c, a highly polymorphic gene that encodes a protein harbouring a HET domain. The *het*-c and *pin*-c loci are closely linked and exist in three different haplotypes. *het*-c and *pin*-c alleles are under balancing selection: individuals carrying *het-c* and *pin-c* haplotypes are equally frequent, show severe linkage disequilibrium (no recombination between *het-c pin-c* haplotypes) and exhibit trans-species polymorphism. *het*-c and *pin*-c homologs are present in many filamentous fungi, but they play a role in heterokaryon incompatibility only in *Neurospora* and closely related genera. Non-allelic interactions also may occur in which alleles at one locus interact with alleles at a second locus to block the formation of a stable heterokaryon. Such, *het* homologs in *P. anserina*, *Fusarium proliferatum* and *Aspergillus niger* do not exhibit allelic variability (Hutchinson and Glass 2012; Leslie 1993).

The genetic basis of vegetative incompatibility in *B. cinerea* is not known. Fournier et al. (2003) cloned and sequenced the *B. cinerea* homolog (Bc-*hch*) of the vegetative-incompatibility loci of *N. crassa* (Nc-*het-c*) and *P. anserina* (Pa-*hch*). Bc-*hch* polymorphism has been observed among *B. cinerea* field isolates. Specifically, two allelic types were found among 117 isolates of *B. cinerea*, and the haplotypes at the Bc-*hch* strictly corresponded to the resistance phenotypes to fenhexamid. The use of Bc-*hch* as a population marker revealed that natural *B. cinerea* populations can be divided into two groups (I and II) (Fournier et al. 2003). The authors supposed that the polymorphism observed at the Bc-*hch* locus corresponds to the fixation of two private alleles in two reproductively isolated populations, with no polymorphism observed within each group at this locus. Barrages were observed in confrontation assays between Group I and Group II isolates in Petri dishes, and barrages also were observed between Group II members, indicating that Group II includes several MCGs, although it is monomorphic at the Bc-*hch* locus. This might indicate that the Bc-*hch* locus does not function as *het* gene in *B. cinerea* (Fournier et al. 2005). Fertile progeny have been obtained from crosses between strains within each group, but not from crosses between the groups, indicating that these groups are reproductively isolated (Walker et al. 2011). A recent analysis of the *B. cinerea* genome identified 52 different genes that encode HET-domain-containing proteins (Aanen et al. 2010).

4.3 Identification of *het* Loci

Three basic techniques have been used to identify *het* loci: direct assessment of heterokaryon formation (usually by complementation of recessive auxotrophic markers), direct assessment of inability to form a heterokaryon (usually through barrage formation) and partial diploid analysis (Leslie 1993). In the last method, a strain with a duplication-generating chromosome rearrangement is crossed with a strain of unknown *het* gene constitution. Homokaryotic, partial-diploid progeny from such a cross that are heterozygous for a given *het* gene within the duplicated segment can be recognized as self-incompatible by phenotypic characteristics that include slow growth, pigmentation and aberrant colony morphology. Partial diploids that are homozygous at the relevant *het* locus look like the wild type. In *N. crassa* and other tested fungi, heterokaryon incompatibility tests and self-incompatible partial diploids tests yielded similar results: congruent incompatibility phenotypes and map locations (Micali and Smith 2003). To our knowledge, no similar study has been conducted in *B. cinerea*. Two other methods for identifying incompatibility loci in *B. cinerea* are described below.

4.4 Vegetative (Heterokaryon) Compatibility Groups (VCGs)

Fungal strains that carry identical alleles of vegetative-incompatibility genes are compatible and form a vegetative compatibility group (VCG). Direct evaluation of heterokaryon formation usually involves the establishment of a stable prototrophic heterokaryon under conditions in which neither of the two auxotrophic components could survive on its own. When a heterokaryon is forced between auxotrophic strains of the same VCG, then a prototrophic heterokaryon results. If the strains are of different VCGs, there will be no prototrophic growth. The introduction of nitrate non-utilising (Nit) mutants has provided a simple method of mutant hunt procedure (Puhalla 1985). Typically, a range of Nit mutant classes are recovered, but the most reliable are usually those deficient in nitrate reductase apoenzyme (*nit1*) and those defective in synthesis of the molybdenum-containing cofactor needed for nitrate reductase and xanthine dehydrogenase activity (NitM mutants) (Bayman and Cotty 1991; Beever and Weeds 2004; Korolev et al. 2000; Tomsett and Garrett 1980).

There have been a number of reports of *nit* mutants in *B. cinerea* (Beever and Parkes 2003; Delcan and Melgarejo 2002; Levis et al. 1997; Weeds et al. 1998; White et al. 1998). These mutants are resistant to chlorate and are selected on media amended with that substance, which is the toxic analogue of nitrate. Selenate is the toxic analogue of another common nutrient, sulphate, and can be used to select selenate-resistant mutants. Selenate-resistant mutants with altered sulphur transport and assimilation have been characterized in several fungal genera (Vialta et al. 1999; Harp and Correll 1998; Correll and Leslie 1987; Arst 1968). Different classes of sulphate non-utilising mutants have been described, one comprised of chromate-resistant mutants and several chromate-sensitive genotypes (Arst 1968; Marzluf 1970). Selenate-resistant mutants, both resistant and sensitive to chromate, have been described in *B. cinerea* (Weeds et al. 1998; Korolev et al. 2006, 2008).

Weeds et al. (1998) presented the first evidence of self-compatibility of a *B. cinerea* strain and incompatibility between two different strains, as demonstrated by forced heterokaryosis between Nit mutants. When several *nit1* mutants from the same strain were paired with each other, some pairings complemented and this behaviour was attributed to intragenic complementation. Complementation was also demonstrated between sulphate non-utilising mutants and *nit1* mutants selected from the same parent, but not between mutants generated from different strains.

Delcan and Melgarejo (2002) obtained both *nit1* and NitM *B. cinerea* mutants and demonstrated the formation of heterokaryon between *nit1* and NitM mutants originating from the same parent strain. NitM × *nit1* pairings of mutants from two different isolates resulted in incompatible interaction. Beever and Parkes (2003) devised a method to facilitate the recovery of both *nit1* and NitM pairs and confirmed the existence of multiple VCGs in this species, with all nine strains examined belonging to different groups. None of the parental strains was self-incompatible. Subsequent studies have shown that a large number of VCGs exist within the *B. cinerea* population and, to date, 59 VCGs have been recognized amongst 82 strains from New Zealand, with a few strains assigned to the same group. An additional seven VCGs were found among the progeny of sexual crosses between two strains (Beever and Weeds 2004; Weeds and Beever personal communication). VCGs are determined by unique combinations of *het* genes, such that if six *het* loci with two alleles per locus are segregating in a population, 64 (2^6) VCGs are theoretically possible (Beever and Weeds 2004). The identification of 66 distinct VCGs among field and single-ascospore isolates (isolates derived from a single ascospores, under laboratory condition) of *B. cinerea* is consistent with presence of at least seven *het* genes with two alleles each.

Selenate-resistant sulphate non-utilising mutants (Sul) have been recovered from 21 *B. cinerea* strains found in Israel. These mutants exhibit two different phenotypes: chromate-resistant and chromate-sensitive (Korolev et al. 2008). Vigorous heterokaryotic growth was obtained in all of the intra-strain pairings of the different sulphate non-utilising phenotypes (Fig. 4.1).

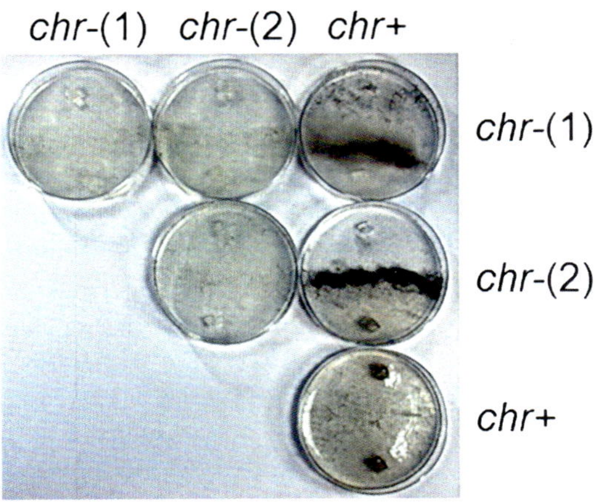

Fig. 4.1 Pairings of chromate-sensitive (*chr*−) and chromate-resistant (*chr*+) sulphate non-utilising mutants of *B. cinerea*. Heterokaryons formed between chromate-sensitive (*chr*−) and chromate-resistant (*chr*+) sulphate non-utilising mutants of *Botrytis cinerea*, but not when mutants were paired with themselves or with mutants carrying the same type of mutation. The photo shows results for two chromate-sensitive and one chromate-resistant mutant of strain B4, paired in all possible combinations on minimal medium in 5-cm-diameter plates kept at 20–22 °C in the dark (Photo was taken 12 days after inoculation (Korolev et al. 2008))

This demonstrates that: (i) chromate-sensitive and chromate-resistant Sul mutants complement each other and could be used for vegetative compatibility studies; (ii) heterokaryon self-compatibility is common among *B. cinerea* strains from Israel, similar to the self-compatibility observed between Nit mutants (Beever and Parkes 2003; Delcan and Melgarejo 2002) and in pairings of Sul and Nit mutants derived from the same strain (Weeds et al. 1998); and (iii) inter-strain incompatibility is widespread in *B. cinerea*. The 22 strains were assigned to 12 single-member VCGs and 3 multi-member VCGs. Twenty-two VCGs, including an average of 1.8 isolates per group, were identified in a strawberry field population of *B. cinerea* (Habib et al. unpublished).

Some fungi carry mutations that prevent them from anastomosing with themselves. Such isolates have been described in many fungal genera, including *Fusarium* spp. (Leslie 1993), *Neurospora* spp. (Glass et al. 2000) and *Verticillium* spp. (Korolev et al. 2000). Correll et al. (1989) determined that heterokaryon self-incompatibility (HSI) in *Gibberella fujikuroi* is controlled by a single segregating gene (*hsi-1*). The frequency of HSI mutants among field isolates of *Fusarium moniliformae* and *Verticillium dahliae* was about 1–2 % (Leslie 1993; Korolev et al. 2000). The high proportion (>50 %) of HSI mutants observed in a collection of *Verticillium lecanii* (Korolev and Gindin 1999) suggests that, in this entomopathogenic species, HSI mutations may have some advantage for survival. All tested isolates of *B. cinerea* were found to be self-compatible (Beever and Parkes 2003; Delcan and Melgarejo 2002; Korolev et al. 2008). Self-compatibility was observed both as heterokaryon formation between complementary Sul mutants and as the absence of any strong barrage reaction between mycelia of field isolates belonging to the same VCG, although weak antagonistic interactions did sometimes occur (Korolev et al. 2008).

4.5 Mycelial Compatibility Groups (MCGs)

Three main types of interaction can occur when fungal hyphae from different mycelia of the same species grow towards each other: mutual intermingling, inhibition and mutual intermingling plus inhibition. In cases of mutual intermingling, the hyphae intermingle in the zone of contact and form numerous hyphal fusions via anastomosis. After a time, the border between the two mycelia becomes unrecognizable. In the case of inhibition, when the different hyphae approach each other, an inhibition zone free of hyphae is formed between the two mycelia. This phenomenon may be caused by unilateral or mutual interaction based on the secretion and diffusion of inhibitory substances. In the case of mutual intermingling plus inhibition, a barrage is formed. In that situation, as the two mycelia grow and intermingle, there is an antagonistic reaction. In contrast to inhibition by diffusible substances, the barrage reaction requires cytoplasmic contact via hyphal fusion. Nuclear exchange is not inhibited, but the two types of mycelia form abnormal and even lethal fusions. A line of contact appears as the culture ages. Barrages may be

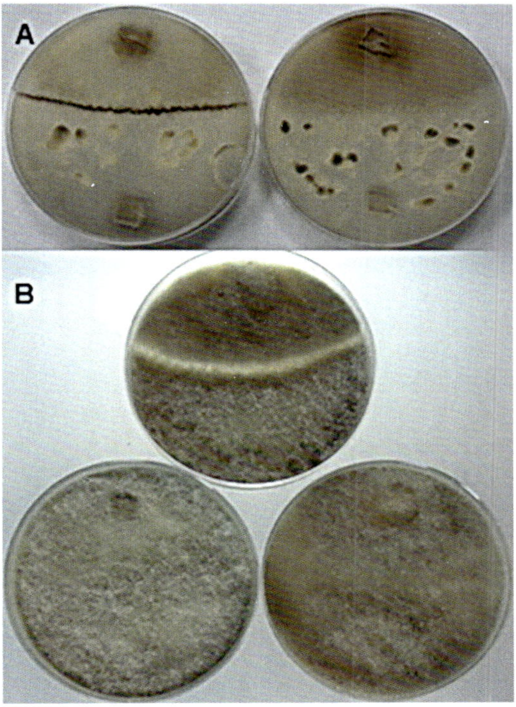

Fig. 4.2 Formation of barrages between incompatible strains of *B. cinerea*. (**a**) Pigmented interaction zone between incompatible strains B4 and B21 (*left*) and the absence of a barrage between compatible strains B7 and B21 (*right*). (**b**) Interaction zone in the form of sparse mycelium between incompatible strains B4 and B7 (*top*) and the absence of a barrage when the strains were paired with themselves (*bottom*) on PDA in 5-cm-diameter plates maintained at 20–22 °C in the dark (Photo was taken 10 days after inoculation (Korolev et al. 2008))

colourless or pigmented (Esser 2006). Mutual intermingling describes the interaction between vegetatively compatible strains and the two other types of interaction (inhibition and mutual intermingling plus inhibition) are incompatible interactions. All three types of interaction are seen in *B. cinerea* (Fig. 4.2).

Often, the term 'barrage' is used in connection with both types of incompatible interactions and, broadly, a barrage is the line or zone of demarcation that may develop at the point of contact between genetically different fungi. In *N. crassa*, at least three types of barrages have been identified: a dark line, a clear zone and a raised aggregate of hyphae (Micali and Smith 2003). Two main types of barrages have been described in *B. cinerea*: a dark line and a clear zone (Beever and Parkes 1993; Delcan and Melgarejo 2002; Korolev et al. 2008).

Beever and Parkes (1993, 2003) described barrages formed between field isolates and between ascospore isolates in *B. cinerea*. Those authors found that strong interaction zones often formed between ascospore isolates grown on malt extract agar amended with 0.68 M NaCl. A distinct zone formed between the parents, as well as in 141 of 144 pairings of randomly selected progeny. The zones took the form of a discrete narrow black line or pair of black lines, which was sometimes associated with a brownish zone spreading out from the interaction line for a few mm. No such zones were formed when strains were paired with themselves. Based on the observation of compatibility in only 2 % of the interactions, those authors suggested that the two parent strains differ at five or six vegetative-compatibility *loci* (Beever and Parkes 1993). In that work, the scoring of zone formation between field isolates proved difficult; many pairings resulted in a faint response that was difficult to distinguish from the merging reaction observed when isolates are paired with themselves. The formation of strong barrage lines between field isolates of *B. cinerea* has been relatively infrequent in other study as well. Korolev et al. (2008) reported that 30 % of the pairings in their study did not yield a barrage and an additional 34 % of pairings were characterised by a weak or inconsistent antagonistic interaction (Korolev et al. 2008).

In their study of barrages in *N. crassa*, Micali and Smith (2003) did not observe any barrage formation when wild-collected strains were paired. However, an increase in the frequency of pairings that produced barrages was observed among strains obtained by back-crossing wild strains with laboratory strains, or through successive rounds of inbreeding of wild-derived strains. Those authors suggested that wild strains might possess genes that suppress the barrage reaction. Similar barrage-suppressive mechanisms could exist in field isolates of *B. cinerea*.

The barrage phenomenon is often regarded as the opposite of prototrophic vegetatively compatible heterokaryon formation. In early work with *Cryphonectria parasitica* and *Podospora anserina*, a correlation was observed between vegetative incompatibility, expressed as barrage, and heterokaryon incompatibility as assayed through complementation (Huber 1996; Rizet 1952). However, in *N. crassa*, barrages were sometimes observed between strains that were identical at all major heterokaryon incompatibility (*het*) loci and possessed the same mating-type gene, *mat1*, which acts as a heterokaryon incompatibility locus during the vegetative phase of *N. crassa*. Examples of cases in which barrages did not form between strains that had genetic differences at *het-6*, *het-c*, and *mat* were also reported. Taken together, these results suggest that the genetic control of barrage formation in *N. crassa* can operate independently from that of heterokaryon incompatibility and mating type (Micali and Smith 2003). Similarly, barrage formation is not necessarily an indication of heterokaryon incompatibility in *Sclerotinia sclerotiorum* (Ford et al 1995) or *C. parasitica* (Smith et al. 2006). This observation may not be surprising considering that, through chemical signalling, a barrage can form between hyphae separated by some distance. A barrage can also form during cell-cell contact through surface molecules, or after cell fusion through heterokaryon incompatibility mechanisms. In some species, *het* genes may represent only a subset of all of the factors involved in vegetative incompatibility. Barrage formation and heterokaryon incompatibility are not different manifestations of the same process. Rather, evidence indicates that heterokaryon incompatibility is one component of a vegetative incompatibility system that may also use cell-surface or extracellular factors to trigger programmed cell death to modulate non self-recognition in fungi (Micali and Smith 2003; Smith et al 2006). In *B. cinerea*, all pairings between single-ascospore tester strains representing nine VCGs recognized by complementation between *nit* mutants resulted in the formation of barrage zones, indicating congruence between mycelial incompatibility and heterokaryon incompatibility (Beever and Parkes

1993). However, in another study (Beever and Parkes 2003), single-ascospore strains of *B. cinerea*, after a series of backcrosses, were complemented by *nit* mutants, but still produced a dark interaction line.

Classifying field strains into distinct groups using the mycelial compatibility test was often difficult in that work, with the number and intensity of dark lines varying depending on the combination. Beever and Parkes (2003) concluded that it was premature to equate mycelial and vegetative incompatibility in *B. cinerea*. In another work, barrage formation between *B. cinerea* field isolates, in the form of a zone of dark pigmentation or sparse mycelia with or without dark pigmentation along the line of confrontation, was observed in 36 % of the combinations and a weak antagonistic reaction was observed for an additional 34 % of the pairings. No visible adverse reaction was observed in 30 % of the pairings. There was no correspondence between MCGs and VCGs (complementation between *sul* mutants), since pairings of strains from different VCGs did not necessarily produce barrages. However, self-compatibility was observed both as heterokaryon formation between complementary *sul* mutants and the absence of barrage between mycelia. Field isolates belonging to the same VCG did not exhibit strong antagonistic reactions to one another; only weak antagonistic interactions were sometimes observed (Korolev et al. 2008).

4.6 Population Implications and Biological Functions of Vegetative Incompatibility

VCGs serve as a natural means to subdivide fungal populations, and subdivisions induced by VCGs within a sexual population are quite different from those expected for an asexual population. In an asexual population, differences at the *het* loci effectively limit the exchange of genetic information to those individuals that belong to the same VCG. Since sexual recombination does not occur, members of the same VCG will form a genetically isolated subpopulation (Leslie 1993). VCGs have been very useful for identifying such subpopulations in fungi that are largely asexual and, in those species, VCGs often correlate with pathogenicity and other traits (Korolev et al. 2000).

Sexually reproducing populations of ascomycetes are expected to have high levels of VCG diversity. Typically, filamentous ascomycetes whose *het* loci have been mapped have eight to ten loci, with two to three alleles at each locus. *het* loci are not clustered, but scattered throughout the genome. Thus, for a species with eight *het* loci and two alleles at each locus, there would be 256 (2^8) different allele combinations and successful fusions between strains that are "self" or at least very closely related are likely to be almost nonexistent in nature (Hutchinson and Glass 2012). *B. cinerea* has a high level of vegetative incompatibility. A large number of VCGs have been identified in this species and only a small proportion of those VCGs include more than one strain, suggesting that sexual recombination occurs in *B. cinerea* populations (Beever and Weeds 2004; Habib et al. (unpublished); Korolev et al. 2008).

Population studies using molecular markers have indicated that the *B. cinerea* population is genetically very diverse with no indication of widespread clonal lineages. The most common haplotypes accounted for only 5–8 % of all strains (Giraud et al. 1997, 1999). This extensive genotypic diversity indicates limited clonal propagation and the significant role of recombination (for more details see Chap. 6). *B. cinerea* is the anamorph of *Botryotinia fuckeliana*, and sexual crossing studies indicate that most strains are heterothallic and capable of crossing. However, apothecia are seldom found in nature and have never been found in areas with a warm dry climate, such as Israel or southern Spain (Raposo et al. 2001; Yunis and Elad 1989). The large proportion of field strains infected by mycoviruses also suggests that, despite its critical role, sexual reproduction is infrequent. Beever and Weeds (2004) suggested that in the absence of the frequent sexual stage, the main mechanism for heterokaryon generation and the maintenance of high genetic diversity in *B. cinerea* is mutation within long-lived somatic lineages. It is interesting to note that the new cryptic species *Botrytis pseudocinerea*, which lives in sympatry with *B. cinerea* (Walker et al. 2011), has only one MCG while there are numerous *B. cinerea* MCGs. No MCG overlapped the two groups (Fournier et al. 2003, 2005).

Vegetative incompatibility has significant biological implications, primarily in preventing the transfer of deleterious genetic elements between colonies, including mycoviruses and senescence plasmids. Thus, differences in *het* loci significantly reduce viral transmission in *C. parasitica* and completely block viral transfer in the black *Aspergilli*. Genetic differences at *het* loci also restrict the transfer of hypovirulence-associated dsRNAs in *Sclerotinia homoecarpa*. In addition to affecting viral transmission, vegetative incompatibility has been shown to restrict the transfer of a mitochondrial senescence phenotype in *P. anserina* and the transfer of mitochondrial senescence plasmids in *N. crassa* (Hutchinson and Glass 2012). Highly variable virus-like dsRNA is common in *B. cinerea*. Over 70 % of isolates were found to be infected with viruses located in cytoplasm and associated with some cellular degeneration (Howitt et al. 1995; Rodriguez-Garcia et al. 2014). Hypovirulence-associated dsRNA and ssRNA mycoviruses, which reduce the pathogenicity of their host fungi, are of special interest due to their potential for use as biological control agents. Such viruses have been identified in *B. cinerea* (Cao et al. 2011; Castro et al. 2003; Pearson and Bailey 2013). However, the high level of vegetative incompatibility in *B. cinerea* may prevent the transmission of these viruses and be an obstacle for successful biocontrol. In some fungi, incompatibility barriers are suppressed during conidial anastomosis, which (conidial anastomosis) has been described in 73 species of filamentous fungi, including *B. cinerea* (Roca et al. 2005, 2012).

4.7 Conclusions

The genetic basis of vegetative incompatibility in *B. cinerea* is not known, but is presumed conform to the system found in other ascomycetes. Using nitrate non-utilizing mutants, many distinct VCGs were identified among field and single ascospore isolates of *B. cinerea*, an observation that is consistent with the presence of at least six *het* genes. Sulphate non-utilizing mutants have also been used to identify multiple VCGs in this species. The large number of VCGs and the limited occurrence of isolates displaying the same VCG suggest that sexual recombination plays an important role in field populations of *B. cinerea*. This is surprising in view of the rare reports of actual apothecia presence in places where *B. cinerea* is an important pathogen.

Mycelial compatibility groups recognized by the formation of interaction lines between isolates paired on agar media (barrages), indicate incompatibility in *B. cinerea*, however, there is no direct correlation with VCGs recognized by complementation. Sexual crosses have been shown to generate new *B. cinerea* VCGs. In practice, vegetative incompatibility has significant biological implications, primarily in preventing the transfer of deleterious genetic elements between colonies, including mycoviruses and senescence plasmids.

References

Aanen DK, Debets AJM, Glass NL, Saupe SJ (2010) Biology and genetics of vegetative incompatibility in fungi. In: Borkovich KA, Ebbole DJ (eds) Cellular and molecular biology of filamentous fungi. ASM Press, Washington, DC, pp 274–288

Amselem J, Cuomo CA, Van Kan JAL et al (2011) Genomic analysis of the necrotrophic fungal pathogens *Sclerotinia sclerotiorum* and *Botrytis cinerea*. PLoS Genet 7:e1002230

Arst HN (1968) Genetic analysis of the first steps of sulphate metabolism in *Aspergillus nidulans*. Nature 219:268–270

Bayman P, Cotty PJ (1991) Vegetative compatibility and genetic diversity in the *Aspergillus flavus* population of a single field. Can J Bot 69:1707–1711

Beever RE, Parkes SL (1993) Mating behavior and genetics of fungicide resistance of *Botrytis cinerea* in New Zealand. N Z J Crop Hortic Sci 21:303–310

Beever RE, Parkes SL (2003) Use of nitrate non-utilizing (Nit) mutants to determine vegetative compatibility in *Botryotinia fuckeliana* (*Botrytis cinerea*). Eur J Plant Pathol 109:607–613

Beever RE, Weeds PL (2004) Taxonomy and genetic variation of *Botrytis* and *Botryotinia*. In: Elad Y, Williamson P, Tudzinski P, Delen N (eds) *Botrytis*: biology pathology and control. Kluwer Academic, Dordrecht, pp 29–52

Cao JB, Zjou Y, Zhang L et al (2011) dsRNA-free transmissible hypovirulence associated with formation of intra-hyphal hyphae in *Botrytis cinerea*. Fungal Biol 115:660–661

Castro M, Kramer K, Valdivia L, Ortiz S, Castillo A (2003) A double-stranded RNA mycovirus confers hypovirulence-associated traits to *Botrytis cinerea*. FEMS Microbiol Lett 228:87–91

Correll JC, Leslie JH (1987) Recovery of spontaneous selenate-resistant mutants from *Fusarium oxysporum* and *Fusarium moniliforme*. Phytopathology 77:1710

Correll JC, Klittich CJR, Leslie JF (1989) Heterokaryon self-incompatibility in *Gibberella fujikuroi* (*Fusarium moniliforme*). Mycol Res 93:21–27

Delcan J, Melgarejo P (2002) Mating behaviour and vegetative compatibility in Spanish populations of *Botryotinia fuckeliana*. Eur J Plant Pathol 108:391–400

Esser K (2006) Heterogenic incompatibility in fungi. In: Kues U, Fisher R (eds) The mycota I. Growth, differentiation and sexuality. Springer, Berlin, pp 141–165

Faretra F, Pollastro S (1993) Genetics of sexual compatibility and resistance to benzimidazole and dicarboximide fungicides in isolates of *Botryotinia fuckeliana* (*Botrytis cinerea*) from nine countries. Plant Pathol 42:48–57

Faretra F, Antonacci E, Pollastro S (1988) Sexual behavior and mating system of *Botryotinia fuckeliana* (*Botrytis cinerea*). J Gen Microbiol 134:2543–2550
Fisher-Harman V, Jackson KJ, Munoz A et al (2012) Evidence for tryptophan being a signal molecule that inhibits conidial anastomosis tube fusion during colony initiation in *Neurospora crassa*. Fungal Genet Biol 49:896–902
Ford EJ, Miller RV, Gray H, Sherwood JE (1995) Heterokaryon formation and vegetative compatibility in *Sclerotinia sclerotiorum*. Mycol Res 99:241–247
Fournier E, Levis C, Fortini D et al (2003) Characterization of Bc-*hch*, the *Botrytis cinerea* homolog of the *Neurospora crassa het-c* vegetative incompatibility locus, and its use as a population marker. Mycologia 95:251–261
Fournier E, Giraud T, Albertini K, Brygoo Y (2005) Partition of the *Botrytis cinerea* complex in France using multiple gene genealogies. Mycologia 97:1251–1267
Giraud T, Fortini D, Levis C, Brygoo Y (1997) RFLP markers show genetic recombination in *Botryotinia fuckeliana* (*Botrytis cinerea*) and transposable elements revealed two sympatric species. Mol Biol Evol 14:1177–1185
Giraud T, Fortini D, Levis C et al (1999) Two sibling species of the *Botrytis cinerea* complex, *transposa* and *vacuma*, are found in sympatry on numerous host plants. Phytopathology 89:967–973
Glass NL, Kuldau GA (1992) Mating type and vegetative incompatibility in filamentous ascomycetes. Annu Rev Phytopathol 30:201–224
Glass NL, Jacobson DJ, Shiu PKT (2000) The genetics of fungal fusion and vegetative incompatibility in filamentous fungi. Annu Rev Genet 34:165–186
Glass NL, Rasmussen C, Roca MG, Read ND (2004) Hyphal homing, fusion and mycelial interconnectedness. Trends Microbiol 12:135–141
Grindle M (1979) Phenotypic differences between natural and induced variants of *Botrytis cinerea*. J Gen Microbiol 111:109–120
Hansen HN (1938) The dual phenomenon in imperfect fungi. Mycologia 30:442–445
Hansen HN, Smith RE (1932) The mechanism of variation in imperfect fungi: *Botrytis cinerea*. Phytopathology 22:953–964
Harp TL, Correll JC (1998) Recovery and characterization of spontaneous selenate-resistant mutants of *Magnaporthe grisea*, the rice blast pathogen. Mycologia 90:954–963
Heale JB (1988) *Verticillium* spp., the cause of vascular wilts in many species. Adv Plant Pathol 6:291–312
Howitt RLJ, Beever RE, Pearson MN, Forster RLS (1995) Presence of double-stranded RNA and virus-like particles in *Botrytis cinerea*. Mycol Res 99:1472–1478
Huber DH (1996) Genetic analysis of vegetative compatibility polymorphisms and horizontal transmission in the chestnut blight fungus *Cryphonectria parasitica*. PhD dissertation, Michigan State University, East Lansing
Hutchinson EA, Glass NL (2012) Programmed cell death and heterokaryon incompatibility in filamentous fungi. In: Wytzany G (ed) Biocommunication of fungi. Springer, Dordrecht, pp 115–138. doi:10.1007/978-94-007-4264-2_8
Jarvis WR (1977) *Botryotinia* and *Botrytis* species: taxonomy, physiology, and pathogenicity. Research Branch, Canada Department of Agriculture, Ottawa
Jinks JL (1959) Lethal suppressive cytoplasms in aged clones of Aspergillus glaucus. J Gen Microbiol 21:397–409
Korolev N, Gindin G (1999) Vegetative compatibility in the entomopathogen *Verticillium lecanii*. Mycol Res 103:833–840
Korolev N, Katan J, Katan T (2000) Vegetative compatibility groups of *Verticillium dahliae* in Israel: their distribution and association with pathogenicity. Phytopathology 90:529–536
Korolev N, Katan T, Elad Y (2006) Use of selenate-resistant strains as markers for the spread and survival of *Botrytis cinerea* under greenhouse conditions. Phytopathology 96:1195–1203
Korolev N, Elad Y, Katan T (2008) Vegetative compatibility grouping in *Botrytis cinerea* using sulphate non-utilizing mutants. Eur J Plant Pathol 122:369–383

Leslie JF (1993) Fungal vegetative compatibility. Annu Rev Phytopathol 31:127–150

Levis C, Fortini D, Brygoo Y (1997) Transformation of *Botrytis cinerea* with the nitrate reductase gene (*niaD*) shows a high frequency of homologous recombination. Curr Genet 32:157–162

Lorbeer JW (1980) Variation in *Botrytis* and *Botryotinia*. In: Coley-Smith JR, Verhoeff K, Jarvis WR (eds) The biology of *Botrytis*. Academic, London, pp 19–40

Mamiev M, Korolev N, Elad Y (2013) Resistance to polyoxin AL and other fungicides in *Botrytis cinerea* collected from sweet basil crops in Israel. Eur J Plant Pathol 137:79–91

Marzluf GA (1970) Genetic and metabolic controls for sulfate metabolism in *Neurospora crassa*: isolation and study of chromate-resistant and sulfate transport-negative mutants. J Bacteriol 102:716–720

Menziger W (1966) Zurvariabilität and Taxonomie von Arten and Formen der Gattung Botrytis Mich. Zentralblätt Bakteriologie Parasitenkunde Infektionskr Hyg 120:179–196

Micali CO, Smith ML (2003) On the independence of barrage formation and heterokaryon incompatibility in *Neurospora crassa*. Fungal Genet Biol 38:209–219

Pearson MN, Bailey AM (2013) Viruses of *Botrytis*. Adv Virus Res 86:249–272

Pollastro S, Faretra F, Di Canio V, De Guido A (1996) Characterization and genetic analysis of field isolates of *Botryotinia fuckeliana* (*Botrytis cinerea*) resistant to dichlofluanid. Eur J Plant Pathol 102:607–613

Puhalla JE (1985) Classification of strains of *Fusarium oxysporum* on the basis of vegetative compatibility. Can J Bot 63:179–183

Raposo R, Gomez V, Urrutia T, Melgarejo P (2001) Survival of *Botrytis cinerea* in southern Spanish greenhouses. Eur J Plant Pathol 107:229–236

Read ND, Goryachev AB, Lichius A (2012) The mechanistic basis of self-fusion between conidial anastomosis tubes during fungal colony initiation. Fungal Biol Rev 26:1–11

Rizet G (1952) Les phenomenes de barrage chez *Podospora anserine*: analyse genetique des barrages entre souchess et S. Rev Cytologiquede Biol Veg 13:51–92

Roca MG, Read ND, Wheals AE (2005) Conidial anastomosis tubes in filamentous fungi. FEMS Microbiol Lett 249:191–198

Roca MG, Weichert M, Siegmund U, Tudzinski P, Fleiβner A (2012) Germling fusion via conidial anastomosis tubes in the grey mould *Botrytis cinerea* requires NADPH oxidase activity. Fungal Biol 116:379–387

Rodriguez-Garcia C, Medina V, Alonso A, Ayllon MA (2014) Mycoviruses of *Botrytis cinerea* isolates from different hosts. Ann Appl Biol 164:46–61

Smith ML, Gibbs CC, Milgroom MG (2006) Heterokaryon incompatibility function of barrage-associated vegetative incompatibility genes (*vic*) in *Cryphonectria parasitica*. Mycologia 98:43–50

Summers RW, Heaney SP, Grindle M (1984) Studies of a dicarboximide resistant heterokaryon of *Botrytis cinerea*. Br Crop Protect Conf Pests Dis 2:453–458

Tomsett AB, Garrrett RH (1980) The isolation and characterization of mutants defective in nitrate assimilation in *Neurospora crassa*. Genetics 95:649–660

Tong LCH, Silar P, Lalucque H (2014) Genetic control of anastomosis in *Podospora anserina*. Fungal Genet Biol 70:94–103

Tsukasaki W, Maruyama J, Katsuhiko K (2014) Establishment of a new method to quantitatively evaluate hyphal fusion ability in *Aspergillus orysae*. Biosci Biotechnol Biochem 78:1254–1262

Tudzynski P, Heller J, Siegmund U (2012) Reactive oxygen species generation in fungal development and pathogenesis. Curr Opin Microbiol 15:653–659

Vialta A, Catani CF, Junior RB, Azevedo JL (1999) Isolation and characterization of selenate-resistant mutants of *Acremonium chrysogenum*. Bras Arch Biol Technol 42:369–374

Vittal R, Yang HC, Hartman GL (2012) Anastomosis of germ tubes and migration of nuclei in germ tube networks of the soybean rust pathogen, *Phakopsora pachyrhizi*. Eur J Plant Pathol 132:163–167

Walker AS, Gautier A, Confais J et al (2011) *Botrytis pseudocinerea*, a new cryptic species causing gray mold in French vineyards in sympatry with *Botrytis cinerea*. Phytopathology 101:1433–1445

Weeds PL, Beever RE, Long PG (1998) New genetic markers for *Botrytis cinerea* (*Botryotinia fuckeliana*). Mycol Res 102:791–800

White GJ, Dobinson K, Traquair JA (1998) Selection of nitrate-nonutilizing mutants in *Verticillium*, *Alternaria*, and *Botrytis*. Can J Plant Pathol 20:340

Yourman LF, Jeffers SN (1999) Resistance to benzimidazole and dicarboximide fungicides in greenhouse isolates of *Botrytis cinerea*. Plant Dis 83:569–575

Yunis H, Elad Y (1989) Survival of dicarboximide-resistant strains of *Botrytis cinerea* in plant debris during summer in Israel. Phytoparasitica 17:13–21

Chapter 5
RNA Mycoviruses and Their Role in *Botrytis* Biology

Mingde Wu, Jing Zhang, Long Yang, and Guoqing Li

Abstract Mycoviruses are viruses inhabiting and replicating in cells of filamentous fungi, yeasts, and oomycetes. Some mycoviruses can attenuate pathogenicity of their host fungi, thus having the potential for biocontrol of plant fungal diseases. RNA mycoviruses infecting *Botrytis* were first recorded in 1995. So far, several species of RNA mycoviruses (dsRNA or ssRNA) belonging to *Alphaflexiviridae*, *Gammaflexiviridae*, *Narnaviridae*, *Partitiviridae*, *Totiviridae*, and an unassigned viral family have been detected in *Botrytis*, mostly in *B. cinerea*. Genomes of a few RNA mycoviruses including *Botrytis cinerea mito virus 1* (BcMV1), *Botryotinia fuckeliana totivirus 1* (BfTV1), *Botryotinia fuckeliana partiti virus 1* (BfPV1), *Botrytis cinerea* CCg378 virus 1 (Bc378V1), *Botrytis virus F* (BVF), *Botrytis virus X* (BVX) and *Botrytis porri RNA virus 1* (BpRV1) have been fully or partially sequenced. Three mycoviruses (BcMV1, BVF and BVX) in *B. cinerea* have a worldwide distribution. BpRV1 was detected in *B. porri* and *B. squamosa*. Role of infection of *Botrytis* by BcMV1, BpRV1 and the 6.8-kb dsRNA mycovirus in attenuation of mycelium growth and pathogenicity of their host fungi has been determined. Vertical transmission of BcMV1 and BpRV1 from hyphae to conidia and horizontal transmission of the two mycoviruses from BcMV1- or BpRV1-infected hypovirulent isolates to virulent isolates of the same and the different VCGs were detected. The combined information suggests that RNA mycoviruses are widespread in *Botrytis* and some can cause virulence (pathogenicity) attenuation of *Botrytis*. Further studies on characterization of novel hypovirulence-causing (or hypovirulence-associated) mycoviruses are needed in order to use mycoviruses to control *Botrytis*.

Keywords *Botrytis cinerea* • dsRNA • Hypovirulence • Mycovirus • Biological control

---

M. Wu • J. Zhang • L. Yang • G. Li (✉)
The State Key Laboratory of Agricultural Microbiology and the Key Laboratory of Plant Pathology of Hubei Province, Huazhong Agricultural University, Wuhan 430070, China
e-mail: guoqingli@mail.hzau.edu.cn

S. Fillinger, Y. Elad (eds.), *Botrytis – the Fungus, the Pathogen and its Management in Agricultural Systems*, DOI 10.1007/978-3-319-23371-0_5

5.1 Introduction

Mycoviruses or fungal viruses are a group of viruses infecting filamentous fungi, yeasts and oomycetes (Ghabrial and Suzuki 2009; Pearson et al. 2009). Mycovirus was first recorded in 1948 as the causal agent of La France disease of a cultivated mushroom (*Agaricus bisporus*). Three different virus particles were observed in the abnormal sporophores of *A. bisporus* and this case was regarded the beginning of mycovirology (Hollings 1962). Then, a substance called "statolon" was detected in the cultural filtrates of *Penicillium stoloniferum*. Statolon was found capable of inhibiting viral infection in animals through enhancing animal resistance to viral infection, as it can improve production of interferon in animals. The nature of statolon was subsequently determined to be a mycovirus (Banks et al. 1968). So far, mycoviruses have been detected in all major taxonomic groups of plant pathogenic fungi including *Botrytis cinerea* and *B. porri* (Ghabrial and Suzuki 2009; Pearson et al. 2009; Wu et al. 2012, 2014).

In most cases, mycoviral infection does not cause visible changes in cultural, morphological and biological characteristics of the host fungi. However, infection by some mycoviruses causes abnormal symptoms in their host fungi including reduced mycelium growth and sporulation, suppressed production of secondary metabolites (pigments, toxins) and debilitated virulence/hypovirulence, compared to the mycovirus-free individuals (Ghabrial and Suzuki 2009; Nuss 2005; Nuss and Koltin 1990; Pearson et al. 2009).

Most reported mycoviruses have the genomes of double-stranded (ds) RNA or positive single-stranded (ss) RNA. Mycoviruses in seven families including *Chysoviridae*, *Megabirnaviridae*, *Endornaviridae*, *Partitiviridae*, *Quadriviridae*, *Reoviride*, and *Totiviridae* accommodate dsRNA genomes (Fukuhara and Gibbs 2011; Ghabrial and Suzuki 2009). The mycovirses in the families *Chysoviridae*, *Megabirnaviridae*, *Partitiviridae*, *Quadriviridae* and *Reoviride* have divided dsRNA genomes, whereas the mycoviruses in the family *Totiviridae* and *Endornaviridae* have undivided dsRNA genome (Fukuhara and Gibbs 2011; Ghabrial and Suzuki 2009). Mycoviruses in five other families including *Alphaflexiviridae*, *Barnaviridae*, *Gamma-flexiviridae*, *Hypoviridae*, and *Narnaviridae* accommodate + ssRNA genomes. In addition, Kondo et al. (2013) provided evidence indicating possible existence of negative single-stranded RNA (–ssRNA) mycoviruses in fungi.

Besides RNA mycoviruses, Yu et al. (2010) reported a single-stranded DNA virus, *Sclerotinia sclerotiorum hypovirulence-associated DNA virus 1* (SsHADV-1), in *S. sclerotiorum*, the causal agent of Sclerotinia stem rot of oilseed rape (*Brassica napus*). Besides *S. sclerotiorum*, SsHADV-1 has been detected in various environments including urban river sediments (Kraberger et al. 2013), near-surface atmosphere (Whon et al. 2012) and in dragonflies (Rosario et al. 2012).

Mycoviruses can be used as biocontrol agents for control of plant fungal diseases. The classical example is use of the RNA mycovirus *Cryphonectria hypovirus 1* (CHV1) to control chestnut blight caused by *Cryphonectria (= Endothia) parasitica* (Anagnostakis 1982; MacDonald and Fulbright 1991) in Europe. However, attempts to control this disease with CHV1 in USA were not as unsuccessful as did in Europe, as there are more complex vegetative compatible groups (VCGs) in populations of *C. parasitica* in Northern America than the VCGs in populations of this plant pathogen in Europe (Liu and Milgroom 1996; Milgroom and Cortesi 2004).

Numerous basic studies conducted by Nuss's group using the CHV1-*C. parasitica* system indicated that mycoviruses can be used as tools for deciphering molecular mechanisms involved in interaction between viruses and hosts (Nuss 2011). Successful construction of infectious viral cDNA clone/transcript of CHV1 in 1990s made it possible to establish a reverse genetic system for manipulation of the genes in CHV1. Now, interaction between CHV1 and *C. paracitica* becomes a model system, which can be used for determining the function of related genes in other RNA mycoviruses (Nuss 2011).

Botrytis is a genus accommodating more than 22 species (Hennebert 1973; Staats et al. 2005; Chap. 6). *B. cinerea* is the generalist in this genus causing grey mould disease on more than 1,400 plant species, including many economically important crops such as cucumber, strawberry, table grapes and tomato (Williamson et al. 2007; Chap. 20). The first detailed description of mycoviruses infecting *B. cinerea* was based on observation of virus-like particles (VLPs) and on detection of dsRNAs from mycelial extracts (Howitt et al. 1995). The VLPs appearing in *B. cinerea* exhibited various shapes (isometric, bacilliform and flexuous) (Howitt et al. 1995). In most cases, the presence of dsRNAs in mycelia of *B. cinerea* is co-purified with the corresponding VLPs (Howitt et al. 1995), indicating that the dsRNAs might be the genomes of the mycoviruses. With rapid development of cDNA sequencing technology, numerous species of RNA mycoviruses, including BVF (*Botrytis virus F*), BVX (*Botrytis virus X*), BcMV1 (*Botrytis cinerea mitovirus 1*), BfTV1 (*Botryotinia fuckeliana totivirus 1*), BfPV1 (*Botryotinia fuckeliana partitivirus 1*) and Bc378V1 (*Botrytis cinerea CCg 378 virus 1*) infecting *B. cinerea*, and BpRV1 (*Botrytis porri RVA virus 1*) infecting *B. porri*, the causal agent of garlic leaf blight, have been characterized at the molecular level (Fig. 5.1; Table 5.1). This chapter is to review the research progress so far made in characterization of molecular features of a few RNA mycoviruses infecting *Botrytis* (BcMV1, Bc378V1, BfTV1, BfPV1, BVF, BVX, BpRV1), and of their effects on biology of *Botrytis*. Information about genomes of some mycoviruses infecting *B. cinerea* has been reviewed by other authors (Ghabrial and Suzuki 2009; Pearson et al. 2009; Pearson and Bailey 2013).

5.2 RNA Mycoviruses in *Botrytis*

5.2.1 Genome Non-sequenced RNA Mycoviruses

Howitt et al. (1995) first reported that 143 out of 200 isolates of *B. cinerea* from various host plants (blackberry, cucumber, French bean, grapes, kiwifruit, strawberry, tomato) in New Zealand were detected to accommodate dsRNAs

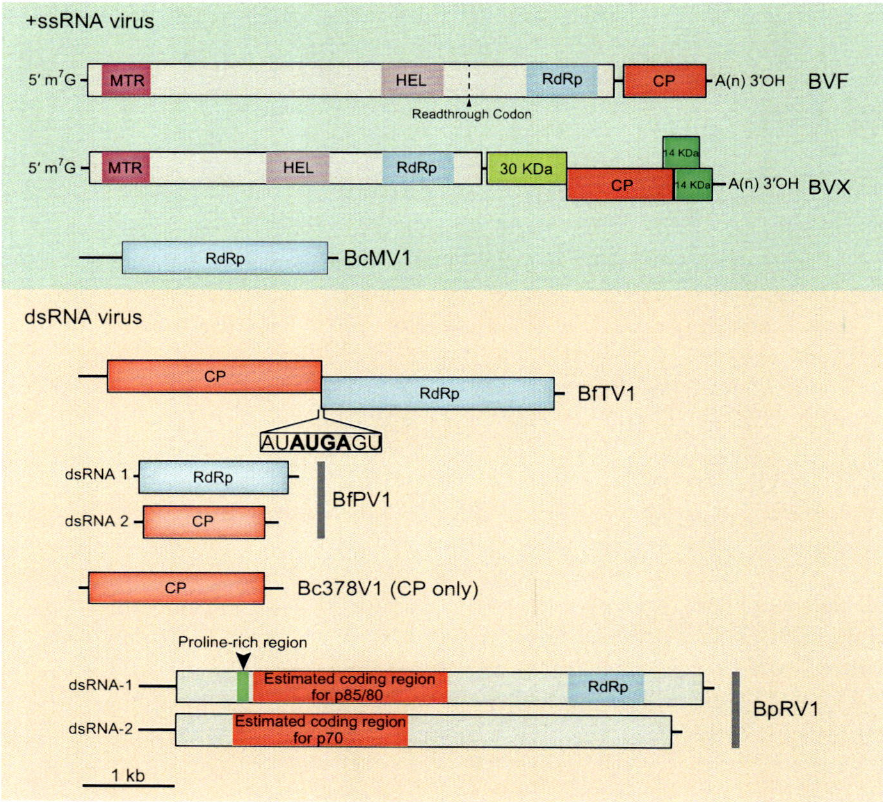

Fig. 5.1 Schematic diagrams of the genome organization of fully or partially sequenced RNA mycoviruses infecting *Botrytis*. *BVF* Botrytis virus F, *BVX* Botrytis virus X, *BcMV1* Botrytis cinerea mito virus 1, *BfTV1* Botryotinia fuckeliana toti virus 1, *BfPV1* Botryotinia fuckeliana partiti virus 1, *Bc378V1* Botrytis cinerea CCg 378 virus 1, *BpRV1* Botrytis porri RNA virus 1, *CP* coat protein, *Hel* helicase, *MTR* methyltransferase, *RdRp* RNA dependent RNA polymerase

ranging from 0.8 to 15 kb in size. Virus-like particles (VLPs) of isometric shape (approximately 30, 35, 40 and 45 nm in diameter) and bacilliform shape (approx. 25 × 63 nm) were observed in three of the five dsRNA-positive isolates of *B. cinerea* (Howitt et al. 1995). Comparison of 12 dsRNA-positive isolates and 12 dsRNA-negative isolates indicated that *B. cinerea* of the two groups differed slightly in radial growth rate, conidial yield, sclerotial yield and weight on agar media, and virulence or pathogenicity on plant tissues (Howitt et al. 1995). Therefore, mycovirus infection does not always cause harmful effects on *B. cinerea*. Mycovirus infection of *B. cinerea* was confirmed by the later studies (Vilches and Castillo 1997; Castro et al. 1999, 2003) with the isolates CVg25, 55k and CCg425 of *B. cinerea* collected from grapes in Chile. The isolate CVg25 had three dsRNA molecules of

Table 5.1 Viruses and dsRNAs reported in *Botrytis*

Classification							
Family	Genus	Species	Viron	Genome	Host fungus	GenBack accession no.	Reference
Gammaflexiviridae	*Mycoflexivirus*	Botrytis virus F	Flexuous, ~720 nm	+ssRNA, ~6.8 kb	*B. cinerea*	AF238884	Howitt et al. (2001)
Alphaflexiviridae	*Botrexvirus*	Botrytis virus X	Flexuous, ~720 nm	+ssRNA, ~6.9 kb	*B. cinerea*	AY055762	Howitt et al. (2006)
Narnaviridae	*Mitovirus*	Botrytis cinerea mitovirus 1	Unencapsidated	+ssRNA, ~2.8 kb	*B. cinerea*	AM087550	Wu et al. (2007, 2010); Rodríguez-García et al. (2013)
Partitiviridae	*Partitivirus*	Botryotinia fuckeliana partitivirus1	Isometric, ~40 nm	dsRNAs, ~1.8+1.5 kb	*B. cinerea*	dsRNA1, AM491609; dsRNA2, AM491610	De Guido et al. (2007)
		Botrytis cinerea CCg378 virus 1	Isometric, ~32 nm	dsRNAs, ~2.2+1.95 kb	*B. cinerea*	KF201714	Potgieter et al. (2013)
Totiviridae	*Victorivirus*	Botryotinia fuckeliana totivirus1	Unkown	dsRNA, 5.2 kb	*B. cinerea*	AM491608	De Guido et al. (2007)
Unassigned	Unassigned	Botrytis porri RNA virus 1	Isometric, ~35 nm	dsRNAs, 6.2+5.9 kb	*B. porri*	dsRNA-1, JF716350; dsRNA-2, JF716351	Wu et al. (2012)

(continued)

Table 5.1 (continued)

Classification							
Family	Genus	Species	Viron	Genome	Host fungus	GenBack accession no.	Reference
Unassigned	Unassigned	Unassigned	Isometric, 30–45 nm	dsRNA	*B. cinerea*	ND[a]	Howitt et al. (1995)
			Bacilliform, 25×63 nm	dsRNA		ND	
			Unencapsidated	dsRNA		ND	
Unassigned	Unassigned	Unassigned	Isometric, ~40 nm	dsRNA, 8.3 kb	*B. cinerea*	ND	Vilches and Castillo (1997)
			Unencapsidated	dsRNA, 2.0 kb		ND	
			Unencapsidated	dsRNA, 1.4 kb		ND	
Unassigned	Unassigned	Unassigned	Isometric, 28 nm	dsRNA, 1.8 kb	*B. cinerea*	ND	Castro et al. (1999)
Unassigned	Unassigned	Unassigned	Isometric, 33 nm	dsRNA, 6.8 kb	*B. cinerea*	ND	Castro et al. (2003)

[a]No sequence data is available

1.4 kb (S-dsRNA), 2.0 kb (M-dsRNA) and 8.3 kb (L-dsRNA), and isometric VLPs (approximately 40 nm in diameter) in hyphal cytoplasm forming clusters within membraneous structures (Vilches and Castillo 1997). L-dsRNA was encapsidated, whereas S-dsRNA and M-dsRNA were not unencapsidated (Vilches and Castillo 1997). The isolate 55k was infected by a mycovirus with the genome of 1.8 kb dsRNA and a coat protein of 68 kDa forming isometric virus particles of approximately 28 nm in diameter (Castro et al. 1999). The hyphal cells of *B. cinerea* infected by this mycovirus exhibited degeneration symptoms characterized by formation of abundant membranous structures and degeneration of cellular organelles (Castro et al. 1999). The isolate CCg425 contained a 33-nm isometric mycovirus with the 6.8-kb dsRNA genome (Castro et al. 2003). The isolate displayed lower conidial production, lower laccase activity and less aggressiveness than the dsRNA-free virulent isolate CKg54 of *B. cinerea* (Castro et al. 2003). The causal effect of the mycovirus in CCg425 was successfully established by introducing the purified virus particles into CKg54, as the resulting transfected isolate CKg54vi425 displayed similar phenotypes to those of the hypovirulent parental isolate CCg425 (Castro et al. 2003). These studies suggest that RNA mycoviruses are widespread in populations of *B. cinerea*, although the taxonomic status of these RNA mycoviruses was not clearly elucidated.

5.2.2 Positive ssRNA Mycoviruses

Three mycoviruses (BVF, BVX, BcMV1) with +ssRNA genomes have been fully cDNA-sequenced (Table 5.1). *Botrytis virus F* (BVF) is the first fully sequenced mycovirus in *B. cinerea* (Howitt et al. 2001). It belongs to the family *Gammaflexiviridae* (Table 5.1). The genome of BVF is 6,827 nucleotides (nt) long with a poly (A) tail at the 3′-terminus (Fig. 5.1). Two open reading frames (ORFs), ORF1 and ORF2, were deduced in the genome of BVF. ORF1 encodes a 212-kDa polypeptide, whereas ORF2 encodes a 32-kDa polypeptide. The 212-kDa polypeptide shows homologous to RNA dependent RNA polymerase (RdRp) of plant 'tymo-' and 'potex-like' viruses. The ORF1 for BVF appears different from the ORF1 for plant viruses, as the BVF ORF1 has an opal putative readthrough codon between the helicase and RdRp regions. The 32-kDa polypeptide was predicted to be the coat protein (CP), which is also homologous to those of plant 'potex-like' viruses in amino acid (aa) sequence. Moreover, there are three untranslated regions (UTR), 5′-UTR of 63 nt long, 3′-UTR of 70 nt long preceding the poly (A) tail and an internal UTR of 93 nt long located between ORFs 1 and 2. The purified virus particles of BVF are flexuous and rod-shaped (approximately 720×13 nm, length×diameter). A new virus genus, *Mycoflexivirus*, was established to accommodate BVF (Adams et al. 2011a).

Botrytis virus X (BVX) is a +ssRNA virus co-infested with BVF in the isolate RH106-10 of *B. cinerea* (Howitt et al. 2006). The virus particles are flexuous and rod-shaped with a modal length of approximately 720 nm (Table 5.1). The virus particles encapsidate the ssRNA genome of 6,966 nt long with a poly (A) tail at the 3′ terminus (Fig. 5.1). There are five deduced ORFs in the genome of BVX. Two ORFs, ORF1 and ORF3, encode polypeptides with homologues in the public database NCBI (e.g. http://www.ncbi.nlm.nih.gov/), ORF1 encoding a polypepetide with expected molecular mass of 158 kDa containing conserved domains for methyltransferase, helicase and RdRp, and ORF3 encoding a 20-kDa polypeptide homologous to the CPs of plant 'potex-like' viruses. The remaining three ORFs (ORFs 2, 4, 5) encode three unknown polypeptides. BVX belongs to a new genus, *Botrexvirus* in the family *Alphaflexiviridae* (Adams et al. 2011b).

Botrytis cinerea mito virus 1 (BcMV1), formerly known as *Botrytis cinerea debilitation-related virus* (BcDRV), was detected in the hypovirulent isolate CanBc-1 of *B. cinerea* (Wu et al. 2007, 2010). BcMV1 is a mitovirus in the family *Narnaviridae* (Table 5.1). The genome of BcMV1 is 2,804 nt long and AU-rich (66.8 %). It was deduced to have one large ORF coding for RdRp based on the mitochondrial codon usage. The nucleotide sequence from 447 to 2,778 nt and the RdRp sequence of BcMV1 share 95 and 96 % identity to the full-length nucleotide sequence (1–2,332 nt) and the RdRp sequence in *Ophiostoma novo-ulmi mitovirus 3b* (OnuMV3b), respectively. Therefore, BcMV1 and OnuMV3b might be the same mycoviral species. However, the sequence of BcMV1 is 472 nt (26 and 446 nt long at 3′- and 5′-terminal regions, respectively) longer than the nucleotide sequence of OnuMV3b. The BcMV1 RdRp contains six conserved amino acid sequence motifs (I–VI) and the motif I was considered to be a characteristic motif for mitoviruses (Hong et al. 1999).

A small defective mitovirus designated as BcMV1-S was detected in the isolate CanBc-1c-78, a single-conidium isolate of CanBc-1 (Wu et al. 2007, 2010). BcMV1-S is 2,171 nt long and derived from BcMV1 through a single internal 633-nt in-frame deletion at 1,145–1,777 nt in the genome of BcMV1. A recent report by Rodríguez-García et al. (2013) indicated that BcMV1 infects *B. cinerea* in Europe. Partial sequence analysis revealed the European strain of BcMV1 shares nucleotide sequence identity of 94 % at 312–1,772 nt compared to the corresponding region in the nucleotide sequence of Chinese strain of BcMV1 and shares amino acid (aa) identity of 96 % at 1–432 aa to the corresponding region in the RdRp sequence encoded by the Chinese strain of BcMV1.

5.2.3 dsRNA Mycoviruses

Genomes of a few dsRNA mycoviruses including Bc378V1, BfPV1, BfTV1 and BpRV1 have been cDNA-sequenced. *Botryotinia fuckeliana t oti v irus 1* (BfTV1) is a victorivirus belonging to the family *Totiviridae* (de Guido et al. 2007). The genome of BfTV1 is 5,261 bp long and possesses two large ORFs, ORF1 and ORF2 (Fig. 5.1). ORF1 encodes a 765-aa polypeptide homologous to the coat proteins (CP) of victoviruses. ORF 2 encodes an 837-aa polypeptide homologous to mycoviral RdRps. The stop codon of ORF1 overlaps the start codon of ORF 2 in four nucleotides (AUGA) suggesting a coupled termination-reinitiation mechanism for translation of CP and RdRp. As the termination-reinitiation mechanism is an important feature for differentiation of victoriviruses from totiviruses (Ghabrial and Nibert 2009). Therefore, BfTV1 might be a member of *Victorivirus* (Table 5.1).

Botryotinia uckeliana partiti virus 1 (BfPV1) possesses three dsRNAs including dsRNA1 (1,793 bp), dsRNA2 (1,566 bp) and dsRNA3 (1,383 bp) (de Guido et al. 2007) (Fig. 5.1; Table 5.1). The presence of all these dsRNAs was associated with isometric viral particles (approximately 40 nm in diameter). DsRNA1 encodes a 540-aa polypeptide with highly homologous to RdRps of partitiviruses, whereas dsRNA2 encodes a putative 436-aa CP. DsRNA3 was determined to be a defective RNA of dsRNA2 with an internal deletion of a 210-bp region in the coding region.

Potgieter et al. (2013) reported a species of *Partitivirus* designated as Bc378V1 in the isolate CCg378 of *B. cinerea*. Bc378V1 forms isometric viral particles (approximately 32 nm in diameter) containing two dsRNAs (1.95 and 2.2 kb) (Table 5.1) and a 70-kDa structural protein. The genome of Bc378V1 has not been fully sequenced (Fig. 5.1). Phylogenetic analysis of the sequence for coat protein of Bc378V1 suggests that it is distantly related to BfPV1.

Botrytis porri RNA virus 1 (BpRV1) is the first mycovirus detected in the isolate GarlicBc-72 of *B. porri* (Wu et al. 2012), the causal agent of garlic leaf blight. The genome of BpRV1 comprises two dsRNAs, dsRNA-1 (6,215 bp) and dsRNA-2 (5,879 bp) (Table 5.1). The two dsRNAs share sequence identities of 62 and 95 % at the 3′- and 5′-terminal regions, respectively. Two ORFs, ORF I on dsRNA-1 and ORF II on dsRNA2, were deduced (Fig. 5.1). The polypeptide deduced from the 3′-proximal coding region of ORF I shows sequence identities of 19–23 % with RdRps encoded by mycoviruses in the families *Totiviridae*, *Chrysoviridae*, and *Megabirnaviridae*. However, the remaining partial polypeptide encoded by ORF I and the popypeptide encoded by the whole ORF II lack sequence similarities to any reported virus proteins. Phylogenetic analysis indicated that BpRV1 is closely related to two insect viruses, Spissistilus festinus virus 1 and Cirulifer tenellus virus 1. BpRV1 is distantly related to other known RNA mycoviruses (Wu et al. 2012). BpRV1 forms isometric virions with the diameter of approximately 35 nm (Table 5.1).

Three structural proteins of 70, 80, and 85 kDa encoded by BpRV1 were detected in extracts of the virions of BpRV1. Peptide mass fingerprinting technique was used to locate genes on dsRNA1 and dsRNA2 coding for the structural proteins of BpRV1. The results revealed that the 80- and 85-kDa proteins are encoded by ORF I, whereas the 70-kDa protein is encoded by ORF II (Fig. 5.1). All the combined results suggest that BpRV1 is a novel bipartite dsRNA virus and possibly belongs to a new virus family. Moreover, another strain of BpRV1 was detected in *B. squamosa*, the causal agent of onion leaf blight (Wu et al. 2014). DsRNA1 and dsRNA2 for the *B. squamosa* strain of BpRV1 (Wu et al. 2014) are 91.7 and 97.7 % identical to the corresponding dsRNAs for the *B. porri* strain of BpRV1 (Wu et al. 2012). The two polypeptides encoded by dsRNA-1 and dsRNA-2 of the *B. squamosa* strain of BpRV1 (Wu et al. 2014) shared 97.1 and 98.7 % sequence identity to the corresponding polypeptides encoded by the *B. porri* strain of BpRV1 (Wu et al. 2012).

5.3 Mycovirus Transmission in *Botrytis*

5.3.1 Types of Mycoviral Transmission

Four types of mycoviral transmission have been reported in previous studies. They are extracellular transmission (Yu et al. 2013), vertical transmission (Nuss 2005), horizontal transmission (Nuss 2005) and interspecies transmission (Milgroom and Hillman 2011). Extracellular transmission was only detected in infection of *S. sclerotiorum* by the DNA mycovirus SsHADV-1 (Yu et al. 2013), but has not been detected in RNA mycovirues. However, previous studies showed that purified virus particles of a few RNA mycoviruses such as the 6.8-kb dsRNA mycovirus in *B. cinerea* and BpRV1 in *B. porri* can be used to transfect protoplasts of mycovirus-free isolates of *B. cinerea* and *B. porri*, respectively, under mediation of polyethylene glycol (Castro et al. 2003; Wu et al. 2012).

Vertical transmission refers to transmission of viruses from parents to offsprings through sexual and/or asexual reproduction (Nuss 2005). Generally, vertical transmission of mycoviruses is common through asexual reproduction, but is not common through sexual reproduction (Lecoq et al. 1979). Transmission efficiency varies with different mycoviruses and different host fungi. For example, the transmission efficiency was high (approximately 100 %) for dsRNA mycoviruses infecting *Aspergillus* spp. (Coenen et al. 1997; Van Diepeningen et al. 2006), *Helicobasidium mompa* (Ikeda et al. 2004), the causal agent of plant purple root rot, and *Chalara elegans* (= *Thielaviopsis basicola*), the causal agent of plant black root rot (Park et al. 2006), and for *Cryphonectria mitovirus 1* (CMV1) infecting *C. paracitica* (Polashock and Hillman 1994). However, the transmission efficiency was low (3–55 %) for *Heterobasidion annosum virus* infecting *Heterobasidion annosum*, the causal agent of root rot of many forest trees (Ihrmark et al. 2002).

Wu et al. (2007) reported that 75 out of 94 (79.8 %) single-conidium (SC) isolates of *B. cinerea* CanBc-1 were hypovirulent (HV), whereas the other 19 SC isolates (20.2 %) were virulent (V). Four randomly selected hypovirulent SC isolates were detected positive for BcMV1 (containing 3.0-kb dsRNA) infection, whereas four randomly selected virulent SC isolates were detected negative for BcMV1 infection (Wu et al. 2007). This result suggests that BcMV1 can be transmitted from hyphae to conidia during asexual reproduction with the efficiency of 79.8 %. The transmission efficiency of BVX to conidia of *B. cinerea* was estimated to be about 72 % (Tan et al. 2007). Another study by Wu et al. (2012) showed that 34 out of 35 (97.1 %) SC isolates derived from the hypovirulent isolate GarlicBc-72 of *B. porri* were detected positive for BpRV1 infection, whereas the other one SC isolate was detected negative for BpRV1 infection. This result suggests that BpRV1 can be vertically transmitted from hyphae to conidia in *B. porri* with the transmission efficiency of 97.1 %.

Viral transmission through sexual reproduction was not as efficient as that through asexual reproduction. For example, *Cryphonectria hypovirus 1* (CHV1) are unable to be transmitted to ascospores in *C. parasitica* (Anagnostakis 1982).

In order to improve transmission of CHV1 to ascospores, the hypovirus infectious cDNA of CHV1 was constructed and successfully integrated into the genome of *C. parasitica* (Choi and Nuss 1992). The viral cDNA can be inherited to the ascospore progeny and transcribed to viral RNA of CHV1 (Chen et al. 1993). Most mitoviruses in *O. novo-ulmi* cannot be transmitted to ascospores (Brasier 1983; Rogers et al. 1986). In some cases, transmission of mitoviruses through sexual reproduction is possible. For example, CMV1 was found transmissible to ascospores from the CMV1-infected female parent with the frequency of ~50 % (Polashock et al. 1997). *B. cinerea* and other species of *Botrytis* rarely conduct sexual reproduction (Williamson et al. 2007). In spite of this, evidence showed that BVX in *B. cinerea* can be transmitted to ascospores of this fungus through sexual reproduction (Tan et al. 2007). The transmission frequency was 34 % for the BVX-infected "male" parent, whereas was 53 % for the BVX-infected "female" parent (Tan et al. 2007). Whether other RNA mycoviruses such as BcMV1, BVF and BpRV1 can be vertically transmitted through sexual reproduction or not remains unknown and needs further characterizations.

Horizontal transmission refers to transmission of mycoviruses between two individuals through hyphal contact or anastomosis (hyphal fusion) (Nuss 2005). Wu et al. (2007, 2012) reported that BcMV1 and BpRV1 in the hypovirulent isolates of CanBc-1 (*B. cinerea*) and GarlicBc-72 (*B. porri*) were successfully transmitted to virulent SC isolates CanBc-1c-66 (derived from CanBc-1) and GP72SC35 (derived from GarlicBc-72), respectively, through hyphal contact (possibly through hyphal fusion). However, BcMV1 and BpRV1 failed to transfect the virulent isolates CanBc-2 (*B. cinerea*) and OnionBc-95 (*B. porri*), respectively, through hyphal contact (Wu et al. 2007, 2012). It is well recognized that horizontal transmission of RNA mycoviruses is usually restricted by the vegetative incompatibility or heterokaryon incompatibility, which acts as a barrier (Anagnostakis 1982; Brasier 1983; Charlton and Cubeta 2007; Dalzoto et al. 2006). Fungal vegetative incompatibility is a complex system controlled by *vic* genes or *het* genes (Cortesi and Milgroom 1998; Saup 2000). If two fungal isolates have different alleles at one or more *vic* loci, they are incompatible and considered as different vegetative compatibility groups (VCGs) (Saup 2000; Chap. 3). Liu et al. (2003) reported that there were at least seven anastomosis groups (AG1 to AG7) among 92 isolates of *B. cinerea* from vegetables in Shanxi Province, China, based on analysis of vegetative incompatibility. Beever and Weeds (2004) identified 66 VCGs in *B. cinerea* isolates of their collection. Existence of different VCGs may restrict horizontal transmission of mycoviruses like BcMV1 in *B. cinerea* and BpRV1 in *B. porri*.

Interspecific transmission refers to transmission of mycoviruses between different fungal species through hyphal fusion. Interspecies transmission of mycoviruses has been reported in *Aspergillus niger*/*A. nidulans* (Coenen et al. 1997), *S. sclerotiorum*/*S. minor* (Melzer et al. 2002), and *C. parasitica*/*C. nitschkei* (Liu et al. 2003). Wu et al. (2007, 2010) reported that the nucleotide sequence and the RdRp sequence of BcMV1 share 95 and 96 % identity to the nucleotide sequence of *Ophiostoma novo-ulmi mitovirus 3b* (OnuMV3b), suggesting that there may exist an interspecies mitovirus transmission between *B. cinerea* and *O. novo-ulmi*, the causal agent of Dutch elm disease. Wu et al. (2012, 2014) reported that BpRV1 infects both *B. porri* and *B. squamosa*, suggesting that there may be an interspecies transmission of BpRV1 in these two species of *Botrytis*, which can infect garlic plants.

5.3.2 Geographic Distribution and Incidence of RNA Mycoviruses in Botrytis

Like viruses infecting other hosts (animals and plants), mycoviruses may have a wide geographical distribution although information about epidemiology of mycoviruses is very limited so far. Evidence showed that the RNA mycoviruses BVF, BVX, BcMV1 and BpRV1 have a wide distribution (Pearson and Bailey 2013; Rodríguez-García et al. 2013; Kecskeméti et al. 2014; Wu et al. 2014). Studies conducted by Howitt et al. (2001, 2006) indicated that 17.2 and 18.8 % of *B. cinerea* isolates in New Zealand were infested with BVF and BVX, respectively. BVF was detected to be present alone or together with BVX in the isolates of *B. cinerea* in England, France, and Israel, whereas BVX was detected to be present alone in the isolates of *B. cinerea* in Belgium, Greece, Italy, Portugal, Switzerland and USA (Pearson and Bailey 2013). The average percentages of BVF and BVX in the investigated isolates of *B. cinerea* in the above-mentioned countries were 16.1 and 29 %, respectively. In Germany, none of the two mycoviruses (BVF and BVX) was detected in 53 isolates of *B. cinerea* in 2008, whereas in 2010, 31, 3, and 3 % isolates of *B. cinerea* ($n=100$) were detected to be infested with BVF alone, BVX alone and BVF + BVX, respectively (Kecskeméti et al. 2014).

It seems that infection *B. cinerea* by BVF or BVX varies in different years and different locations. The mycovirus BcMV1 was first reported to infect *B. cinerea* in China (Wu et al. 2007, 2010). Rodríguez-García et al. (2013) reported BcMV1 was detected in approximately 30.2 % isolates of *B. cinerea* ($n=96$) collected from the southern and central Spain. These studies suggest BcMV1 may have a wide geographic distribution. The mycovirus BpRV1 was first reported to infect *B. porri* in China (Wu et al. 2012). It was detected to infect *B. squamosa* (Wu et al. 2014) and *Sclerotinia sclerotiorum* (L. J. Liu and D. H. Jiang, unpublished data) in this country. These results suggest that BpRV1 may have a wide host range.

5.4 Effects of Mycoviral Infection on the Biology of *Botrytis*

Infection of different host fungi by different mycoviruses may result in different effects on the host fungi (beneficial, neutral and harmful) (Nuss and Koltin 1990). For example, killer proteins produced by RNA mycovirus-infected isolates of the yeast *Saccharomyces cerevisiae* and the basidiomycetous fungus *Ustilago maydis*, the causal agent of maize smut, can enhance their survival through inhibiting growth of mycovirus-free isolates of *S. cerevisiae* and *U. maydis*, respectively (Magliani et al. 1997). Interestingly, to better maintain the killer protein production, the mycovirus-infected yeast isolates usually lose the ability to produce the proteins argonaute (AGO1) and dicer (DCR1), two key proteins in the RNA silencing-mediated anti-viral system, and deficiency in this antiviral system may create an optimum intracellular environment for viral proliferation in cells of *S. cerevisiae* (Drinnenberg et al. 2011). In contrast, infection of *A. bisporus* by La France isometric virus causes La France disease on cultivated mushroom (Hollings 1962). Plant pathologists get interested in studies on mycoviruses as infection by mycoviruses may cause fungal virulence (pathogenicity or aggressiveness) debilitation or attenuation (hypovirulence). Hypovirulence-causing (or associated) mycoviruses have been reported in many plant pathogenic fungi including *B. cinerea* and *B. porri*.

5.4.1 *The 6.8-kb dsRNA Mycovirus and Hypovirulence in* B. cinerea

The 6.8-kb dsRNA mycovirus in the isolate CCg425 of *B. cinerea* was the first reported dsRNA mycovirus conferring hypovirulence in *B. cinerea* (Castro et al. 2003). CCg425 caused smaller lesions than the mycovirus-free isolate CKg54 of *B. cinerea* on bean leaves. It was suppressed for growth, sporulation and laccase activity compared to CKg54 in cultural media. In order to confirm that the causal effect of the 6.8-kb dsRNA mycovirus on hypovirulence and the hypovirulence-associated traits, Castro et al. (2003) introduced the purified viral particles from CCg425 to protoplasts of CKg54. The resulting mycovirus-infected isolate CKg54vi425 from protoplast regeneration displayed similar abnormal phenotypes to those exhibited by the parental isolate CCg425.

5.4.2 *BcMV1 and Hypovirulence in* B. cinerea

Studies by Wu et al. (2007, 2010) revealed that infection of *B. cinerea* by BcMV1 is closely associated with hypovirulence of *B. cinerea*. Compared with the virulent isolates of *B. cinerea*, the BcMV1-infected isolate CanBc-1 of *B. cinerea* grew slowly on potato dextrose agar and was nearly avirulent on leaves of oilseed rape (*Brassica napus*) (Fig. 5.2). Wu et al. (2010) observed mitochondrial morphology in hyphal cells of CanBc-1 (hypovirulent) and CanBc-1c-66 (a single-conidum isolate of CanBc-1, virulent) of *B. cinerea* using transmission electron microscope (TEM). They found that in hyphae of CanBc-1c-66, normal oblong-shaped mitochondria with abundant cristae inside were frequently observed. In hyphal cells of CanBc-1, however, besides normal mitochondria, abnormal mitochondria with swollen

Fig. 5.2 Colony morphology (*top two rows*) and pathogenicity (*bottom two rows*) of the isolates CanBc-1 and CanBc-2 of *Botrytis cinerea*, and eight single-conidium isolates of CanBc-1. All the cultures on potato dextrose agar (PDA) were incubated at 20 °C for 2 weeks. Leaves of oilseed rape inoculated with the isolates were incubated at 20 °C for 72 h. (**a–j**) PDA cultures of CanBc-1c-55, CanBc-1c-64, CanBc-1c-66, CanBc-1c-93, CanBc-2, CanBc-1c-50, CanBc-1c-73, CanBc-1c-78, CanBc-1c-82 and CanBc-1, respectively, and leaf lesions caused by these isolates, respectively. Note normal growth of CanBc-1c-55 (**a**), CanBc-1c-64 (**b**), CanBc-1c-66 (**c**), CanBc-1c-93 (**d**) and CanBc-2 (**e**), and necrotic leaf lesions caused these isolates, whereas suppressed growth of CanBc-1c-50 (**f**), CanBc-1c-73 (**g**), CanBc-1c-78 (**h**), CanBc-1c-82 (**i**) and CanBc-1 (**j**), and no lesion formation on leaves inoculated with these isolates. DsRNA$^+$ and dsRNA$^-$ indicate positive detection and negative detection of 3.0-kb dsRNA of BcMV1, respectively, in mycelial extracts of the ten isolates (**a–j**) by agarose gel electrophoresis

appearance, and remnants of degenerated cristae and fibrous matrix materials inside were observed. Wu et al. (2010) also found that BcMV1 is co-purified with mitochondria. These results suggest that mitochondria in hyphal cells of *B. cinerea* CanBc-1 might be the target for the attack by BcMV1.

Zhang et al. (2010) conducted a study to investigate mechanisms involved in hypovirulence of CanBc-1. The hypovirulent isolate CanBc-1 was compared with the virulent isolates CanBc-1c-66 and CanBc-2 of *B. cinerea* for formation of infection cushions on onion bulbs and on leaves of oilseed rape and tomato, as well as for production of pectinase, toxic metabolites, oxalic acid and laccase in different growth media. They found that formation of infection cushions was common on epidermis of onion bulbs and on leaves of oilseed rape and tomato inoculated with the isolates CanBc-1c-66 or CanBc-2, but was rare on these plant tissues inoculated with the isolate CanBc-1. The three isolates were found capable of production of pectinases, toxic metabolites, oxalic acid and laccase in pure cultures. Therefore, deficiency in formation of infection cushions and attenuated mycelium growth are probably responsible for hypovirulence of the BcMV1-infected isolate CanBc-1 of *B. cinerea*.

Additionally, Wu et al. (2010) detected very low titer of BcMV1 in mycelial extracts of the virulent SC isolate CanBc-1c-66 of *B. cinerea* using quantitative real-time PCR (qPCR). However, the 3.0-kb dsRNA (the replication intermediate of BcMV1) in the mycelial extracts of CanBc-1c-66 was not detected by agarose gel electrophoresis. This result suggests that infection of *B. cinerea* by BcMV1 at low titer (latent or cryptic infection) may have no harmful effect on hyphal growth and pathogenicity of *B. cinerea*.

5.4.3 *BpRV1 and Hypovirulence in* B. porri

Wu et al. (2012) reported that BpRV1 is the causal agent for hypovirulence of *B. porri*. The BpRV1-infected isolate GarlicBc-72 of *B. porri* grew slowly on potato dextrose agar with the formation of numerous mycelial sectors at the colony margin and caused smaller lesions on leaves of garlic (*Allium sativum*) compared to virus-free isolates of *B. porri*. The cytoplasm of hypovirulent isolate GarlicBc-72 contained abundant vacuole-like membranous structures and small membranous vesicles and the membranous vacuoles/vesicles. Introduction of the purified virus particles of BpRV1 into the virulent isolate GarlicBc-38 of *B. porri* resulted in a significant pathogenicity attenuation of this isolate (Wu et al. 2012).

5.4.4 *Impacts of BVF and BVX on Biology of* B. cinerea

Evidence so far achieved indicates that BVF and BVX have no significant effects on pathogenicity of *B. cinerea*. Some natural BVX-infected isolates of *B. cinerea* showed a marginal increase in growth rates on malt extract agar plates and a slightly decrease of virulence compared to the mycovirus-free isolates of *B. cinerea* (Pearson and Bailey 2013). Kecskeméti et al. (2014) reported that at 10–30 °C, BVF-infected isolates of *B. cinerea* showed no detectable difference in growth rate from the mycovirus-free isolates of *B. cinerea*. However, at 4 and 7 °C, the BVF-infected isolates were reduced for mycelial growth compared to the mycovirus-free isolates (Kecskeméti et al. 2014).

5.5 Conclusions and Future Perspectives

Infection of *B. cinerea* by RNA mycoviruses was first reported in 1995 (Howitt et al. 1995). So far, RNA mycoviruses in *B. cinerea* have been recorded in southern America (e.g. Chile), Asia (e.g. China and Lebanon), Europe (e.g. Germany and Spain) and Oceania (e.g. New Zealand). Several RNA mycoviruses in *B. cinerea*, including BVF (*Gammaflexiviridae*), BVX (*Alphaflexiviridae*), BcMV1 (*Narnaviridae*), Bc378V1 (*Totiviridae*), BfTV1 (*Partitiviridae*), BfPV1 (*Partitiviridae*), and in *B. porri* such as BpRV1 possibly belonging to a novel family have been fully or partially characterized at the molecular level. Meanwhile, many other RNA mycoviruses reported in previous studies (Castro et al. 1999, 2003; Habib et al. 2013; Howitt et al. 1995; Kecskeméti et al. 2014; Vilches and Castillo 1997) have not been sequenced for their genomes. These results suggest that *B. cinerea* accommodates diversified species of RNA mycoviruses.

Infection of *B. cinerea* either by the 6.8-kb dsRNA mycovirus (Castro et al. 2003) or by BcMV1 (Wu et al. 2007, 2010), and infection of *B. porri* by BpRV1 (Wu et al. 2012) were observed to cause (or to be closely associated with) virulence attenuation (hypovirulence) of the host fungi. Both BcMV1 and BpRV1 were detected vertically transmissible from hyphae to conidia of *B. cinerea* and *B. porri*, respectively, through asexual reproduction, and horizontally transmissible to other virulent isolates of corresponding host fungi through hyphal contact (Wu et al. 2007, 2012). These results suggest that RNA mycoviruses may have potential to be exploited to control *B. cinerea* and *B. porri*.

Previous studies indicated that a high proportion of *B. cinerea* isolates harboring virus-like particles and/or dsRNA elements (Habib et al. 2013; Howitt et al. 1995; Kecskeméti et al. 2014). Besides BpRV1 infecting *B. porri*, mycoviruses in other species of *Botrytis* have not been detected so far. Novel mycoviruses possibly exist in populations of *B. cinerea* and its relatives. From the point of view of fundamental studies, it is necessary to discover and to characterize novel mycoviral species in *Botrytis*, to disclose interaction between mycoviruses and *Botrytis*, and to investigate epidemics of mycoviruses in populations of *Botrytis*.

Mycoviruses can cause hypovirulence in plant pathogenic fungi including *B. cinerea* (Nuss and Koltin 1990; Pearson et al. 2009). Wu et al. (2007, 2010, 2012) reported that BcMV1 and BpRV1 infect *B. cinerea* and *B. porri*, respectively, conferring hypovirulence for the two fungi. However, horizontal transmission of both mycoviruses among different isolates of *B. cinerea* and *B. porri* has been found to be possibly restricted by vegetative incompatibility (Wu et al. 2007, 2012). For example, transmission of BcMV1 in the isolate CanBc-1 of *B. cinerea* was successful from CanBc-1 to the isolate CanBc-1c-66, a single-conidium isolate of CanBc-1, but was unsuccessful from CanBc-1 to the isolate CanBc-2 of *B. cinerea* (Wu et al. 2007). Vegetative incompatibility between the isolates CanBc-1 and CanBc-2 might be responsible for the failure of BcMV1 transmission from CanBc-1 to CanBc-2. Therefore, in order to use mycoviruses to control plant diseases caused by *Botrytis*, it is necessary to screen mycoviruses capable of overcoming vegetative incompatibility in horizontal transmission.

Recent progress in use of the DNA mycovirus SsHADV-1 to control *S. sclerotiorum* provided a good example (Yu et al. 2013). SsHADV-1 in the isolate DT-8 of *S. sclerotiorum* is transmissible to isolates of *S. sclerotiorum* belonging to various vegetative compatibility groups (VCGs) different from the VCG of DT-8. Purified virus particles of SsHADV-1 were detected directly infectious to hyphae of *S. sclerotiorum* belonging to different VCGs, suggesting that SsHADV-1 can extracellularly transfect hyphae of *S. sclerotiorum* (Yu et al. 2013). Whether there exist DNA/RNA mycoviruses in *Botrytis* similar to SsHADV-1 in terms of horizontal transmissibility or not remains unknown and needs further large-scale screening and characterization.

Acknowledgments This research was funded by the Natural Science Foundation of China (Grant Nos. 31070122, 31070122, 31301615) and the R & D Special Fund for Public Welfare Industry (Agriculture) of China (Grant No. 201303025).

References

Adams MJ, Kreuze JF, Pearson MN (2011a) *Gammaflexiviridae*. In: Fauquet CM, Mayo MA, Maniloff J et al (eds) Virus taxonomy, ninth report of the international committee on taxonomy of viruses. Elsevier Press, London, pp 942–943

Adams MJ, Candresse T, Hammond J et al (2011b) *Alphaflexiviridae*. In: Fauquet CM, Mayo MA, Maniloff J, Desselberger U, Ball LA (eds) Virus taxonomy, the ninth report of the international committee on taxonomy of viruses. Elsevier Press, London, pp 904–919

Anagnostakis SL (1982) Biological control of chestnut blight. Science 215:466–471

Banks GT, Buck KW, Chain EB et al (1968) Viruses in fungi and interferon stimulation. Nature 218:542–545

Beever RE, Weeds PL (2004) Taxonomy and genetic variation of *Botrytis* and *Botryotinia*. In: Elad Y, Williamson B, Tudzynski P, Delen N (eds) *Botrytis*: biology, pathology and control. Springer Press, The Netherlands, pp 29–52

Brasier CM (1983) A cytoplasmically transmitted disease of *Ceratocystis ulmi*. Nature 305:220–223

Castro M, Kramer K, Valdivia L et al (1999) A new double-stranded RNA mycovirus from *Botrytis cinerea*. FEMS Microbiol Lett 175:95–99

Castro M, Kramer K, Valdivia L et al (2003) A double-stranded RNA mycovirus confers hypovirulence-associated traits to *Botrytis cinerea*. FEMS Microbiol Lett 228:87–91

Charlton ND, Cubeta MA (2007) Transmission of the M2 double-stranded RNA in *Rhizoctonia solani* anastomosis group 3 (AG-3). Mycologia 99:859–867

Chen B, Choi GH, Nuss DL (1993) Mitotic stability and nuclear inheritance of integrated viral cDNA in engineered hypovirulent strains of the chestnut blight fungus. EMBO J 12:2991–2998

Choi GH, Nuss DL (1992) Hypovirulence of chestnut blight fungus conferred by an infectious viral cDNA. Science 257:800–803

Coenen A, Kevei F, Hoekstra RF (1997) Factors affecting the spread of double-stranded RNA viruses in *Aspergillus nidulans*. Genet Res 69:1–10

Cortesi P, Milgroom MG (1998) Genetics of vegetative incompatibility in *Cryphonectria parasitica*. Appl Environ Microbiol 64:2988–2994

Dalzoto PR, Glienke-Blanco C, Kava-Cordeiro V et al (2006) Horizontal transfer and hypovirulence associated with double-stranded RNA in *Beauveria bassiana*. Mycol Res 110:1475–1481

De Guido MA, Minafra A, Santomauro A, Faretra F (2007) Molecular characterization of mycoviruses in *Botryotinia fuckeliana*. In: 14th international *Botrytis* symposium abstracts book, Cape Town. p 40

Drinnenberg IA, Fink GR, Bartel DP (2011) Compatibility with killer explains the rise of RNAi-deficient fungi. Science 333:1592

Fukuhara T, Gibbs MJ (2011) *Endornaviridae*. In: Fauquet CM, Mayo MA, Maniloff J, Desselberger U, Ball LA (eds) Virus taxonomy, ninth report of the international committee on taxonomy of viruses. Elsevier Press, London, pp 942–943

Ghabrial SA, Suzuki N (2009) Viruses of plant pathogenic fungi. Annu Rev Phytopathol 47:353–384

Ghabrial SA, Nibert ML (2009) *Victorivirus*, a new genus of fungal viruses in the family Totiviridae. Arch Virol 154:373–379

Habib W, De Miccolis Angelini RM, Rotolo C, et al (2013) Genetic variation in *Botryotinia fuckeliana* (*Botrytis cinerea*) populations on greenhouse vegetable crops in Lebanon. Abstracts book: XVI international *Botrytis* symposium (Locorotondo). p 37

Hennebert GL (1973) *Botrytis* and *Botrytis*-like genera. Persoonia 7:183–204

Hollings M (1962) Viruses associated with dieback disease of cultivated mushrooms. Nature 196:962–965

Hong Y, Dover SL, Cole TE et al (1999) Multiple mitochondrial viruses in an isolate of the Dutch elm disease fungus *Ophiostoma novo-ulmi*. Virology 258:118–127

Howitt RLJ, Beever RE, Pearson MN, Forster RLS (1995) Presence of double-stranded RNA and virus-like particles in *Botrytis cinerea*. Mycol Res 99:1472–1478

Howitt RLJ, Beever RE, Pearson MN, Forster RLS (2001) Genome characterization of *Botrytis* virus F, a flexuous rod-shaped mycovirus resembling plant 'potex-like' viruses. J Gen Virol 82:67–78

Howitt RLJ, Beever RE, Pearson MN, Forster RLS (2006) Genome characterization of a flexuous rod-shaped mycovirus, *Botrytis virus X*, reveals high amino acid identity to genes from plant 'potex-like' viruses. Arch Virol 151:563–579

Ihrmark K, Johannesson H, Strenstrom E, Stenlid J (2002) Transmission of double-stranded RNA in *Heterobasidion annosum*. Fungal Genet Biol 36:147–154

Ikeda K, Nakamura H, Arakawa M, Matsumoto N (2004) Diversity and vertical transmission of double-stranded RNA elements in root rot pathogens of trees, *Helicobasidium mompa* and *Rosellinia necatrix*. Mycol Res 108:626–634

Kecskeméti E, Brathuhn A, Kogel KH et al (2014) Presence of transposons and mycoviruses in *Botrytis cinerea* isolates collected from a German grapevine growing region. J Phytopathol 162:582–595

Kondo H, Chiba S, Toyoda K, Suzuki N (2013) Evidence for negative-strand RNA virus infection in fungi. Virology 435:201–209

Kraberger S, Stainton D, Dayaram A et al (2013) Discovery of *Sclerotinia sclerotiorum* hypovirulence-associated virus 1 in urban river sediments of Heathcote and Styx Rivers in Christchurch City, New Zealand. Genome Announc 1:e00559–13

Lecoq H, Boissonnet-Menes M, Delhotal P (1979) Infectivity and transmission of fungal viruses. In: Molitoris HP, Hollings M, Wood HA (eds) Fungal viruses. Springer, Berlin, pp 34–47

Liu HP, Wang JM, Yan XQ et al (2003a) Study on vegetative compatibility and anastomosis group in *Botrytis cinerea* Pers. Acta Phytopathol Sin 33:132–134

Liu YC, Milgroom MG (1996) Correlation between hypovirus transmission and the number of vegetative incompatibility (*vic*) genes different among isolates from a natural population of *Cryphonectria parasitica*. Phytopathology 86:79–86

Liu YC, Hillman BI, Linder-Basso D et al (2003b) Evidence for interspecies transmission of viruses in natural populations of filamentous fungi in the genus *Cryphonectria*. Mol Ecol 12:1619–1628

MacDonald W, Fulbright D (1991) Biological control of chestnut blight: use and limitations of transmissible hypovirulence. Plant Dis 75:656–661

Magliani W, Conti S, Gerloni M et al (1997) Yeast killer systems. Clin Microbiol Rev 10:369–400

Melzer MS, Ikeda SS, Boland GJ (2002) Interspecific transmission of double-stranded RNA and hypovirulence from *Sclerotinia sclerotiorum* to *S. minor*. Phytopathology 92:780–784

Milgroom MG, Cortesi P (2004) Biological control of chestnut blight with hypovirulence: a critical analysis. Annu Rev Phytopathol 42:311–338

Milgroom MG, Hillman BI (2011) The ecology and evolution of fungal viruses. In: Hurst CJ (ed) Studies in viral ecology. Wiley, New York, pp 221–257

Nuss DL, Koltin Y (1990) Significance of dsRNA genetic elements in plant pathogenic fungi. Annu Rev Phytopathol 28:37–58

Nuss DL (2005) Hypovirulence: mycoviruses at the fungal-plant interface. Nat Rev Microbiol 3:632–642

Nuss DL (2011) Mycoviruses, RNA silencing, and viral RNA recombination. Adv Virus Res 80:25–48

Park YJ, Chen XB, Punja ZK (2006) Diversity, complexity and transmission of double-stranded RNA elements in *Chalara elegans* (syn. *Thielaviopsis basicola*). Mycol Res 110:697–704

Pearson MN, Beever RE, Boine B, Arthur K (2009) Mycoviruses of filamentous fungi and their relevance to plant pathology. Mol Plant Pathol 10:115–128

Pearson MN, Bailey AM (2013) Viruses of *Botrytis*. Adv Virus Res 86:249–272

Polashock JJ, Hillman BI (1994) A small mitochondrial double-stranded (ds) RNA element associated with a hypovirulent strain of the chestnut blight fungus and ancestrally related to yeast cytoplasmic T and W dsRNAs. Proc Natl Acad Sci U S A 91:8680–8684

Polashock JJ, Bedker PJ, Hillman BI (1997) Movement of a small mitochondrial double-stranded RNA element of *Cryphonectria parasitica*: ascospore inheritance and implications for mitochondrial recombination. Mol Gen Genet 256:566–571

Potgieter CA, Castillo A, Castro M et al (2013) A wild-type *Botrytis cinerea* strain co-infected by double-stranded RNA mycoviruses presents hypovirulence-associated traits. Virol J 10:220

Rodríguez-García C, Medina V, Alonso A, Ayllón MA (2013) Mycoviruses of *Botrytis cinerea* isolates from different hosts. Ann Appl Biol 164:46–61

Rogers HJ, Buck KW, Brasier CM (1986) Transmission of double-stranded RNA and a disease factor in *Ophiostoma ulmi*. Plant Pathol 35:277–287

Rosario K, Dayaram A, Marinov M, Ware J, Kraberger S, Stainton D, Breitbart M, Varsani A (2012) Diverse circular ssDNA viruses discovered in dragonflies (Odonata: Epiprocta). J Gen Virol 93:2668–2681

Saup SJ (2000) Molecular genetics of heterokaryon incompatibility in filamentous ascomycetes. Microbiol Mol Biol Rev 64:489–502

Staats M, Van Baarlen P, Van Kan JAL (2005) Molecular phylogeny of the plant pathogenic genus *Botrytis* and the evolution of host specificity. Mol Biol Evol 22:333–346

Tan CMC, Pearson MN, Beever RE, Parkes SL (2007) Why fungi have sex? In: 14th international *Botrytis* symposium abstract book, Cape Town, p 26

Van Diepeningen AD, Debets AJM, Hoekstra RF (2006) Dynamics of dsRNA mycoviruses in black *Aspergillus* populations. Fungal Genet Biol 43:446–452

Vilches S, Castillo A (1997) A double-stranded RNA mycovirus in *Botrytis cinerea*. FEMS Microbiol Lett 155:125–130

Whon TW, Kim MS, Roh SW, Shin NR, Lee HW, Bae JW (2012) Metagenomic characterization of airborne viral DNA diversity in the near-surface atmosphere. J Virol 86:8221–8231

Williamson B, Tudzynski B, Tudzynski P, Van Kan JAL (2007) *Botrytis cinerea*: the cause of gray mould disease. Mol Plant Pathol 8:561–580

Wu MD, Zhang L, Li GQ et al (2007) Hypovirulence and double-stranded RNA in *Botrytis cinerea*. Phytopathology 97:1590–1599

Wu MD, Zhang L, Li GQ et al (2010) Genome characterization of a debilitation-associated mitovirus infecting the phytopathogenic fungus *Botrytis cinerea*. Virology 406:117–126

Wu MD, Jin FY, Zhang J, Yang L, Jiang DH, Li GQ (2012) Characterization of a novel bipartite double-stranded RNA mycovirus conferring hypovirulence in the phytopathogenic fungus *Botrytis porri*. J Virol 86:6605–6619
Wu MD, Wang QJ, Zhang J et al (2014) A novel dsRNA mycovirus infecting the phytopathogenic fungus *Botrytis squamosa*. J Huazhong Agric Univ 33:39–46
Yu X, Li B, Fu Y et al (2010) A geminivirus-related DNA mycovirus that confers hypovirulence to a plant pathogenic fungus. Proc Natl Acad Sci U S A 107:8387–8392
Yu X, Li B, Fu Y et al (2013) Extracellular transmission of a DNA mycovirus and its use as a natural fungicide. Proc Natl Acad Sci U S A 110:1452–1457
Zhang L, Wu MD, Li GQ et al (2010) Effect of mitovirus on formation of infection cushions and production of some virulence factors by *Botrytis cinerea*. Physiol Mol Plant Pathol 75:71–80

Chapter 6
Diversity Within and Between Species of *Botrytis*

Anne-Sophie Walker

Abstract The genus *Botrytis* is highly diverse, with numerous species differing in terms of their biology, ecology, morphological features and host range. Progress in molecular genetics, and the development of relevant phylogenetic markers in particular, has resulted in the establishment of ≈ 30 species, a hybrid and a species complex. At least seven new species have been identified in the last decade, albeit with limited scientific support in some cases. *B. cinerea* has long been known to display broad population diversity, possibly due to intense recombination and large population sizes. The introduction of powerful markers, such as SSRs, has provided new insight into the respective contributions of the forces driving this diversity. It has recently been shown that populations may be structured, not only by the host plant as shown in preliminary studies, but also by other factors, such as cropping system, geography and fungicide applications. Evidence of recombination and gene flow, between and within compartments, has also been obtained. Finally, this chapter focuses on the biological and genetic characteristics of *Botrytis* spp. favouring their adaptation to their local environment and speciation. This information is particularly useful for improving the management of diseases on cultivated hosts.

Keywords Adaptation • Phylogeny • Population genetics • Species • Structuring factor

6.1 Introduction

Diversity is the degree of variation within a group. In population biology, diversity makes it possible to distinguish between individuals on the basis of the variation of phenotypic (morphological, physiological or behavioral characteristics) or genotypic (based upon the examination of one or several combined loci) criteria.

A.-S. Walker (✉)
UMR 1290 BIOGER, INRA, AgroParisTech,
BP01, Avenue Lucien Brétignières, F-78850, Thiverval-Grignon, France
e-mail: walker@versailles.inra.fr

S. Fillinger, Y. Elad (eds.), *Botrytis – the Fungus, the Pathogen and its Management in Agricultural Systems*, DOI 10.1007/978-3-319-23371-0_6

Evolution is thus based on variation of the characteristics of organisms at several scales: (1) differences between individuals within a population, and/or (2) between populations and species. At the population scale, the frequencies of the different variants, characterized by their alleles at the loci considered, can be predicted by Mendel's laws in ideal conditions of Hardy-Weinberg equilibrium. Any violation of this equilibrium modifies the frequencies of the variants in populations. The evolutionary path of a given allelic variant may therefore be affected by the effects of four evolutionary forces (*i.e.* genetic drift, mutation, migration, and selection), acting in isolation or in combination, and of the mode of reproduction, which may affect the probability of particular gametes coming together during mating. In some cases, divergence between populations may be driven by ecological forces alone (ecological specialization), with the selection of specific variants in different environments. This may ultimately lead to speciation if reproductive isolation occurs. Speciation thus links microevolution (occurring within populations and species) and macroevolution (diversification of higher taxa). This chapter aims to describe the diversity occurring within the genus *Botrytis* and within individual species of this genus, and the determinants of this diversity. We will focus, in particular, on the factors underlying the partitioning of the population, with the long-term aim of adapting agricultural practices appropriately so as to decrease the impact of grey mould.

6.2 Diversity in the Genus *Botrytis*

6.2.1 General Comments

The definition of the term "species" has long been a matter of debate, mostly because it has been confused with the species criteria (or species concepts) used to delimit species. It is now widely agreed by biologists that species are segments of evolutionary lineages that evolve independently from each other, without exchanging genes any more. Speciation can thus be defined as the splitting of one species into two or more daughter species (Samadi and Barberousse 2006; De Queiroz 1998).

Up to 22 species concepts are available for species delimitation (Mayden 1997), but it is generally accepted that several of these concepts are required for the correct definition of species boundaries (Sites and Marshall 2004). Four of these concepts are recognized as being of particular utility in fungi (Giraud et al. 2008; Taylor et al. 2000): (1) the biological species concept (BSC), focusing on reproductive isolation, (2) the morphological species concept (MSC), focusing on morphological divergence, (3) the ecological species concept (ESC), focusing on adaptation to a particular ecological niche and (4) the phylogenetic species concept (PSC), and its extension, genealogical concordance phylogenetic species recognition (GCPSR), focusing on the lack of gene flow between lineages and nucleotide divergence.

6.2.2 *Taxonomy and Species Identification in the Genus* Botrytis

Botrytis spp. are Ascomycete fungi of class *Leotiomycetes*, order *Heliotiales* and family *Sclerotiniaceae*. The genus *Botrytis* was first described in 1729 by Pier Antonio Micheli, who listed it in the "*Nova Plantarum Genera*". Most of the species were eventually established by Hennebert (1973), Groves and Loveland (1953), and Beever and Weeds (2004). This genus is very closely related to *Sclerotinia*, with the proteins encoded by the genomes of *B. cinerea* and *S. sclerotiorum* displaying 83 % identity (Amselem et al. 2011). The genus *Botrytis* contains more than 30 species (Table 6.1). The life cycle of *Botrytis* spp. can be generalized as follows. There is a somatic (vegetative) stage, in which the mycelium produces asexual macroconidia, sclerotia and microconidia (spermatia). This is essentially the anamorph stage of *Botrytis*. There is then a sexual stage, during which microconidia may fertilize sclerotia to produce apothecia, in which meiosis occurs and ascospores are produced (the *Botryotinia* teleomorph stage, named by Whetzel, 1945). More details about the biology and genetics of *Botrytis* species are provided in Chaps. 2 and 3. Not all *Botrytis* spp. have known teleomorphs. Moreover, in a recent paper, Wingfield et al. (2012) pleaded the cause of a simplification of fungi taxonomy, which has become possible with the advent of molecular tools, which have made the need to split morphs obsolete. At a recent meeting (Bari, Italy, June 2013), the *Botrytis* research community unanimously decided to retain the asexual name, *Botrytis*, which is also the oldest and most widely used name, for fungi from this genus (Chap. 1). The teleomorph name should therefore no longer be used. Johnston et al. (2014) put forward the proposal for this taxonomic change and provided *Botrytis* names for the only two *Botryotinia* species lacking a *Botrytis* equivalent.

6.2.2.1 Identification on the Basis of the Morphological Species Concept

Botrytis spp. have mostly been delimited on the basis of morphological and culture characteristics (MSC species concept). All species of *Botrytis* display the common morphological feature that gave rise to the name of the genus: the botryose, or cluster-of-grape shape of the conidiophores (the conidiophores bear clusters of macroconidia that resemble clusters of grapes). The size and shape of the macroconidia have repeatedly been used as criteria for distinguishing species, as have the number, organization and size of the sclerotia and the morphology of the mycelium on artificial media (Jarvis 1977; Hennebert 1973; Zhang et al. 2010a, b; Lorenzini and Zapparoli 2014; Li et al. 2012). However, many species are morphologically similar (*e.g. B. cinerea* and *B. pseudocinerea* (Walker et al. 2011); *B. aclada* and *Botrytis* sp. B83 (Lorenzini and Zapparoli 2014)). Growing conditions may also significantly influence variation (Grindle 1979; Martinez et al. 2003) Moreover, high levels of morphological diversity may also be recognized in a single species.

Table 6.1 Some species of *Botrytis*

Botrytis sp. (anamorph)[a]	*Botryotinia* sp. (teleomorph)[e]	Mating system	Major hosts	Refs[f]
B. aclada Fresen.	–	–	*Allium*	4,7,8,16
B. allii Munn	–	–	*Allium*	4,8,16
B. anthophila Bondartsev[b,a]	–	–	*Trifolium*	4,8,12
B. byssoidea Walker[c]	–	–	*Allium*	4,7,8,16
B. calthae Hennebert	*Bt. calthae* Hennebert & Elliott	–	*Caltha*	4,6,7,8,9,25
B. caroliniana	–	–	*Rubus*	17
B. cinerea Pers.:Fr	*Bt. fuckeliana* (de Bary) Whetzel	Heterothallic	Polyphagous	4,7,8,9
B. convallariae (Kleb.) Ondřej[a]	–	–	*Convallaria*	8
B. convoluta Whetzel & Drayton	*Bt. convoluta* (Drayton) Whetzel	–	*Iris*	4,7,8,9
B. croci Cooke & Massee	–	–	*Crocus*	7,8,11
B. deweyae		Heterothallic	*Hemerocallis*	18
B. elliptica (Berk.) Cooke	? *Botryotinia* sp.	Heterothallic	*Lilium*	4,7,8,13,14
B. fabae Sardiña	*Bt. fabae* Lu & Wu	–	*Vicia*	7,8,15
B. ficariarum Hennebert	*Bt. ficariarum* Hennebert	–	*Ficaria*	6,7,8,9
B. fabiopsis	–	–	*Vicia*	19
B. galanthina (Berk. & Broome) Sacc.	–	–	*Galanthus*	4,7,8
B. gladiolorum Timmerm.	*Bt. draytonii* (Buddin & Wakef.) Seaver	–	*Gladiolus*	4,7,8,9
B. globosa Raabe	*Bt. globosa* Buchw.	Homothallic	*Allium*	1,7,8,9
B. hyacinthi Westerd. & Beyma	–	–	*Hyacinthus*	4,7,8
B. mali[a]	–	–	*Malus*	24
B. narcissicola Kleb. Ex Westerd. & Beyma	*Bt. narcissicola* (Greg.) Buchw.	–	*Narcissus*	4,7,8,9
B. paeoniae Oudem.	–	–	*Paeonia*, *Allium*	4,7,8,9
B. pelargonii Røed	*Bt. pelargonii* Røed	–	*Pelargonium*	7,8,9
B. polyblastis Dowson	*Bt. polyplastis* (Greg.) Buchw.	–	*Narcissus*	4,7,8,9
B. porri Buchw.	*Bt. porri* (Beyma) Whetzel	Homothallic	*Allium*	3,7,8,9
B. ranunculi Hennebert	*Bt. ranunculi* Hennebert & Groves	Heterothallic	*Ranunculus*	4,6,7,8,9

(continued)

Table 6.1 (continued)

Botrytis sp. (anamorph)[a]	*Botryotinia* sp. (teleomorph)[e]	Mating system	Major hosts	Refs[f]
B. prunorum	–	–	*Prunus*	26
B. pseudocinerea	*Bt. pseudofuckeliana*	Heterothallic	Polyphagous	20
B. ricini Buchw.[a]	*Bt. ricini* (Godfrey) Whetzel	Homothallic	*Ricinus*	4,5,7,8,9
B. sinoallii	–	–	*Allium*	21
B. sinoviticola	–	–	*Vitis*	22
Botrytis fritillarii-pallidiflori (Chen & Li) Seifert & Kohn	*Bt. fritillarii-pallidiflori* Chen & Li	–	–	10
Botrytis sp.[a]	*Sclerotinia spermophila* Noble[b]	Homothallic	*Trifolium*	4,8,9,12
Botrytis sp. B83[d]	–	–	Polyphagous	23
B. sphaerosperma Buchw.	*Bt. sphaerosperma* (Greg.) Buchw.	–	*Allium*	7,8,9
B. squamosa Walker	*Bt. squamosa* Vienn.-Bourg.	Heterothallic	*Allium*	2,4,7,8,9
B. tulipae Lind	–	–	*Tulipa, Allium, Lilium*	4,7,8,9

Modified and updated from Beever and Weeds (2004)

[a]List of species established from published data. Some species could not be recovered from ancient collections, to check their delimitation in the genus phylogeny (Staats et al. 2005; Van Kan, pers. com.) and their status is therefore potentially doubtful. *B. mali* is a herbarium specimen for which no living culture is available

[b]It has been assumed (*e.g.* Farr et al. 1989) that *Sclerotinia spermophila* is the teleomorph of *B. anthophila* but Noble (1948) discussed this possibility and concluded that further evidence was required to confirm this link

[c]It has been assumed (*e.g.* Jarvis (1980) that *Bt. allii* (Sawada) Yamam. is the teleomorph of *B. byssoidea*. However, (Kohn 1979) provided evidence that Yamamoto was wrong to conclude that Sawada's species produced a *Botrytis* anamorph and she regrouped *Sclerotinia allii* (Sawada) as *Ciborina allii* (Sawada) Kohn. Nevertheless, Yamamoto (1949), who worked on Japanese isolates, identified the anamorph as *B. byssoidea*, in which case the teleomorph he describes is that of *B. byssoidea*. However, the relationships of the Japanese fungus require further investigation (Hennebert 1963; Nielsen et al. 2001)

[d]This new entity is actually represented by only one strain collected in Italy, but the available information confirm that it almost certainly corresponds to a new species (Lorenzini and Zapparoli 2014)

[e]Teleomorph names are given wherever possible, to make it possible to refer to older studies. However, *Botrytis* has been accepted over *Botryotinia* for the genus name (Johnston et al. 2014). Therefore, only *Botrytis* (anamorph names) should be used in future publications

[f]1=Buchwald 1953; 2=Bergquist and Lorbeer 1972; 3=Elliott 1964; 4=Farr et al. 1989; 5=Godfrey 1923; 6=Hennebert and Groves 1963; 7=Hennebert 1973; 8=Jarvis 1980; 9=Kohn 1979; 10=Li and Chen 1987; 11=Moore 1959; 12=Noble 1948; 13=Van den Ende and Pennock 1996; 14=Van den Ende and Pennock-Vos 1997; 15=Wu and Lu 1991; 16=Yohalem et al. 2003; 17=Li et al. 2012; 18=Grant Downton et al. 2014; 19=Zhang et al. 2010a; 20=Walker et al. 2011; 21=Zhang et al. 2010b; 22=Zhou et al. 2014; 23=Lorenzini and Zapparoli 2014; 24=O'Gorman et al. 2008; 25=Plesken et al. 2015; 26=Ferrada et al. 2015

For example, in *B. cinerea*, the mycelium has been described as short, aerial or woolly, and the sclerotia have been described as organized randomly, in circles or at the edge of Petri dishes, with conidial density differing between cultures (Martinez et al. 2003). Phenotypic differences can usually be found in complete sets of ascospores from single asci, and in independent cultures established from a single ascospore or conidium from the same wild isolate. Moreover, morphotypes may not remain stable during subculture (Lorenz 1983).

6.2.2.2 Identification on the Basis of the Biological Species Concept

The use of the biological species concept (BSC), which is based on interbreeding, is limited to species for which sexual crosses can be conducted *in vitro*, however laborious. For example, the putative new species *Botrytis* sp. B83 only rarely produces sclerotia, preventing the use of BSC criteria to confirm its existence (Lorenzini and Zapparoli 2014). Respect for the mating system is another essential condition. Some species, such as *B. porri* and *B. globosa*, have been reported to be homothallic (*i.e.* self-fertile) and can produce sexual progeny in a single culture (Elliott 1964; Buchwald 1953). Others are heterothallic (*i.e.* self-sterile) and require strains with a compatible mating type to produce their progeny. Mating is governed by a mating-type locus (two alleles, MAT 1-1 and MAT 1-2; Faretra et al. 1988) in *B. cinerea*, with each locus carrying two different, non-homologous genes (Amselem et al. 2011). The use of the BSC is dependent on mating experiments and is therefore limited and has been largely restricted to elucidation of the sexual system. Nevertheless, the BSC has been successfully used to differentiate between *B. squamosa* and *B. cinerea* (Bergquist and Lorbeer 1972) and between *B. cinerea* and *B. pseudocinerea* (Walker et al. 2011).

6.2.2.3 Identification on the Basis of the Ecological Species Concept

The ESC seems to have been more successfully used for the definition of species. Indeed, most *Botrytis* spp. are host-specific or have a narrow host range. *B. fabae*, for example, infects only plants from the same botanical family *Fabaceae* (Jarvis 1977). *Botrytis* species are, therefore, usually named after their host plants (Table 6.1). Only two species have been reported to have multiple host plants from different botanical families in the wild: *B. cinerea* and *B. pseudocinerea* (Walker et al. 2011; Jarvis 1980; Fekete et al. 2012). *Botrytis* spp. can infect more than 1,400 host species, including not only wild plants, but also major ornamental, greenhouse and field crops, such as tomato, grapevine, lettuce, strawberry, tulip, rose and onion (Chap. 21). They are all necrotrophs, inducing host-cell death, leading to the progressive decay of infected plant tissue. They are also commonly found as saprophytes on dead tissues. However, there is an exception. The newly described species *B. deweyae*, which infects *Hemerocallis*, was recently shown to have an endophytic lifestyle under appropriate conditions (Grant Downton et al. 2014). *B. cinerea* may also be found as an endophytic contaminant of *Primula* and lettuce seeds

(Barnes and Shaw 2003; Sowley et al. 2010). Moreover, ecological preferences may not be governed solely by the host. *B. pseudocinerea*, which is sympatric with *B. cinerea* on the same hosts, is regularly found to be more abundant in the spring than in the autumn. It is also mostly found on dead flower parts from grapevine rather than on living berries. Finally, unlike *B. cinerea*, the distribution of this species is not uniform worldwide, or even within a particular region (Walker et al. 2011; Johnston et al. 2013). This suggests that at least some *Botrytis* species are capable of fine-niche adaptation, but this aspect requires further investigation.

6.2.2.4 Identification on the Basis of the Phylogenetic Species Concept

The PSC is undoubtedly the most widely used concept in *Botrytis* taxonomy. The phylogenetic approach to systematics has been boosted by advances in DNA sequencing and by the identification, over the last 10 years, of genes highly relevant to the resolution of *Botrytis* phylogeny. As a consequence, seven new species have come to the fore in the last four years (Table 6.1). The ITS rDNA region has been extensively used for the species-level discrimination of fungal species. Unfortunately, the number of informative sequence characters is too small in *Botrytis* for the full resolution of relationships between all species. This approach was nevertheless successfully used to infer a preliminary phylogeny of the family *Sclerotiniaceae* (Holst-Jensen et al. 1998). Universal-primed polymerase chain reaction (UP-PCR) fingerprinting, coupled with the restriction of ITS rDNA regions also led to the identification of five groups: *B. cinerea*, *B. squamosa*, *B. byssoidea* and two groups of *B. aclada* (Nielsen et al. 2001), one of which was subsequently recognized as the hybrid between *B. aclada* and *B. byssoidea* and renamed *B. allii* (Nielsen and Yohalem 2001; Yohalem et al. 2003). Moreover, a method coupling ITS rDNA sequencing and the multiplex PCR amplification of two laccase introns can identify several species from the *Sclerotiniaceae*, including *B. cinerea* (Hirschhäuser and Fröhlich 2007). Finally, the intergenic spacer (IGS) rDNA region was successfully used in preliminary studies as an RFLP marker for distinguishing between *B. cinerea* and *B. pseudocinerea* (Giraud et al. 1997).

According to the GCPCSR, as discussed for fungi by Taylor et al. (2000), the best way to demonstrate a lack of gene flow between evolutionary lineages (*i.e.* the definition of a phylogenetic species) is to compare the phylogenies obtained for several independent loci: barriers to gene flow (or species limits) correspond to the nodes from which independent phylogenies become incongruent. The identification of relevant protein-coding genes has greatly improved the resolution of the *Botrytis* phylogeny. Three genes, in particular, are used: the glyceraldehyde 3-phosphate dehydrogenase (*G3PDH*), heat shock protein 60 (*HSP60*) and DNA-dependent RNA polymerase subunit II (*RPB2*) genes. Staats et al. (2005) combined these genes in a comprehensive phylogenetic analysis of the whole genus. The combined analysis of all three loci provides greater resolution than individual gene phylogenies. As a result, these authors were able to confirm the morphological and host plant-based classification of the *Botrytis* genus and divided the genus into two well separated clades. The first of these clades contains only five species (*B. cinerea*,

B. fabae, *B. pelargonii*, *B. calthae*, and *B. pseudocinerea*, added after further investigation) that infect only eudicot plants, whereas the second clade contains all the other species, infecting either eudicot or monocot plants. Interestingly, the host and pathogen topologies were incongruent (particularly for the species infecting the same host, *Allium*, clustered in different clades), suggesting regular host shifts rather than co-speciation. Alternatively, host range may have been underestimated for some species currently thought to be strict single-host pathogens. This study also confirmed the hybrid nature of *B. allii*. Finally, these authors investigated the mode of reproduction in each species. Their results suggested that sexual reproduction had been lost at least three times during the evolution of the genus, due to negative selection. This work probably paved the way for the delimitation of many new species (see next section and Li et al. 2012; Lorenzini and Zapparoli 2014; Zhang et al. 2010a, b; Grant Downton et al. 2014). An updated phylogenetic tree, adapted from the work of Staats et al. (2005) and including these new taxa is presented in Fig. 6.1.

In other studies, one or several of these genes have been used in combination with other genes for the identification of other species. The Funybase phylogenetic database containing 1:1 orthologs from 30 fungal genomes of the highest informative value at the desired taxonomic level, can be considered a good source of genes useful for phylogenetic analyses (Marthey et al. 2008). For example, Walker et al. (2011) used the ATP-dependent RNA helicase gene (*MS547*) to help determine the systematic position of *B. pseudocinerea* in the genus. Andrew et al. (2012) also refined the phylogeny of the *Sclerotiniaceae*, by combining *G3PDH* and *HSP60* topologies with that for the calmodulin gene. Khan et al. (2013) recently distinguished between species causing neck rot disease on onion in New Zealand, using a combination of data for the *ITS* and *IGS* regions and the *G3PDH* sequence, which greatly improved the molecular identification of species. Interestingly, sequences established with the *G3PDH* and β-tubulin genes confirmed the existence of *B. mali* from herbarium apple specimens (O'Gorman et al. 2008), following the initial approximate and, therefore, dubious description of this species (Ruehle 1931). The findings for these genes suggest that this species would probably cluster with *B. paeoniae*. Finally, in addition to the previous markers, two genes encoding the necrosis and ethylene-inducing proteins, *NEP1* and *NEP2*, were found to be highly relevant for establishment of the phylogeny of *Botrytis* phylogeny (Staats et al. 2007a) and for refining the systematic positions of new species (Lorenzini and Zapparoli 2014; Li et al. 2012; Zhang et al. 2010a). The trees established with these genes were entirely congruent with the concatened trees established by Staats et al. (2005), with the exception of *B. gladiolorum*, which was located in a different position in the *NEP2* tree. The *NEP1* and *NEP2* genes are involved in interactions with the plant and have been shown to have evolved under positive selection (Staats et al. 2007a). Caution is therefore required in their use and their neutrality should be checked before their use for phylogenetic purposes, particularly if they are to be used for dating.

Recent developments in molecular phylogenetics have generated a set of relevant genes for delimiting *Botrytis* spp. They have been shown to be of significant interest

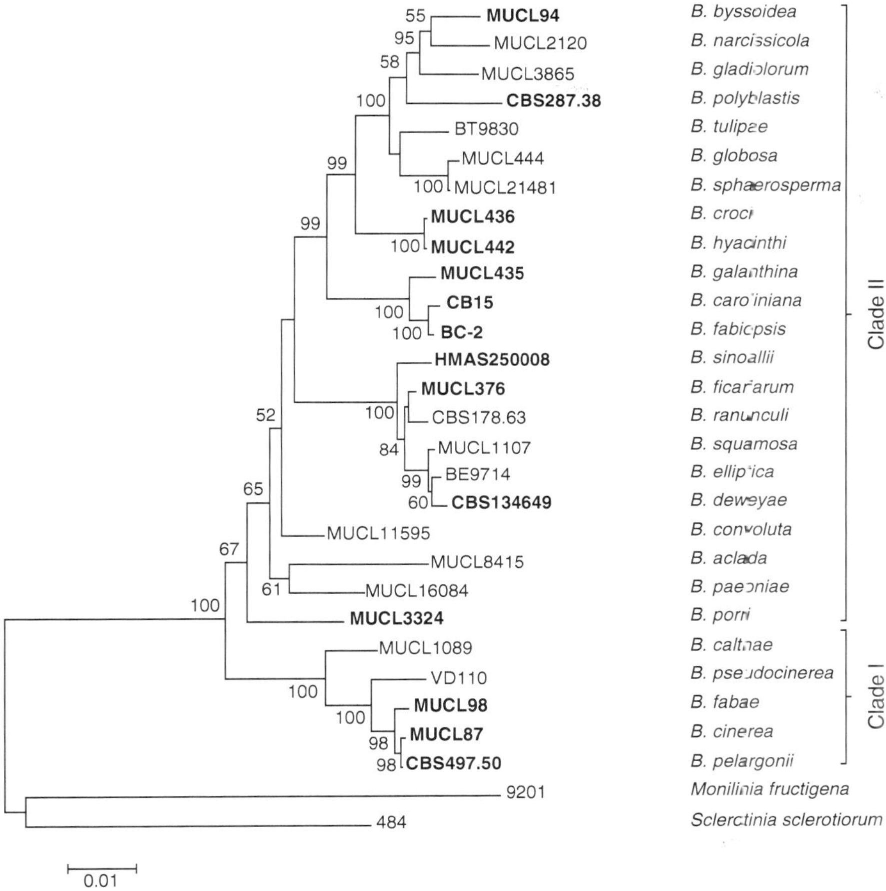

Fig. 6.1 Updated phylogenetic tree of *Botrytis* sp., based on concatenated sequences of the G3PDH, HSP60 and RPB2 genes (From Hyde et al. 2014). The tree was constructed with the maximum likelihood method. Some species from Table 6.1 could not be integrated in this tree, due to a lack of published sequences or an inability to obtain fresh cultures from collections for the establishment of these sequences

in at least one phylogenetic study and it is, therefore, difficult to rank them. However, it has now been established that the most appropriate approach for correctly delimiting distinct evolutionary lineages (*i.e.* for placing species boundaries) is to use several independent genes and to evaluate the congruence of their phylogenies. Unfortunately, the congruence of gene topologies is only rarely quantified, although at least one useful congruence index has been published (de Vienne et al. 2007, 2009). Moreover, given these limitations, new species should not be delimited exclusively on the basis of molecular phylogenetic data (a tempting approach as it would minimize the amount of laboratory work required), but on the basis of a body of evidence accumulated from as many of the species concepts as possible.

6.2.3 *Recently Discovered* Botrytis *Species*

B. fabiopsis was first collected from 35 locations in Hubei Province (Central China), between 2006 and 2009. It was sympatric with *B. cinerea* and *B. fabae* on broad bean (*Vicia faba*) (Zhang et al. 2010a). It was the second most frequently identified species in this sampling campaign after *B. cinerea*, accounting for 36.4 % of the strains. It may therefore have a significant epidemiological impact on this crop. The "chocolate spot" symptoms were similar to those usually observed with the other two species. Host range was not investigated further. This fungus formed pale grey colonies with short aerial mycelia and produced grey to black sclerotia in concentric rings on PDA medium. It differed from *B. cinerea* and *B. fabae* in terms of sclerotium production and the size of its conidia. Phylogenetic analysis with *G3PDH*, *HSP60*, *RPB2*, *NEP1*, and *NEP2* sequences placed this species in clade II of the most recent phylogeny (Staats et al. 2005), well separated from *B. cinerea* and *B. fabae* (clade I). Instead, it was found to be a sister species to *B. galanthina*, which infects snowdrop (*Galanthus nivalis*).

B. caroliniana was first isolated in 2010 from infected blackberries (*Rubus fruticosus*) on several farms in South Carolina, USA (Li et al. 2012). It was sympatric with *B. cinerea*. It was shown to be pathogenic on broad bean leaves and tomato in laboratory conditions, and to produce lesions, without conidiation, on apple, pear, orange lemon, table grape, and raspberry (Schnabel and Li, pers. com.). Phylogenetic analysis with *G3PDH*, *HSP60*, *RPB2*, *NEP1* and *NEP2* sequences showed this species to be most closely related to *B. fabiopsis* (infecting broad bean) and to *B. galanthina* (infecting snowdrop), and to belong to clade II of the most recent phylogeny (Staats et al. 2005). This finding contrasts with what would be expected on the basis of host plant data, which would suggest phylogenetic relatedness to *B. cinerea*. It forms white to pale grey colonies with a short, tufted aerial mycelium and produces black sclerotia on PDA at 20 °C. Its conidia are similar to those of *B. cinerea*, but smaller than those of *B. fabiopsis* and *B. galanthina*. Little is known about its ecological requirements, and its frequency in *Botrytis* populations from blackberry. It is therefore difficult to draw any firm conclusions about its relevance for blackberry production.

B. deweyae was isolated in 2009 from daylilies (*Hemerocallis* hybrids) in the UK, as the causal agent of "spring sickness", an emerging disease of this ornamental plant. It produces slightly fluffy, whitish to pale brown colonies on artificial medium, but its principal characteristic is the absence of macroconidial conidiation, particularly on plants. Two alleles were detected at the mating-type locus homologous to that of *B. cinerea*, suggesting heterothallism. This new species is genetically related to *B. elliptica* (*ITS*, *G3PDH*, *NEP1*), which can also affect daylilies. This suggests that the emergent pathogen *B. deweyae* may have arisen from *B. elliptica* after a host shift. The distribution and abundance of this species in populations are unknown, as only a few specimens have been isolated. Given its morphological features, the absence of macroconidia and the nature of disease development, *B. deweyae* may be an endophyte undergoing the transition to a more aggressive pathogenic state (Grant Downton et al. 2014).

B. sinoalli was first collected in 2006 from green onion (*Allium fistulosum*) and then from other *Allium* crops at seven sites in Hubei Province (Central China; Zhang et al. 2010b). Within populations, it was found in association with *B. cinerea*, *B. squamosa*, *B. porri*, or *B. byssoidea*, but accounted for only 2.2 % of the population. This species produces white mycelia and abundant small sclerotia and can be distinguished from the other species on the basis of its conidium production on living host tissues, but not on PDA. Phylogenetic analysis with the *G3PDH*, *HSP60*, and *RPB2* genes showed that *B. sinoallii* formed a single lineage within clade II that was closely related to *B. squamosa*, but only distantly related to other species growing on *Allium*, including *B. cinerea*, *B. porri*, *B. aclada*, *B. allii*, *B. byssoidea*, *B. globosa* and *B. sphaerosperma*.

B. pseudocinerea is a new cryptic species recently distinguished from *B. cinerea* (Walker et al. 2011). It is described in more detail below, in the section on the *Botrytis* species complex.

B. sinoviticola was isolated between 2004 and 2012 from *Botrytis*-infected table grapes collected in two Chinese provinces, but little is known about its relevance in natural populations (Zhou et al. 2014). It produces whitish colonies and abundant small sclerotia. Its conidia appear to be thinner than those produced by *B. cinerea* and they carry characteristic villiform appendages on their surface. Phylogenetic analysis (*G3PDH*, *HSP60*, and *RPB2*) placed this species in a single lineage of clade I, separate from *B. cinerea* and *B. pseudocinerea*, although closely related to these species. *B. sinoviticola* has an intermediate level of sensitivity to the fungicide fenhexamid, between those of *B. cinerea* (sensitive) and *B. pseudocinerea* (naturally resistant). However, resistance to the full range of modes of action of fungicides, as in *B. pseudocinerea* (as described in Leroux et al. 1999), has not been assessed. Finally, the inoculation of grapevine leaves with mycelium showed this species to be less virulent than *B. cinerea* and *B. pseudocinerea*, suggesting that it may also occur on other plants.

A new species was recently collected from withered grapes (*Vitis vinifera*) in a fruit drying room in Valle dei Laghi (Trentino, Italy) (Lorenzini and Zapparoli 2014). This species is currently represented by a single strain, *Botrytis* sp. B83, and further collections and analyses are thus required to confirm the status of this new entity and to identify its diagnostic features. The morphology of *Botrytis* sp. B83 colonies on PDA was different from that of *B. cinerea*, which usually produces colonies of various shades of grey. *Botrytis* sp. B83 instead produces a moderately deep and floccose aerial white-cream mycelium on PDA. Sclerotia are produced rarely, and only under specific conditions. More generally, several characters indicate that this strain is different from but closely related to *B. aclada* in clade II, as confirmed by phylogenetic analysis with *G3PDH*, *HSP60*, *RPB2*, *NEP1*, and *NEP2* sequences. Conversely, its host range indicates that *Botrytis* sp. B83 behaves like *B. cinerea*, acting as a pathogen of many different plant hosts exclusively infected with grey mould (*i.e.* grapevine, green bean, kiwifruit, pepper, cucumber, strawberry, hortensia, blackcurrant, sage, and tomato). Its presence in a fruit drying room may therefore be incidental.

At last, *B. prunorum* was recently isolated from asymptomatic flowers of Japanese plums (*Prunus salicina*) in the Central Valley of Chile (Ferrada et al. 2015). They were supposed to cause blossom blight, together with *B. cinerea*. The authors described low-sporulating isolates, developing a white cottony mycelium and scarce conidiation on several media. Sclerotia were obtained only in restricted conditions. Molecular analysis with *G3PDH*, *HSP60*, *RPB2*, *NEP1*, and *NEP2* sequences established this entity as a new species within clade II, related to *B. paeoniae* and *B. aclada*. No further description was provided about the ecology and host range of this new species.

In conclusion, given the large number of newly established species, mostly identified through the use of recently developed relevant markers, *Botrytis* taxonomy seems to be at an early stage. Many more species may remain undiscovered, some parasitic but others endophytic, and therefore much harder to detect (Van Kan et al. 2014). For example, six 'new' *Botrytis* spp. from *Centaurea stoebe* (*Asteraceae*) have been identified as endophytes and their further characterization is planned in the near future (Shipunov et al. 2008). One of these species is very closely related to some *B. pseudocinerea* strains from a specific New-Zealand clade (P. Johnston, pers. com.). A foliar pathogen of *Hosta* that is closely related to *B. tulipae* is also currently awaiting full identification (Laundon 1978). In addition, the quality of new species descriptions is highly heterogeneous. Biological and ecological characteristics are often not recorded and few studies describe the mode of reproduction of the new species and provide intra- and interspecific crossing data (BSC). Moreover, population surveys are also often required to delimit geographical and host ranges. Recently developed tools should also prove useful for refining the descriptions of long-known species, such as *B. calthae* (Plesken et al. 2015; Hennebert and Groves 1963).

6.2.4 *The* Botrytis *Species Complex*

Cryptic species have identical morphological features but can be distinguished on the basis of their molecular sequences. They often inhabit the same environment. Thus, cryptic fungal species often parasitize the same host, produce similar symptoms and, therefore, form what is called a species complex. In the genus *Botrytis*, several species may correspond to this description. *B. pelargonii* and *B. cinerea* produce identical symptoms on *Pelargonium*; molecular phylogenetics studies do not support the separation of these two species (Staats et al. 2005, 2007a; Walker et al. 2011) and the existence of *B. pelargonii* is therefore doubtful.

Within *B. cinerea*, a subdivision into two distinct genetic groups has long been proposed, based on the presence or absence of the transposable elements *Boty* (Diolez et al. 1995) and *Flipper* (Levis et al. 1997). The presence/absence of these two elements has been used to describe four transposon types in populations: *vacuma* (strains with neither of these elements), *transposa* (strains with both elements), *Boty* and *Flipper* (strains with one or other of the two elements). For a limited number of strains, this subdivision coincided with two sympatric species, "group I" (only *vacuma*) or "group II" (the other three types), but this pattern was subsequently shown to be invalid in independent studies of larger datasets. The identification of "group I" as a new species was corroborated by the polymorphism of many genes: *Bc-hch* (encoding the homolog of the *Neurospora crassa* het-c vegetative incompatibility locus; Fournier et al. 2003), *cyp51* (encoding the eburicol 14α-demethylase, target of azole fungicides; Albertini et al. 2002), *erg27* (encoding the 3-cetoreductase, target of the fungicide fenhexamid; Albertini and Leroux 2004), *sdhA/B/C/D* (genes encoding the four subunits of succinate dehydrogenase; Leroux et al. 2010), for example. All these loci display many fixed polymorphisms within groups I and II, and an absence of shared polymorphisms between groups I and II, strongly suggesting an ancient divergence, with a complete absence of gene flow between these two groups. This was also confirmed by population genetics studies based on PCR-RFLP markers (Giraud et al. 1997) or SSRs (Walker et al. 2011). This body of evidence led to the establishment of *Botrytis* "group I" as the new species *B. pseudocinerea* (Walker et al. 2011; Fournier and Giraud 2008; Fournier et al. 2005). The classification based upon transposable elements became obsolete, because some *transposa* strains were found in *B. pseudocinerea*, albeit at low frequencies (Walker et al. 2011; Johnston et al. 2013; Fekete et al. 2012).

Several phylogenetic studies using *Bc-hch*, *cyp51*, *63-R* and/or *β-tub* (Fournier et al. 2005; Johnston et al. 2013), or *G3PDH*, *HSP60* and/or *MS547* (Walker et al. 2011; Johnston et al. 2013) clearly identified strains from groups I and II as belonging to different species within clade I of the *Botrytis* phylogeny. However, these studies showed that these species were not sisters, because *B. cinerea* was closer to *B. fabae* than to *B. pseudocinerea*. Indeed, a molecular clock analysis demonstrated that these two species diverged between 7 and 18 million years ago. Until recently, *B. pseudocinerea* had been detected only in European countries (Walker et al. 2011; Fekete et al. 2012) and was thought to originate from this part of the world. However, strains have recently been found in the Southern hemisphere, in Chile, New Zealand and South Africa (Johnston et al. 2013; Wessels et al. 2013; Munoz et al. 2015) and also in Central China on tomato (Li et al. 2014) and on blueberry in North America (Saito et al. 2014), suggesting that human activities may have mediated the migration of this species. Interestingly, two clades were resolved within *B. pseudocinerea* samples from New Zealand, only one of which has been reported in European vineyards, although this clade contained only a small number of strains (Johnston et al. 2013). *B. cinerea* and *B. pseudocinerea* have identical morphological features. Mating between these two species generates no progeny or a sterile progeny, suggesting the possible existence of a reproductive barrier based on prezygotic isolation. Moreover, the two species have different patterns of susceptibility to fungicides. Among many other fungicide markers, *B. pseudocinerea* is naturally resistant to fenhexamid and hypersensitive to morpholines (Albertini et al. 2002; Leroux et al. 1999). Both are polyphagous (Plesken et al. 2015; Giraud et al. 1999; Walker et al. 2011) and they are sympatric on the same hosts, but *B. pseudocinerea* is more abundant on dead flower parts in the spring (Johnston et al. 2013; Walker et al. 2011).

Recent studies of German strawberry fields identified a predominant new entity called *Botrytis* group S (for strawberry). These strains are present on other crops, such as grapevine, but at a lower frequency (Walker unpublished data; Leroch et al. 2013; Johnston et al. 2013). They have a morphology similar to that of *B. cinerea*. They display a 21-bp deletion in a transcription factor-encoding gene, *mrr1*, which is routinely used for their identification. Some group S strains express an original multidrug-resistant phenotype (MDR1h), conferring a higher resistance factor than the previously described MDR1 phenotype (Leroux and Walker 2013; Kretschmer et al. 2009), due to an additional 3-bp deletion in the same *mrr1* gene (Leroch et al. 2013). The phylogenetic tree established with *mrr1*, *MS547*, *FG1020*, *HSP60*, and *NEP2* did not clearly resolve this clade, strains from group S being grouped together with *B. cinerea* or considered a different species more closely related to *B. fabae*, depending on the gene considered (Johnston et al. 2013; Leroch et al. 2013). More information, from the sequencing of these genomes, is required for firm conclusions to be drawn about whether this new entity can be considered to be a separate species. Population genetics studies may also help to determine whether *Botrytis* group S corresponds to a subpopulation particularly adapted to strawberry and other plants or whether it is a sympatric new species present in grey mould populations.

6.3 Diversity of *B. cinerea* Populations

6.3.1 General Comments

B. cinerea is the *Botrytis* species for which by far the most information on diversity at the population level has been provided. In association with taxonomists (see previous section), diversity was first assessed and quantified together with morphological features in populations. Isolates were classified by several authors and quantified in populations as being of a “mycelial type”, with various categories, based upon the absence of sclerotia, conidiation and the appearance of the mycelium on synthetic medium, or of a “sclerotial type”, based on the number, size and organization of sclerotia in Petri dishes (Martinez et al. 2003; Mirzaei et al. 2009). Diversity is also evident if metabolic criteria are considered. For example, the production of secondary metabolites may differ between strains (Chap. 15). In particular, only a very small number of strains in populations produce the pink polyketide bikaverin pigment, due to the acquisition of a fully functional six-gene cluster possibly by horizontal gene transfer from *Fusarium* sp., probably before the divergence of the genus, this cluster being partially inactivated in most strains (Schumacher et al. 2013). Finally, diversity in acquired resistance to fungicides may also be observed in populations, with strains displaying adaptation to most of the available modes of action of fungicides and the selection of original resistance mechanisms (Chap. 10).

Over the last two decades, diversity within populations has mostly been assessed with a number of neutral molecular markers. Population genetics studies have flourished over the last decade (Table 6.2) and have helped to improve our knowledge of *B. cinerea* populations in various environments, to identify the factors shaping population structure and to understand the relationships between the various compartments. Outdoor populations are often shown to have high levels of genic and genotypic diversity, consistent with the large size of these populations, strong gene flow and regular recombination events. Haplotypic diversity may be low in some greenhouse populations, due to the predominance of selected genotypes. Diversity indices estimated with SSR markers were recently compiled and compared for various situations (Leyronas et al. 2015a). Genetic diversity has also been observed at a very fine scale: lesions from a single grapevine plant have been shown to be caused by different haplotypes, and up to five haplotypes have been distinguished from a single lesion on a berry (Giraud et al. 1997). Finally, population genetics studies may also provide information about the mode of reproduction at work in populations. Some indoor tomato populations favor clonal reproduction (Walker et al. 2015), but regular recombination, consistent with cryptic sexual reproduction, is observed in outdoor populations, as shown by the low proportion of clones and low level of linkage disequilibrium (Fournier and Giraud 2008; Giraud et al. 1997, 1999; Walker et al. 2015; Vaczy et al. 2008). Nevertheless, a key exception has been reported, for outdoor populations of rooibos seedlings in the Western Cape, South Africa, which displayed higher clonal fractions, accompanied by disequilibrium of mating type ratios (Wessels et al. 2013).

6.3.2 *Molecular Markers Available for Population Genetics*

Some studies have used random amplified polymorphic DNA (RAPD) to assess genetic diversity in populations (Table 6.2), mostly because this technique has been shown to be a powerful tool for genetic analysis and has been applied to a wide range of organisms. Commercial random decamer primers are easily obtained, making it relatively easy to gain rapid access to this technique. However, Moyano et al. (2003) showed that these markers were less polymorphic than amplified-fragment length polymorphism (AFLP) markers for *B. cinerea* (Vos et al. 1995). Surprisingly, AFLP analyses have been carried out in only a small number of studies (Table 6.2). In addition to these multi-locus techniques, various studies have used restriction fragment length polymorphism after PCR (PCR-RFLP) to measure diversity, because this single-locus technique is more reproducible and precise for the estimation of genetic parameters (high levels of polymorphism). Many genes have been used for this technique: the IGS, nitrate reductase, ATP synthase, ADP-ATP translocase (Giraud et al. 1997; Baraldi et al. 2002; Munoz et al. 2002, 2010; Giraud et al. 1999; Kretschmer and Hahn 2008) and *Bc-hch* (Fournier et al. 2003) genes.

Table 6.2 *Botrytis* spp. studies assessing diversity with various molecular markers

Botrytis species and reference	Host	Region/ country	Sample size (no. of isolates)	Neutral markers				Structuring factor exploited[a]						
				RAPD	RFLP/ AFLP	TE	SSR	Year	Season	Geography	Host plant	Cropping system	Species delim.	Other
B. cinerea														
Van Der Vlugt-Bergmans et al. (1993)	Various hosts	The Netherlands	10	●										
Giraud et al. (1997)	Grapevine	Champagne/ France	356		●	●			●	●			●	
Kerssies et al. (1997)	Rose, *Gerbera*	The Netherlands	30	●				●	●	●		●		
Choi et al. (1998)	9 different hosts	Korea	34	●				●		●	●			
Giraud et al. (1999)	21 wild hosts	Champagne/ France	107		●	●				●	●		●	
Thompson and Latorre (1999)	7 hosts, including grapevine	Chile	29	●							●			
Alfonso et al. (2000)	6 greenhouse crops, air	Almeria/ Spain	185	●					●	●		●		
Yourman et al. (2000)	Various ornamentals	South Carolina/USA	56	●										●[c]
Baraldi et al. (2002)	Kiwi fruit	Italy	90		●									●[c]

Munoz et al. (2002)	Grapevine, tomato, kiwi fruit, blueberry	Chile	58	●	●	●					●			
Rigotti et al. (2002)	Strawberry and others	Switzerland	34	●									●	
Fournier et al. (2003)	Various	France, Germany, Japan	132		●	●							●	
Moyano et al. (2003)	Unknown greenhouse crops, air	Almeria/ Spain	44	●	●[b]							●		●[d]
Topolovec-Pintaric et al. (2004)	Grapevine	Croatia	30			●								
Ma and Michailides (2005)	Fig, grapevine, kiwi fruit, pea, squash	California/ USA	234			●	●			●	●			
Martinez et al. (2005)	Grapevine	Bordeaux area/France	741			●			●					●[e,f]
Calpas et al. (2006)	20 plants in greenhouse, including tomato	Alberta/ Canada	91	●						●	●	●		
Isenegger et al. (2008b)	Chickpea	Bangladesh	146				●			●				

(continued)

Table 6.2 (continued)

Botrytis species and reference	Host	Region/ country	Sample size (no. of isolates)	Neutral markers				Structuring factor exploited[a]						
				RAPD	RFLP/ AFLP	TE	SSR	Year	Season	Geography	Host plant	Cropping system	Species delim.	Other
Isenegger et al. (2008a)	Chickpea, lentil, grape and various flowers	Bangladesh, Australia	266			●	●			●				
Karchani-Balma et al. (2008)	Tomato, grapevine, strawberry, bean, others	Tunisia	153			●				●	●	●		
Kretschmer and Hahn (2008)	Grapevine	Germany	209		●	●			●					
Martinez et al. (2008)	Grapevine	France	1,172			●			●					●[f]
Vaczy et al. (2008)	Grapevine	Eger area/ Hungary	109			●	●	●		●				
Decognet et al. (2009)	Greenhouse tomato, air	Provence/ France	897				●					●		
Mirzaei et al. (2009)	36 wild or cultivated plants	Iran	44	●					●	●	●			
Tanovic et al. (2009)	Blackberry, strawberry, raspberry	Serbia	26				●				●			
Munoz et al. (2010)	Grapevine	Argentina	35		●	●					●			

Rajaguru and Shaw (2010)	Strawberry, blackberry, dandelion, primrose	Southern England	243				●			●	●			
Valiuskaite et al. (2010)	Several horticultural plants	Lithuania	18		●[b]					●	●			
Esterio et al. (2011)	Grapevine	Chile	457	●		●			●	●				●[f]
Walker et al. (2011)	Grapevine, blackberry	Champagne/ France	1,088				●				●		●	
Fekete et al. (2012)	Rapeseed, strawberry, raspberry	Hungary	92			●	●						●	
Kuzmanovska et al. (2012)	Tomato	Macedonia	123			●				●				
Samuel et al. (2012)	Strawberry, grapevine, apple, tomato	Greece	334			●					●			
Asadollahi et al. (2013)	Strawberry, raspberry	Hungary	490		●	●	●	●	●		●			
Fournier et al. (2013)	Grapevine	France	164				●			●				●[g]
Leroch et al. (2013)	Grapevine, strawberry	Germany	323		●						●			
Johnston et al. (2013)	Grapevine	New Zealand	1,331			●			●	●				

(continued)

Table 6.2 (continued)

Botrytis species and reference	Host	Region/ country	Sample size (no. of isolates)	Neutral markers				Structuring factor exploited[a]						
				RAPD	RFLP/ AFLP	TE	SSR	Year	Season	Geography	Host plant	Cropping system	Species delim.	Other
Munoz and Campos (2013)	Pine, eucalyptus	Bio-Bio/Chile	55		●					●				
Wessels et al. (2013)	Rooibos	South Africa	207				●			●				
Bardin et al. (2014)	Greenhouse tomato, air	Southern France	712				●			●				
Kumari et al. (2014)	Chikpea, marigold and others	India, Nepal	79	●		●		●			●			
Adjebli et al. (2015)	Greenhouse tomato	Algeria	174				●					●		
Leyronas et al. (2015a)	Greenhouse tomato and lettuce	Provence/ France	86				●				●			
Walker et al. (2015)	Grapevine, blackberry, tomato, litter	France	3,546				●	●	●	●	●	●		

Leyronas et al. (2015b)	Air	Provence/ France	683				●						●[h]	*B. elliptica* and *B. tulipae*
Huang et al. (2001)	Lilies	USA, Taiwan	69	●						●	●			
Staats et al. (2007b)	Tulip, lily	The Netherlands	344		●[b]				●		●			
Botrytis species from onion														
Nielsen et al. (2001)	Onion	Denmark	51		●								●	
Valiuskaite et al. (2010)	Several horticultural plants, including onion	Lithuania	18		●[b]				●	●			●	

[a]Factor identified by the authors as possibly structuring their dataset and therefore tested in their analysis
[b]Studies using AFLP markers
[c]Adaptation to storage conditions (cold)
[d]Molecular epidemiology in greenhouses; redistribution of genotypes by the air in the greenhouse
[e]Influence of fungicide treatment
[f]Adaptation to the host organ
[g]Differentiation between noble rot and grey mould
[h]Structure of airborne populations

The transposable elements (TEs) *Boty* and *Flipper* (Sect. 2.4) are probably the most widely used markers in population studies (Table 6.2). Indeed, they were among the first markers to be identified and are rapid and convenient to use in any laboratory. However, even though the subdivision between *vacuma* and *transposa* led to the discovery of *B. pseudocinerea*, new results from genome sequencing have called into question the utility of these markers for population studies. TEs are thought to account for 0.4–0.6 % of the *B. cinerea* genome and are highly diverse (Chap. 3; Amselem et al. 2011). A single genome may contain not only various numbers of full-length copies of TEs, but also truncated copies or, for retrotransposons, solo-LTRs, resulting from recombination between the LTRs of ancestral TE copies. Consequently, diversity studies based on TE classification may principally reflect transposition dynamics within a genome, rather than exclusively population processes, as expected for canonical nuclear "neutral" markers. Moreover, several techniques are available for detecting *Boty* and *Flipper*: dot-blot hybridization (Giraud et al. 1997) and PCR detection with transposon-specific primers. Several pairs of primers are available for PCR detection. Some flank the transposon whereas others partially overlap it (Munoz et al. 2002; Ma and Michailides 2005; Martinez et al. 2008; Kretschmer and Hahn 2008; Johnston et al. 2013). These partially overlapping primers detect transposon copies in a specific genomic environment. The results obtained for a given strain may therefore differ according to the technique used, making comparisons of results difficult, as demonstrated by some reported unsatisfactory concordance rates (Martinez et al. 2008).

More recent studies have been based on the use of microsatellite or single sequence repeat (SSR) markers. These markers are thought to be neutral, and it is easy to test for departure from expectations for neutrality. They can be combined in multiplex PCR and are often highly polymorphic. This technique is easy to use and highly reproducible. A set of nine SSRs, named Bc1 to Bc10 (Fournier et al. 2002), has been used in all but one study. Bc9 and Bc10 may be linked, as they were isolated from the same clone. This SSR set has been shown to be highly discriminating for use in population studies (Karchani-Balma et al. 2008). Another set of 16 MP-PCR markers amplifying microsatellite motifs was also developed for the analysis Californian populations (Ma and Michailides 2005).

6.3.3 Confirmation of Species Boundaries Through the Use of Population Genetics Markers

In addition to phylogenetic genes, population genetics markers have also proved useful for the confirmation and recognition of species boundaries, because some of these markers may have specific ('private') alleles for some species and/or highly contrasting frequencies of shared alleles in natural populations. This property has been used to identify fungi, including several *Botrytis* spp., from symptomless strawberries by RAPD fingerprinting (Rigotti et al. 2002). However, it has chiefly been used to confirm the existence of a barrier to gene flow between *B. cinerea* and *B. pseudocinerea* in populations, with either RFLP (Fekete et al. 2012; Fournier et al. 2003; Giraud et al. 1997, 1999) or SSR markers (Walker et al. 2011; Fekete et al. 2012). Allelic diversity was found to be lower at locus Bc4 and a private allele (size 86 bp) was identified at locus Bc6 in *B. pseudocinerea*, in a study based on SSR markers (Walker et al. 2011; Walker unpublished). These differences confirmed the delimitation of the two species after population genetics analysis, with indices of differentiation between the two species increasing and significantly greater than those usually observed between populations from the same species. Bayesian analysis with no prior assumptions about the species of the strain unambiguously assigned individuals into two distinct, independent clusters, confirming the absence of gene flow between them (Walker et al. 2011). Population genetics may thus constitute a powerful tool for the discovery of new cryptic species, but the taxonomic status of these entities as species must subsequently be confirmed by phylogenic approaches (see above). The specificity of the available SSR markers should nevertheless be checked in *Botrytis* sp.

6.3.4 Geography as a Structuring Factor

Studies of geographic distance as a factor potentially structuring population diversity are justified by the ability of pathogens to disperse gradually over space, due to their biological features and the properties of their spores and the presence of natural elements (*e.g.* mountain chains, oceans or rivers) that may prevent dispersal. This has important consequences for disease propagation. Indeed, this factor was explored in 51 % of the papers reviewed here (Table 6.2). The effect of geographic distance is mostly dependent on the scale studied. Very little, or no differentiation has been reported at the regional (Kerssies et al. 1997; Giraud et al. 1999; Alfonso et al. 2000; Moyano et al. 2003; Ma and Michailides 2005; Calpas et al. 2006; Vaczy et al. 2008; Rajaguru and Shaw 2010; Wessels et al. 2013), and national scales (Choi et al. 1998; Isenegger et al. 2008b; Valiuskaite et al. 2010; Walker et al. 2015; Fournier and Giraud 2008; Munoz et al. 2010; Mirzaei et al. 2009; Esterio et al. 2011). By contrast, differentiation has been reported for some Tunisian populations separated by the Great Dorsal, a mountain chain constituting a geographic barrier to gene flow (Karchani-Balma et al. 2008). Differentiation is observed at the continental scale, between populations collected on different hosts in Australia and South Asia (Bangladesh, Nepal, India) (Isenegger et al. 2008a). High levels of genetic differentiation have also been observed between grapevine populations in France and Argentina (Munoz et al. 2010). These findings suggest that effective population sizes are large and that migration rates are high at the regional and national scales. This intense migration may also be facilitated by the large number of hosts that can potentially be contaminated, constituting as many transient reservoirs within an agricultural or natural landscape.

6.3.5 Host as a Structuring Factor

Adaptation to the host is a key issue for generalist pathogens such as *B. cinerea*, because it sets crucial levels for disease management (role of different hosts as within- *vs*. between-crop reservoirs or in the diffusion of fungicide resistance alleles, for example). Indeed, wild (*e.g.* Giraud et al. 1999; Rajaguru and Shaw 2010) or cultivated (*e.g.* Leyronas et al. 2015a; Karchani-Balma et al. 2008) host plants were found to have been studied as a structuring factor in 42 % of the population studies available (Table 6.2). Studies carried out with TE markers have shown that *transposa* strains predominate in populations collected from grapes, strawberries and tomatoes, whereas *vacuma* strains dominate *B. cinerea* populations from kiwi fruit and apples (Esterio et al. 2011; Johnston et al. 2013; Martinez et al. 2005; Munoz et al. 2002; Samuel et al. 2012). However, there is currently no clear explanation for these observations. Moreover, in studies based on the use of SSRs, differentiation indices or analysis recognize the host as the most powerful structuring factor for populations, ahead of geography (*e.g.* Fournier and Giraud 2008; Walker et al. 2015). Thus, populations collected from the same host in distant regions may be more genetically similar than populations collected from different hosts within the same region. An exception has been reported, for Californian populations, for which no genetic differentiation was observed between populations collected from different host plants (various fruit spp.) (Ma and Michailides 2005). However, this study used an original set of markers, ruling out comparisons with other studies. Similarly, no differentiation was observed between populations collected from grapevine and those collected from the surrounding litter (dead and decaying plant tissues lying on the ground) (Walker et al. 2015).

The coexistence in sympatry of subdivided populations (*i.e.* with restricted gene exchange) developing on different hosts in agricultural and natural environments suggests an effect of ecological divergence due to adaptation to the host. Indeed, fungal life cycles are unusual in that there is no migration between development on a host and reproduction: a given individual can reproduce with another individual that developed on the same host. Thus, the probability of a particular individual mating is dependent solely on its genotype, which favours or disfavours its development on a particular host. Specialization (*i.e.*, selection for genotypes with the best fitness on one host) therefore acts as a "magic trait" (Gavrilets 2004) in some fungi, pleiotropically allowing both adaptation to the host and reproductive isolation, thus facilitating sympatric divergence (Giraud et al. 2008).

Many studies highlighting the existence of sympatric subdivided *B. cinerea* populations on different hosts, thus suggesting ecological adaptation to the host, also showed that ecological divergence was not "complete", with some gene flow continuing between the different demes. Indeed, significant proportions of spillover individuals (*i.e.* individuals collected on one host but belonging to a genetic group specialized on another host) are regularly recorded (Fournier and Giraud 2008; Leyronas et al. 2015a; Walker et al. 2015). This may be because host adaptation in *B. cinerea* is mostly based on quantitative, rather than in "all or nothing" mechanisms.

Thus, gene flow and mating can still occur between host-specialized sub-populations and individuals are able to infect several host plants, although differences in aggressiveness may be recorded. As discussed by Anderson et al. (2004), ecological divergence, even if incomplete, should be accompanied by a trade-off in the aggressiveness of the pathogen on a different host (*i.e.* a local adaptation pattern for quantitative traits). This issue has seldom been addressed and further investigations are warranted. Leyronas et al. (2015a) showed that the host specialization of *B. cinerea* on tomato and lettuce grown in the same greenhouses was not always accompanied by significant measurable differences in aggressiveness. This may reflect the large, diverse and sophisticated biochemical arsenal of *B. cinerea*, with numerous alternative pathways deployed during the various steps of penetration, development, and decomposition processes (Chap. 12; Amselem et al. 2011; Van Kan 2006). This would allow basal growth to occur on a large range of hosts, with very small quantitative differences.

Moreover, although host adaptation may be incomplete in *B. cinerea*, recent studies have suggested that it may operate at a very fine scale. Indeed, Walker et al. (2015) demonstrated the surprising subdivision of grapevine-adapted populations into three genetic clusters co-existing on this host and only partially subsisting over time, due to recurrent recombination between them and/or with strains from clusters specialized on other hosts. The reasons for this fine-scale adaptation are unclear, but may involve specific host tissues, organs or elements of host physiology. Indeed, preliminary studies have shown that the frequencies of *vacuma* and *transposa* isolates differ significantly as a function of grapevine phenological stage and organ (Martinez et al. 2005, 2008), however, these markers may not always be relevant (see discussion above). Moreover, this clustering may reflect differences in the lifestyle of *B. cinerea*, as it has been suggested that this species can display both facultative cryptic endophytic behavior and classical necrotrophic behavior (Van Kan et al. 2014).

6.3.6 Cropping System as a Structuring Factor

The cropping system may significantly affect the genetics and dynamics of *Botrytis* populations. Indeed, many *B. cinerea* hosts are not cultivated outdoors, where gene flow is thought to be affected by host adaptation and geography. Instead, they are grown indoors, in greenhouses. The diversity of populations sampled from the air, within greenhouses or outdoors, has been shown to be high (Decognet et al. 2009; Kerssies et al. 1997; Bardin et al. 2014), consistent with the hypothesis that substantial amounts of inoculum from the external environment frequently find their way into the greenhouse (Chap. 7). Nevertheless, a comprehensive SSR-based study assessing diversity in populations collected from various ecological niches demonstrated that isolates collected from greenhouse tomatoes in several French regions clustered together and displayed a high degree of differentiation from isolates collected outdoors on other crops. This suggests that, together with host specialization, cropping system could be a powerful structuring factor (Walker et al. 2015). The proportion of clonal genotypes in populations was low or rising, depending on the greenhouse considered and, particularly, its technological equipment and the prophylaxis measures implemented (*i.e.* the efforts made to limit gene flow from the outside).

Similarly, high levels of diversity were also detected by RAPD analysis in greenhouses from the Almeria region in south-eastern Spain, but the comparison may have been biased by the collection of only a small number of strains per greenhouse (Alfonso et al. 2000; Moyano et al. 2003). However, in this region of intensive indoor cropping concentrated in a limited area and over a short time period, the diversity of the crops grown, together with the associated epidemics and poor prophylaxis (Alfonso et al. 2000), may favor the regular migration of genotypes within and between greenhouses. Considerable genetic diversity was also detected in tomato populations from northern Algerian greenhouses, in which it was suggested that recombination might occur (Adjebli et al. 2015). By contrast, in some tomato greenhouses in south-eastern France, a limited number of genotypes, or even a single well adapted genotype, may colonize the whole greenhouse. This highlights the importance of the secondary inoculum produced within the crop during long growing seasons and suggests that the polycyclic development of grey mould epidemics may occur in such situations (Bardin et al. 2014; Decognet et al. 2009). It may also explain why some genotypes collected on various indoor hosts could be grouped together, after RAPD analysis, according to the greenhouse of origin (Calpas et al. 2006). Finally, as host specialization is thought to be limited, successive crops (*e.g.* tomato and lettuce) may be infected by endogenous inoculum, maintaining the epidemic throughout the year. Nevertheless, differences in cropping practices for these two hosts may result in differences in the epidemiological features of the disease and may account for the more frequent production of secondary inoculum on tomato than on lettuce, resulting in lower levels of genotypic diversity on this host (Leyronas et al. 2015a).

6.3.7 Fungicides as a Structuring Factor

Fungicides may exert an intense selection pressure on populations, logically leading to an increase in the frequency of resistance within populations. However, little is known about their ability to shape population structure. A first set of studies (Table 6.2) investigated whether resistant strains clustered together in analyses with neutral markers. Resistant strains collected in Bordeaux vineyards (France) tended to be of the *transposa* type (Martinez et al. 2005), although the limitations of this marker must be borne in mind, as discussed previously. Yourman et al. (2000) found that all isolates from greenhouses in South Carolina, USA were different, but observed some clustering according to fungicide sensitivity, whereas Moyano et al. (2003) barely detected this pattern in Spanish populations collected in Almerian greenhouses. This suggests that, in some situations, resistance is selected in a limited number of neutral genetic backgrounds that recombine poorly with other genetic backgrounds.

A second set of studies (Table 6.2) investigated whether fungicides could induce the differentiation of populations subjected to various spraying programs. Fungicide treatments were found to have no effect on neutral subdivision, in Chilean populations collected from table grapes and subjected to RAPD analysis (Esterio et al. 2011) or in populations from Champagne vineyards subjected to SSR analysis (Walker and Fournier 2014). Moreover, in this last study, diversity (measured as gene diversity) and the mode of reproduction (measured as clonal richness), were similar in treated and untreated plots. Nevertheless, significant differences in allelic richness and private allele richness between the untreated and treated plots were observed for two of the three sites studied. This is consistent with greater genetic drift, leading to the loss of alleles, particularly for the rarest alleles, in populations that regularly shrink due to fungicide applications (bottlenecks). Moreover, as expected, resistance frequency was significantly higher in treated plots, but only for loci under contemporary selection, not for loci selected by ancient modes of action. Interestingly, the frequency of resistance decreased at all sites, for all four loci, probably due to the combined effect of winter migration and negative selective pressure (resistance cost). Fungicide-mediated selection was also demonstrated by the detection of clines (patterns of resistance organized into a spatial gradient between the treated and untreated plots), especially at vintage time for loci under contemporary selection (Walker and Fournier 2014). Resistance frequencies could therefore be used to model the evolution of these traits over a period of years and to provide estimates of the magnitude of the evolutionary forces at work in vineyards (*e.g.* migration, positive selection, negative selection of resistance cost).

6.3.8 Other Hypotheses Relating to the Structuring of Populations

Three additional studies explored the differentiation of the population according to additional factors. Baraldi et al. (2002) detected no genetic differentiation in RFLP analyses of isolates from kiwi fruit that had and had not been stored in the cold, despite the demonstration of a clear adaptation to cold in cold-stored populations. Fournier et al. (2013) used microsatellite genotyping and clustering methods to determine whether isolates sampled from plants displaying grey mould and isolates from plants displaying noble rot symptoms in three French regions belonged to genetically differentiated populations. The inferred population structure matched geography rather than the type of symptom. Noble rot symptoms therefore do not seem to be caused by a specific *B. cinerea* population, depending instead essentially on microclimatic conditions, with implications for the production of sweet wines. At last, Leyronas et al. (2015b) identified eight strongly differentiated genetic clusters from *B. cinerea* populations collected from the air, in southern France. Cluster abundance showed temporal variation and was linked to the season, the climatic parameters and the origin of air masses. *B. pseudocinerea* strains were also isolated and their frequency varied over time, tending to be greater in winter. The relationship between airborne and terrestrial clusters remains to establish but may help understanding the emergence of grey mould epidemics on crops, as well as the distribution of fungicide resistant alleles in populations from distinct compartments.

In conclusion, considerable amount of information about population genetics has been obtained for *B. cinerea* over the last decade, due to sophisticated marker use and data analysis. In the future, efforts should be made to harmonize the sampling and analysis of isolates in different situations, to improve the comparability of results and to synergize knowledge acquisition by different research groups. Some studies are clearly based on a limited number of isolates (Table 6.2), and poor or false information may be extracted from populations of fewer than 20 or 30 individuals, particularly if these individuals are collected from different hosts or sites. Sampling schemes should dissociate structuring factors, to measure their respective impacts and to prevent situations such as the comparison of populations from tomatoes grown in a greenhouse with populations from outdoor crops or the comparison of populations from one host with those from another host in another country, for example.

6.4 Diversity the Populations of Other Botrytis spp.

Much less is known about the population genetics and genetic diversity of other species of *Botrytis* than about *B. cinerea*.

6.4.1 Botrytis elliptica *and* B. tulipae

Population genetics studies have been carried out on *B. elliptica* with 69 isolates from Taiwan and the USA and RAPD markers (Huang et al. 2001). In total, 43 unique haplotypes were identified, falling into two clusters according to country of origin. Further investigations compared the abilities of three molecular typing methods (multi-locus sequencing, restriction analysis of IGS region, AFLP) to assess variability within *B. cinerea*, *B. elliptica*, and *B. tulipae*. AFLP analysis was found to be the most effective method and was therefore used for the subsequent population genetics studies. In total, 105 genotypes were identified for 174 *B. elliptica* isolates sampled from lilies in the Netherlands. Linkage disequilibrium scores were low and clonal genotypes were detected only within the growing season, at a single site (Staats et al. 2007b). These results suggest that sexual reproduction plays a significant role in determining population diversity in this species, although apothecia have been observed in the field only in the Netherlands, with few details reported (Van den Ende and Pennock-Vos 1997).

By contrast, only 25 genotypes could be distinguished for 170 *B. tulipae* isolates, and clonal genotypes were frequently found in different growing seasons and different locations from the Netherlands. Higher linkage disequilibrium indices confirmed that the *B. tulipae* population was mostly clonal, with some recombination (Staats et al. 2007b).

6.4.2 Botrytis *Species from Onion*

Morphological mutants of *B. squamosa* have been recovered by chemical mutagenesis, and mutants resistant to the fungicide botran have also been obtained, with this resistance segregating as a single gene (Bergquist and Lorbeer 1972). Variation has been investigated in a number of onion-associated species, by UP-PCR (Nielsen et al. 2001). *B. squamosa* was found to be highly diverse, with 10 of the 11 isolates studied having unique haplotypes, consistent with the known heterothallic sexual reproduction of this species (Carisse et al. 2011). By contrast, *B. aclada* and *B. allii* displayed little variation, consistent with a high degree of clonality and the absence of known teleomorphs for these species. Only three isolates of *B. byssoidea* were examined. All were identical, despite originating from different sites in the USA, the Netherlands and the UK, suggesting that the populations of this species may be mostly clonal (Nielsen et al. 2001).

6.4.3 Botrytis pseudocinerea

The formal description of this species was accompanied by a population study (Walker et al. 2011). Indeed, sympatric populations of *B. cinerea* and *B. pseudocinerea* from Champagne vineyards clustered separately, demonstrating the lack of gene flow between these two species. Genotypic diversity was high, of the same order of magnitude as in *B. cinerea*; the clonal fraction was small, and weak linkage disequilibrium was observed, suggesting that recombination regularly occurs in *B. pseudocinerea* populations, through sexual reproduction. The two species are thought to have similar population genetics, but the influence of possible structuring factors has not yet been evaluated in *B. pseudocinerea*.

6.5 Conclusions

Our knowledge of *Botrytis* diversity has greatly increased over the last decade, following the development of relevant techniques and markers and the sequencing of two *B. cinerea* genomes. This knowledge is not only of academic interest, but may also be very useful for disease management. Delimiting species causing similar symptoms and characterizing their biological features and epidemiological importance will undoubtedly contribute to the appropriate adaptation of control strategies. Accurate and complete characterization is thus required, together with depositions in international collections and a relevant multi-criterion determination tool covering all known species. Identifying barriers to gene flow between the various population compartments of *B. cinerea* in natural and agricultural environments should facilitate their manipulation, improving prophylaxis, host deployment or appropriate antifungal compound applications, for example. Finally, although *B. cinerea* is the model species within this genus, similar developments, particularly in population genetics, would also be useful to improve disease control on major crops affected by other *Botrytis* species.

Acknowledgments I thank Ross E. Beever and Pauline L. Weeds for paving the way to the writing of this chapter, through their contributions to the previous edition of this book. Their writing has greatly influenced this updated version, although much has been discovered about *Botrytis* diversity since 2004. I am also grateful to Kevin D. Hyde and coworkers for authorizing the publication of Fig. 6.1. Finally, I warmly thank Elisabeth Fournier, Jean-Marc Pradier, Pierre Leroux, Peter Johnston and Jan van Kan for their helpful comments on early drafts of this chapter.

References

Adjebli A, Leyronas C, Aissat K, Nicot PC (2015) Comparison of *Botrytis cinerea* populations collected from tomato greenhouses in Northern Algeria. Plant Pathol 163(2):124–132

Albertini C, Leroux P (2004) A *Botrytis cinerea* putative 3-keto reductase gene (ERG27) that is homologous to the mammalian 17 beta-hydroxysteroid dehydrogenase type 7 gene (17 beta-HSD7). Eur J Plant Pathol 110(7):723–733

Albertini C, Thebaud G, Fournier E, Leroux P (2002) Eburicol 14 alpha -demethylase gene (CYP51) polymorphism and speciation in *Botrytis cinerea*. Mycol Res 106(10):1171–1178

Alfonso C, Raposo R, Melgarejo P (2000) Genetic diversity in *Botrytis cinerea* populations on vegetable crops in greenhouses in south-eastern Spain. Plant Pathol 49(2):243–251

Amselem J, Cuomo CA, Van Kan JAL et al (2011) Genomic analysis of the necrotrophic fungal pathogens *Sclerotinia sclerotiorum* and *Botrytis cinerea*. Plos Genet 7(8). doi:e1002230

Anderson PK, Cunningham AA, Patel NG et al (2004) Emerging infectious diseases of plants: pathogen pollution, climate change and agrotechnology drivers. Trends Ecol Evol 19(10):535–544

Andrew M, Barua R, Short SM, Kohn LM (2012) Evidence for a common toolbox based on necrotrophy in a fungal lineage spanning necrotrophs, biotrophs, endophytes, host generalists and specialists. PLoS ONE 7(1). doi:e29943

Asadollahi M, Fekete E, Karaffa L et al (2013) Comparison of *Botrytis cinerea* populations isolated from two open-field cultivated host plants. Microbiol Res 168(6):379–388

Baraldi E, Bertolini P, Chierici E et al (2002) Genetic diversity between *Botrytis cinerea* isolates from unstored and cold stored kiwi fruit. J Phytopathol 150(11–12):629–635

Bardin M, Decognet V, Nicot P (2014) Remarkable predominance of a small number of genotypes in greenhouse populations of *Botrytis cinerea*. Phytopathology 104(8):859–864

Barnes SE, Shaw MW (2003) Infection of commercial hybrid primula seed by *Botrytis cinerea* and latent disease spread through the plants. Phytopathology 93(5):573–578

Beever R, Weeds P (2004) Taxonomic and genetic variation of *Botrytis* and *Botryotinia*. In: Elad Y, Williamson B, Tudzynski P, Delen N (eds) *Botrytis*: biology, pathology and control. Kluwer Academic Publisher, Dordrecht, pp 29–52

Bergquist RR, Lorbeer JW (1972) Apothecial production, compatibility and sex in *Botryotinia squamosa*. Mycologia 64:1270–1281

Buchwald N (1953) *Botryotinia* (*Sclerotinia*) *globosa* sp. n. on *Allium ursinum*, the perfect stage of *Botrytis globosa* Raabe. J Phytopathol 20:241–254

Calpas JT, Konschuh MN, Toews CC, Tewari JP (2006) Relationships among isolates of *Botrytis cinerea* collected from greenhouses and field locations in Alberta, based on RAPID analysis. Can J Plant Pathol 28(1):109–124

Carisse O, Tremblay D-M, McDonald MR et al (2011) Management of *Botrytis* leaf blight of onion: the Québec experience of 20 years of continual improvement. Plant Dis 95(5):504–514

Choi IS, Kim DH, Lee CW et al (1998) Analysis of genetic variation in *Botrytis cinerea* isolates using random amplified polymorphic DNA markers. J Microbiol Biotechnol 8(5):490–496

De Queiroz K (1998) The general lineage concept of species, species criteria, and the process of speciation: a conceptual unification and terminological recommendations. In: Howard DJ, Berlocher SH (eds) Endless forms: species and speciation. Oxford University Press, New York, pp 57–75

de Vienne DM, Giraud T, Martin OC (2007) A congruence index for testing topological similarity between trees. Bioinformatics 23(23):3119–3124

de Vienne DM, Giraud T, Martin OC (2009) In response to comment on 'A congruence index for testing topological similarity between trees'. Bioinformatics 25(1):150–151

Decognet V, Bardin M, Trottin-Caudal Y, Nicot PC (2009) Rapid change in the genetic diversity of *Botrytis cinerea* populations after the introduction of strains in a tomato glasshouse. Phytopathology 99(2):185–193

Diolez A, Marches F, Fortini D, Brygoo Y (1995) Boty, a long-terminal-repeat retroelement in the phytopathogenic fungus *Botrytis cinerea*. Appl Environ Microbiol 61(1):103–108

Elliott ME (1964) Self-fertility in *Botryotinia porri*. Can J Bot 42(10):1393–1395

Esterio M, Munoz G, Ramos C et al (2011) Characterization of *Botrytis cinerea* isolates present in Thompson seedless table grapes in the central valley of Chile. Plant Dis 95(6):683–690

Faretra F, Antonacci E, Pollastro S (1988) Sexual behaviour and mating system of *Botrytotinia fuckeliana*, teleomorph of *Botrytis cinerea*. J Gen Microbiol 134:2543–2550

Farr DF, Bills GF, Chamuris GP, Rossman AY (1989) Fungi on plants and plant products in the United States. American Phytopathological Society Press, St Paul

Fekete E, Irinyi L, Karaffa L et al (2012) Genetic diversity of a *Botrytis cinerea* cryptic species complex in Hungary. Microbiol Res 167(5):283–291

Ferrada EE, et al (2015) Identification and characterization of Botrytis Blossom Blight of Japanese plums caused by *Botrytis cinerea* and *B. prunorum* sp. nov. in Chile. Phytopathology (in press)

Fournier E, Giraud T (2008) Sympatric genetic differentiation of a generalist pathogenic fungus, *Botrytis cinerea*, on two different host plants, grapevine and bramble. J Evol Biol 21(1):122–132

Fournier E, Giraud T, Loiseau A et al (2002) Characterization of nine polymorphic microsatellite loci in the fungus *Botrytis cinerea* (Ascomycota). Mol Ecol Notes 2(3):253–255

Fournier E, Levis C, Fortini D et al (2003) Characterization of Bc-hch, the *Botrytis cinerea* homolog of the *Neurospora crassa* het-c vegetative incompatibility locus, and its use as a population marker. Mycologia 95(2):251–261

Fournier E, Giraud T, Albertini C, Brygoo Y (2005) Partition of the *Botrytis cinerea* complex in France using multiple gene genealogies. Mycologia 97(6):1251–1267

Fournier E, Gladieux P, Giraud T (2013) The "Dr Jekyll and Mr Hyde fungus": noble rot versus gray mold symptoms of *Botrytis cinerea* on grapes. Evol Appl 6(9):960–969

Gavrilets S (2004) Fitness landscapes and the origin of species. Princeton University Press, Princeton

Giraud T, Fortini D, Levis C et al (1997) RFLP markers show genetic recombination in *Botryotinia fuckeliana* (*Botrytis cinerea*) and transposable elements reveal two sympatric species. Mol Biol Evol 14(11):1177–1185

Giraud T, Fortini D, Levis C et al (1999) Two sibling species of the *Botrytis cinerea* complex, transposa and vacuma, are found in sympatry on numerous host plants. Phytopathology 89(10):967–973

Giraud T, Refregier G, Le Gac M et al (2008) Speciation in fungi. Fungal Genet Biol 45(6):791–802

Godfrey GH (1923) Gray mold of castor bean. J Agric Res 23:679–716

Grant Downton RT, Terhem RB, Kapralov MV et al (2014) A novel *Botrytis* species is associated with a newly emergent foliar disease in cultivated *Hemerocallis*. PLoS One 9(6):e89272

Grindle M (1979) Phenotypic differences between natural and induced variants of *Botrytis cinerea*. J Gen Microbiol 111(1):109–120

Groves J, Loveland C (1953) The connection between *Botryotinia fuckeliana* and *Botrytis cinerea*. Mycologia 45:415–425

Hennebert GL (1963) Les *Botrytis* des *Allium*. Medelingen Van de Landbouwhogeschool En de Opzoekingsstations Van de Staat Te Gent 28:851–876

Hennebert G (1973) *Botrytis* and *Botrytis*-like genera. Persoonia 7:183–204

Hennebert GL, Groves JW (1963) Three new species of *Botryotinia* on *Ranunculaceae*. Can J Bot 41(3):341–370

Hirschhäuser S, Fröhlich J (2007) Multiplex PCR for species discrimination of *Sclerotiniaceae* by novel laccase introns. Int J Food Microbiol 118(2):151–157

Holst-Jensen A, Vaage M, Schumacher T (1998) An approximation to the phylogeny of *Sclerotinia* and related genera. Nord J Bot 18(6):705–719

Huang JB, Hsieh TF, Chastagner GA, Hsiang T (2001) Clonal and sexual propagation in *Botrytis elliptica*. Mycol Res 105:833–842

Hyde KD, Nilsson RH, Alias SA et al (2014) One stop shop: backbones trees for important phytopathogenic genera: I (2014). Fungal Divers 67(1):21–125

Isenegger DA, Ades PK, Ford R, Taylor PWJ (2008a) Status of the *Botrytis cinerea* species complex and microsatellite analysis of transposon types in South Asia and Australia. Fungal Divers 29:17–26

Isenegger DA, Macleod WJ, Ford R, Taylor PWJ (2008b) Genotypic diversity and migration of clonal lineages of *Botrytis cinerea* from chickpea fields of Bangladesh inferred by microsatellite markers. Plant Pathol 57(5):967–973

Jarvis WR (1977) *Botryotinia* and *Botrytis* species: taxonomy, physiology and pathogenicity. Research Branch, Canada Department of Agriculture, Ottawa

Jarvis WR (1980) Taxonomy. In: The biology of *Botrytis*. Coley-Smith, J.R. Verhoeff, K. Jarvis, W.R., Academic, London

Johnston PR, Hoksbergen K, Park D, Beever RE (2013) Genetic diversity of *Botrytis* in New Zealand vineyards and the significance of its seasonal and regional variation. Plant Pathol 63(4):888–898

Johnston PR, Seifert KA, Stone JK et al (2014) Recommendation on the generic names competing for use in Leotiomycetes. IMA Fungus 5(1):91–120

Karchani-Balma S, Gautier A, Raies A, Fournier E (2008) Geography, plants and growing systems shape the genetic structure of Tunisian *Botrytis cinerea* populations. Phytopathology 98(12):1271–1279

Kerssies A, BoskervanZessen AI, Wagemakers CAM, Van Kan JAL (1997) Variation in pathogenicity and DNA polymorphism among *Botrytis cinerea* isolates sampled inside and outside a glasshouse. Plant Dis 81(7):781–786

Khan MI, Marroni V, Keenan S et al (2013) Enhanced molecular identification of *Botrytis* spp. from New Zealand onions. Eur J Plant Pathol 136(3):495–507

Kohn LM (1979) Delimitation of the economically important plant pathogenic *Sclerotinia* species. Phytopathology 69(8):881–886

Kretschmer M, Hahn M (2008) Fungicide resistance and genetic diversity of *Botrytis cinerea* isolates from a vineyard in Germany. J Plant Dis Prot 115(5):214–219

Kretschmer M, Leroch M, Mosbach A et al (2009) Fungicide-driven evolution and molecular basis of multidrug resistance in field populations of the grey mould fungus *Botrytis cinerea*. PLoS Pathog 5(12):e1000696

Kumari S, Tayal P, Sharma E, Kapoor R (2014) Analyses of genetic and pathogenic variability among *Botrytis cinerea* isolates. Microbiol Res 169(11):862–872

Kuzmanovska B, Rusevski R, Jankuloski L et al (2012) Phenotypic and genetic characterization of *Botrytis cinerea* isolates from tomato. Gen-Belgrade 44(3):633–647

Laundon GF (1978) New host records of plant disease fungi in New-Zealand. N Z J Agric Res 21(4):705–707

Leroch M, Plesken C, Weber RWS et al (2013) Gray mold populations in German strawberry fields are resistant to multiple fungicides and dominated by a novel clade closely related to *Botrytis cinerea*. Appl Environ Microbiol 79(1):159–167

Leroux P, Walker A (2013) Activity of fungicides and modulators of membrane drug transporters in field strains of *Botrytis cinerea* displaying multidrug resistance. Eur J Plant Pathol 135(4):683–693

Leroux P, Chapeland F, Desbrosses D, Gredt M (1999) Patterns of cross-resistance to fungicides in *Botryotinia fuckeliana* (*Botrytis cinerea*) isolates from French vineyards. Crop Prot 18(10):687–697

Leroux P, Gredt M, Leroch M, Walker A-S (2010) Exploring mechanisms of resistance to respiratory inhibitors in field strains of *Botrytis cinerea*, the causal agent of gray mold. Appl Environ Microbiol 76(19):6615–6630

Levis C, Fortini D, Brygoo Y (1997) Flipper, a mobile Fot1-like transposable element in *Botrytis cinerea*. Mol Gen Genet 254(6):674–680

Leyronas C, Bryone F, Duffaud M et al (2015a) Assessing host specialization of *Botrytis cinerea* on lettuce and tomato by genotypic and phenotypic characterization. Plant Pathol 64(1):119–127

Leyronas C, Halkett F, Nicot PC (2015b) Relationship between the genetic characteristics of *Botrytis* sp. airborne inoculum and meteorological parameters, seasons and the origin of air masses. Aerobiologia 31(3):367–380

Li JL, Chen QT (1987) The taxonomic study on *Botryotinia fritillarii- pallidiflori* Q. T. Chen et J. L. Li (sp. nov.). Acta Mycol Sin 6(1):15–19

Li X, Kerrigan J, Chai W, Schnabel G (2012) *Botrytis caroliniana*, a new species isolated from blackberry in South Carolina. Mycologia 104(3):650–658

Li N, Zhang J, Yang L, Wu MD, Li GQ (2014) First report of *Botrytis pseudocinerea* causing gray mold on tomato (*Lycopersicon esculentum*) in Central China. Plant Dis 99(2):283–283

Lorenz DH (1983) Investigations on the morphological variability and the pathogenicity of *Botrytis cinerea* and *Botryotinia fuckeliana* Whetz. J Plant Dis Prot 90(6):622–633

Lorenzini M, Zapparoli G (2014) An isolate morphologically and phylogenetically distinct from *Botrytis cinerea* obtained from withered grapes possibly represents a new species of *Botrytis*. Plant Pathol 63(6):1326–1335

Ma ZH, Michailides TJ (2005) Genetic structure of *Botrytis cinerea* populations from different host plants in California. Plant Dis 89(10):1083–1089

Marthey S, Aguileta G, Rodolphe F et al (2008) FUNYBASE: a FUNgal phYlogenomic dataBASE. BMC Bioinformatics 9:456. doi:10.1186/1471-2105-9-456

Martinez F, Blancard D, Lecomte P et al (2003) Phenotypic differences between vacuma and transposa subpopulations of *Botrytis cinerea*. Eur J Plant Pathol 109(5):479–488

Martinez F, Dubos B, Fermaud M (2005) The role of saprotrophy and virulence in the population dynamics of *Botrytis cinerea* in vineyards. Phytopathology 95(6):692–700

Martinez F, Corio-Costet MF, Levis C et al (2008) New PCR primers applied to characterize distribution of *Botrytis cinerea* populations in French vineyards. Vitis 47(4):217–226

Mayden RL (1997) A hierarchy of species concepts: the denouement in the saga of the species problem. In: Claridge MF, Dawah HA, Wilson MR (eds) Species: the units of biodiversity. Chapman and Hall, London, pp 381–424

Mirzaei S, Goltapeh EM, Shams-Bakhsh M et al (2009) Genetic and phenotypic diversity among *Botrytis cinerea* isolates in Iran. J Phytopathol 157(7–8):474–482

Moore WC (1959) British parasitic fungi. Cambridge University Press, Cambridge, UK

Moyano C, Alfonso C, Gallego J et al (2003) Comparison of RAPD and AFLP marker analysis as a means to study the genetic structure of *Botrytis cinerea* populations. Eur J Plant Pathol 109(5):515–522

Munoz G, Campos F (2013) Genetic characterization of *Botrytis cinerea* isolates collected from pine and eucalyptus nurseries in Bio-Bio Region, Chile. For Pathol 43(6):509–512

Munoz G, Hinrichsen P, Brygoo Y, Giraud T (2002) Genetic characterisation of *Botrytis cinerea* populations in Chile. Mycol Res 106:594–601

Munoz C, Talquenca SG, Oriolani E, Combina M (2010) Genetic characterization of grapevine-infecting *Botrytis cinerea* isolates from Argentina. Rev Iberoam Micol 27(2):66–70

Munoz G, Campos F, Salgado D et al (2015) Molecular identification of *Botrytis cinerea, B. paeoniae* and *B. pseudocinerea* associated to gray mold in peony (*Paeonia lactiflora* Pall.) in southern Chile. Rev Iberoam Micol, p ii: S1130–1406(15):00023–6. doi:10.1016/j.riam.2015.02.002

Nielsen K, Yohalem DS (2001) Origin of a polyploid *Botrytis* pathogen through interspecific hybridization between *Botrytis aclada* and *B. byssoidea*. Mycologia 93(6):1064–1071

Nielsen K, Justesen AF, Jensen DF, Yohalem DS (2001) Universally primed polymerase chain reaction alleles and internal transcribed spacer restriction fragment length polymorphisms distinguish two subgroups in *Botrytis aclada* distinct from B-byssoidea. Phytopathology 91(6):527–533

Noble M (1948) Seed-borne diseases of clover. Trans Br Mycol Soc 30:84–91

O'Gorman DT, Sholberg PL, Stokes SC, Ginns J (2008) DNA sequence analysis of herbarium specimens facilitates the revival of *Botrytis mali*, a postharvest pathogen of apple. Mycologia 100(2):227–235

Plesken C, Westrich LD, Hahn M (2015) Genetic and phenotypic characterization of *Botrytis calthae*. Plant Pathol 64(1):128–136

Rajaguru BAP, Shaw MW (2010) Genetic differentiation between hosts and locations in populations of latent *Botrytis cinerea* in southern England. Plant Pathol 59(6):1081–1090

Rigotti S, Gindro K, Richter H, Viret O (2002) Characterization of molecular markers for specific and sensitive detection of *Botrytis cinerea* Pers.: Fr. in strawberry (*Fragaria* × *ananassa* Duch.) using PCR. FEMS Microbiol Lett 209(2):169–174

Ruehle GD (1931) New apple-rot fungi from Washington. Phytopathology 21:1141–1152

Saito S, Michailides TJ, Xiao CL (2014) First report of *Botrytis pseudocinerea* causing gray mold on blueberry in North America. Plant Dis 98(12):1743–1743

Samadi S, Barberousse A (2006) The tree, the network, and the species. Biol J Linn Soc 89(3):509–521

Samuel S, Veloukas T, Papavasileiou A, Karaoglanidis GS (2012) Differences in frequency of transposable elements presence in *Botrytis cinerea* populations from several hosts in Greece. Plant Dis 96(9):1286–1290

Schumacher J, Gautier A, Morgant G et al (2013) A functional bikaverin biosynthesis gene cluster in rare strains of *Botrytis cinerea* is positively controlled by *VELVET*. PLoS One 8(1):e53729

Shipunov A, Newcombe G, Raghavendra AKH, Anderson CL (2008) Hidden diversity of endophytic fungi in an invasive plant. Am J Bot 95(9):1096–1108

Sites JW, Marshall JC (2004) Operational criteria for delimiting species. Ann Rev Ecol Evol Syst 35:199–227

Sowley ENK, Dewey FM, Shaw MW (2010) Persistent, symptomless, systemic, and seed-borne infection of lettuce by *Botrytis cinerea*. Eur J Plant Pathol 126(1):61–71

Staats M, Van Baarlen P, Van Kan JAL (2005) Molecular phylogeny of the plant pathogenic genus *Botrytis* and the evolution of host specificity. Mol Biol Evol 22(2):333–346

Staats M, Van Baarlen P, Schouten A et al (2007a) Positive selection in phytotoxic protein-encoding genes of *Botrytis* species. Fungal Genet Biol 44(1):52–63

Staats M, Van Baarlen P, Van Kan JAL (2007b) AFLP analysis of genetic diversity in populations of *Botrytis elliptica* and *Botrytis tulipae* from the Netherlands. Eur J Plant Pathol 117(3):219–235

Tanovic B, Delibasic G, Milivojevic J, Nikolic M (2009) Characterization of *Botrytis cinerea* from small fruits and grapevine in Serbia. Arch Biol Sci 61(3):419–429

Taylor JW, Jacobson DJ, Kroken S et al (2000) Phylogenetic species recognition and species concepts in fungi. Fungal Genet Biol 31(1):21–32

Thompson JR, Latorre BA (1999) Characterization of *Botrytis cinerea* from table grapes in Chile using RAPD-PCR. Plant Dis 83(12):1090–1094

Topolovec-Pintaric S, Milicevic T, Cvjetkovic B (2004) Genetic diversity and dynamic of pyrimethanil-resistant phenotype in population of *Botrytis cinerea* Pers.: Fr. in one wine-growing area in Croatia. J Plant Dis Prot 111(5):451–460

Vaczy KZ, Sandor E, Karaffa L et al (2008) Sexual recombination in the *Botrytis cinerea* populations in Hungarian vineyards. Phytopathology 98(12):1312–1319

Valiuskaite A, Surviliene E, Baniulis D (2010) Genetic diversity and pathogenicity traits of *Botrytis* spp. isolated from horticultural hosts. Zemdirbyste-Agriculture 97(4):85–90

Van den Ende E, Pennock I (1996) The perfect stage of *Botrytis elliptica*. In: XI international *Botrytis* symposium, Wageningen, p 16

Van den Ende J, Pennock-Vos IMG (1997) Primary sources of inoculum of *Botrytis elliptica* in lily. Acta Horticult 430:1193–1200

Van der Vlugt-Bergmans CJB, Brandwagt BF, Vant't Klooster JW et al (1993) Genetic variation and segregation of DNA polymorphisms in *Botrytis cinerea*. Mycol Res 97(10):1193–1200

Van Kan JAL (2006) Licensed to kill: the lifestyle of a necrotrophic plant pathogen. Trends Plant Sci 11(5):247–253

Van Kan JAL, Shaw MW, Grant Downton RT (2014) *Botrytis* species: relentless necrotrophic thugs or endophytes gone rogue? Mol Plant Pathol 15(9):957–961

Vos P, Hogers R, Bleeker M et al (1995) AFLP – a new technique for DNA-fingerprinting. Nucleic Acids Res 23(21):4407–4414

Walker AS, Fournier E (2014) Habitat- and host-specific differentiation in the multihost pathogen *Botrytis cinerea* and evidence for fungicide selection in populations. In: Dehne DW et al (ed) 17th International Reinhardsbrunn Conf, Friedrichroda, Germany, April 21–25, 2013 2014. DPG Spectrum Phytomedizin

Walker A-S, Gautier A, Confais J et al (2011) *Botrytis pseudocinerea*, a new cryptic species causing gray mold in French vineyards in sympatry with *Botrytis cinerea*. Phytopathology 101(12):1433–1445

Walker AS, Gladieux P, Decognet V et al (2015) Population structure and temporal maintenance of the multihost fungal pathogen *Botrytis cinerea*: causes and implications for disease management. Environ Microbiol 17(4):1261–1274

Wessels BA, Lamprecht SC, Linde CC et al (2013) Characterization of the genetic variation and fungicide resistance in *Botrytis cinerea* populations on rooibos seedlings in the Western Cape of South Africa. Eur J Plant Pathol 136(2):407–417

Whetzel H (1945) A synopsis of the genera and species of the *Sclerotiniaceae*, a family of stromatic inoperculate discomycetes. Mycologia 37(6):648–714

Wingfield MJ, De Beer ZW, Slippers B et al (2012) One fungus, one name promotes progressive plant pathology. Mol Plant Pathol 13(6):604–613

Wu TH, Lu JY (1991) A new species of *Botryotinia* – the teleomorph of *Botrytis fabae* Sardina. Acta Mycol Sin 10(1):27–30

Yamamoto W (1949) Species of the *Sclerotiniaceae* in Japan. Trans Mycol Soc Jpn 2:2–8

Yohalem DS, Nielsen K, Nicolaisen M (2003) Taxonomic and nomenclatural clarification of the onion neck rotting *Botrytis* species. Mycotaxon 85:175–182

Yourman LF, Jeffers SN, Dean RA (2000) Genetic analysis of isolates of *Botrytis cinerea* sensitive and resistant to benzimidazole and dicarboximide fungicides. Phytopathology 90(8):851–859

Zhang J, Wu M-D, Li G-Q et al (2010a) *Botrytis fabiopsis*, a new species causing chocolate spot of broad bean in central China. Mycologia 102(5):1114–1126

Zhang J, Zhang L, Li G-Q et al (2010b) *Botrytis sinoallii*: a new species of the grey mould pathogen on *Allium* crops in China. Mycoscience 51(6):421–431

Zhou YJ, Zhang J, Wang XD et al (2014) Morphological and phylogenetic identification of *Botrytis sinoviticola*, a novel cryptic species causing gray mold disease of table grapes (*Vitis vinifera*) in China. Mycologia 106:43–56

Chapter 7
Epidemiology and Aerobiology of *Botrytis* spp.

Odile Carisse

Abstract A thorough understanding of the epidemiology and aerobiology of *Botrytis*-induced diseases is essential for the design of rational management strategies. *B. cinerea* and other *Botrytis* spp. are important pathogens of ornamental, fruit, vegetable, and legume crops, and *Botrytis*-induced diseases can be found in nurseries, fields, greenhouses, and in storage. Although a generic outline of the life cycle of *Botrytis* spp. can be drawn, there are large variations depending on the host, agricultural practices, environmental conditions, and the geographical region. A *Botrytis* epidemic comprises a sequence of processes (perennation and infection, colonization, conidiation, and conidia dispersal), each of which is influenced by the host and the surrounding environment. The survival of *Botrytis* spp. between cropping seasons and the dispersal of conidia to uninfected plants are critical phases of the *Botrytis*-induced disease cycle, because if either phase is prevented, the disease will be delayed or will not occur. For most *Botrytis*-induced diseases, the pathogen reproduces profusely and asexually by air-dispersed conidia. Hence aerobiology plays a key role in understanding the epidemiology of *Botrytis* spp., which have developed a variety of strategies to infect and colonize their host. With a view to the design of best management practices, this chapter reviews the epidemiology and aerobiology of some important *Botrytis*-induced diseases and discuss the influence of the host, the environment, and agricultural practices on disease development.

Keywords Airborne inoculum • Fruitrot • Fungal dispersal • Grey mould

---

O. Carisse (✉)
Agriculture and Agri-Food Canada, 430 Gouin Boulevard,
St-Jean-sur-Richelieu, Quebec, Canada, J3B 3E6
e-mail: odile.carisse@agr.gc.ca

S. Fillinger, Y. Elad (eds.), *Botrytis – the Fungus, the Pathogen and its Management in Agricultural Systems*, DOI 10.1007/978-3-319-23371-0_7

7.1 Introduction

Botrytis-incited diseases are among the most important diseases worldwide (Dean et al. 2012). This ranking can be explained by a number of factors: the exceptionally wide host range of *Botrytis* spp., particularly *B. cinerea*; the endemic nature of *Botrytis* epidemics and the extent of crop losses that they cause; the ability of *Botrytis* spp. to cause quiescent infections; the remarkable adaptation of *Botrytis* spp. to various environments; the genetic plasticity of *Botrytis* spp., including adaptation to fungicides; the lack of host resistance; and the flexible lifestyle of *Botrytis* spp. (from saprophyte to pathogen). Because *B. cinerea* is one of the most important plant pathogens, its entire genome sequence has been determined (Williamson et al. 2007).

The ecology of *Botrytis* spp. and the epidemiology of the diseases that they cause were reviewed in a previous edition of this textbook (Elad et al. 2004), which followed the textbook published in 1980 by Coley-Smith et al. (1980). In addition, several reviews and reports have addressed the epidemiology of *Botrytis* diseases in specific pathosystems, including grape (Elmer and Michailides 2007; Nair and Nadtotchei 1987; Nair et al. 1988, 1995; Seyb 2004), strawberry (Boff 2001; Boff et al. 2001; Braun and Sutton 1987; Stromeng et al. 2009; Sutton 1990; Sutton et al. 1988), kiwifruit (Michailides and Elmer 2000), pear (Spotts and Cervantes 2001; Spotts and Serdani 2006), boysenberry (Walter et al. 1997, 2004), sweet cherry (Borve and Stensvand 2004; Borve et al. 2005), and onion (Carisse et al. 2009; Sutton 1990).

Because of the wide host range of *Botrytis* spp., reviewing the epidemiology of *Botrytis* diseases in only the most important pathosystems would be a huge task, Nevertheless, such a review would contain repetition, because several of these diseases share common features. *Botrytis* epidemics involve a sequence of events, starting with the production and dispersal of initial inoculum, followed by primary infection and by the production and dispersal of secondary inoculum, and completed by the production of survival structures. This chapter takes a broad approach, with the disease cycle as a conceptual framework for the discussion of *Botrytis* diseases (Fig. 7.1). Hence, this chapter presents the epidemiology of *Botrytis* diseases according to the major components of the life cycle: perennation, inoculum, and infection. In addition, special attention is given to the aerobiology of *Botrytis* spp., as inoculum production and dispersal are key phases in the development of these airborne diseases.

An epidemiology-based approach was chosen because, in most cases, management programs that are based on a thorough knowledge of the epidemiology and disease cycle are more effective (de Wolf and Isard 2007). Each stage in disease development is influenced by environmental conditions such as temperature, rain, humidity, and wind; by cultural practices such as fertilization and fungicide usage; and by host factors such as tolerance and phenology. To illustrate the epidemiology of *Botrytis* diseases, this chapter will look mainly at *Botrytis* grey mould caused by *B. cinerea* as an example of a fruit pathogen (Fig. 7.2) and *Botrytis* leaf blight caused by *B. squamosa* as an example of a leaf pathogen (Fig. 7.3). See also Chap. 2, Fig. 2.1.

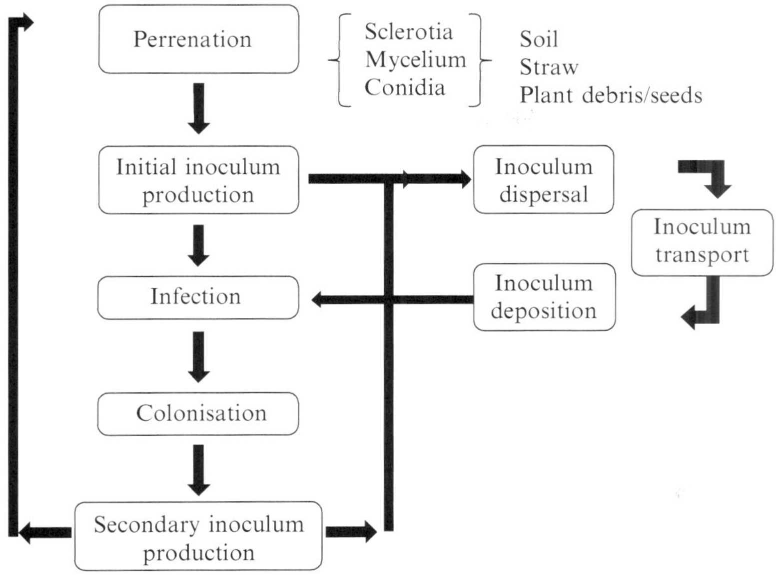

Fig. 7.1 Generic *Botrytis* disease cycle

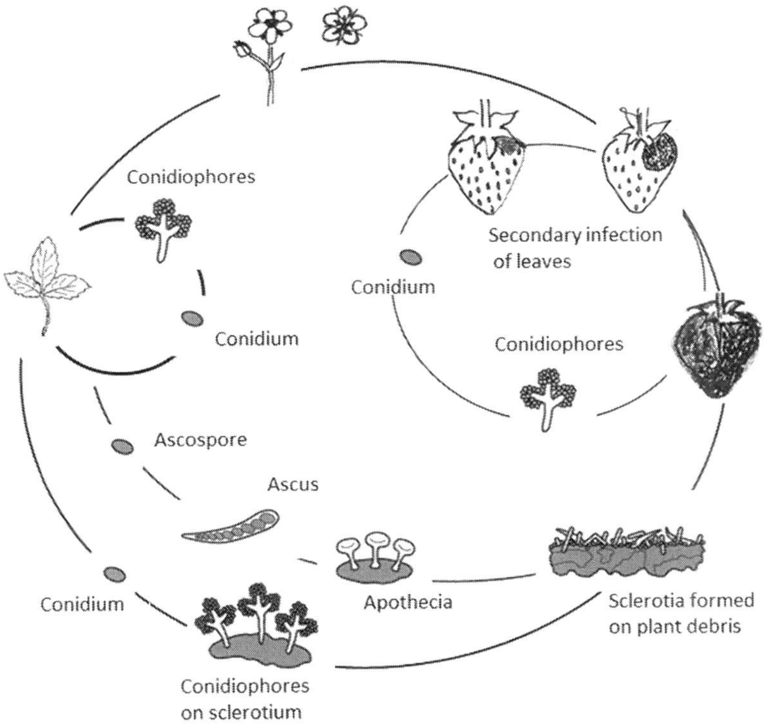

Fig. 7.2 Example of a life cycle of flower-infecting *Botrytis* spp.

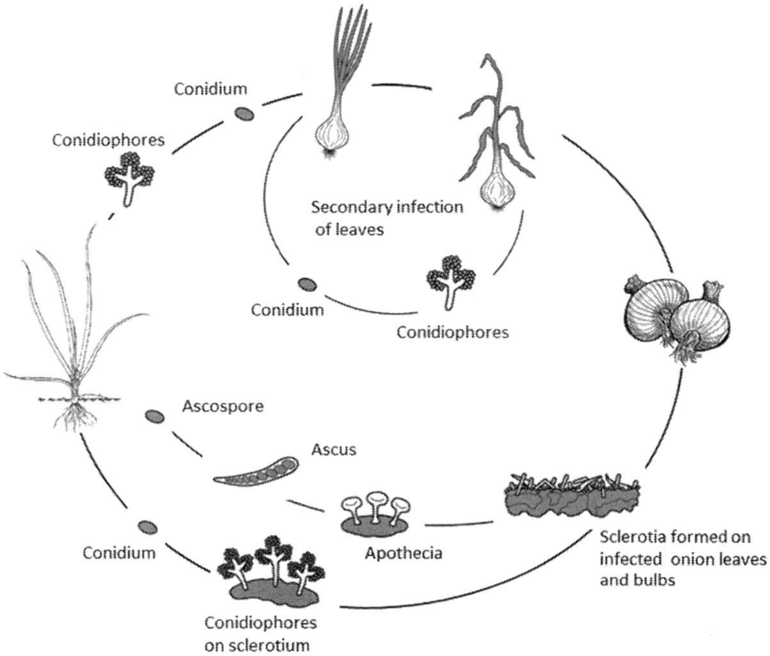

Fig. 7.3 Example of a life cycle of leaf-infecting *Botrytis* spp.

7.1.1 **Botrytis** *Diseases*

Botrytis diseases are among the most common diseases of berry and fruit crops, vegetables, ornamentals (including cut flowers), and some field crops. *Botrytis* diseases are found in various environments, including fields, nurseries, greenhouses, storage rooms, and transit houses. The most common disease is probably fruit rot or grey mould of grapes, berries, fruits, and tomatoes. However, *Botrytis* spp. also cause blossom blight, stem cankers, leaf spot, damping-off, and tuber, corm, bulb, and root rot. Examples of such diseases are grey mould of strawberry, *Botrytis* bunch rot of grape, *Botrytis* leaf blight of onion, onion neck rot, calyx end rot of apple, and bulb rot of gladiolus. Once fruits, berries, vegetables, and flowers have been harvested, *Botrytis* spp. may cause secondary soft rot during storage or transit or even in the marketplace. Damping-off diseases may occur on seedlings grown in cold frames but are also possible in the field when seeds are contaminated with *Botrytis* sclerotia (Fig. 7.4).

Fig. 7.4 Symptoms of *Botrytis cinerea* on raspberry (**a**), strawberry (**b**) grapes (**c**), and tomato stem (**d**), and of leaf spot (**e**) and leaf blight (**f**) on onion leaves caused by *Botrytis squamosa*

7.1.2 Effect on Yield

Despite the worldwide importance of *Botrytis* diseases, reliable estimates of crop losses are scarce in the scientific literature. Elmer and Michailides (2007) reported that internationally, the economic impact of *Botrytis* bunch rot of grape is estimated at up to US$2 billion. Depending on the pathosystem, *Botrytis* spp. may cause direct crop losses when the agricultural product, such as berries, fruits, flowers, or bulbs, is affected to such an extent that it is unmarketable. In Florida, it was found that direct losses due to *Botrytis* strawberry fruit rot could range from 0.5 to 13 % in fungicide-sprayed plots and that in unsprayed plots, pre-harvest losses could reach 35 % and post-harvest losses could be significant (Mertely et al. 2000). Shoemaker and Lorbeer (1971) reported onion yield losses caused by *B. squamosa* ranging from 7 to 30 % in unsprayed plots. Similarly, yield losses of 26 % were reported in the Netherlands (De Visser 1996).

Botrytis spp. also cause indirect losses. In grapes, *Botrytis*-infected berries may be unsuitable for wine production because of a laccase enzyme that is produced by *B. cinerea* and affects the flavour, colour, and storage stability of wine. The presence of *Botrytis* may also dictate that crops be harvested before they have reached the maturity required for commercialization for a particular purpose.

In addition, plant parts infected by *Botrytis* can provide entry for or increase susceptibility to other organisms such as *Penicillium* (blue-green mould), which produces geosmin, a compound that causes strong, damp, earthy aromas in wine. The threat of *Botrytis* may also include economic costs resulting from a longer time required for harvest (related to different sorting requirements) and from the disposal of damaged products. Losses caused by *Botrytis* are often reported in terms of the cost of control. For example, in New Zealand's wetter regions, direct crop losses caused by *Botrytis* bunch rot are up to NZ$5000/ha, while the cost of control is NZ$1500/ha (Hoksbergen 2010).

7.2 Perrenation of *Botrytis* Diseases

Most *Botrytis* spp. overwinter or over-season in the soil as sclerotia in decayed plant material. In general, more sclerotia are produced during wet periods. Sclerotia of *B. squamosa* form on blighted onion leaves near and after harvest, as well as on onion bulb necks, seed stalks, and the inner leaf sheaths in the neck region of bulbs. The sclerotia of different *Botrytis* species vary in size but are all composed of mycelium surrounded by a melanized rind and β-glucans that protect the sclerotia from desiccation, UV radiation, and microbial parasitism (Backhouse and Willets 1984). Even though sclerotia are structures adapted to survive adverse conditions, some conditions may affect survival and only a small proportion of the sclerotia population may maintain viability. Nevertheless, depending on the size of the population, even a small number of surviving sclerotia will generally have considerable inoculum potential.

Survival periods of sclerotia ranging from several weeks to several years were reported for different *Botrytis* spp. (Coley-Smith and Cooke 1971; Entwistle 1987), including 15 months for *B. tulipae* (Coley-Smith and Cooke 1971; Coley-Smith and Javed 1970; Javed 1977), 5–9 months for *B. cinerea* (Thomas et al. 1983), and up to 21 months for *B. squamosa* (Ellerbrock and Lorbeer 1977a). However, survival is influenced by how deeply sclerotia are buried in the soil, with survival percentages of 30 % after 8 months when *B. squamosa* sclerotia overwintered at the soil surface and survival after 21 months of 7 and 66 % when *B. squamosa* sclerotia overwintered at soil depths of 3 or 15 cm, respectively (Ellerbrock and Lorbeer 1977a).

In perennial strawberry, *B. cinerea* overwinters as sclerotia but also as mycelium in leaves, petioles, crop debris, straw mulches, and weeds (Braun and Sutton 1987). Conidia produced from mycelium that overwinters are the main source of primary inoculum (Jarvis 1962a, b, c; Braun and Sutton 1987, 1988). In Scotland, however, Jarvis 1962a, b, c reported that sclerotia were found in barley-straw mulch (bearing sclerotia) and that mummified berries and weeds were important sources of primary inoculum (conidia). For greenhouse crops such as roses, sclerotia are not considered an important source of *B. cinerea* initial inoculum, in part because sclerotia have not been found but mainly because there is a constant, year-round supply of infected, diseased tissue (Araujo et al. 2005; Raposo et al. 2001). In organic soil, Ellerbrock and Lorbeer (1977b) reported that *B. squamosa* mycelium does not survive in onion leaf debris. However, mycelium of that species was found to survive for up to 17 months in association with onion seeds, although the percentage of infected seeds was low, with a maximum of 6 %. In warm regions survival during summer is usually as mycelium in plant debris (Yunis and Elad 1989).

7.3 Production of Inoculum

As for most fungal diseases, inoculum is a key component of *Botrytis* epidemics, because the inoculum source, amount, and type significantly influence the onset of an epidemic, the rate of disease progress and, indirectly, yield losses. *Botrytis* inoculum can be divided into initial (or primary) and secondary inocula, where initial inoculum is the inoculum that has survived the winter or non-cropping period and that triggers new epidemics, and secondary inoculum is the inoculum produced as a result of the completion of the first infection–conidiation cycle (Fig. 7.1).

7.3.1 Initial Inoculum Production

For most *Botrytis* spp., sclerotia can undergo carpogenic germination: sclerotia are spermatized, and apothecia that contain asci are produced, with each ascus containing eight binucleate ascospores. However, ascospores are considered a minor source of primary inoculum. In eastern Canada, apothecia of *B. fuckeliana* are reported to be observed rarely in strawberry fields (Braun and Sutton 1987). In addition, under field conditions from November to June, non-spermatized sclerotia of *B. cinerea* were found to fail to produce apothecia, although apothecia could be produced from several crosses under laboratory conditions (Braun and Sutton 1987). In most instances, initial inoculum is made up of conidia that formed from myceliogenic germination of sclerotia, that formed directly from mycelium, or that survived the non-cropping period.

For several *Botrytis* spp., such as *B. squamosa*, several cohorts of conidia are produced on each sclerotium, and thus initial inoculum is produced over an extended period from the spring until the early summer (Ellerbrock and Lorbeer 1977b). Similarly, *B. cinerea* sclerotia conidiate continuously for up to 12 weeks after the production of the first cohort of conidia (Nair and Nadtotchei 1987). When sclerotia germination occurs, temperature is generally not a limiting factor; for example, *B. squamosa* conidia are produced on sclerotia at temperatures ranging from 3 to 27 °C. In the absence of onions, *B. squamosa* conidia were found to survive for up to 2 months in natural soil under both controlled and field conditions. However, high temperatures reduced survival, and the amount of conidia declined rapidly in soil that was alternately dried and remoistened.

7.3.2 Secondary Inoculum Production

The formation of conidia is stimulated by specific wavelengths of light (Epton and Richmond 1980), although some isolates conidiate in darkness. For example, *B. squamosa* conidia are produced on necrotic tissues primarily at night, when onion leaves are wet for at least 12 h and the mean temperature during the period of leaf wetness is between 8 and 22 °C (Alderman and Lacy 1984a). Conidiation occurs during shorter periods of leaf wetness only if the preceding day was humid (relative humidity above 70 % for at least 6 h) and if the necrotic tissues are moist. In the *B. cinerea*–grape pathosystem, conidiation on grape rachides occurs at temperatures of 10–30 °C, provided that the rachides are wet for a minimum of 3 h. Conidiation increases as the length of the wetness period increases, with optimum conidiation after 36 h at 10 °C and 24 h at 25 °C. On strawberry, the optimum temperature for conidiation on wet dead leaves is around 18 °C, and only a few conidia are produced at 30 °C.

7.3.3 Inoculum Sources and Dispersal

The scientific literature contains little information on *Botrytis* spp. inoculum dispersal. One explanation for that paucity of information is the ubiquitous nature of *B. cinerea* and the general assumption that inoculum is always present, even though that is not the case for all crops or for all *Botrytis* spp. Nevertheless, from an epidemiological standpoint, inoculum dispersal is an essential step in *Botrytis* disease progress, as dispersal ensures population survival, disease progress through the colonization of new host tissues (healthy areas), and reproduction. For most *Botrytis* spp., inoculum is made up of conidia, and dispersal can be broken down, as for most airborne diseases, into release, transport, and deposition on host tissues.

7.3.4 Inoculum Release

The release of conidia is regulated by a hygroscopic mechanism as well as by leaf vibrations, air movements, and rain splashes (Jarvis 1962a, b, c, 1980b; Sutton 1990; Thomas et al. 1988). For most *Botrytis* spp., there is a circadian periodicity in airborne conidia concentration: the maximum occurs at about midday, which corresponds to the highest wind speed and turbulence near the ground (Fig. 7.5). In general, this periodicity is correlated positively with increasing temperature and wind velocity and negatively with increasing percentage of relative humidity (RH) and the presence of dew. In raspberry plantings, the maximum numbers of conidia trapped during the day were found to coincide with increasing temperatures and decreasing RH and were sometimes followed by a smaller peak in the evening (Jarvis 1962b, c). Dispersal occurs mostly during the day but also happens at night in conjunction with rain (Jarvis 1962a). Lorbeer (1966) reported that 80 % of airborne *B. squamosa* conidia were observed between 8:00 and 16:00, an observation that was supported by the report by Sutton et al. (1978) that most conidia were observed between 9:00 and 12:00.

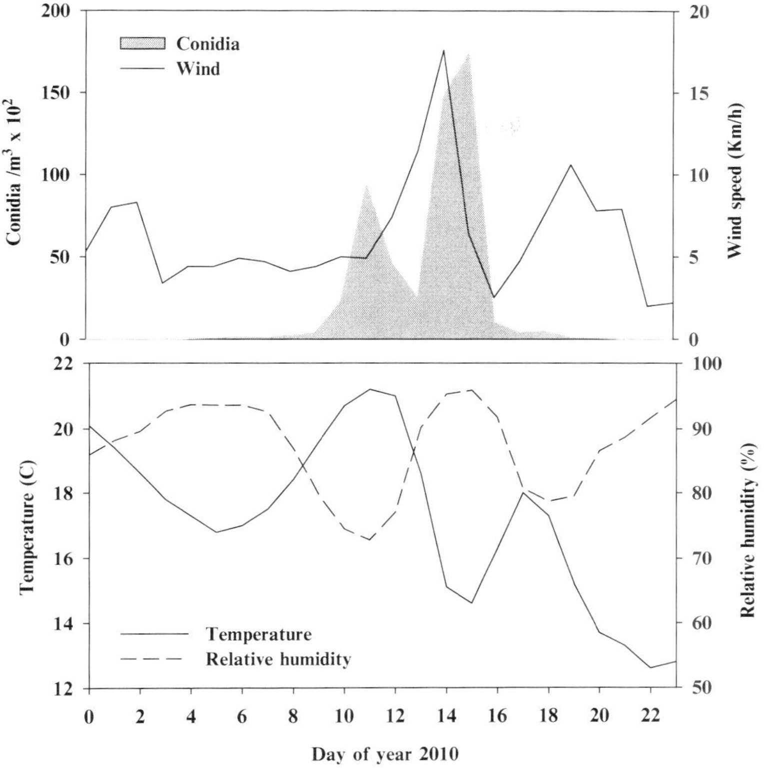

Fig. 7.5 Periodicity of airborne concentration of *Botrytis cinerea* conidia above a strawberry field monitored using rotating arm sampler and a qPCR assay for quantification of conidia

The release of *Botrytis* conidia from conidiating material is influenced by wind speed, and the removal of conidia generally increases with increasing wind speed. However, depending on how strongly conidia are attached to conidiophores, not all conidia may be removed from conidiating material at commonly occurring wind speeds (Harrison and Lowe 1987). The release of conidia is also influenced by the size of the spores or spore clumps, which affects the drying rate required for release. Infected wind-blown flower petals containing *B. cinerea* mycelium serve as dispersal propagules (Johnson and Powelson 1983).

7.3.5 Inoculum Transport

The next step following the release of conidia from the source is transport of the inoculum. The size and weight of *Botrytis* conidia suggest that the scale of aerial dispersal is probably measurable in kilometres. Despite this dispersal potential, most inoculum comes from within the field for perennial crops and from nearby sources for annual crops, because the dead tissues of numerous plants, seeds, and soil are potential sources of inoculum (Coley-Smith 1980; Jarvis 1980a, b).

Inoculum is transported either within or outside the plant canopy. Harrison and Lowe (1987) reported that the dispersal of *B. fabae* and *B. cinerea* increased with increasing wind speed up to 2.8 km/h. However, the proportion of conidia that escape the canopy, assuming that the source is within it, depends on the equilibrium between deposition and turbulent transport and on the vertical position of the inoculum source. When conidia are produced on a source near the ground or in the lower canopy, they are exposed to slow wind speeds, low turbulence, and rapid rates of sedimentation, conditions that are conducive to short-distance transport. Chastagner et al. (1978) reported that in tomato fields, wind velocities above the canopy of up to 9.6 km/h corresponded to wind velocities within the canopy, where stem cankers are located, of less than 0.5 km/h. The same author reported that grey mould was observed within 8 m of the inoculum source. A similar pattern was reported for *B. cinerea* dispersal in snap bean fields (Johnson and Powelson 1983). In a study using laboratory-grown inoculum placed on the ground, inoculum dispersal was monitored as the number of viable *B. cinerea* conidia washed from bean foliage. Overall, the results showed that dispersal during blooming was limited to 3 m from the source. Similar observations were reported for grapevines in New Zealand: Seyb (2004) reported that 95 % of airborne *B. cinerea* conidia landed within 1.6 m of their source, and thus the most important source of inoculum is within the vine canopy in close proximity to grape bunches.

These observations are supported by the results of a recent study on the relative importance of trash type as a source of inoculum for *Botrytis* bunch rot of grapes caused by *B. cinerea*. Jaspers et al. (2013) reported that the most important source of *B. cinerea* inoculum was infected rachides, followed by tendrils, petioles, and cane pieces, which had mean conidiation potentials of 1.0×10^4, 5.6×10^3, 3.3×10^3, and 1.7×10^3 conidia per quadrat (0.25 m^2), respectively. The same authors studied the conidiation potential of necrotic grape tissue over time and showed that potential diminished from flowering to harvest, with mean conidia numbers decreasing from 3.9×10^5 to 2.7×10^3 per ground rachis and from 3.5×10^5 to 2.6×10^4 per canopy rachis. Similarly, although strawberry leaves infected by *B. cinerea* are generally symptomless, they serve as an inoculum reservoir, and large amounts of inoculum could be produced for several months on these infected leaves when they senesce and die (Braun and Sutton 1987). In addition to leaves, other important sources of inoculum are senescent sepals, petals, and stamens (Powelson 1960). These reports are supported by a recent study analyzing airborne *B. cinerea* inoculum curves from strawberry, raspberry, and grape plantings (Carisse et al. 2015). That quantitative analysis showed that inoculum is continuously produced and released and that the rate of inoculum progress varies depending on the crop (Carisse et al. 2015).

Following release and dispersal, *Botrytis* conidia become airborne and can be monitored (Carisse et al. 2015). In flower-infecting *Botrytis* pathosystems, the seasonal progress of airborne *B. cinerea* in berry plantings is characterized generally by an airborne conidia concentration that is initially low but then increases during fruit ripening and until the last fruit harvest (Fig. 7.6). In these systems, the airborne conidia concentration is curtailed by harvest (Carisse et al. 2015). In *Botrytis* leaf blight systems, however, the airborne conidia concentration increases logarithmically with the accumulation of dead leaf tissue in the crop unless the soil temperature is above 30 °C (Carisse et al. 2011).

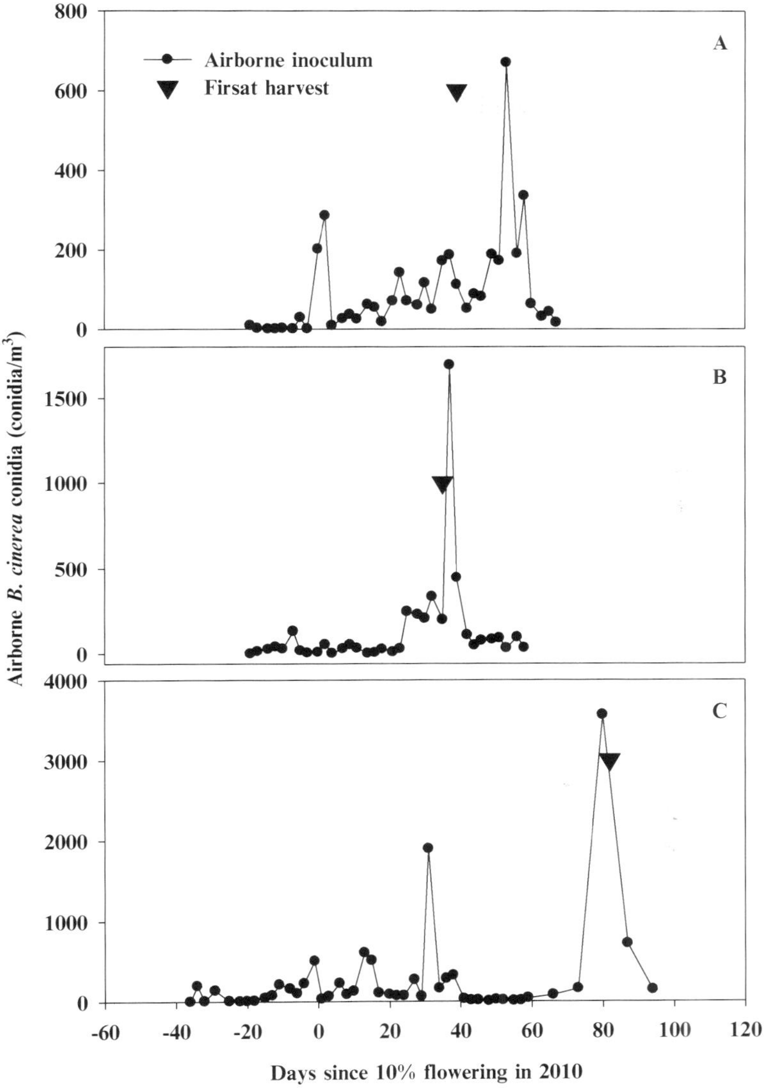

Fig. 7.6 Dynamics of *B. cinerea* airborne conidia monitored in raspberry (**a**), strawberry (**b**), and grape (**c**) plantings in Frelighsburg, Quebec, Canada, in 2010

7.3.6 Inoculum Deposition

Most of what is known about the deposition of *Botrytis* spp. conidia was derived from controlled or simulated wind conditions experiments, with single conidia or clumps consisting of a few conidia each (Harrison and Lowe 1987). Knowledge on the deposition of airborne conidia under uncontrolled or field conditions on different plant surfaces such as leaves, shoots, and fruits is scarce. *B. cinerea* inoculum is deposited on grape berries mostly as single airborne conidia (Jarvis 1962a, b, c). That observation suggests that infection by solitary conidia, and not by conidial clusters, likely plays a major role in the epidemiology of *Botrytis* diseases. However, some conidia are dispersed either within or at the surface of raindrops.

Spotts and Holz (1996) showed that conidia stick more strongly when they are applied in water suspension or to the wet surface of a grape berry than when they are dry and are applied to a dry surface. The results of these studies with dry and wet *Botrytis* conidia suggest that the types of inoculum (wet or dry) influence not only conidial germination and growth on plant surfaces but also the resulting symptom development. For example, on gerbera flowers, typical necrotic lesions such as those found in commercial productions were obtained only following inoculation with dry *B. cinerea* conidia (Salinas et al. 1989). On grape berries, a 14-h wetness period was required for 63 % disease incidence at 23 °C, whereas dry conidia did not produce any symptoms on berries even after 96 h of wet conditions (Coertze and Holz 1999; Coertze et al. 2001). Based on these finding, it is expected that under humid or wet conditions, airborne conidia have an equal potential to infect dry and wet berry surfaces.

7.4 Infection, Colonization, and Symptom Development

7.4.1 Infection Pathways

For *B. cinerea*, floral organs are important infection courts. Depending on the host, *B. cinerea* can infect various flower organs. For example, in blackcurrant, *B. cinerea* infects flowers through the style or the carpels (McNicol and Williamson 1989), whereas in pear, grape, and strawberry, the style may not be an important penetration site (Bristow et al. 1986; De Kock and Holz 1992; Holz et al. 2003). In pear, the most important infection sites are the stamens. In grape, conidia infect mainly the flower receptacle and, to a lesser extent, the stigma and style (Viret et al. 2004). For several *Botrytis* spp., conidia can penetrate undamaged host tissues directly or enter through natural openings. This infection pathway is common to several *Botrytis* leaf blight pathogens such as *B. squamosa*. That species penetrates onion leaves through stomata or the cuticle but, like many necrotrophs, produces lesions more readily on old leaves than on young ones (Alderman and Lacy 1984a, 1984b). On Oriental lily leaves, *B. elliptica* conidia were found to germinate on both the adaxial and abaxial surfaces, but penetration occurred only on the abaxial surface (Hsieh et al. 2001). Similarly, *B. cinerea* can penetrate bean leaves, plum fruits, and nectarine fruits directly. Lastly, most *Botrytis* spp. can infect their host through wounded tissues. Following penetration, *B. cinerea* often remains in a quiescent state for varying periods of time before rotting leaves, flowers, or mature fruits.

7.4.2 Influence of the Environment on Infection

Regardless of the infection route, the germination of a *Botrytis* conidium is influenced by the presence of free surface water or high relative humidity (above 93 %), which is essential for both germination and penetration. Shoots, flowers, leaves, and fruits can become infected by *B. cinerea* under a wide range of temperature and humidity conditions, with optimal infection generally occurring at 15–20 °C, provided that free water or RH above 90 % is present for at least 4 h. Slow drying conditions in conjunction with high RH (above 90 %) favour the development of *Botrytis*. As for several other fungal pathogens, there is an interaction between temperature and length of the wet or humid period: the wet period required becomes longer as the temperature departs further from the optimum.

Under controlled conditions, *B. cinerea* can infect grape berries at temperatures from 12 to 30 °C, but infection was found to develop in only 9 and 37 % of berries after 4 h of berry wetness at 12 and 30 °C, respectively. The incidence of infected berries was found to increase with increasing duration of wetness, up to 90 % at 12–20 °C and 24 h of wetness (Broom et al. 1995). The optimum temperature for strawberry flower infection is around 20 °C, provided that the flowers are wet for 24 h (Bulger et al. 1987). Significantly fewer infected flowers were observed at temperatures below 15 °C or above 25 °C, and the incidence of infected flowers increased with increasing duration of flower wetness.

The infection of onion leaves by *B. squamosa* is influenced by temperature, duration of leaf wetness, epicuticular wax, leaf age, and conidia concentration. In controlled experiments, regardless of the duration of leaf wetness, the number of lesions increased as the temperature increased from 10 to 20 °C, but the number then decreased at 25 °C, and only a few lesions developed at 30 °C. Between 10 and 25 °C, the number of lesions per square centimetre of leaf area was found to increase gradually as the duration of leaf wetness increased from 12 to 72 h (Carisse et al. 2012). When the leaf wetness period is interrupted, the number of lesions is reduced proportionally to the length of the dry period, and as little as 0.3–1.7 h of dryness was found to be sufficient to significantly reduce the number of lesions produced (Alderman et al. 1985). Dry conidia applied to dry onion leaves survived for up to 2 days in the absence of leaf wetness and retained the potential to induce lesions when the leaves were subsequently wetted (Alderman et al. 1985).

7.4.3 Symptom Development

In perennial strawberry, *B. cinerea* infects young leaves as they appear in the spring. The pathogen then remains latent and, as the leaves senesce and die, colonizes the tissues and conidiates. These conidia are an important source of inoculum for flower infection. Following flower infection, the rate at which *B. cinerea* colonizes fruits depends on temperature and fruit maturity. *B. cinerea* conidiates profusely on diseased fruits, and the conidia formed on these tissues contribute to fruit rot epidemics in day-neutral and annual strawberries (Fig. 7.4b). Unlike in grape, direct infection of strawberry and raspberry fruits by conidia is not considered important (Elmer and Michailides 2007; Marois et al. 1986), although healthy fruit could be infected by contact with adjacent infected fruits.

After *B. squamosa* penetrates onion leaves, colonization is limited to a small area around the infection site, resulting in small lesions that appear 24–48 h after inoculation. These lesions are whitish and 1–5 mm in length and are usually bordered by a greenish-white halo that first appears water-soaked (Fig. 7.4e). This phase of disease development is often called the "leaf-spotting phase." As the leaf ages, *B. squamosa* progressively invades the leaf's tissues. Under optimum conditions, partial or complete leaf blighting occurs within 5–12 days after initial lesion development (Fig. 7.4f). Prolonged leaf wetness, high temperatures, and increased leaf age favour colonization and leaf dieback (Alderman and Lacy 1983; Shoemaker and Lorbeer 1977). Blighted leaves are the main sites for secondary conidia production. The optimum temperature for disease development is 18–20 °C, with a maximum temperature for lesion development of 24 °C.

7.4.4 Relationship between Airborne Inoculum and Disease Development

The influence of inoculum on disease progress or crop yield has been studied mostly for soil-borne diseases, and very little is known about the relationship between *Botrytis* inoculum and disease development. The interaction between inoculum production and dispersal on one hand and infection on the other is poorly understood for most *Botrytis* diseases. The influence of airborne conidia concentration in a tomato greenhouse on both flower and stem-wound infections by *B. cinerea* was studied (Carisse et al. 2015). Regardless whether the temperature was 15, 20, or 25 °C, the proportion of infected flowers remained low when the airborne conidia concentration was below 10 conidia/m^3. Above that level, flower infection increased with increasing airborne conidia concentration. For stem infections, no infected wounds were observed when the airborne conidia concentration was below 100 conidia/m^3. Above that level, infection increased linearly with increasing *B. cinerea* airborne conidia concentration (Carisse et al. 2015).

Xu et al. (2000) studied the relationship between the incidence of strawberry flower infection by *B. cinerea*, meteorological data, and numbers of airborne conidia. These data were used to develop mathematical models describing flower infection from weather data only, inoculum data only, and both types of data. Models using both weather variables and inoculum data gave the best predictions; however, including inoculum in the models resulted in only a small improvement over the models based on weather variables only. The weather variables that best predicted flower infection were daytime vapour pressure deficit and night time temperature: infection was favoured by a low daytime vapour pressure deficit and a high nighttime temperature. In a similar experiment, Blanco et al. (2006) reported a positive but non-significant correlation between incidence of flower infection and airborne conidia concentration. However, a positive significant correlation between *Botrytis* fruit rot incidence and accumulated number of conidia over 7 days was observed (Blanco et al. 2006).

Carisse et al. (2005) reported a linear relationship between *B. squamosa* airborne conidia concentration and number of lesions per onion leaf. Concentrations of 10–15 and 25–35 conidia/m^3 were associated with 1 and 2.5 lesions per leaf, respectively. A significant spatial association between *B. squamosa* airborne conidia concentration and lesion density measured after various lag times was reported at the scale of an onion field (Carisse et al. 2008b). This relationship was used to design risk indicators for the timing of fungicide applications to manage onion leaf blight (Carisse et al. 2008a; Van Der Heyden et al. 2012).

7.5 Analysis of *Botrytis* Disease and Inoculum Progress Curves

An improved understanding of epidemic processes should guide the design of effective disease management programs. Hence, a thorough understanding of the epidemiology and aerobiology of *Botrytis*-induced diseases should be the foundation of integrated management so that the various control practices available can be assessed. Considering that each control practice, including physical, cultural, genetic, biological, and chemical ones, has its specific mode of action, it is essential to apply these practices optimally according to the epidemic processes. For example, biological control agents that antagonize sclerotia will have an effect on the survival of sclerotia and the production of initial inoculum, resulting in delayed epidemic development. Synthetic fungicides that inhibit conidia germination will prevent infection, whereas those that affect conidiation will influence the reproduction rate (production of secondary inoculum), resulting in a slower rate of disease progress.

Host genetic resistance may affect several processes such as infection, colonization, and length of the latency period and thus may reduce the rate of disease progress. All control practices affect the progress of the disease, although their efficiency varies. Hence, the analysis of disease progress curves can be used to evaluate the efficacy of individual or combined control practices, the level of host resistance, the effectiveness of fungicides, or the appropriateness of tactical and strategic management decisions (Van der Plank 1963).

Several parameters are used to describe and compare disease progress curves, the most common parameters being time of disease onset, initial amount of disease, rate of disease progress, area under the disease progress curve, shape of the disease progress curve, maximum and final amount of disease, and overall duration of the epidemic. The analysis of disease progress curves implies that disease is assessed at several times during the course of a growing season. That method is not always possible for *Botrytis* grey mould, because diseased fruits are often present only during a short period of time and damage is commonly measured at harvest as the percentage of diseased fruits. Although this method allows yield losses to be evaluated, it is of limited use for comparing epidemics. One way to circumvent this limitation is to "force" the expression of disease symptoms. Xu et al. (2000) collected strawberry flowers and incubated them under optimum laboratory conditions for disease expression. The *Botrytis*-infected flowers were then assessed and expressed as the percentage of infected flowers. Nevertheless, during the few weeks when mature fruits are present, disease progress curves can be built (Fig. 7.7). For example, in June-bearing strawberry, there are significant differences in the area under the *Botrytis* progress curve and in the rate of disease progress measured in plantings with different cultivars grown under the same conditions (Fig. 7.7).

For leaf diseases caused by *Botrytis* spp. such as *B. squamosa*, building disease progress curves is easier, because diseased leaves are potentially present all season long. Van Der Heyden et al. (2012) compared three indicators for the initiation of fungicide spray programs for managing *Botrytis* leaf blight of onion. The spray programs were initiated when the first *B. squamosa* airborne conidia were detected, when the cumulative airborne conidia concentration reached 15 conidia/m^3, or when the first lesions were detected. The interval between subsequent fungicide applications was determined by taking into account conidiation potential, airborne conidia concentration, and lesion density. The effectiveness of the spray programs was evaluated based on observed maximum disease severity, area under the disease progress curve, and rate of disease progress. The fungicide spray programs initiated when the first *B. squamosa* airborne conidia were detected resulted in the lowest maximum disease severity and the lowest area under the disease progress curve. In that study, initiating fungicide sprays when the first conidia were detected prevented disease severity from reaching the critical level for curative fungicide applications.

Airborne inoculum progress curves similar to curves for disease progress can be built (Carisse et al. 2015). Advances in molecular biology mean than airborne inoculum can now be monitored in a timely fashion (Carisse et al. 2009). For a recent analysis of *B. cinerea* airborne inoculum curves, airborne conidia concentrations were monitored over a few years in strawberry, raspberry, and grape plantings that were not being managed for grey mould (Carisse et al. 2015). Airborne inoculum was always present, a fact that is suggestive of successive infection–conidiation cycles. Overall, the absolute rate of inoculum progress (proportional increase in airborne inoculum per one unit of time) was smaller in the grape plantings than in the raspberry or strawberry plantings, with mean values of 0.008, 0.014, and 0.047, respectively. Over the course of that study, the grape plantings showed a wide range of variation in the amount of airborne inoculum at the beginning of the flowering period (10 % flowering) as well as in the period when epidemics reached 50 % of the maximum (Carisse et al. 2015). These results suggest that epidemics of *Botrytis* bunch rot of grape differ from 1 year to another and hence that monitoring of inoculum could be used as a disease risk indicator. These inoculum progress descriptors could be used to critically compare *B. cinerea* epidemics.

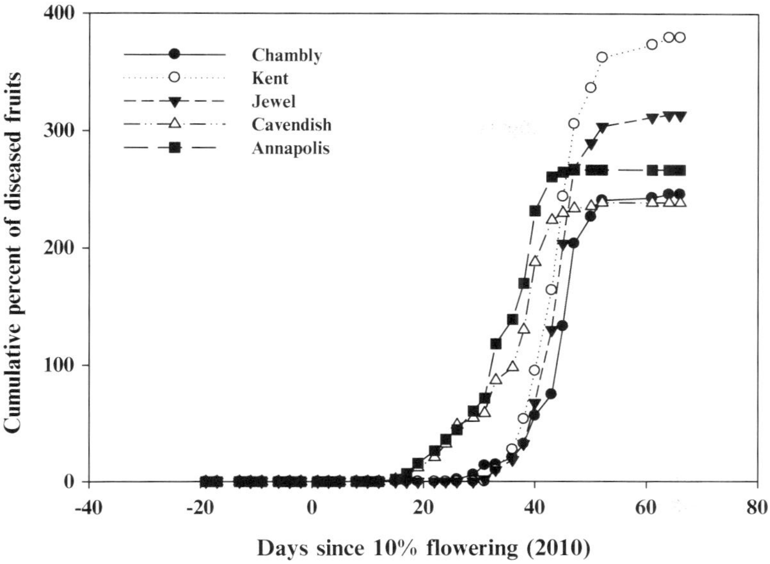

Fig. 7.7 Progress of *Botrytis* fruit rot in strawberry plantings with various cultivars at the Agriculture and Agri-Food Canada experimental farm in 2010. The area under the disease progress curves was 280.0, 480.5, 389.0, 272.0, 306.5, for the cultivars Chambly, Kent, Jewel, Cavendish, and Annapolis, respectively. The rate of disease progress was determined by fitting the Richards model (Carisse et al. 2015) and was 0.25, 0.23, 0.28, 0.15, and 0.17, for the cultivars Chambly, Kent, Jewel, Cavendish, and Annapolis, respectively

7.6 Sampling for *Botrytis*

Improving epidemiological knowledge in order to evaluate management practices requires more monitoring tools. Accurate quantification depends on the availability of both monitoring tools and sampling procedures. Improved detection and quantification tools for *B. cinerea* and *B. squamosa* have been developed during the past decade (Carisse et al. 2009, 2015) and can be used to detect *B. cinerea* in an active or quiescent state in various plant materials or to quantify airborne inoculum. Only a few sampling procedures are available, however, with the exception of some for *Botrytis* leaf blight of onion. The presence of an action threshold for timing either the first fungicide spray or the interval between sprays probably motivated the development of sampling procedures.

In 1984, Boivin and Sauriol (1984) studied the spatial distribution of *B. squamosa* lesions on onion leaves and concluded that the population of lesions was aggregated and that the spatial pattern was not influenced by fungicide treatments. Based on the pattern of lesion distribution within onion fields, a sequential sampling procedure was developed with a threshold of one-half lesion per leaf. Under that procedure, on average 54 leaves needed to be assessed before it could be determined if the action threshold was reached or not. That sampling procedure was improved on by Vincelli and Lorbeer (1987) so that disease level could be estimated based on a sample size of only 15–50 onion leaves.

More recently, Carisse et al. (2008b) studied the spatial and temporal relationships between *B. squamosa* airborne inoculum and *Botrytis* blight intensity in both experimental fields that were not being managed for *Botrytis* blight and commercial fields that were being managed according to standard grower practices. At the beginning of the onion-growing season, lesion density was randomly distributed, but the level of aggregation increased as disease progressed. For most sampling dates, a significant spatial correlation between airborne inoculum and lesion density assessed 1 week after inoculum assessment was observed in both the unmanaged and managed onion fields. Using a computer simulation with 1 % and 3 % berry disease incidence, different clustering levels (random, low, and high), and small (100 bays) or large (2,500 bays) hypothetical vineyard sizes, McKay et al. (2012) evaluated various sampling procedures for *Botrytis* bunch rot of grape. The sampling procedures were evaluated based on sample size and sample efficiency, and the results suggest that inverse sampling and simple random sampling could be used to accurately and efficiently estimate the amount of *Botrytis* bunch rot in vineyards.

7.7 Conclusion

Botrytis diseases develop as the result of the successful completion of several processes (conidiation, dispersal, and infection) that are influenced by a number of biological, environmental, and agricultural factors. Consequently, *Botrytis* spp. display a wide variety of behaviours and interactions with a broad range of hosts and conditions. Even with a much-studied, cosmopolitan plant pathogen such as *B. cinerea*, improved knowledge of epidemiology is required. We do not fully understand specific aspects of the epidemiology of *Botrytis* spp., such as the spatial distribution of disease and inoculum (including sclerotia and initial and secondary inoculum) and disease progress (especially in some minor crops such as sweet basil).

References

Alderman SC, Lacy ML (1983) Influence of dew period and temperature on infection of onion leaves by dry conidia of *Botrytis squamosa*. Phytopathology 73:1020–1023

Alderman SC, Lacy ML (1984a) Influence of temperature and moisture on growth and sporulation of *Botrytis squamosa*. Can J Bot 62:2793–2797

Alderman SC, Lacy ML (1984b) Influence of leaf position and maturity on development of *Botrytis squamosa* in onion leaves. Phytopathology 74:1461–1463

Alderman SC, Lacy MI, Everts KL (1985) Influence of interruptions of dew period on numbers of lesions produced on onion by *Botrytis squamosa*. Phytopathology 75:808–811

Araújo AE, Maffia LA, Mizubuti ESG et al (2005) Survival of *Botrytis cinerea* as mycelium in rose crop debris and as Sclerotia in soil. Fitopatol Bras 30:516–521

Backhouse D, Willetts HJ (1984) A histochemical study of sclerotia of *Botrytis cinerea* and *Botrytis fabae*. Can J Microbiol 30:171–178

Boff P (2001) Epidemiology and biological control of grey mould in annual strawberry crops. PhD thesis, Wageningen University, Wageningen, The Netherlands, 128 p. doi.org/10.1590/S0100-41582003000100011

Boff P, Kastelein P, De Kraker J et al (2001) Epidemiology of grey mould in annual waiting-bed production of strawberry. Eur J Plant Pathol 107:615–624

Boivin G, Sauriol P (1984) Dispersion statistics and sequential sampling plan for leaf blight caused by *Botrytis squamosa* in onions. Phytopathology 74:1385–1387

Børve J, Stensvand A (2004) Non-abscised aborted sweet cherry fruits are vulnerable to fruit decaying fungi and may be sources of infection for healthy fruits. Acta Agric Scand Sect B Soil Plant Sci 54:31–36

Børve J, Flatland S, Vangdal E, Stensvand A (2005) Non-abscised aborted fruits are a source of inoculum for fungal pathogens causing fruit decay in sweet cherry. Acta Horticult 667:499–501

Braun PG, Sutton JC (1987) Inoculum sources of *Botrytis cinerea* in fruit rot of strawberries in Ontario. Can J Plant Pathol 9:1–5

Braun PG, Sutton JC (1988) Infection cycles and population dynamics of *Botrytis cinerea* in strawberry leaves. Can J Plant Pathol 10:133–141

Bristow PR, McNicol RJ, Williamson B (1986) Infection of strawberry flowers by *Botrytis cinerea* and its relevance to grey mould development. Ann Appl Biol 109:545–554

Broome JC, English JT, Marois JJ et al (1995) Development of an infection model for Botrytis bunch rot of grapes based on wetness duration and temperature. Phytopathology 85:97–102

Bulger MA, Ellis MA, Madden LV (1987) Influence of temperature and wetness duration on infection of strawberry flowers by *Botrytis cinerea* and disease incidence of fruit originating from infected flowers. Phytopathology 77:1225–1230

Carisse O, Van Der Heyden H (2015) Relationship of airborne *Botrytis cinerea* conidium concentration to tomato flower and stem infections. Plant Dis 99:137–142

Carisse O, McCartney HA, Gagnon JA, Brodeur L (2005) Quantification of airborne inoculum as an aid in the management of leaf blight of onion caused by *Botrytis squamosa*. Plant Dis 89:726–733

Carisse O, McRoberts N, Brodeur L (2008a) Comparison of monitoring- and weather-based risk indicators of botrytis leaf blight of onion and determination of action thresholds. Can J Plant Pathol 30:442–456

Carisse O, Savary S, Willocquet L (2008b) Spatiotemporal relationships between disease development and airborne inoculum in unmanaged and managed botrytis leaf blight epidemics. Phytopathology 98:38–44

Carisse O, Tremblay DM, Lévesque CA et al (2009) Development of a TaqMan real-time PCR assay for quantification of airborne conidia of *Botrytis squamosa* and management of botrytis leaf blight of onion. Phytopathology 99:1273–1280

Carisse O, Tremblay DM, McDonald MR, Brodeur L, McRoberts N (2011) Management of botrytis leaf blight of onion: the Quebec experience of 20 years of continual improvement. Plant Dis 95:504–514

Carisse O, Levasseur A, Van der Heyden H (2012) A new risk indicator for botrytis leaf blight of onion caused by *Botrytis squamosa* based on infection efficiency of airborne inoculum. Plant Pathol 61:1154–1164

Carisse O, Tremblay DM, Lefebvre A (2015) Comparison of *Botrytis cinerea* airborne inoculum progress curves from raspberry, strawberry and grape plantings. Plant Pathol 63:983–993

Chastagner GA, Ogawa JM, Manji BT (1978) Dispersal of *Botrytis cinerea* in tomato fields. Phytopathology 68:1172–1176

Coertze S, Holz G (1999) Surface colonization, penetration, and lesion formation on grapes inoculated fresh or after cold storage with single airborne conidia of *Botrytis cinerea*. Plant Dis 8:917–924

Coertze S, Holz G, Sadie A (2001) Germination and establishment of infection on grape berries by single airborne conidia of *Botrytis cinerea*. Plant Dis 85:668–675

Coley-Smith JR (1980) Sclerotia and other structures of survival. In: Coley-Smith JR, Verhoeff K, Jarvis WR (eds) The biology of *Botrytis*. Academic, London, pp 85–114

Coley-Smith JR, Cooke RC (1971) Survival and germination of fungal sclerotia. Annu Rev Phytopathol 9:65–92

Coley-Smith JHR, Javed ZUR (1970) Testing the viability of fungal sclerotia. Ann Appl Biol 65:59–63

De Kock SL, Holz G (1992) Blossom-end rot of pears: systemic infection of flowers and immature fruit by *Botrytis cinerea*. J Phytopathol 135:317–327

de los Blanco C, Santos B, Romero F (2006) Relationship between concentrations of Botrytis cinerea conidia in air, environmental conditions, and the incidence of grey mould in strawberry flowers and fruits. Eur J Plant Pathol 114:415–425

De Visser CLM (1996) Field evaluation of a supervised control system for botrytis leaf blight in spring sown onions in the Netherlands. Eur J Plant Pathol 102:795–805

Dean R, Van Kan JAL, Pretorius ZA et al (2012) The top 10 fungal pathogens in molecular plant pathology. Mol Plant Pathol 13:414–430

DeWolf ED, Isard SA (2007) Disease cycle approach to plant disease prediction. Annu Rev Phytopathol 45:203–220

Elad Y, Williamson B, Tudzynski P, Delen N (2004) *Botrytis* spp. and diseases they cause in agricultural systems – an introduction. In: Elad Y, Williamson B, Tudzynski P, Delen N (eds) *Botrytis*: biology, pathology and control. Kluwer Academic Publishers, Dordrecht, pp 1–8

Ellerbrock LA, Lorbeer JW (1977a) Survival of sclerotia and conidia of *Botrytis squamosa*. Phytopathology 67:219–225. doi: 0.1094/Phyto-67-219.

Ellerbrock LA, Lorbeer JW (1977b) Sources of primary inoculum of *Botrytis squamosa*. Phytopathology 67:363–372. doi: 10.1094/Phyto-67-363.

Elmer PAG, Michailides TJ (2007) Epidemiology of *Botrytis cinerea* in orchard and vine crops. In: Elad Y, Williamson B, Tudzynski P, Delen N (eds) Botrytis: biology, pathology and control. Kluwer Academic Publishers, Dordrecht, pp 243–272

Entwistle AR (1987) Inoculum production and survival in fungi which form sclerotia. In: Pegg GF, Ayres GA (eds) Fungal infection of plants. Symposium of the British Mycological Society, Cambridge, pp 349–365

Epton HAS, Richmond DV (1980) Formation, structure and germination of conidia. In: Coley-Smith JR, Verhoeff K, Jarvis WR (eds) The biology of *Botrytis*. Academic, London

Harrison JG, Lowe R (1987) Wind dispersal of conidia of *Botrytis* spp. pathogenic to *Vicia faba*. Plant Pathol 36:5–15

Hoksbergen T (2010) Impact of *Botrytis cinerea* on vineyard returns and wine quality. Grape Day seminar, Proceedings, Marlborough, New Zealand Wine growers

Holz G, Gütschow M, Coertze S, Calitz FJ (2003) Occurrence of *Botrytis cinerea* and subsequent disease expression at different positions on leaves and bunches of grape. Plant Dis 87:351–358

Hsieh TF, Huang JW, Hsiang T (2001) Light and scanning electron microscopy studies on the infection of oriental lily leaves by *Botrytis elliptica*. Eur J Plant Pathol 107:571–581

Jarvis WR (1962a) The infection of strawberry and raspberry fruits by *Botrytis cinerea* Fr. Ann Appl Biol 50:569–575

Jarvis WR (1962b) The dispersal of spores of *Botrytis cinerea* Fr. in a raspberry plantation. Trans Br Mycol Soc 45:549–559

Jarvis WR (1962c) The epidemiology of *Botrytis cinerea* Pers. in strawberries. In: Proceedings of the 16th International Horticultural Congress, pp 258–262

Jarvis WR (1969) The phenology of flowering in strawberry and raspberry in relation to grey mould control. Hortic Res 9:8–17

Jarvis WR (1980a) Taxonomy. In: Coley-Smith JR, Verhoeff K, Jarvis WR (eds) The biology of *Botrytis*. Academic, London, pp 1–18

Jarvis WR (1980b) Epidemiology. In: Coley-Smith JR, Verhoeff K, Jarvis WR (eds) The biology of *Botrytis*. Academic, London, pp 219–250

Jaspers MV, Seyb AM, Trought MCT, Balasubramaniam R (2013) Overwintering grapevine debris as an important source of *Botrytis cinerea* inoculum. Plant Pathol 62:130–138

Javed ZUR (1977) Studies of the behaviour in soil of five sclerotium-forming fungi. Trans Mycol Soc Jpn 18:111–118

Johnson KB, Powelson ML (1983) Analysis of spore dispersal gradients of *Botrytis cinerea* and gray mold disease gradient in snap bean. Phytopathology 73:741–746

Lorbeer JW (1966) Diurnal periodicity of *Botrytis squamosa* conidia in the air. Phytopathology 56:887 (abstr)

Marois JJ, Nelson JK, Morrison JC et al (1986) The influence of berry contact within grape clusters on the development of *Botrytis cinerea* and epicuticular wax. Am J Enol Viticult 37:293–296

McKay AH, Hill GN, Beresford RM (2012) Comparison of sampling procedures for estimating botrytis bunch rot incidence in New Zealand vineyards. NZ Plant Protect 65:241–248

McNicol RJ, Williamson B (1989) Systemic infection of black currant flowers by *Botrytis cinerea* and its possible involvement in premature abscission of fruits. Ann Appl Biol 114:243–254

Mertely JC, Chandler CK, Xiao CL, Legard DE (2000) Comparison of sanitation and fungicides for management of Botrytis fruit rot of strawberry. Plant Dis 84:1197–1202

Michailides TJ, Elmer PAG (2000) Botrytis gray mold of kiwifruit caused by *Botrytis cinerea* in United States and New Zealand. Plant Dis 84:208–223

Nair NG, Nadtotchei A (1987) Sclerotia of *Botrytis* as a source of primary inoculum for bunch rot of grapes in New-South-Wales, Australia. J Phytopathol 119:42–51

Nair NG, Emmett RW, Parker FE (1988) Some factors predisposing grape berries to infection by *Botrytis cinerea*. NZ J Exp Agric 16:257–263

Nair NG, Guilbaud-Oulton S, Barchia I, Emmett R (1995) Significance of carry over inoculum, flower infection and latency on the incidence of *Botrytis cinerea* in berries of grapevines at harvest in New South Wales. Austr J Exp Agric 35:1177–1180

Powelson RL (1960) Initiation of strawberry fruit rot caused by *Botrytis cinerea*. Phytopathology 50:491–494

Raposo R, Gomez V, Urrutia T, Melgarejo P (2001) Survival of *Botrytis cinerea* in southeastern Spanish greenhouses. Eur J Plant Pathol 107:229–236

Salinas J, Glandorf DCM, Picavet ED, Verhoeff K (1989) Effects of temperature, relative humidity and age of conidia in the incidence of spotting on gerbera flowers caused by *Botrytis cinerea*. Neth J Plant Pathol 95:51–64

Seyb AM (2004) *Botrytis cinerea* inoculum sources in the vineyard system. PhD thesis, Lincoln University, Lincoln, New Zealand, 227 p

Shoemaker PB, Lorbeer JW (1971) Spray volume, interval, and fungicide rate for control of botrytis leaf blight of onion. Plant Dis Rep 55:565–569

Shoemaker PB, Lorbeer JW (1977) The role of dew and temperature in the epidemiology of botrytis leaf blight of onions. Phytopathology 67:1267–1272

Spotts RA, Cervantes LA (2001) Disease incidence-inoculum dose relationships for *Botrytis cinerea* and *Penicillium expansum* and decay of pear fruit using dry, airborne conidia. Plant Dis 85:755–759

Spotts RA, Holz G (1996) Adhesion and removal of conidia of *Botrytis cinerea* and *Penicillium expansum* from grape and plum fruit surfaces. Plant Dis 80:688–691

Spotts RA, Serdani M (2006) Inoculum sources of *Botrytis cinerea* important to pear orchards in Oregon. Plant Dis 90:750–754

Stromeng GM, Hjeljord LG, Stensvand A (2009) Relative contribution of various sources of *Botrytis cinerea* inoculum in strawberry fields in Norway. Plant Dis 93:1305–1310

Sutton JC (1990) Epidemiology and management of botrytis leaf blight of onion and gray mold of strawberry: a comparative analysis. Can J Plant Pathol 12:100–110

Sutton JC, Swanton CJ, Gillespie TJ (1978) Relation of weather variables and host factors to incidence of airborne spores of *Botrytis squamosa*. Can J Bot 56:2460–2469

Sutton JC, James TDW, Dale A (1988) Harvesting and bedding practices in relation to grey mould of strawberries. Ann Appl Biol 113:167–175

Thomas AC, Kotze JM, Matthee FN (1983) Development of a technique for the recovery of soilborne sclerotia of *Botrytis cinerea*. Phytopathology 73:1374–1376

Thomas CS, Marois JJ, English JT (1988) The effects of wind speed, temperature and relative humidity on development of aerial mycelium and conidia of *Botrytis cinerea* on grape. Phytopathology 78:260–265

Van der Heyden H, Carisse O, Brodeur L (2012) Comparison of monitoring based indicators for initiating fungicide spray programs to control botrytis leaf blight of onion. Crop Prot 33:21–28

Van der Plank JE (1963) Plant diseases: epidemics and control. Academic, New York, p 349

Vincelli PC, Lorbeer JW (1987) Sequential sampling plan for timing initial fungicide application to control botrytis leaf blight of onion. Phytopathology 77:1301–1303

Viret O, Keller M, Jaudzems VG, Cole FM (2004) *Botrytis cinerea* infection of grape flowers: light and electron microscopical studies of infection sites. Phytopathology 94:850–857

Walter M, Boyd-Wilson KSH, Stanley J, et al (1997) Epidemiology of *Botrytis cinerea* in boysenberry (*Rubus* spp.). In: Proceedings of the 50th New Zealand plant protection rot Conf, pp 93–100

Walter M, Harris-Virgin P, Waipara NW et al (2004) From the laboratory to the field: litter management for control of *Botrytis cinerea* in boysenberry gardens. Fruits (Paris) 59:291–299

Williamson B, Tudzynski B, Tudzynski P, Van Kan JAL (2007) *Botrytis cinerea*: the cause of grey mould disease. Mol Plant Pathol 8:561–580

Xu X-M, Harris DC, Berrie AM (2000) Modeling infection of strawberry flowers by *Botrytis cinerea* using field data. Phytopathology 90:1367–1413

Yunis H, Elad Y (1989) Survival of dicarboximide-resistant strains of *botrytis cinerea* in plant debris during summer in Israel. Phytoparasitica 17:13–21

Chapter 8
Cultural and Integrated Control of *Botrytis* spp.

Yigal Elad

Abstract Epidemics of *Botrytis*-incited grey mould are common in open fields, orchards and greenhouses. These infections are promoted by high humidity and the presence of a film of water on susceptible plant organs and those conditions may be manipulated to prevent infection. Traditionally, heating greenhouses was a popular means of controlling the humidity in those structures and this practice remains popular in some temperate regions. However, the cost of active heating has forced farmers in some regions to abandon this disease-management strategy, which has led to the increased incidence and severity of grey mould, as susceptible organs of the crop plants remain wet for longer periods of time. Cultural methods for controlling *Botrytis*-incited disease include reducing the planting density, managing the crop canopy to allow for the aeration of the crop or susceptible organs via passive and active ventilation, fertigation with increased levels of potassium and calcium and reduced amounts of nitrogen, the use of soil mulch and passive solar heating of unheated greenhouses, avoiding harvesting on rainy days, and timing fungicide applications for the optimal protection of fresh harvest wounds and so that the crop remains wet for shorter periods of time. Row and field positioning and direction can affect grey mould as these factors affect local air movement and temporal temperature changes. A combination of treatments can provide better grey mould suppression than individual treatments and appropriate integration of control measures can provide sufficient disease control with minimal use of chemical fungicides.

Keywords Agrotechnical disease control • Integrated disease management • Plant nutrition • Resistance

---

Y. Elad (✉)
Department of Plant Pathology and Weed Research, Agricultural Research Organization, The Volcani Center, Bet Dagan 50250, Israel
e-mail: elady@volcani.agri.gov.il

S. Fillinger, Y. Elad (eds.), *Botrytis – the Fungus, the Pathogen and its Management in Agricultural Systems*, DOI 10.1007/978-3-319-23371-0_8

8.1 Introduction

Although the limiting factors for the germination of *Botrytis* spp. conidia and grey mould infection are generally the presence of liquid water and high humidity, the management of *Botrytis*-induced diseases can vary in different agro-systems. The initial *Botrytis* spp. inoculum in various crops may originate from inside or outside the field, as shown for *B. cinerea* in greenhouses (Korolev et al. 2006). In some crops, grey mould epidemics have been described as polycyclic (Ben Kalifa et al. 2012; Decognet et al. 2009; Elmer and Michailides 2004). Additionally, the common targets for efforts aimed at suppressing *Botrytis* spp. under field conditions in various crops are the possibility of limiting the presence of water, growing cultivars/crops with greater resistance to the pathogen and manipulating the availability of nutrients to the plants to decrease host susceptibility. However, agro-technical means for suppression of epidemics may vary according to limitations imposed by the local growing conditions and what host plants can tolerate without any significant yield loss.

Practices that allow air to move within the crop canopy, reduce the relative humidity around the plant, limit the presence of a film of water and increase resistance will be described below. General strategies for reducing the general humidity within a crop include planting on relatively high ground, allowing the wind to move along the rows and mulching with polyethylene or straw. Heating and ventilating greenhouses can be a powerful technique for controlling humidity inside those structures and suppressing disease suppression. But, this may not be economically viable when energy prices are high. Nutrition can affect the susceptibility of plants to *Botrytis* spp. and those effects will also be discussed in this chapter. In our discussion below, the plant hosts of *Botrytis* spp. are divided into two groups: open-field (and orchard) crops and greenhouse-grown crops. In this chapter, we describe cultural means of *Botrytis* control as stand-alone practices in various crops, as well as combined means of control in some specific pathosystems.

8.2 Host Characteristics

Botrytis cinerea is a major pathogen in orchards and vineyards. Various means of combating fruit infections have been explored and implemented; these methods have generally relied upon host characteristics and canopy management (Elmer and Michailides 2004). The morphological, anatomical and chemical characteristics of grape cultivars were measured in an attempt to correlate grey mould susceptibility with host characteristics. In that work, the number of naturally occurring pores in a leaf was found to be negatively correlated with resistance to *B. cinerea* (Eibach 1994; Mlikota Gabler et al. 2003). Cultivar resistance was also associated with higher cuticle and wax contents, but not induced or constitutive antifungal host defence mechanisms (Mlikota Gabler et al. 2003). In some raspberry cultivars, the stigmatic fluid inhibited *B. cinerea*. This mechanism allows those plants to avoid latent infections, which might develop into acute infections (Williamson and Jennings 1992). This observation suggests that cultivar selection can play a role in *Botrytis* management strategies.

The effects of root depth and water uptake on skin splitting and grey mould have been studied on wine grapes in France (Ribèreau-Gayon et al. 1980). Berries on deep-rooted vines were found to be much less susceptible to splitting and grey mould than those on shallow-rooted vines. Rootstocks also had a significant effect on the extent of fruit micro-cracking in sweet cherries and this rootstock effect is believed to be due to differences in their uptake of soil moisture (Cline et al. 1995). Thus, not only genetic resistance of cultivars, but also the nature of the rootstock can affect the outcome of the interaction between *Botrytis* and its host.

8.3 Canopy Management and Sanitation

Dense grape canopies are associated with increased wetness and more severe grey mould epidemics. Improving air circulation in the bunch zone, reducing relative humidity in the canopy and exposing more fruit bunches to light (Savage and Sall 1982) reduced bunch rot (Smithyman et al. 1997). Leaf removal affects the microclimate around the bunches of grapes. The removal of leaves from the fruiting zone of vines significantly reduced the incidence of *Botrytis* epidemics in grapes (English et al. 1989, 1993; Gubler et al. 1987; Percival et al. 1994; Thomas et al. 1988; Zoecklein et al. 1992). Following leaf removal, exposed Riesling grapes developed large amounts of epicuticular wax and cuticle, which resulted in significantly lower levels of grey mould (Percival et al. 1993).

Vines trained along horizontal bilateral cordons had improved exposure to light and a lower incidence of *Botrytis* than vines in a 'high' cordon system supported by a one-wire trellis (Redl 1988). A higher incidence of grey mould was observed in the Pergola system, whereas vines pruned and trained to the Guyot system had the lowest levels of disease (Cargnello et al. 1991). Infection of Cabernet Sauvignon clusters by *B. cinerea* after veraison was significantly influenced by cluster compactness (Fermaud et al. 2001a, b; Vail and Marois 1991). Reduction of berry-to-berry contact within the bunch reduced the development of *B. cinerea* (Martin 1990).

Botrytis spp. also infect various legume crops (Davidson et al. 2004). Grey mould of lentils is caused by both *B. cinerea* and *B. fabae* (Bailey et al. 2000). Fava beans are infected by both *B. fabae* and *B. cinerea*, but *B. fabae* is more important in causing chocolate spot in this crop (Sundheim 1973; Mansfield 1980). There are some effective cultural methods for controlling humidity-promoted diseases in annual open-field legume crops (Davidson et al. 2004). A dense canopy encourages the development of lentil grey mould, but there are several agronomic practices that can be used to avoid a dense canopy (Bretag and Mebalds 1987), including the adjustment of sowing dates and sowing density, increasing the amount of space between rows and employing appropriate weed control (Bayaa and Erskine 1998). Dense plantings of fava bean create a high-humidity environment, which should also be avoided (Jellis et al. 1998). An association between early sowing and chocolate spot was noted in Syria (Hanounik and Robertson 1988). In common bean, cultural practices can provide disease reduction i.e. reduced planting densities, careful use of fertilisers to prevent the development of a dense crop and judicious watering all help to reduce the onset of disease (Jarvis 1991).

The removal of necrotic tissue before onset of epidemics led to significant reductions in the number of *Botrytis* spp. conidia in the air and incidence *Botrytis* infections (Köhl et al. 1992). Similarly, the removal of kiwifruit flowers from male vines (an important source of *B. cinerea* inoculum) led to reduced levels of stem-end rot in neighboring vines (Michailides and Elmer 2000). Removal of senescent floral tissues and aborted berries was found to reduce *B. cinerea* by approx. 30 % in Merlot grapes (Jermini et al. 1986). Similarly, pruning infection foci in tomato leaves stopped epidemics in greenhouses (Shtienberg et al. 1998). These results reveal an effective means of epidemic management that is, unfortunately, quite labour intensive.

8.4 Managing Environmental Conditions Inside the Greenhouse

8.4.1 Heated Greenhouses

Botrytis cinerea infects many vegetable, ornamental and other horticultural crops (Elad et al. 2004). The temperature range for grey mould development in those various crops is from 12 to 30 °C (Jarvis 1992), with an optimal range of 15–20 °C, although the pathogen is active at temperatures as low as 0 °C (Droby and Lichter 2004). Greenhouses may or may not be actively heated. The use of supplemental heating has been shown to result in lower levels of grey mould in greenhouse-grown vegetable and ornamental crops (Dik and Wubben 2004; Elad and Shtienberg 1995; Hausbeck et al. 1996; Jarvis 1992).

Studies on grey mould have suggested the use of heating or aeration and ventilation of the greenhouse (Dik and Wubben 2004; Eden et al. 1996; Elad and Shtienberg 1995; Jarvis 1992; Morgan 1984) to reduce humidity levels and the amount of dew on susceptible plant tissues. Heating greenhouses not only results in higher temperatures, but can also lead to higher vapour pressure deficits (lowered relative humidity; Dik and Wubben 2004). In heated geranium (*Pelargonium*) crops, it was demonstrated that periods of wetness lasting more than 4 h should be avoided (Sirjusingh and Sutton 1996) and that forced heated air can reduce the incidence of stem blight and sporulation (Hausbeck et al. 1996). In heated tomato greenhouses with three climate regimes, more grey mould was found in crops in greenhouses kept at 16 °C with no ventilation. Less tomato grey mould was observed in crops kept at 13 °C with ventilation and even less in those kept at 16 °C with ventilation (Morgan 1984). These results demonstrate the importance of ventilation for reducing humidity and preventing the conditions that promote infection.

8.4.2 Light Quality Management in Greenhouses

Crops susceptible to grey mould may be grown in polyethylene tunnels. Different types of polyethylene have different effects on the sporulation of *B. cinerea* (Reuveni et al. 1989). Light, especially near-UV (nUV) light, increases the sporulation of *B. cinerea*. In *Primula* and strawberry crops grown under polyethylene tunnels, the incidence of infection over two seasons was reduced when nUV blocking film (up to 405 nm) was used instead of a standard film (West et al. 2000). Several weeks after the inoculation of tomato plants with *B. cinerea*, conidia production was low under polyethylene films that absorbed nUV light (Nicot et al. 1996).

Spectrally modified polyethylene films were found to have a similar effect on the development of *B. cinerea* on greenhouse-grown tomato plants (Reuveni and Raviv 1992) and UV-absorbing polyvinyl chloride film had a similar effect in cucumber (Honda et al. 1977). Elad (1997) demonstrated an effect of far-red light on the production of conidia by field isolates of *B. cinerea*. A green pigmented polyethylene film reduced conidia production and infection in commercial greenhouses. However, Elad (1997) identified some field isolates of *B. cinerea* that were capable of producing conidia in the dark and were not affected by light quality. Since similar *B. cinerea* fungi exist under field conditions, it was concluded that the disease-suppressive effect of light-filtering polyethylene covers was definitely due to a change in the physiological status of the plants, that is, induced resistance in plants grown under conditions of change in light quality (Elad 1997).

8.4.3 Crop Density in Greenhouses

Increasing the air circulation in the field may reduce grey mould by reducing the amount of time that a water film is present on susceptible plant parts (Elad and Shtienberg 1995). Cultural methods such as reduced planting density and reduced canopy density can help to improve the environmental conditions in the field and limit the development of grey mould (Legard et al. 2000; Trolinger and Strider 1984). Similarly, reducing leaf wetness by moving from open-field production to unheated greenhouses dramatically reduced the incidence of strawberry grey mould (Xiao et al. 2001).

Crop density affects the movement of air in the canopy and grey mould severity. More cucumber grey mould was found on fruits in a cucumber crop with two stems per plant than in a crop with one stem per plant (Elad and Shtienberg 1995). Higher levels of grey mould were observed on floral parts of *Exacum affine* grown under crowded conditions as compared to less crowded conditions (Trolinger and Strider 1984) and similar results were observed in densely planted strawberry (Legard et al. 2000).

8.4.4 Stem Infection and Its Management

Grey mould develops at relative humidity (RH) levels of >90 % and *Botrytis* conidia germinate in liquid water (Jarvis 1992). However, grey mould can develop on tomato and sweet basil stem wounds under lower RH conditions. In these crops, optimal disease development has been documented at 75–85 % RH (Sharabani et al. 1999; O'Neill et al. 1997). *B. cinerea* infection at lower RH is possible due to moisture on the wound surface. This moisture originates from within the plant and promotes infection. The infection further develops inside the host tissue, where it is wet (Eden et al. 1996).

Lisianthus (*Eustoma grandiflorum*) is a cut flower crop that is susceptible to infection by *B. cinerea* (Shpialter et al. 2009), which causes grey mould on the harvested flowers and stem bases and on the stem stubs after harvest (Shpialter et al. 2009; Wegulo and Vilchez 2007), especially in unheated greenhouses. The lower leaves of lisianthus plants are often in close contact with the wet soil. When a barrier, in the form of a polyethylene sheet, was placed between the lower leaves and the wet soil, the severity of disease symptoms on those leaves was significantly reduced (Shpialter et al. 2009).

Infection of the stem base is likely to be affected by soil moisture, due to the proximity of the stem base to the soil. The location of irrigation drippers affects the moisture level of the soil around the stem base and is also likely to affect the infection of plants by *B. cinerea*. In lisianthus, when drippers were placed 20 cm below the soil surface instead of at the soil surface, the incidence of stem infection was drastically reduced. The sub-surface irrigation plots were characterized by lower RH and shorter dew periods than those observed in plots in which an aboveground drip irrigation system was used (Shpialter et al. 2009).

In a study of early blight (*Alternaria solani*) and Septoria leaf spot (*Septoria lycopersici*) in fresh market tomato crops, Mills et al. (2002) concluded that an observed reduction in foliar disease in mulched plots was associated with reduced splash dispersal and, in one of two trials, reduced leaf wetness. Strawberry plants grown in beds covered with black polyethylene had lower incidences of *B. cinerea-incited* rot than those grown in beds covered with either clear polyethylene or paddy straw (Saks et al. 1996). Reducing ambient humidity by covering the soil with polyethylene has also been suggested as a means for restricting grey mould development in greenhouse crops (Elad 2000; Elad and Shtienberg 1995).

In greenhouse-grown lisianthus, polyethylene mulch reduced RH and dew duration and drastically reduced the incidence of disease (Shpialter et al. 2009). The mechanisms by which lisianthus stem base infections are reduced through the use of polyethylene soil cover may be complex. The reductions in the RH and the amount of dew in the canopy contribute to the inhibition of the germination of *B. cinerea* conidia and limit the fungus's penetration of the host plant (Holz et al. 2004). Also, as shown in laboratory experiments, a physical barrier between the lower leaves and the wet soil affects conditions for the development of infection on the leaves near the stem base.

Grey mould of lisianthus stems can also serve as an example of a situation in which cultural methods and their combination can affect disease development. In-bed aeration in greenhouse-grown lisianthus, based on the use of perforated polyethylene sleeves and blowing unheated air, reduced disease incidence by changing the microclimate around the stem base (Elad et al. 2009b). As mentioned above, soil cover suppressed disease. Moreover, small planting holes in the polyethylene were superior to larger holes. This is probably due to the fact that when large holes are used, the lower leaves may come into contact with the wet soil and this moisture may promote *B. cinerea* infection of those leaves and the spread of that rot towards the stem. In contrast, when the planting holes are smaller, the lower leaves are less likely to come into contact with the wet soil. The disease control provided by whole-plant applications of chemical fungicides together with soil cover is not superior to that provided by soil cover alone. A polyethylene soil cover with small planting holes was as effective as chemical control. The combination of a polyethylene soil cover with small holes and chemical sprays directed at the plant stubs provided better disease suppression than either of the treatments alone (Elad et al. 2009b).

8.4.4.1 Effects of Polyethylene Soil Cover and Deep Irrigation on Foliar *B. cinerea* Infection

The use of polyethylene soil cover can suppress grey mould (Elad 2000), as mentioned above. For instance, the use of plastic sheets that have a white upper surface and a black lower surface increased yield and fruit size in strawberry and reduced the incidence of *B. cinerea* infection in that crop, as compared with non-mulched soil (Laugale et al. 2012). Polyethylene soil cover is associated with increased yields, thanks to the accumulation of heat in the root zone and increased soil temperatures (Flores Velasquez and Ibarra 1998).

Apart from the physical barrier between the soil and the aerial plant parts and the prevention of evaporation from the soil, Cohen et al. (2006) suggested a detailed explanation for the effect of polyethylene mulch on foliar late blight (*Phytophthora infestans*) of tomato and *Pseudoperonospora cubensis* in cucumber. They found that, in the early winter, covering the soil surface with polyethylene dramatically reduced the rate of infection, as compared with that observed in greenhouses in which the crops were grown in bare soil. Microclimate measurements showed that night-time soil and air temperatures were higher in the mulched greenhouses. On days when infection occurred in the control plots, air temperature reached the dew-point temperature in the control greenhouses, but not in the mulched greenhouses. Night-time soil temperatures and air temperatures near the ground were higher under the mulch and the difference between the air temperature and the dew point in the greenhouse was higher above the mulch, indicating a lower incidence of condensation on the leaves (Shtienberg et al. 2010; Cohen et al. 2006).

This temperature–dew effect explains part of the effect of polyethylene mulch on grey mould, but not the effect of deep irrigation, which also affected the duration of wetness of the lower leaves and the RH. Interestingly, plant tissues sampled in the mulched plots had relatively high levels of N, P, K, Ca, and Mg, as compared to tissue samples collected from the unmulched plots. These nutritional differences might contribute to disease control by affecting the susceptibility of lisianthus tissues to *B. cinerea* infection (Shpialter et al. 2009).

8.4.4.2 The Sweet Basil–Grey Mould Test Case

Sweet basil (*Ocimum basilicum*) is an annual herb crop grown in greenhouses. Branches are harvested several times a season and this harvesting creates wounds that are susceptible to infection by *B. cinerea*. Stem infection can develop until the entire plant dies and a large amount of conidia are discharged into the air (Sharabani et al. 1999). Rot can also develop on the crop postharvest (Aharoni et al. 2010). Sweet basil is grown year-round in unheated polyethylene greenhouses, but the majority of grey mould damage is observed during the winter which is the rainy season in the region. In the absence of active heating, the conditions in sweet basil greenhouses are favorable for *B. cinerea*. Indeed, it was found that the limiting factor for grey mould development was the need for a high level of humidity in the greenhouse rather than the presence of inoculum. Regardless of crop age, the severity of grey mould observed on harvesting wounds was found to be related to seasonal microclimatic conditions.

Sharabani et al. (1999) found that rain outside the sweet basil greenhouses, which is naturally associated with increased RH in the greenhouse, was the dominant factor influencing the timing of grey mould epidemics. The cut ends of stems were found to be highly susceptible to infection soon after harvest; susceptibility diminished gradually and stem cuts inoculated 2 days after harvest were rarely diseased. In field experiments (Sharabani et al. 1999), disease severity did not change much when harvesting was done on dry days. However, disease outbreaks were found to coincide with harvests carried out on rainy days. It was found that if harvest was carried out during or immediately after a rain event, then the application of one fungicidal spray soon after harvest is needed and will provide adequate disease suppression. Avoiding harvesting during rain events is another good way to prevent stem grey mould (Sharabani et al. 1999). Applying the fungicide treatment immediately after harvesting (wounding) may help to reduce the amount of chemical fungicide needed, as well as lower the risk of contaminating the marketable crop with pesticide residues.

Plant spacing and foliage management can affect the RH inside the crop canopy and denser foliage is associated with higher levels of grey mould incidence (Legard et al. 2000; Trolinger and Strider 1984). Observations in commercial sweet basil plots have revealed that higher planting densities and restricted aeration contribute to epidemics (Elad et al. 2014b). In experiments, a planting density that was half that of the common practice was associated with lower disease levels, with no significant negative yield effect. This could be attributed to better aeration within the canopy at the initial stages of growth and to reduced amounts of receptive host tissue. At a later stage, the branches harvested from the lower-density plots were less susceptible to *B. cinerea* infection, indicating that a higher level of plant tissue resistance was induced by the lower-density field conditions (Elad et al. 2014b). A combination of increased plant spacing and the use of a polyethylene soil cover synergistically improved the yield of sweet basil branches. Similarly, increasing the distance between plants in strawberry plots reduced the incidence of grey mould in those plots. Nevertheless, the fruit yield was negatively affected by this practice (Laugale et al. 2012; Daugaard 2003; Legard et al. 2000).

Aeration is an accepted practice for reducing the RH inside greenhouses. For instance, nocturnal ventilation can significantly reduce the level of humidity in a greenhouse and suppress tomato *B. cinerea* grey mould (Baptista et al. 2012). Aeration of sweet basil tunnels decreased the incidence of grey mould in those tunnels. Moreover, the combination of passive aeration and wider plant spacing improved the yield of sweet basil, as did the combination of wider plant spacing and polyethylene soil cover (Elad et al. 2014b).

8.4.5 Passive Heating of Polyethylene Greenhouses

We have developed a method to suppress foliar diseases by increasing the temperature inside greenhouses during day. Elevated day-time temperatures are obtained by closing the side walls of the greenhouses, which allows the temperature inside the greenhouse to rise above 25 °C in winter. The initial information on the effect of day-time elevated temperatures was published in the context of the tomato–*O. neolycopersici* pathosystem (Elad et al. 2008; Jacob et al. 2007) and the sweet pepper–*O. sicula* pathosystem (Elad et al. 2009a).

Temperatures in polyethylene-covered sweet basil greenhouses that were kept closed for 6 h each day reached 42 °C on winter days. The incidence of grey mould in these greenhouses was significantly reduced and basil yield increased as a result of the passive day-time heating. A significant negative correlation was observed between the duration of temperatures above 30 °C and the incidence of grey mould. The suppressive effects of high day-time temperatures on the three diseases in the three plant species can be attributed to a direct effect of temperature on the pathogen inoculum, as well as an indirect effect involving the induction of resistance in the plants. New results from work with sweet basil (Elad et al. 2014b) indicate that the heat treatment has an indirect effect involving the induction of resistance in the host plants. In this case, plants exposed to a day-time heat treatment and later infected by *B. cinerea* or *S. sclerotiorum* exhibited less severe infections than inoculated plants that had not received the heat treatment (Elad et al. 2014a). Thus, it may be concluded that high day-time temperatures may reduce foliar disease. However, one needs to carefully verify that this heating has no deleterious effect on fruiting or negative interaction with other biotic stresses, such as humidity-promoted pathogens.

8.5 Plant Nutrition

Mineral nutrients are essential for plant growth, development and reproduction. Nutrients can also affect plants' susceptibility to pathogens (Engelhard 1989). Calcium is important for fruit and vine production (Ferguson 1984), affects various cell functions (Conway 1982) and increases resistance to disease (Volpin and Elad 1991; Conway et al. 1991). Calcium in plant tissues has been shown to reduce the severity of grey mould (Yermiyahu et al. 2006; Bar-Tal et al. 2001; Chardonnet and Doneche 1995; Volpin and Elad 1991).

Calcium reduces the leakage of exudates to the host surface, reducing the availability of those exudates to the pathogen (Volpin and Elad 1991). It can also directly inhibit the production of polygalacturonase by *B. cinerea* (Biggs et al. 1997). Ca^{2+} deficiency increases susceptibility to *B. cinerea* (Schwab et al. 1993). Incubating *B. cinerea* conidia in increasing concentrations of $CaCl_2$ decreased conidial germination and germ tube length (Chardonnet et al. 2000). Possible mechanisms by which high levels of calcium in plant tissues may reduce disease include the reduced emission of ethylene from the plant tissue, decreased susceptibility of plant membranes, particularly pectin constituents, to pathogens' hydrolytic enzymes and reduced leakage of nutrients to plant surfaces (Elad and Evensen 1995).

In strawberry, increasing fruit calcium content reduced the incidence of *Botrytis* infection (Cheour et al. 1990; Karp and Starast 2002; Wojcik and Lewandowski 2003). Nevertheless, when conventional fungicides were replaced with $CaCl_2$ to protect strawberries from *B. cinerea*, the $CaCl_2$ treatments did not reduce the epidemic (Erincik et al. 1998). When Ca^{2+} was applied to table grapes in the vineyard, resistance to *B. cinerea* increased and this resistance was correlated with increased levels of cellulose and both oxalate- and alkali-soluble pectins (Miceli et al. 1999).

High nitrogen rates enhance plant growth and foliage density. In general, increasing the nitrogen level is believed to result in increased susceptibility to *B. cinerea*. However, the results in this area are sometimes contradictory. Nitrogen supplied at 1.5, 3.8 and 6.0 g/m^2 resulted in a quadratic increase in grey mould incidence on chrysanthemum flowers (Hobbs and Waters 1964). In contrast, Verhoeff (1968) found that the susceptibility of soil-grown tomatoes to grey mould increased with decreasing levels of nitrogen in the soil. Hoffland et al. (1999) found a linear correlation between the leaf C:N ratio and the susceptibility of tomato to *B. cinerea*. They attributed that correlation to levels of available soluble carbohydrates.

Increasing calcium concentrations in fertigation fluid reduced the intensity of conidiation of *B. cinerea* on sweet basil stems and the incidence of infected plants (Yermiyahu et al. 2006). Furthermore, increased nitrogen increased the susceptibility of sweet basil to *B. cinerea*; this effect was related to the indirect negative influence of N on Ca in the canopy (Yermiyahu et al. 2006). Similarly, increased N levels have been shown to increase the severity of grey mould in chrysanthemum (Hobbs and Waters 1964), but decrease the incidence of the disease in tomato (Verhoeff 1968). High levels of nitrogen result in increased plant growth and the increased incidence *B. cinerea* on fruits in storage (Pertot and Perin 1999; Prasad and Spiers 1991). In sweet basil, higher nitrogen levels in the fertigation solution resulted not only in more grey mould infection, but also in increased severity of white mould (*Sclerotinia sclerotiorum*; Elad and co-workers, unpublished). These findings indicate that limiting nitrogen fertilization can minimize grey mould, but that would also be expected to affect yields.

Similar nutrition guidelines are commonly suggested for open-field crops such as legumes. Avoiding a dense canopy in lentil by avoiding high N levels decreases grey mould (Bayaa and Erskine 1998). A higher incidence of chocolate spot was observed in plants fava bean grown with increased applications of N fertiliser (Hegab and Beshir 1994). However, in fava bean, the use of adequate fertiliser, particularly the application of adequate amounts of K and N, is necessary to prevent early senescence, which leads to dead leaves upon which the fungus sporulates (Jellis et al. 1998). The use of adequate amounts of fertilizer, particularly K, has also been shown to reduce susceptibility to infection in field pea (Biddle 2001). K is absorbed by plants in its cation form and it is the major cation in the cytoplasm. It is readily translocated in the phloem and xylem (Marschner 1986). A large number of enzymes are either completely dependent upon or stimulated by K (Suelter 1970). Interestingly, we found that fertigation with low levels of K is associated with higher levels of grey mould in sweet basil; whereas fertigation with comparably higher levels of K can suppress the disease under field conditions (Israeli et al. 2011; Yermiyahu et al. 2013). In grapevine, increased applications of K have been associated with less severe outbreaks of grey mould (Kiraly 1964) and, in pumpkin, increased applications of K have been associated with less severe white mould infections (Abia and Smith 1980).

8.6 Conclusions

As examples drawn from some *Botrytis*-based pathosystems could demonstrate (mentioned above), integrated disease management have advantages over using single means of control and longer durability. A strategy was developed for the integrated biological and chemical control of *B. cinerea* in vegetable crops grown in unheated greenhouses. The integrated strategy, named BOTMAN (Botrytis Manager) has been used in greenhouse-grown tomato and cucumber and achieved a drastic reduction in chemical fungicide use (Shtienberg and Elad 1997) while promoting cultural management and biocontrol in greenhouse-grown vegetable crops. This system takes into consideration not only chemical and biocontrol agents, but also environmental conditions and their potential effects on grey mould epidemics (Shtienberg and Elad 1997). Decisions of whether to spray a biological agent or a fungicide are made in relation to weather forecasts. When cultural means of control seem to be sufficient and slow or no disease progress is expected, the system does not suggest any spray treatments. In contrast, when an outbreak is expected, the application of a chemical fungicide is recommended. In all other cases when the risk is of mid-level, the application of a biocontrol agent is recommended. In this system, disease warnings are based on weather forecasts.

Infections caused by pathogenic *Botrytis* species are more common under high-humidity conditions and cultural techniques aimed at decreasing humidity can suppress those infections. In some cases, cultural methods actually increase the resistance of the crop to the disease. Different cultural methods can be combined for improved disease suppression. Moreover, cultural methods can be combined with chemical botryticides or biocontrol treatments (Chap. 9). Integrated crop management, including disease control, may involve the use a decision-support system, such as BOTMAN.

References

Abia JA, Smith BN (1980) Mineral nutritional status of pumpkin and infection by *Sclerotinia sclerotiorum*. Plant Physiol 65:106

Aharoni N, Kenigsbuch D, Chalupowicz D et al (2010) Reducing chilling injury and decay in stored sweet basil. Israel J Plant Sci 58:167–180

Baas R, Van Oers S, Silber A et al (2003) Calcium distribution in cut roses as related to transpiration. J Hortic Sci Biotechnol 78:1–9

Bailey KL, Gossen BD, Derksen DA, Watson PR (2000) Impact of agronomic practices and environment on diseases of wheat and lentil in south-eastern Saskatchewan. Can J Plant Sci 80:917–927

Baptista FJ, Bailey BJ, Meneses JF (2012) Effect of nocturnal ventilation on the occurrence of *Botrytis cinerea* in Mediterranean unheated tomato greenhouses. Crop Prot 32:144–149

Bar-Tal A, Baas R, Ganmore-Neumann R et al (2001) Rose flower production and quality as affected by Ca concentration in the petal. Agronomie 21:393–402

Bayaa B, Erskine W (1998) Diseases of lentils. In: Allen DJ, Lenné JM (eds) The pathology of food and pasture legumes. CAB International, Wallingford, pp 423–471

Ben Kalifa H, Rav David D, Borenshtein M et al (2012) Climate change effect on plant–pathogen–beneficial microorganism interaction in high humidity-promoted tomato diseases. IOBC/WPRS Bull 78:15–18

Biddle AJ (2001) Botrytis gray mold. In: Kraft JM, Pfleger FL (eds) Compendium of pea diseases and pests, 2nd edn. American Phytopathological Society Press, St. Paul, pp 31–32

Biggs AR, El-Kholi MM, El-Neshawy S, Nickerson R (1997) Effects of calcium salts on growth, polygalacturonase activity, and infection of peach fruit by *Monilinia fructicola*. Plant Dis 81:399–403

Bretag TW, Mebalds MI (1987) Pathogenicity of fungi isolated from *Cicer arietinum* (chickpea) grown in north-western Victoria. Aust J Exp Agric 27:141–148

Cargnello G, Forno S, Terzuolo S (1991) Research on the influence of agricultural techniques on epidemic patterns: investigations of grape training systems. Vignevini 18:53–57

Chardonnet C, Doneche B (1995) Relation between calcium content and resistance to enzymatic digestion of the skin during grape ripening. Vitis 34:95–98

Chardonnet CO, Sams CE, Trigiano RN, Conway WS (2000) Variability of three isolates of *Botrytis cinerea* affects the inhibitory effects of calcium on this fungus. Phytopathology 90:769–774

Cheour F, Willemot C, Arul J et al (1990) Foliar application of calcium chloride delays postharvest ripening of strawberry. J Am Soc Hortic Sci 115:789–792

Cline JA, Sekse L, Meland M, Webster AD (1995) Rain-induced fruit cracking of sweet cherries: I. Influence of cultivar and rootstock on fruit water absorption, cracking and quality. Acta Agric 45:213–223

Cohen S, Ziv G, Grava A et al (2006) Influence of polyethylene mulch on night microclimate, dew point and *Phytophthora infestans* infection in non-heated tomato greenhouses in southern Israel. Acta Horticult 718:277–282

Conway WS (1982) Effect of postharvest calcium treatment on decay of delicious apples. Plant Dis 66:402–403

Conway WS, Sams CE, Abbott JA, Bruton BD (1991) Postharvest calcium treatment of apple fruit to provide broad-spectrum protection against postharvest pathogens. Plant Dis 75:620–622

Daugaard H (2003) Effect of plant spacing on yield and incidence of *Botrytis cinerea* Pers. in strawberry. IOBC/WPRS Bull 26(2):147–151

Davidson JA, Pande S, Bretag TW et al (2004) Biology and management of *Botrytis* spp. in legume crops. In: Elad Y, Williamson B, Tudzynski P, Delen N (eds) *Botrytis*: biology, pathology and control. Kluwer Academic Publishers, Dordrecht, pp 295–318

Decognet V, Bardin M, Trottin-Caudal Y, Nicot PC (2009) Rapid change in the genetic diversity of *Botrytis cinerea* populations after the introduction of strains in a tomato glasshouse. Phytopathology 99:185–193

Dik AJ, Wubben JP (2004) Epidemiology of *Botrytis cinerea* diseases in greenhouses. In: Elad Y, Williamson B, Tudzynski P, Delen N (eds) *Botrytis*: biology, pathology and control. Kluwer Academic Publishers, Dordrecht, pp 319–333

Droby S, Lichter A (2004) Post-harvest *Botrytis* infections: etiology, development and management. In: Elad Y, Williamson B, Tudzynski P, Delen N (eds) *Botrytis*: biology, pathology and control. Kluwer Academic Publishers, Dordrecht, pp 349–367

Eden MA, Hill RA, Beresford R, Stewart A (1996) The influence of inoculum concentration, relative humidity, and temperature on infection of greenhouse tomatoes by *Botrytis cinerea*. Plant Pathol 45:795–806

Eibach R (1994) Defense mechanisms of the grapevine to fungus disease. Am Vineyard 1:8–10

Elad Y (1997) Effect of filtration of solar light on the production of conidia by field isolates of *Botrytis cinerea* and on several diseases of greenhouse-grown vegetables. Crop Prot 16:635–642

Elad Y (2000) Changes in disease epidemics on greenhouse grown crops. Acta Horticult 534:213–220

Elad Y, Evensen K (1995) Physiological aspects of resistance to *Botrytis cinerea*. Phytopathology 85:637–643

Elad Y, Shtienberg D (1995) *Botrytis cinerea* in greenhouse vegetables: chemical, cultural, physiological and biological controls and their integration. Integr Pest Manag Rev 1:15–29

Elad Y, Williamson B, Tudzynski P, Delen N (2004) *Botrytis* spp. and diseases they cause in agricultural systems – an introduction. In: Elad Y, Williamson B, Tudzynski P, Delen N (eds) *Botrytis*: biology, pathology and control. Kluwer Academic Publishers, Dordrecht, pp 1–8

Elad Y, Jacob D, Rav David D et al (2008) Development of climate control methods for integrated management of powdery mildew of tomato caused by *Oidium neolycopersici*. Acta Horticult 807:727–732

Elad Y, Brand M, Messika Y et al (2009a) Spray treatments combined with climate modification for the management of *Leveillula taurica* in sweet pepper. Eur J Plant Pathol 124:309–329

Elad Y, Shpialter L, Rav David D et al (2009b) Suppression of gray mold in lisianthus by passive means of greenhouse environment management. Acta Horticult 893:1277–1283

Elad Y, Fogel M, Rav David D et al (2014a) Daytime solar heat treatment for the suppression of foliar plant pathogens in polyethylene-covered greenhouses. Acta Horticult 1015:89–94, http://www.actahort.org/books/1015/1015_9.htm

Elad Y, Israeli L, Fogel M et al (2014b) Conditions influencing the development of sweet basil grey mould and cultural measures for disease management. Crop Prot 64:67–77

Elmer PAG, Michailides TJ (2004) Epidemiology of *Botrytis cinerea* in orchards and vine crops. In: Elad Y, Williamson B, Tudzynski P, Delen N (eds) *Botrytis*: biology, pathology and control. Kluwer Academic Publishers, Dordrecht, pp 243–272

Engelhard W (1989) Soilborne plant pathogens: management of diseases with macro- and microelements. American Phytopathological Society Press, St. Paul

English JT, Thomas CS, Marois JJ, Gubler WD (1989) Microclimates of grapevine canopies associated with leaf removal and control of botrytis bunch rot. Phytopathology 79:395–401

English JT, Kaps ML, Moore JF et al (1993) Leaf removal for control of *Botrytis* bunch rot of wine grapes in the midwestern United States. Plant Dis 77:1224–1227

Erincik O, Madden LV, Scheerens JC, Ellis MA (1998) Evaluation of foliar applications of calcium chloride for control of *Botrytis* fruit rot on strawberry and effects on strawberry fruit quality. Adv Strawberry Res 17:7–17

Ferguson IB (1984) Calcium in plant senescence and fruit ripening. Plant Cell Environ 7:477–489

Fermaud M, Liminana JM, Froidefond G, Pieri P (2001a) Grape cluster microclimate and architecture affect severity of *Botrytis* rot of ripening berries. IOBC WPRS Bull 24(7):7–9

Fermaud M, Pieri P, Liminana JM (2001b) *Botrytis* and micro-climates: propagation of *Botrytis cinerea* in grapes in controlled climatic conditions. Phytoma 543:40–43

Flores Velasquez J, Ibarra JL (1998) Cultivation of peppers using plastic mulch with coloured films and nutrient irrigation. Plasticulture 116:16–26

Gubler WD, Marois JJ, Bledsoe AM, Bettiga LJ (1987) Control of *Botrytis* bunch rot of grape with canopy management. Plant Dis 71:599–601

Hanounik SB, Robertson LD (1988) New sources of resistance in *Vicia faba* to chocolate spot caused by *Botrytis fabae*. Plant Dis 72:696–698

Hausbeck MK, Pennypacker SP, Stevenson RE (1996) The use of forced heated air to manage Botrytis stem blight of geranium stock plants in a commercial greenhouse. Plant Dis 80:940–943

Hegab MT, Beshir MA (1994) Effect of nitrogen and fertilizer and application of fungicides on chocolate spot, rust diseases and yield components of field bean under calcareous soil. Ann Agric Sci 32:717–729

Hobbs EL, Waters WE (1964) Influence of nitrogen and potassium on susceptibility of *Chrysanthemum morifolium* to *Botrytis cinerea*. Phytopathology 54:674–676

Hoffland E, Van Beusichem ML, Jeger MJ (1999) Nitrogen availability and susceptibility of tomato leaves to *Botrytis cinerea*. Plant Soil 210:263–272

Holz G, Coertze S, Williamson B (2004) The ecology of *Botrytis* on plant surfaces. In: Elad Y, Williamson B, Tudzynski P, Delen N (eds) *Botrytis*: biology, pathology and control. Kluwer Academic Publishers, Dordrecht, pp 9–27

Honda Y, Toki T, Yunoki T (1977) Control of gray mold of greenhouse cucumber and tomato by inhibiting sporulation. Plant Dis Rep 61:1041–1048

Israeli L, Yermiyahu U, Rav-David D et al (2011) The influence of nutritional elements on sweet basil diseases incited by *Sclerotinia sclerotiorum* and *Botrytis cinerea*. Phytoparasitica 39:245

Jacob D, Rav David D, Elad Y (2007) Biology and biological control of tomato powdery mildew (*Oidium neolycopersici*). IOBC/WPRS Bull 30(6–1):329–332

Jarvis WR (1991) Gray mold. In: Hall R (ed) Compendium of bean diseases. American Phytopathological Society Press, St. Paul, pp 21–22

Jarvis WR (1992) Managing diseases in greenhouse crops. The American Phytopathological Society Press, St. Paul

Jellis GJ, Bond DA, Boulton RE (1998) Diseases of faba bean. In: Allen DJ, Lenné JM (eds) The pathology of food and pasture legumes. CAB International, Wallingford, pp 371–422

Jermini M, Jelmini G, Gessler C (1986) Control of *Botrytis cinerea* on Merlot grapevine in Ticino. Role of latent infections. Rev Suisse Vitic Arboric Hortic 18:161–166

Karp K, Starast M (2002) Effects of springtime foliar fertilization on strawberry yield in Estonia. Acta Horticult 594:501–505

Kiraly Z (1964) Effect of nitrogen fertilization on phenol metabolism of selected plants species and its ecological implications. Bot Acta 105:355–361

Köhl J, Molhoek WML, Van der Plas CH et al (1992) Biological control of botrytis leaf blight of onions: significance of sporulation suppression. In: Verhoeff K, Malathrakis NE, Williamson B (eds) Recent advances in *Botrytis* research. Pudoc Scientific Publishers, Wageningen, pp 192–196

Korolev N, Katan T, Elad Y (2006) The use of selenate-resistant strains as markers for the spread and survival of *Botrytis cinerea* under greenhouse conditions. Phytopathology 96:1195–1203
Laugale V, Lepse L, Strautina S et al (2012) Effect of planting density and plastic soil mulch on strawberry plant development, yield and fruit quality. Acta Horticult 926:517–523
Legard DE, Xiao CL, Mertely JC, Chandler CK (2000) Effects of plant spacing and cultivar on incidence of botrytis fruit rot in annual strawberry. Plant Dis 84:531–538
Mansfield JW (1980) Mechanism of disease resistance in *Botrytis*. In: Coley-Smith JR, Verhoeff K, Jarvis WR (eds) The biology of *Botrytis*. Academic, London, pp 181–218
Marschner H (1986) Mineral nutrition of higher plants. Academic, London, pp 229–448
Martin SR (1990) Systematic management to minimize botrytis bunch rot in three Victorian vineyards. Aust NZ Wine Ind J 5:235–237
Miceli A, Ippolito A, Linsalata V, Nigro F (1999) Effect of preharvest calcium treatments on decay and biochemical changes in table grape during storage. Phytopathol Mediterr 38:47–53
Michailides TJ, Elmer PAG (2000) Botrytis gray mold of kiwifruit caused by *Botrytis cinerea* in the United States and New Zealand. Plant Dis 84:208–223
Mills DJ, Coffman CB, Teasdale JR et al (2002) Factors associated with foliar disease of staked fresh market tomatoes grown under differing bed strategies. Plant Dis 86:356–361
Mlikota Gabler F, Smilanick JL, Mansour M et al (2003) Correlations of morphological, anatomical, and chemical features of grape berries with resistance to *Botrytis cinerea*. Phytopathology 93:1263–1273
Morgan WM (1984) The effect of night temperature and glasshouse ventilation on the incidence of *Botrytis cinerea* in a late-planted tomato crop. Crop Prot 3:243–251
Nicot PC, Mermier M, Vaissière BE, Lagier J (1996) Differential spore production by *Botrytis cinerea* on agar medium and plant tissue under near-ultraviolet light-absorbing polyethylene film. Plant Dis 80:555–558
O'Neill TM, Shtienberg D, Elad Y (1997) Effect of some host and microclimate factors on infection of tomato stems by *Botrytis cinerea*. Plant Dis 81:36–40
Percival DC, Sullivan JA, Fisher KH (1993) Effect of cluster exposure, berry contact and cultivar on cuticular membrane formation and occurrence of bunch rot (*Botrytis cinerea* Pers.: Fr.) with 3 *Vitis vinifera* L. cultivars. Vitis 32:87–97
Percival DC, Fisher KH, Sullivan JA (1994) Use of fruit zone leaf removal with *Vitis vinifera* L. cv. Riesling grapevines. II. Effect on fruit composition, yield, and occurrence of bunch rot (*Botrytis cinerea* Pers.:Fr.). Am J Enol Viticult 45:133–140
Pertot I, Perin L (1999) Influence of N-fertilization on rot caused by *Botrytis cinerea* on kiwifruit in cold store. Notiziario dall'Ente Regionale per lo Sviluppo e la Promozione dell'Agricoltura del Friuli Venezia Giulia (ERSA) 12(6):39–41
Prasad M, Speirs TM (1991) The effect of nutrition on the storage quality of kiwifruit (A review). Acta Horticult 297:579–585
Redl H (1988) Results of a ten-year study on the suitability of one-wire training for wide-spaced, high-stemmed grapevine plantations. Vitis 27:33–40
Reuveni R, Raviv M (1992) The effect of spectrally modified polyethylene films on the development of *Botrytis cinerea* in greenhouse grown tomato plants. Biol Agric Hortic 9:77–86
Reuveni R, Raviv M, Bar R (1989) Sporulation of *Botrytis cinerea* as affected by photoselective sheets and filters. Ann Appl Biol 115:417–424
Ribéreau-Gayon J, Ribéreau-Gayon P, Seguin G (1980) *Botrytis cinerea* in enology. In: Coley-Smith JR, Verhoeff K, Jarvis WR (eds) The biology of *Botrytis*. Academic, London, pp 251–274
Saks Y, Copel A, Barkai Golan R (1996) Improvement of harvested strawberry quality by illumination: colour and *Botrytis* infection. Postharvest Biol Technol 8:19–27
Savage SD, Sall MA (1982) Vineyard cultural practices may help reduce *Botrytis* bunch rot caused by *Botrytis cinerea*. Calif Agric 36(2&3):8–9
Schwab M, Noga G, Barthlott W (1993) Influence of water and nutrient deficiency on epicuticular waxes of kohlrabi. Angew Bot 67:186–191

Sharabani G, Shtienberg D, Elad Y, Dinoor A (1999) Epidemiology of *Botrytis cinerea* in sweet basil and implications for disease management. Plant Dis 83:554–560

Shpialter L, Rav David D, Dori I et al (2009) Cultural methods and environmental conditions affecting gray mold and its management in lisianthus. Phytopathology 99:557–570

Shtienberg D, Elad Y (1997) Incorporation of weather forecasting in integrated, biological-chemical management of *Botrytis cinerea*. Phytopathology 87:332–340

Shtienberg D, Elad Y, Niv A et al (1998) Significance of leaf infection by *Botrytis cinerea* in stem rotting of tomatoes grown in non-heated greenhouses. Eur J Plant Pathol 104:753–763

Shtienberg D, Elad Y, Borenshtein M et al (2010) Polyethylene mulch modulates greenhouse microclimate and reduces infection of *Phytophthora infestans* in tomato and *Pseudoperonospora cubensis* in cucumber. Phytopathology 100:97–104

Sirjusingh C, Sutton JC (1996) Effects of wetness duration and temperature on infection of geranium by *Botrytis cinerea*. Plant Dis 80:160–165

Smithyman RP, Howell GS, Miller DP (1997) Influence of canopy configuration on vegetative development, yield, and fruit composition of Seyvalblanc grapevines. Am J Enol Viticult 48:482–491

Suelter CH (1970) Enzymes activated by monovalent cations. Science 168:789–795

Sundheim L (1973) *Botrytis fabae*, *B. cinerea* and *Ascochyta fabae* on broad bean (*Vicia faba*) in Norway. Acta Agric Scand 23:43–51

Thomas CS, Marois JJ, English JT (1988) The effects of wind speed, temperature, and relative humidity on development of aerial mycelium and conidia of *Botrytis cinerea* on grape. Phytopathology 78:260–265

Trolinger JC, Strider DL (1984) Botrytis blight of *Exacum affine* and its control. Phytopathology 74:1181–1188

Vail ME, Marois JJ (1991) Grape cluster architecture and the susceptibility of berries to *Botrytis cinerea*. Phytopathology 81:188–191

Verhoeff K (1968) Studies on *Botrytis cinerea* in tomatoes. Effect of soil nitrogen level and of methods of deleafing upon occurrence of *B. cinerea* under commercial conditions. Neth J Plant Pathol 74:184–192

Volpin H, Elad Y (1991) Influence of calcium nutrition on susceptibility of rose flowers to *Botrytis* blight. Phytopathology 81:1390–1394

Wegulo SN, Vilchez M (2007) Evaluation of lisianthus cultivars for resistance to *Botrytis cinerea*. Plant Dis 91:997–1001

West JS, Pearson S, Hadley P et al (2000) Spectral filters for the control of *Botrytis cinerea*. Ann Appl Biol 136:115–120

Williamson B, Jennings DL (1992) Resistance to cane and foliar diseases in red raspberry (*Rubus idaeus*) and related species. Euphytica 63:59–70

Wojcik P, Lewandowski M (2003) Effect of calcium and boron sprays on yield and quality of "Elsanta" strawberry. J Plant Nutr 26:671–682

Xiao CL, Chandler CK, Price JF et al (2001) Comparison of epidemics of Botrytis fruit rot and powdery mildew of strawberry in large plastic tunnel and field production systems. Plant Dis 85:901–909

Yermiyahu U, Shamai I, Peleg R et al (2006) Reduction of *Botrytis cinerea* sporulation in sweet basil by altering the concentrations of nitrogen and calcium in the irrigation solution. Plant Pathol 55:544–552

Yermiyahu U, Israeli L, Fogel M, Elad Y (2013) Enhanced potassium status in sweet basil reduced diseases caused by *Botrytis cinerea* and *Sclerotinia sclerotiorum*. Proc Int Plant Nutr Colloq XVII:469–470

Zoecklein BW, Wolf TK, Duncan NW et al (1992) Effects of fruit zone leaf removal on yield, fruit composition, and fruit rot incidence of Chardonnay and White Riesling (*Vitis vinifera* L.) grapes. Am J Enol Viticult 43:139–148

Chapter 9
Biological Control and Biopesticide Suppression of *Botrytis*-Incited Diseases

Philippe C. Nicot, Alison Stewart, Marc Bardin, and Yigal Elad

Abstract Recent years have seen the development of many biological control agents and other biopesticides, such as plant extracts, minerals and organic compounds, against *Botrytis*-incited diseases. This chapter presents significant examples of such commercially available products and reviews our increasing comprehension of mechanisms implicated in biological control, including recent breakthroughs on complex interactions between the biocontrol agent, the host plant and the pathogen. Its highlights progress made in characterizing the determinants of efficacy and documents the growing body of knowledge on natural biological control and the potential for its enhancement. Finally, future prospects and challenges are presented, including issues on the durability of biocontrol methods.

Keywords Biocontrol • Mode of action • Efficacy

9.1 Introduction

Botrytis-incited diseases have long been the target of efforts to develop biological control and successful protection of vegetable crops with applications of microbial preparations was already reported in the 1950s (Dubos 1992; Newhook 1957; Wood 1951). Due to difficulties associated with chemical control (Chap. 10) and growing societal pressure to reduce pesticide use, alternative methods and strategies of integrated protection (Chap. 8) have gained increased attention worldwide. Although anti-*Botrytis* biocontrol products long remained rare despite abundant research efforts (Nicot et al. 2011), and biological control of plant diseases was

P.C. Nicot (✉) • M. Bardin
INRA, UR407 Pathologie végétale, CS 60094, 84143 Montfavet cedex, France
e-mail: philippe.nicot@avignon.inra.fr

A. Stewart
Marrone Bio Innovations, 1540 Drew Avenue, Davis, CA 95618, USA

Y. Elad
Department of Plant Pathology and Weed Research, Agricultural Research Organization, The Volcani Center, Bet Dagan 50250, Israel

S. Fillinger, Y. Elad (eds.), *Botrytis – the Fungus, the Pathogen and its Management in Agricultural Systems*, DOI 10.1007/978-3-319-23371-0_9

often considered as a weak link in integrated management strategies in certain crops (Nicot and Bardin 2012), recent years have seen a boost in the development and use of biopesticides against grey mould. The term "biopesticide" is meant here to include preparations based on living microorganisms (referred to as biocontrol agents or "BCA" in the present chapter) and substances of natural origin such as plant and microbial extracts, minerals and organic compounds. In parallel to this commercial development of biocontrol agents, tremendous progress has been achieved in our comprehension of their modes of action and in the characterisation of factors that determine their efficacy. Much interest has also developed on natural biological control and the potential for its enhancement.

9.2 Types of Products and Commercial Availability

A wide range of different plant and microbe groups have been reported to exhibit inhibitory activity against *Botrytis cinerea* in laboratory and greenhouse trials but only few have shown consistent field performance and even fewer have been commercialized (Nicot et al. 2011). This section describes representative examples of three groups of commercially available biopesticides (Table 9.1). For each group, we highlight a number of key examples of products with differing active ingredients and/or modes of action and discuss aspects related to extent of commercialization. Many of them share the following key features: (i) registration for use very close to harvest (due to absence of toxicity and residue issues), (ii) approval for use in organic farming in various countries, for example through the Organic Materials Review Institute (OMRI; http://www.omri.org/) or the Ecocert certification body (http://www.ecocert.com/en/). There are several registered commercial products based on plant extracts but, to date, none based on microbial extracts.

9.2.1 Plant Extracts

Melaleuca alternifolia An extract of the tea tree is commercialized as Timorex Gold®24EC. It has broad-spectrum activity against various fungal pathogens including *B. cinerea* in vegetables, herbs, grapevines and orchards. In field trials on grapes in California, it gave >90 % suppression of grey mould incidence which was equal to the best fungicide treatments (Nguyen et al. 2013). It is registered in over 25 countries including USA, Canada, most countries in Central and South America and several countries in Europe and South-east Asia. It exhibits both prophylactic and curative efficacy, which acts against various stages of the infection process. Tea tree extract has multi-target fungicidal effects and inhibits conidial germination, germ tube growth and mycelial growth of *B. cinerea* (Antonov et al. 1997).

Table 9.1 Examples of commercial biopesticides with activity against *Botrytis*-incited diseases

Active ingredient	Product	Mode of action[a]	Company
Plant extracts			
Melaleuca alternifolia	Timorex Gold®	Antifungal	STK Stockton Group www.stockton-ag.com
Reynoutria sachalinensis	Regalia®	ISR	Marrone Bio Innovations www.marronebioinnovations.com
Micro-organisms			
Aureobasidium pullulans strains 14940/14941	Botector®	Competitive exclusion	Bio-ferm www.bio-ferm.com
Bacillus amyloliquefaciens	Double Nickel 55WDG/LC™	Antimicrobial	Certis USA www.certisusa.com
B. subtilis QST 713	Serenade® Max	Antimicrobial, induction of plant defences	Bayer Crop Science www.bayercropscience.com
B. subtilis GB03	Companion®	Antibiosis (iturins), ISR	Growth Products www.growthproducts.com
B. megaterium	Bio Arc	Enzymatic action	Sphere Bio-Arc PVT Ltd
Chlonostachys rosea	Endofine®	Competition	Adjuvants Plus www.adjuvantsplus.com
Gliocladium catenulatum J1446	Prestop®	Competition, hyperparasitism	Verdera Oy www.agbio-inc.com
Pseudomonas syringae ESC-10	Bio-save®	Competition	Jet Harvest Solutions www.jetharvest.com
Streptomyces griseoviridis K61	Mycostop®	Competition	Verdera Oy www.verdera.fi

(continued)

Table 9.1 (continued)

Active ingredient	Product	Mode of action[a]	Company
S. lydicus WYCD108	Actinovate®	Competition, antibiosis	Novozymes
			www.bioag.novozymes.com
Trichoderma atroviride LC52	Sentinel®	Competitive exclusion	Agrimm Technologies Ltd
			www.tricho.com
T. harzianum + *T. polysporum*	Binab TF	Antibiosis, SAR	Binab Bio Innovation AB
			www.binab.com
T. harzianum	Supresivit	Competition	Borregaard Bioplant
			www.bioplant.com
Ulocladium novo-zealandiae	Botryzen®	Competitive exclusion	Botryzen NZ Ltd
			www.botryzen.com
Mineral oils			
Neem oil	Trilogy®	Fungicidal	Certis USA
	Triact® 70EC		www.certisusa.com
Paraffinic oil	JMS Stylet Oil®	Fungistatic	JMS Flower Farms Inc
			www.stylet-oil.com
Paraffinic oil	Suffoil-X®	Fungistatic	Bioworks
			www.bioworks.com
Thyme oil	Promax™	Fungicidal	Bio Huma Netics Inc
			www.bhn.us

[a]*ISR* Induced systemic resistance, *SAR* Systemically acquired resistance

Reynoutria sachalinensis An extract of the giant knotweed is commercialized as Regalia®. It has a broad spectrum of activity against both foliar and soil-borne fungal and bacterial pathogens, including *B. cinerea* on tomato, grape, pepper, cucurbit, strawberry, walnut, and turf. The product was first registered in the USA in 2006 with subsequent registrations in Canada, Turkey, South Africa and several South American countries.

Several active compounds have been identified such as emodin and physcion and these have been shown to trigger Induced Systemic Resistance (ISR) (Schmitt et al. 1996, 2005). Regalia® treatment increases the production and accumulation of specialized proteins and other compounds known to inhibit *Botrytis* (Marrone 2002). It also increases papillae formation at pathogen penetration sites as well as the lignification of plant cell walls. The active ingredient provides translaminar protection (Quarles 2009; Su 2012). Since induction of ISR takes 1–2 days to exhibit, the product must be used preventively.

9.2.2 Living Microorganisms

This group contains the largest number of commercial botryticide products with active ingredients from a wide range of microbial groups including bacteria (e.g. *Bacillus*, *Pseudomonas*), actinomycetes (e.g. *Streptomyces*), yeasts (e.g. *Aureobasidium*, *Candida*) and fungi (e.g. *Trichoderma*, *Gliocladium*, *Chlonostachys*, *Ulocladium*). Representative examples are described below.

Aureobasidium pullulans Two strains of this yeast (14940 and 14941, isolated from untreated apple trees) constitute the active ingredient of Botector®. This product is registered for grape protection in the USA and several countries in Europe and also has a label claim for tomato and ornamentals. It can be applied shortly before harvest and does not affect the colour or taste of the treated grapes. The product works through natural competition for space and nutrients on the berries. Due to the high proliferation rate of *A. pullulans* after application, the pathogen cannot infect the plant (Schilder 2013).

Bacillus subtilis Serenade®Max contains strain QST 713 of this bacterium. It has a wide spectrum of activity against various fungal and bacterial pathogens of numerous crops including *Botrytis* spp. on grapes, citrus, berry fruit, kiwifruit, persimmon, onion, garlic, tomato, pepper, eggplant, cucumber, squash, and bean. It is suitable for organic production and can be used close to harvest. The product claims >2 year storage stability and compatibility with standard fungicides. Its mode of action is a combination of nutrient competition, site exclusion, direct antifungal activity and induction of the plant's natural systemic resistance. The bacterium has been reported to produce three different classes of lipopeptides that disrupt fungal cell membranes, causing the cells to collapse and killing the pathogen (Ongena et al. 2010). In addition, upon contact, Serenade® Max triggers the plant's internal defences and physiological responses. The effect is systemic and appears to activate a pathway (related to induction of PR1) distinct from those of systemically acquired resistance (SAR) and induced systemic resistance (ISR) (Ongena et al. 2010).

Gliocladium catenulatum Prestop® contains mycelium and conidia of strain J1446 of *G. catenulatum*. The product has a shelf life of 12 months in cool storage conditions (<8 °C) and is safe for beneficials and compatible with most chemical pesticides. The product is registered in the USA, Canada and in various European countries. It is particularly active against grey mould and stem canker caused by *B. cinerea* on tomato, pepper, cucumber, lettuce, herbs, and ornamental plants. A specific formulation of the same active ingredient is applied through pollinator-assisted distribution (delivered via honeybees) for the control of grey mould on strawberry and raspberry. For both products, the antagonist deprives the pathogen of space and nourishment by colonizing the plant surfaces and outcompeting the pathogen. It also acts as a hyperparasite by producing enzymes which disrupt the cell walls of the pathogen.

Streptomyces **sp.** Strain K61 of the actinomycete *S. griseoviridis* is the active ingredient of Mycostop®. This product can control or suppress many root rot and wilt fungi as well as *B. cinerea*. It is suitable for organic production and is registered for use on a wide range of crops such as vegetables, herbs, and ornamentals. It has been registered in the USA since 1994 with additional registrations in more than 15 countries. After application, the microbe grows on the plant surface, creating a biological defence against plant pathogenic fungi.

Another actinomycete, *S. lydicus* strain WYEC108 is the active ingredient of Actinovate® AG, a product commercialized for the control of a wide range of foliar diseases including grey mould. It is registered in the USA, Canada, Turkey, India, New Zealand, Vietnam and Trinidad, with EU registration currently in progress. The product is approved for organic farming. It has a zero-day pre-harvest interval and can be tank mixed or dry mixed with a wide range of fungicides, insecticides and fertilizers. The microbe grows on the surface of the foliage and acts against the pathogen through a combination of competitive exclusion and secretion of antifungal compounds, including lytic enzymes such as chitinases, glucanases and peroxidases (Crawford et al. 2005; Lichatowich 2007). The microbe has also been reported to act through direct parasitism.

Trichoderma atroviride Strain LC52 of this fungus is the active ingredient of Sentinel®, a bio-fungicide commercialized against grey mould in grape, strawberry, and tomato. It was registered in New Zealand in 2006, with registrations pending in several other countries. The fungus aggressively colonizes leaves, flowers and truss/leaf plucking wound sites, where it actively competes with and repels invasion by *B. cinerea* (Card 2005). Sentinel acts as a protectant bio-fungicide and has no curative effect. It can be applied until harvest.

9.2.3 Mineral Oils and Organic Acids

A number of oil based products have claims for control of *Botrytis*-incited diseases. The two most common are paraffinic oil and neem oil.

Paraffinic Oil JMS Stylet-Oil® is a 1 % paraffinic oil used to control certain fungal diseases including Botrytis bunch rot. It may smother fungal growth and reduce conidia germination on treated surfaces. It is mostly fungistatic, stopping fungal growth rather than killing the pathogen. To combat fungal pathogens, the oil generally must be applied prophylactically prior to infection. Repeated applications of the oil may be needed to achieve desired levels of control.

Neem Oil Trilogy®90EC is a commercial product based on clarified hydrophobic extract of neem oil. This product is generally regarded as being most active against powdery mildew diseases but is also reported to have activity against *B. cinerea* on citrus, cucurbits, bulbs, vegetables, small fruit and nuts. Thorough coverage of the plant is essential and care must be taken with time of application to prevent the risk of leaf burn. The product has a re-entry interval of 4 days and can be applied up until the day of harvest. It is suitable for organic production (OMRI listed) but applications are restricted during the bloom period due to its toxicity to bees. The product works by preventing fungal attack of the plant tissue.

9.3 Modes of Actions

Botrytis spp. are affected by BCAs that utilize various strategies during their interaction while suppression of the pathogen is achieved. *Botrytis* spp. conidia need the presence of water and exogenous nutrients for germination and host penetration. On the plant surface the germinating conidia can be affected by inhibitory substances, competition for nutrients and space, limitation of saprophytic ability. Following infection, the invading pathogen is affected by host characteristics that are typical to defending host tissues as occurs with induced resistance (Elad and Stewart 2004). In many cases, it can be assumed that biocontrol is a result of several combined modes of action of the BCA.

Reduction of the concentration of nutrients such as nitrogen, carbon, macro-elements and micro-elements results in a reduced rate of germination of the pathogen's conidia and thus the extent of infection. This can be achieved by competition for nutrients (Blakeman 1993). Examples of competition for nutrients or space include yeasts such as *Candida oleophila* applied on apple fruits (Mercier and Wilson 1994) or *Aureobasidium pullulans* on grapes (Schilder 2013), as well as bacteria and filamentous fungi (Zimand et al. 1996).

Modification of plant surface properties may change pathogen behavior on plant surfaces. For example, *Bacillus brevis* applied to Chinese cabbage caused water drops to spread and dry, and generally changed the wettability of plant surfaces

(Edwards and Seddon 1992). Furthermore, the attachment of microorganisms to the pathogen can be implicated in biocontrol. The attachment of the yeasts *Rhodotorula glutinis* and *Cryptococcus albidus* to conidia of *B. cinerea* is associated with the formation of a fibrillar material. This material is likely to be the pathogen's polysaccharide extracellular matrix (ECM) and the involvement of lectin-like compounds was suggested (Elad 1996). Scanning electron microscopy showed attachments of the *Trichoderma* conidia to the ECM of *B. cinerea* (Meyer et al. 2001).

The inhibition of *Botrytis* spp. by microbial compounds was thoroughly studied (Elad and Stewart 2004). For instance, *Penicillium chrysogenum* produced inhibitory products, which reduced conidial germination of *B. fabae* and lesion development on faba bean (Jackson et al. 1994), and peptaibol antibiotics from *T. harzianum* and gliotoxin from *G. virens* inhibited conidia of *B. cinerea* (Schirmböck et al. 1994). Control of *B. cinerea* by *Bacillus subtilis* and *Bacillus pumilus* was related to antibiotic production (Leifert et al. 1995); *Bacillus brevis* secretes gramicidin S that is very inhibitory on *B. cinerea* (Edwards and Seddon 1992) and a UV-induced antibiotic deficient mutant of *B. subtilis* was not active. Other bacteria such as *Serratia plymuthica* and several species of *Pseudomonas* described as BCAs are known to produce pyrrolnitrin, an antibiotic which inhibits mycelial growth of *B. cinerea* (Ajouz et al. 2011). Volatile compounds may also be involved. Inhibitory volatile-producing fungus *T. hamatum* reduced grey mould on snap bean (Nelson and Powelson 1988). It should be noted that the antimicrobial compounds may affect not only the plant pathogen but also non-target organisms.

Hyperparasitism of fungi by microorganisms is a well-known phenomenon (Elad and Stewart 2004). For instance *Trichoderma, Gliocladium* and *Pythium* spp. are among the best-known mycoparasites able to attack the mycelium of *B. cinerea* (Elad 1996). Parasitism of sclerotia has also been described (Köhl and Schlösser 1989; Dubos et al. 1982). Several enzymes have been implicated in mycoparasitism, including cell wall degrading enzymes (CWDEs) such as proteinases, mannanases, laminarinases and chitinases (Labudova and Gogorova 1988), and genes coding for certain of these enzymes have been identified (Geremia et al. 1993; Viterbo et al. 2001; Kamensky et al. 2003). Further progress has been achieved recently in the understanding of the *Botrytis*-BCA interactions with the analysis of secreted protein patterns of *T. harzianum* ETS 323 in laboratory conditions (Yang et al. 2009). One L-amino acid oxidase (LAAO) and two endochitinases of the BCA were uniquely induced in the medium that contained deactivated *B. cinerea* mycelium as the sole carbon source.

Activities of the CWDEs, β-1,3-glucanases, β-1,6-glucanases, chitinases, proteases and xylanases, were significantly higher in media with deactivated *B. cinerea* mycelium than in other media, suggesting that the cell wall of *B. cinerea* is indeed the primary target of the BCA in the biocontrol mechanism (Yang et al. 2009). The effect of *T. harzianum* ETS 323 against *B. cinerea* during the mycoparasitic process was evaluated in culture using a biexponential equation (Cheng et al. 2012). The secretion of LAAO by *T. harzianum* ETS 323 was increased when the BCA was grown with deactivated hyphae of *B. cinerea* and this oxidase inhibited *B. cinerea* growth in vitro and on apple fruit and tobacco leaves. Furthermore, an apoptosis-like response, including the generation of reactive oxygen species, was observed in

B. cinerea after treatment with LAAO, suggesting that it triggers programmed cell death in *B. cinerea*. Cheng et al. (2012) suggested a two-step antagonism of the BCA against *B. cinerea*. In practice, however, a BCA's ability to act as a mycoparasite or to produce CWDEs does not necessarily guarantee effective biocontrol under field conditions since the activity of mycoparasites is generally regarded as too slow to reduce effectively the fast penetration process of *Botrytis* into the host tissue.

Interference with the Pathogenicity Processes. The *T. harzianum* strain T39 prevents penetration of *B. cinerea* into the host tissue and interference with the pathogenicity processes was reported as being responsible for this effect (Zimand et al. 1996). T39 reduced the activities of exo- and endo-polygalacturonase, pectin methyl esterase and pectate lyase (Zimand et al. 1996), chitinase, β-1,3-glucanase and cutinase produced by *B. cinerea* (Kapat et al. 1998). It was shown that *T. harzianum* T39 produced a cysteine protease that reduced the activity of pathogenicity-related enzymes of *B. cinerea* and subsequent disease development (Elad and Kapat 1999) as part of its biocontrol mechanism; this was further proven by the fact that a specific inhibitor of the T39 protease nullified biocontrol activity (Elad et al. 1998). Another pathogenicity factor that is produced by *B. cinerea* and *Sclerotinia sclerotiorum* is oxalic acid (Chap. 12). Schoonbeek et al. (2007) have found oxalate-degrading bacteria that have protective activity on *Arabidopsis thaliana* cucumber, grapevine, and tomato leaves against *B. cinerea*. Two yeasts with higher (*Cryptococcus laurentii* LS-28) or lower (*Rhodotorula glutinis*) antagonistic activity against the postharvest pathogens *B. cinerea* and *Penicillium expansum* were tested. LS-28 was more resistant to ROS-generated oxidative stress. Castoria et al. (2003) suggested that resistance to oxidative stress may be a mechanism of biocontrol yeasts antagonism against postharvest wound pathogens.

Reduction in Inoculum Production. As epidemics created by *Botrytis* spp. are commonly polycyclic, reduction in inoculum production may create a cumulative effect over several disease cycles (Köhl and Fokkema 1993). Several microorganisms suppress conidiation of *B. cinerea* on strawberry (Peng and Sutton 1991) and other crops (Morandi et al. 2000). *Ulocladium atrum* reduced sporulation of *B. cinerea* on dead leaves of lily and onion exposed to field conditions (Köhl et al. 1995). Colonisation of necrotic tissue by *U. atrum* prevents saprophytic colonisation of those leaves by *B. cinerea*. Two antibiotic-producing isolates of *Serratia liquefaciens* reduced conidiation on irradiated grape leaves and the degree of suppression was affected by the antibiotic production by the bacteria (Whiteman and Stewart 1998).

Induced resistance (IR) was recognized in recent years as an important mode of disease suppression. While several types of micro-organisms have long been known to induce resistance to *B. cinerea*, including *Pseudomonas aeruginosa* (Audenaert et al. 2001) and *Trichoderma* spp. (De Meyer et al. 1998), tremendous progress has recently been made in the comprehension of implicated mechanisms. Studying the effect of metabolite complexes secreted by *T. atroviride* on plant cells, Navazio et al. (2007) found that secreted fungal molecules are sensed by plant cells through intracellular Ca^{2+} changes and that plant cells have the ability to discriminate signals originating in the single or two-fungal partner interaction and modulate defence responses. Indeed, plant-*Trichoderma* spp. interaction is known to correlate with changes in the plant proteome and transcriptome (Shoresh et al. 2010; Hermosa et al. 2012).

Recently, *T. harzianum* T39 was shown to be involved in a complex transcriptional reprogramming in grapevine (Palmieri et al. 2012), affecting proteins associated with stress responses, photosynthesis, redox signalling and energy metabolism (Perazzolli et al. 2012). The hormonal and molecular IR pathways involved in *Trichoderma* spp.-plant interactions are usually dependent on the jasmonate (JA) and/or ethylene (ET) signaling pathways typical of ISR (Shoresh et al. 2005, 2010; Korolev et al. 2008; Perazzolli et al. 2008). In some cases, *Trichoderma* spp. induced the expression of defence-related genes associated with both the salicylate (SA) and the JA/ET pathways (Hermosa et al. 2012). In addition, the effects of *Trichoderma-based* treatments on the expression levels of defence-related genes vary with the type of treatment, the type of tissue analysed, and plant age. For instance, SA-induced gene expression was observed in tomato foliage after long-term *Trichoderma* spp. treatment, soil incorporated or applied as seed coating (Alfano et al. 2007; Tucci et al. 2011), but opposite results were observed when T39 was sprayed onto grapevine foliage (Perazzolli et al. 2011, 2012).

Expression levels of defence-related genes in plants often reflect the state of the plant's awareness of biotic and abiotic stresses. Observations of *B. cinerea* infection in leaves harvested from plants grown in the treated soils revealed that drenching with a T39 suspension induced systemic resistance against *B. cinerea* and primed SA- and ET-related gene expression in a manner proportional to the concentration of *Trichoderma* (Meller Harel et al. 2014). Similarly, up-regulation of *TomloxA* and, to a lesser extent, *TomloxC* was observed in tomato plants grown from seeds coated with *T. harzianum* T22 (Tucci et al. 2011). *T. harzianum* T39 induced strong priming effects on *PR1a*, *Chi9*, and *GluB* expression. The priming effect of T39 on defence- and microbial recognition-related gene expression upon infection with *B. cinerea* is expected for a BCA triggering the ISR pathway (Ahn et al. 2007) and has been reported (Malmierca et al. 2012; Shoresh et al. 2005; Palmieri et al. 2012). Expression of the ET-related genes *EFR1* and *ACO1* was also primed by the T39 treatments. Opposite results were reported in regards of gene expression induced by strain T22, showing up-regulation of SA-marker genes before infection with *B. cinerea* and the down-regulation of those genes following infection (Tucci et al. 2011). In the case of *P. viticola*-grapevine interactions, induction of *Lox9* genes transcription was observed in T39-treated plants (Perazzolli et al. 2011). Prior to inoculation, the SA-responsive genes were inhibited by T39. Following inoculation with *B. cinerea*, T39 induced strong priming of the SA-responsive genes expression. Finally, the research with *T. atroviride* and *T. harzianum* and various tomato lines revealed that induced systemic resistance was evident for some, but not all, the tested tomato lines. This means that the genetic background of the host plant is important for response to the resistance induction (Tucci et al. 2011).

Gene expression was followed also in fruits at postharvest stage. In apple fruits treated by *Candida oleophila* and *B. cinerea*, PR-8 expression was significantly elevated in response to both fungi. The PR-8 gene was cloned into a *Pichia pastoris* expression system revealing that the PR protein it codes for may play a role in the inhibition of *B. cinerea* by *C. oleophila* in apple fruit, namely by the induction of this specific host PR gene (Liu et al. 2013).

9.4 Determinants of Efficacy

Although examples of steady efficacy are not uncommon (Calvo-Garrido et al. 2013; Ilhan and Karabulut 2013), biological control is generally considered as inconsistent in field conditions, either from one growing season to the next or over different sites (Nicot et al. 2011). This is generally blamed on its vulnerability to changeable environmental conditions and/or on an unstable quality of the products. It may also result from variability in the sensitivity of plant pathogens to BCAs.

9.4.1 Environmental Conditions and Stability of the Active Ingredient

Fluctuating environmental conditions in commercial production systems can influence the survival, establishment and activity of BCAs (Morandi et al. 2008). Many studies have evaluated the effect of microclimate on the efficacy of BCAs in controlled or semi-controlled conditions. In most cases, both temperature and relative humidity were key factors. For example, *T. harzianum* T39 was more effective in controlling *B. cinerea* on cucumber when temperature was above 20 °C and relative humidity in the range of 80–97 % (Elad et al. 1993). Similarly, high day-time temperature and high relative humidity overnight were associated with a reduction in efficacy of *A. pullulans*, *C. albidus* and *T. harzianum* to control *B. cinerea* on cucumber and tomato (Dik and Elad 1999). Even small-amplitude variations in climate can strongly influence biocontrol efficacy. Among seven BCAs tested by Hannusch and Boland (1996), six were highly dependent on climatic conditions. Changes by 4 °C in temperature or 5 % in relative humidity (in the range of 20–28 °C and 90–100 % RH) were sufficient to cause variability in grey mould suppression on bean that ranged from 15 to 100 %.

In contrast, consistent biocontrol efficacy over wide ranges of temperatures and highly fluctuating relative humidity may be achieved in some situations. In the context of postharvest for example, *Rhodotorula glutinis* significantly reduced the development of *B. cinerea* on strawberry during 5 days of storage at 4 °C followed by 3 days at 20 °C (Zhang et al. 2007). In specific cases, such as the protection of pruning wounds of tomato, temperature but not relative humidity (between 50 and 90 %) affected the efficacy of *Microdochium dimerum* and *U. atrum* to reduce the incidence of pruning wound infection by *B. cinerea* (Nicot et al. 2002). However, a complicating factor may be that the effect of microclimate is not necessarily the same for all components of biocontrol efficacy. For example, in contrast to the protection against wound infection, other components of efficacy such as delay of infection, expansion of stem lesions and inhibition of sporulation were little influenced by the microclimatic conditions studied (Nicot et al. 2002).

In addition to a necessary adaptation to their physical environment, the persistence of BCAs on the plant after application requires that they can also compete with or blend in communities of microorganisms that may change overtime. In a recent study, strawberry plants were repeatedly sprayed with three BCAs (*Bacillus amyloliquefaciens*, *T. harzianum* and *Beauveria bassiana*) to suppress *B. cinerea* infections. Both *Bacillus* and *Trichoderma* spp. persisted in the strawberry phyllosphere throughout the strawberry season, suggesting a high competitive ability on the leaves. In parallel, no significant impact of the treatments was detected on the leaf microbiota (Sylla et al. 2013). In some cases the natural microflora can be beneficial for efficacy, as shown on apple wounds for *C. oleophila* against *B. cinerea* (Mercier and Wilson 1994).

9.4.2 Dose Effect and "Quality" of the Preparations

The protective efficacy of a BCA depends on features of the product that influence its persistence on the plant. In this matter, microbial products present particular challenges when their effect depends on the presence of a sufficient amount of living cells (Zhang et al. 2009; Kowalska et al. 2012; Mari et al. 2012; Nicot et al. 2002). Commercial products have to be engineered to contain a minimum concentration of living cells and retain it through the chain of distribution, so it is sufficient at the time of application by the farmer. Formulations have been designed to preserve both the survival of the BCA and its protective properties, thus ensuring a satisfactory shelf life as shown for *Pseudomonas* (Janisiewicz and Jeffers 1997), *A. pullulans* (Mounir et al. 2007) or various *Trichoderma* sp. (Ruocco et al. 2011). Production processes also play an important role (Jackson et al. 1991) and substantial gains in protection efficacy can be obtained by optimizing the nutrient substrate composition as shown for *Rhodotorula* (Calvente et al. 2001) and *Pichia carribbica* (Zhao et al. 2012, 2013). Although much innovation is needed on this topic, currently published information is quite limited.

9.4.3 Delivery Systems and Timing of Applications

Timely and accurate coverage of the plant tissue to be protected may be a key factor for successful biological control, requiring the target to be precisely identified so as to define the pertinent application method. In heated tomato greenhouses for example, grey mould symptoms develop mainly on stems, following infection of pruning wounds. The BCA *Fusarium* sp., developed for its efficacy in protecting pruning wounds of tomato plants was equally efficacious when applied with a sprayer or through pruning shears equipped with a small nozzle which sprayed a conidia suspension of the BCA directly on the blades (Decognet et al. 1999). Similarly, spray directed on the disease lesion on tomato enhanced the efficacy of *G. catenulatum-based* product Prestop® (Lahdenpera and Korteniemi 2008).

Timing of BCA application, often as a preventive treatment, is also an important factor. For instance, *Trichoderma* sp. is highly effective when applied to blossoms or fruits for control of *B. cinerea* on strawberry. Even low levels of the organism applied to strawberry blossoms by bee delivery or by sprays of liquid formulations are effective (Ruocco et al. 2011). In a greenhouse experiment, treatment of blueberry blossoms with *Clonostachys rosea* significantly reduced establishment of *B. cinerea*, while application of the BCA after establishment of the pathogen was not effective (Reeh and Cutler 2013). The greatest inhibition of grey mould on apple, using *A. pullulans*, was observed when fruits were treated with the antagonist no later than 6 h after inoculation (Mari et al. 2012). Conversely, *G. catenulatum* was able to stop further development of infection of *B. cinerea* even in situations where symptoms already occured (Lahdenpera and Korteniemi 2008).

Increased biocontrol efficacy may also be achieved by combining various compounds together with a given BCA. Many compounds such as calcium salts, ammonium molybdate, chitosan, amino acids, salicylic acid, carbohydrates and essential oils were shown to enhance the efficacy of BCAs against *B. cinerea* (Nunes et al. 2002a; El-Ghaouth et al. 2000a, b; Guetsky et al. 2002a; Yu et al. 2012; Zamani-Zadeh et al. 2013; Zhang et al. 2010). Combining BCAs is also generally reported as beneficial (Magnin-Robert et al. 2013; Nunes et al. 2002b; Guetsky et al. 2002b) although this is not always the case, as shown for resistance-inducing *T. harzianum* Tr6 and *Pseudomonas* sp. Ps14 (Alizadeh et al. 2013).

9.4.4 Host-Dependent Effects

As *B. cinerea* has a broad host-range, demonstrating the efficacy of a BCA on several of these plant species can make their commercial development more attractive (Dik and Wubben 2007; Köhl 2004). To evaluate the potential market size of *U. atrum*, tests were conducted on strawberry, onion, cyclamen, pelargonium, pot roses, tomato, grapevine, and lily, with convincing efficacy for most of these species (Köhl 2004). Conversely, post-harvest rot control efficacy of *Metschnikowia fructicola* was strongly dependent on the soft fruit type tested (Ferrari et al. 2007). Recent studies have also indicated that the efficacy of BCAs may be highly different depending on the variety of a given plant species, for example as mentioned above for resistance-inducing *Trichoderma* spp. on tomato (Tucci et al. 2011).

Little is known on the possible role of mineral plant nutrition on the efficacy of BCAs, although fertilisation was shown to affect the impact of *B. cinerea* on various crops (Lecompte et al. 2010, 2013). Recent studies have reported an effect of nitrogen fertilisation on the efficacy of defence-stimulating biocontrol products (chitosan, *B. subtilis*-based Serenade Max®) against *B. cinerea* on strawberry plants (Nicot et al. 2013) and on *T. atroviride* and *M. dimerum* against *B. cinerea* of tomato (Abro et al. 2014). Interestingly, high levels of nitrogen enhanced biocontrol efficacy in the tomato experiments, while the opposite effect was observed on strawberry. Further work will be needed to fully understand the underlying mechanisms.

9.4.5 Variability in the Susceptibility of **Botrytis** *spp.*

Few studies have evaluated differences in sensitivity to BCAs among populations of *B. cinerea*. Recently, 204 isolates were tested for their sensitivity to pyrrolnitrin, an antibiotic produced by various BCAs (Ajouz et al. 2011). A broad range of diversity was observed, with an 8.4-fold difference between the most and the least sensitive isolates of *B. cinerea*. Despite these differences, the isolates were equally well controlled on tomato by a pyrrolnitrin-producing strain of *Pseudomonas chlororaphis*. Diversity was also found in the sensitivity of 29 *B. cinerea* isolates to *R. glutinis* PM4 on geranium leaf discs (Buck and Jeffers 2004). Some isolates were totally insensitive while others produced significantly reduced symptoms in the presence of the BCA. Similar results were found with 40 isolates of *B. cinerea* on tomato and lettuce leaves in the presence of *B. subtilis* QST713 (Bardin et al. 2013a). The protective effect of the bacterium ranged from 40 to 86 % on tomato leaves and 0–80 % on lettuce leaves, depending on the *B. cinerea* isolate. In another study, in contrast, all isolates were equally well controlled by a preparation of *M. dimerum* used at the recommended dose (Bardin et al. 2013b). However, differences in protective efficacy were observed among isolates of *B. cinerea* when the BCA was used at a ten-fold reduced dose. In this case, a correlation was noticed between the level of aggressiveness of *B. cinerea* and the level of protection provided by the BCA (Bardin et al. 2013b). This correlation was presumably the consequence of the putative mode of action of this BCA, which is not based on a direct effect but considered to be essentially competition for nutrients.

Taken together, these results show that *B. cinerea* populations have a high level of diversity in their sensitivity to BCAs regardless of their mode of action. Gradual selection of less sensitive isolates and reduced effectiveness of BCAs against *B. cinerea* in the field should thus not be regarded as an impossible outcome of selection pressure, if biocontrol became widely and intensively used by farmers. Indeed, studies of Li and Leifert (1994) have shown that *B. cinerea* was able to gradually build up resistance to *B. subtilis* CL27 although this BCA produces three different antifungal antibiotics, two of which were effective in vitro against *B. cinerea* (Leifert et al. 1995). It only took ten successive treatments with the BCA to observe a complete loss of efficacy (Li and Leifert 1994).

In evolutionary experiments conducted in vitro, strains of *B. cinerea* consistently became resistant to pyrrolnitrin after successive generations produced under selection pressure (Ajouz et al. 2010). This buildup of resistance was accompanied with a loss of susceptibility in vitro to the inhibitory effect of pyrrolnitrin-producing *P. chlororaphis* but also a reduction in aggressiveness of the more resistant strains. Furthermore, selection pressure exerted by the fungicide iprodione in similar experiments also led to the build-up of resistance to pyrrolnitrin and to *P. chlororaphis* due to similar modes of action (Fillinger et al. 2012). These results suggest that the use of chemical methods in parallel to or in combination with biological control need to be carefully thought out in order to avoid any impact on the durability of the efficacy of certain BCAs.

9.5 Natural Biological Control and Its Enhancement

Traditionally, agricultural patho-systems implicating *Botrytis* spp. are described with three levels of biological interactions (BCA – *Botrytis*, *Botrytis* – plant, BCA – plant) within a general context of possible effect of the abiotic environment. However, an additional component has to be considered in this system, namely the indigenous microflora on the plant surface. This microflora may interact with the pathogen and the plant and affect the outcome of the plant – pathogen interaction. Interestingly, the microflora composition is affected by changes in the plant growth medium, as reflected in rhizosphere microflora, especially as a result of various soil amendments.

Composts produced with various types of organic matter are commonly incorporated to soil and substrates in which plants are grown. There is evidence that foliar diseases including grey mould can be reduced by such amendments (Segarra et al. 2007; Vallad et al. 2003; Yogev et al. 2010; Zhang et al. 1998). A compost of olive origin was shown to induce resistance against *B. cinerea* in *Arabidopsis thaliana*, in parallel with triggering the expression of various genes related to plant response to both microbial and abiotic stress stimuli (Segarra et al. 2013). This highlighted the possible role of soil microflora in providing natural biocontrol against *Botrytis*.

Soil solarisation is effective in controlling a variety of soilborne pathogens (Katan and DeVay 1991; Stevens et al. 2003). Mechanisms of control are both physical and biological as solarisation stimulates rhizobacteria, such as fluorescent pseudomonads and *Bacillus* spp. (Stapleton and DeVay 1984; Gamliel and Katan 1991). These microorganisms are potential inducers of plant resistance and their enhancement in the solarised soil could contribute indirectly to induced resistance (Katan and Gamliel 2012). Strawberry, cucumber and common bean grown on solarised substrates showed significant reduction in disease after leaf inoculation with *B. cinerea*, hence indicating induced resistance that is systemic (Okon Levy et al. 2004). Furthermore, denaturizing gradient gel electrophoresis (DGGE) revealed changes in the patterns of rhizosphere and phyllosphere populations and some bacilli, pseudomonads and actinobacteria were detected. Culturable bacilli and pseudomonads were isolated from the rhizosphere of plants grown in solarised soil, applied to soil and found to induce systemic resistance to grey mould in cucumber (Elad et al. unpublished results). The JA marker gene *PI2* and the *Pti4* gene, which codes for trans-acting factors of PR genes, were up-regulated by soil solarisation.

Following *B. cinerea* infection, up-regulation of the PR genes PR1a, GluB and CHI9 in treated plants revealed a priming effect of soil solarisation (Okon Levy et al. 2015). Thus, the soil treatment involves not only a direct effect on the plant but also an indirect effect via stimulation of beneficial microorganisms in the rhizosphere that by themselves induce resistance.

Biochar is a product of pyrolysis of waste biomass in the absence of oxygen. It can be applied to the soil as a conditioner, where it remains in an essentially permanent form and leads to net removal of carbon from the atmosphere (Laird 2008; Lehmann 2007) while improving soil properties and plant productivity (Glaser et al. 2002). Biochar also induces systemic resistance to *B. cinerea* and other foliar diseases (Elad et al. 2010). Phylogenetic characterization of bacterial isolates from biochar-amended growing mixtures (based on partial 16S rRNA gene analysis) showed that out of the 21 unique identified isolates, 16 were affiliated with previously described plant growth promoting or biocontrol agents, including *Pseudomonas*, *Mesorhizobium*, *Microbacterium*, *Brevibacillus*, *Bacillus*, and *Streptomyces spp.* (Graber et al. 2010). Molecular fingerprinting (DGGE and T-RFLP) of 16S rRNA gene fragments of the root-associated rhizosphere community showed a clear differentiation between biochar-amended and non-amended plants. Pyrosequencing of 16S rRNA amplicons showed a relative abundance of members of the *Bacterioidetes* phylum increased from 12 to 30 % in the biochar-amended samples, while that of the *Proteobacteria* decreased from 71 to 47 %. The *Bacteroidetes*-affiliated *Flavobacterium* was the strongest biochar-induced genus. Additional biochar-induced genera included degraders of chitin (*Chitinophaga*), cellulose (*Cellvibrio*) and aromatic compound degraders (*Hydrogenophaga* and *Dechloromonas*). It was suggested that the biochar-augmented microflora is at least partially responsible for the beneficial effect of biochar amendment on plant growth and disease resistance, including against *B. cinerea* (Kolton et al. 2011). Selected rhizosphere microbial isolates from biochar treated soil were found as effective BCAs.

Interestingly, application to leaves or roots of *T. harzianum*, a BCA by itself, resulted in increased variability in the bacterial population inhabiting the leaves. It was suggested that some of the effect exerted by *T. harzianum* is associated with microbial changes (Okon Levy et al. 2005). It is possible that the improved plant growth and disease control observed in the *T. harzianum* treatment is due to a direct effect of *T. harzianum* or due to an increase in populations of beneficial microbes (Elad et al. 2004) as revealed by the recent investigation of resistance pathways and related genes that are involved in *B. cinerea* suppression by induced resistance in tomato leaves (Meller Harel et al. 2014), DGGE analyzes, and activity of isolated microorganisms from bean plants.

Finally, naturally occurring endophytic microorganisms are known to produce inhibitory compounds. For example, an endophytic *Aspergillus clavatonanicus*, isolated from *Taxus mairei*, produces clavatol and patulin which both exhibited inhibitory activity in vitro against *B. cinerea* and other fungi. Zhang et al. (2008) suggested that these compounds may be involved in the protection of *T. mairei* against attack by plant pathogens.

9.6 Future Prospects and Challenges

Much change has occurred in the last decade concerning *Botrytis* control with biopesticides. One remarkable evolution is the increase in the number of commercialised preparations and the range of their area of registration. This has been accompanied by improvements in production and manufacturing processes, as well as in formulation technology, resulting in more robust and cost effective products. Additionally, biopesticides which were often marketed as complete replacements for chemical control are increasingly promoted as components of an integrated disease management system and as useful tools for the management of fungicide resistance. Together with their suitability when disease pressure is low or when chemicals cannot be used (e.g. over bloom time or just before harvest), this situation is likely to assist their widespread adoption by farmers.

Future years will also likely entail continued progress in our comprehension of the mechanisms of biocontrol, especially those related to induced resistance and direct interference of the BCA with pathogenesis. Much promise is shown by RNA interference technology, both as a research tool to understand the role of specific genes and as a potential tool to construct new biocontrol methods or to devise screening strategies for the selection of new BCAs. Progress in the comprehension of host-dependent effects will be needed to guide the deployment of BCAs according to the varieties used by the growers. It may also lead to the inclusion in varietal selection schemes of criteria related to the capacity of the plant to be induced or primed for *Botrytis*-resistance or to enhance the efficacy of biocontrol. Finally, with anticipated wider field use of BCAs against *Botrytis*, vigilance will be needed in future years to ascertain that durability does not become an issue for biocontrol as it has been for chemical control.

References

Abro MA, Lecompte F, Bardin M et al (2014) Nitrogen fertilization impacts biocontrol of tomato gray mold. Agron Sustain Dev 4:641–648

Ahn IP, Lee SW, Suh SC (2007) Rhizobacteria-induced priming in *Arabidopsis* is dependent on ethylene, jasmonic acid, and NPR1. Mol Plant Microbe Interact 20:759–768

Ajouz S, Nicot PC, Bardin M (2010) Adaptation to pyrrolnitrin in *Botrytis cinerea* and cost of resistance. Plant Pathol 59:556–566

Ajouz S, Walker AS, Fabre F et al (2011) Variability of *Botrytis cinerea* sensitivity to pyrrolnitrin, an antibiotic produced by biological control agents. Biocontrol 56:353–363

Alfano G, Ivey MLL, Cakir C et al (2007) Systemic modulation of gene expression in tomato by *Trichoderma hamatum* 382. Phytopathology 97:429–437

Alizadeh H, Behboudi K, Ahmadzadeh M et al (2013) Induced systemic resistance in cucumber and *Arabidopsis thaliana* by the combination of *Trichoderma harzianum* Tr6 and *Pseudomonas* sp Ps14. Biol Control 65:14–23

Antonov A, Stewart A, Walter M (1997) Inhibition of conidium germination and mycelial growth of *Botrytis cinerea* by natural products. In: Proceedings 50th NZ plant protection conference 159–164

Audenaert K, Pattery T, Cornelis P et al (2001) Induced resistance to *Botrytis cinerea* by *Pseudomonas aeruginosa*: role of siderophores and pyocyanin. IOBC WPRS Bull 24(3):37–41

Bardin M, Comby M, Lenaerts R et al (2013a) Diversity in susceptibility of *Botrytis cinerea* to biocontrol products inducing plant defence mechanisms. IOBC WPRS Bull 88:45–49

Bardin M, Comby M, Troulet C et al (2013b) Relationship between the aggressiveness of *Botrytis cinerea* on tomato and the efficacy of biocontrol. IOBC WPRS Bull 86:163–168

Blakeman JP (1993) Pathogens in the foliar environment. Plant Pathol 42:479–493

Buck JW, Jeffers SN (2004) Effect of pathogen aggressiveness and vinclozolin on efficacy of *Rhodotorula glutinis* PM4 against *Botrytis cinerea* on geranium leaf disks and seedlings. Plant Dis 88:1262–1268

Calvente V, de Orellano ME, Sansone G et al (2001) Effect of nitrogen source and pH on siderophore production by *Rhodotorula* strains and their application to biocontrol of phytopathogenic moulds. J Ind Microbiol Biotechnol 26:226–229

Calvo-Garrido C, Elmer PAG, Vinas I et al (2013) Biological control of botrytis bunch rot in organic wine grapes with the yeast antagonist *Candida sake* CPA-1. Plant Pathol 62:510–519

Card SD (2005) Biological control of *Botrytis cinerea* in lettuce and strawberry crops. PhD thesis. Lincoln University

Castoria R, Caputo L, de Curtis F et al (2003) Resistance of postharvest biocontrol yeasts to oxidative stress: a possible new mechanism of action. Phytopathology 93:564–572

Cheng CH, Yang CA, Peng KC (2012) Antagonism of *Trichoderma harzianum* ETS 323 on *Botrytis cinerea* mycelium in culture conditions. Phytopathology 102:1054–1063

Crawford DL, Kowalski M, Roberts MA et al (2005) Discovery, development and commercialization of a microbial biocontrol agent *Streptomyces lydicus* WYEC108: history of a decade long endeavor. SIMS News 55:88–95

De Meyer G, Bigirimana J, Elad Y et al (1998) Induced systemic resistance in *Trichoderma harzianum* T39 biocontrol of *Botrytis cinerea*. Eur J Plant Pathol 104:279–286

Decognet V, Trottin-Caudal Y, Fournier C et al (1999) Protection of stem wounds against *Botrytis cinerea* in heated tomato greenhouses with a strain of *Fusarium* sp. IOBC WPRS Bull 22(1):53–56

Dik AJ, Elad Y (1999) Comparison of antagonists of *Botrytis cinerea* in greenhouse-grown cucumber and tomato under different climatic conditions. Eur J Plant Pathol 105:123–137

Dik AJ, Wubben JP (2007) Epidemiology of *Botrytis cinerea* diseases in greenhouses. In: Elad Y, Williamson B, Tudzynski P, Delen N (eds) *Botrytis*: biology, pathology and control. Springer, Dordrecht

Dubos B (1992) Biological control of *Botrytis*: state-of-the-art. In: Verhoeff K, Malathrakis NE, Williamson B (eds) Recent advances in *Botrytis* research. Pudoc Scientific Publishers, Wageningen, pp 267–271

Dubos B, Jailloux F, Bullit J (1982) Protection du vignoble contre la pourriture grise: les propriétés antagonistes du *Trichoderma* à l'égard du *Botrytis cinerea*. Les Colloques de l'INRA 11:205–219

Edwards SG, Seddon B (1992) *Bacillus brevis* as a biocontrol agent against *Botrytis cinerea* on protected Chinese cabbage. In: Verhoeff K, Malathrakis NE, Williamson B (eds) Recent advances in *Botrytis* research. Pudoc Scientific Publishers, Wageningen, pp 267–271

Elad Y (1996) Mechanisms involved in the biological control of *Botrytis cinerea* incited diseases. Eur J Plant Pathol 102:719–732

Elad Y, Kapat A (1999) Role of *Trichoderma harzianum* protease in the biocontrol of *Botrytis cinerea*. Eur J Plant Pathol 105:177–189

Elad Y, Stewart A (2004) Microbial control of *Botrytis* spp. In: Elad Y, Williamson B, Tudzynski P, Delen N (eds) *Botrytis*: biology, pathology and control. Kluwer Academic Publishers, Dordrecht, pp 223–241

Elad Y, Zimand G, Zaqs Y et al (1993) Use of *Trichoderma harzianum* in combination or alternation with fungicides to control cucumber grey mould (*Botrytis cinerea*) under commercial greenhouse conditions. Plant Pathol 42:324–332

Elad Y, Rav David D, Levi T et al (1998) *Trichoderma harzianum* T39 – mechanisms of biocontrol of foliar pathogens. In: Lyr H, Russell PE, Dehne H-W, Sisler HD (eds) Modern fungicides and antifungal compounds II. Intercept, Aandover, pp 459–467

Elad Y, Baker S, Faull JL et al (2004) Multi trophic relationships – interaction of a biocontrol agent and a pathogen with the indigenous micro-flora on bean leaves. IOBC WPRS Bull 27(8):151–154

Elad Y, Rav David D, Meller Harel Y et al (2010) Induction of systemic resistance in plants by biochar, a soil-applied carbon sequestering agent. Phytopathology 100:913–921

El-Ghaouth A, Smilanick JL, Wilson CL (2000a) Enhancement of the performance of *Candida saitoana* by the addition of glycolchitosan for the control of postharvest decay of apple and citrus fruit. Postharvest Biol Technol 19:103–110

El-Ghaouth A, Smilanick JL, Wisniewski M et al (2000b) Improved control of apple and citrus fruit decay with a combination of *Candida saitoana* and 2-deoxy-D-glucose. Plant Dis 84:249–253

Ferrari A, Sicher C, Prodorutti D et al (2007) Potential new applications of Shemer, a *Metschnikowia fructicola* based product, in post-harvest soft fruit rots control. IOBC WPRS Bull 30(6):43–46

Fillinger S, Ajouz S, Nicot PC et al (2012) Functional and structural comparison of pyrrolnitrin- and iprodione-induced modifications in the class III histidine-kinase Bos1 of *Botrytis cinerea*. PLoS ONE 7(8):e42520

Gamliel A, Katan J (1991) Involvement of fluorescent pseudomonads and other microorganisms in increased growth response of plants in solarized soils. Phytopathology 81:494–502

Geremia RA, Goldman GH, Jacobs D et al (1993) Molecular characterization of the proteinase-encoding gene prb1, related to mycoparasitism by *Trichoderma harzianum*. Mol Microbiol 8:603–613

Glaser B, Lehmann J, Zech W (2002) Ameliorating physical and chemical properties of highly weathered soils in the tropics with charcoal – a review. Biol Fertil Soils 35:219–230

Graber ER, Meller Harel Y, Kolton M et al (2010) Biochar impact on development and productivity of pepper and tomato grown in fertigated soilless media. Plant Soil 337:481–496

Guetsky R, Elad Y, Shtienberg D et al (2002a) Improved biocontrol of *Botrytis cinerea* on detached strawberry leaves by adding nutritional supplements to a mixture of *Pichia guilermondii* and *Bacillus mycoides*. Biocontrol Sci Technol 12:625–630

Guetsky R, Shtienberg D, Elad Y et al (2002b) Improving biological control by combining biocontrol agents each with several mechanisms of disease suppression. Phytopathology 92:976–985

Hannusch DJ, Boland GJ (1996) Interactions of air temperature, relative humidity and biological control agents on grey mold of bean. Eur J Plant Pathol 102:133–142

Hermosa R, Viterbo A, Chet I et al (2012) Plant-beneficial effects of *Trichoderma* and of its genes. Microbiology 158:17–25

Ilhan K, Karabulut OA (2013) Efficacy and population monitoring of bacterial antagonists for gray mold (*Botrytis cinerea* Pers. ex. Fr.) infecting strawberries. Biocontrol 58:457–470

Jackson AM, Whipps JM, Lynch JM et al (1991) Effects of some carbon and nitrogen sources on spore germination, production of biomass and antifungal metabolites by species of *Trichoderma* and *Gliocladium virens* antagonistic to *Sclerotium cepivorum*. Biocontrol Sci Technol 1:43–51

Jackson AJ, Walters DR, Marshall G (1994) Antagonistic interactions between the foliar pathogen *Botrytis fabae* and isolates of *Penicillium brevicompactum* and *Cladosporium cladosporioides* on faba beans. Biol Control 8:97–106

Janisiewicz WJ, Jeffers SN (1997) Efficacy of commercial formulation of two biofungicides for control of blue mold and gray mold of apples in cold storage. Crop Prot 16:629–633

Kamensky M, Ovadis M, Chet I et al (2003) Soil-borne strain IC14 of *Serratia plymuthica* with multiple mechanisms of antifungal activity provides biocontrol of *Botrytis cinerea* and *Sclerotinia sclerotiorum* diseases. Soil Biol Biochem 35:323–331

Kapat A, Zimand G, Elad Y (1998) Biosynthesis of pathogenicity hydrolytic enzymes by *Botrytis cinerea* during infection of bean leaves and *in vitro*. Mycol Res 102:1017–1024

Katan J, DeVay JE (eds) (1991) Soil solarization. CRC Press, Boca Raton

Katan J, Gamliel A (2012) Mechanisms of pathogen and disease control and plant-growth improvement involved in soil solarization. In: Gamliel A, Katan J (eds) Soil solarization. APS Press, St. Paul, pp 135–145

Köhl J (2004) Biological control of *Botrytis* spp. by *Ulocladium atrum* through competitive colonisation of necrotic tissues. Plant Research International B.V, Wageningen

Köhl J, Schlösser F (1989) Decay of sclerotia of *Botrytis cinerea* by *Trichoderma* spp. at low temperatures. J Phytopathol 125:320–326

Köhl J, Fokkema NJ (1993) Fungal interactions on living and necrotic leaves. In: Blakeman JP, Williamson B (eds) Ecology of plant pathogens. CABI, Oxon, pp 321–334

Köhl J, Van der Plas CH, Molhoek WML et al (1995) Effect of interrupted leaf wetness periods on suppression of sporulation of *Botrytis allii* and *Botrytis cinerea* by antagonists on dead onion leaves. Eur J Plant Pathol 101:627–637

Kolton M, Meller Harel Y, Pasternak Z et al (2011) Impact of biochar application to soil on the root-associated bacterial community structure of fully developed greenhouse pepper plants. Appl Environ Microbiol 77:4924–4930

Korolev N, Rav David D, Elad Y (2008) The role of phytohormones in basal resistance and *Trichoderma*-induced systemic resistance to *Botrytis cinerea* in *Arabidopsis thaliana*. Biocontrol 53:667–683

Kowalska J, Drożdżyński D, Remlein-Starosta D, Sas-Paszt L, Malusá E (2012) Use of *Cryptococcus albidus* for controlling grey mould in the production and storage of organically grown strawberries. J Plant Dis Prot 119:174–178

Labudova I, Gogorova L (1988) Biological control of phytopathogenic fungi through lytic action of *Trichoderma* species. FEMS Microbiol Lett 52:193–198

Lahdenpera ML, Korteniemi M (2008) Application methods for commercial biofungicides in greenhouses. IOBC WPRS Bull 32:111–114

Laird DA (2008) The charcoal vision: a win–win–win scenario for simultaneously producing bioenergy, permanently sequestering carbon, while improving soil and water quality. Agron J 100:178–181

Lecompte F, Abro MA, Nicot PC (2010) Contrasted responses of *Botrytis cinerea* isolates developing on tomato plants grown under different nitrogen nutrition regimes. Plant Pathol 59:891–899

Lecompte F, Abro MA, Nicot PC (2013) Can plant sugars mediate the effect of nitrogen fertilization on lettuce susceptibility to two necrotrophic pathogens: *Botrytis cinerea* and *Sclerotinia sclerotiorum*? Plant Soil 369:387–401

Lehmann J (2007) Bio-energy in the black. Front Ecol Environ 5:381–387

Leifert C, Li H, Chidburee S, Hampson S et al (1995) Antibiotic production and biocontrol activity by *Bacillus subtilis* CL27 and *Bacillus pumilus* CL45. J Appl Bacteriol 78:97–108

Li H, Leifert C (1994) Development of resistance in *Botryotinia fuckeliana* (de Bary) Whetzel against the biological control agent *Bacillus subtilis*. J Plant Dis Prot 101:414–418

Lichatowich T (2007) The plant growth enhancing and biocontrol mechanisms of *Streptomyces lydicus* WYEC108 and its use in nursery and greenhouse production. USDA Forest Service Proceedings RMRS-P-50, 61–62

Liu J, Wisniewski M, Artlip T et al (2013) The potential role of PR-8 gene of apple fruit in the mode of action of the yeast antagonist, *Candida oleophila*, in postharvest biocontrol of *Botrytis cinerea*. Postharvest Biol Technol 85:203–209

Magnin-Robert M, Quantinet D, Couderchet M et al (2013) Differential induction of grapevine resistance and defense reactions against *Botrytis cinerea* by bacterial mixtures in vineyards. Biocontrol 58:117–131

Malmierca MG, Cardoza RE, Alexander et al (2012) Involvement of *Trichoderma* trichothecenes in the biocontrol activity and induction of plant defense related genes. Appl Environ Microbiol 78:4856–4868

Mari M, Martini C, Spadoni A et al (2012) Biocontrol of apple postharvest decay by *Aureobasidium pullulans*. Postharvest Biol Technol 73:56–62

Marrone PG (2002) An effective biofungicide with novel modes of action. Pestic Outlook 13:193–194

Meller Harel Y, Haile Mehari Z, Rav-David D et al (2014) Induced systemic resistance against gray mold in tomato (*Solanum lycopersicum*) by benzothiadiazole and *Trichoderma harzianum* T39. Phytopathology 104:150–157

Mercier J, Wilson CL (1994) Colonization of apple wounds by naturally occurring microflora and introduced *Candida oleophila* and their effect on infection by *Botrytis cinerea* during storage. Biol Control 4:138–144

Meyer UM, Fischer E, Barbul O et al (2001) Effect of biocontrol agents on antigens present in the extracellular matrix of *Botrytis cinerea*, which are important for pathogenesis. IOBC WPRS Bull 24(3):5–9

Morandi MAB, Sutton JC, Maffia LA (2000) Effects of host and microbial factors on development of *Clonostachys rosea* and control of *Botrytis cinerea* in rose. Eur J Plant Pathol 106:439–448

Morandi MAB, Mattos LPV, Santos ER et al (2008) Influence of application time on the establishment, survival, and ability of *Clonostachys rosea* to suppress *Botrytis cinerea* sporulation on rose debris. Crop Prot 27:77–83

Mounir R, Durieux A, Bodo E et al (2007) *Aureobasidium pullulans* (1113-5) microbial antagonist for the control of post-harvest decay on apple fruit: development of active biomass formulation at a lab scale. IOBC WPRS Bull 30(6):575–578

Navazio L, Baldan B, Moscatiello R et al (2007) Calcium-mediated perception and defense responses activated in plant cells by metabolite mixtures secreted by the biocontrol fungus *Trichoderma atroviride*. BMC Plant Biol 7:41

Nelson ME, Powelson ML (1988) Biological control of gray mold of snap beans by *Trichoderma hamatum*. Plant Dis 72:727–729

Newhook FJ (1957) The relationship of saprophytic antagonism to control of *Botrytis cinerea* Pers. on tomatoes. N Z J Sci Technol 38:473–481

Nguyen T, Bay IS, Abramians AA, Gubler WD (2013) Evaluation of fungicide programs for management of *Botrytis* bunch rot of grapes: 2013 field trial. http://plantpathology.ucdavis.edu/Cooperative_Extension/Gubler/2013_Fruit_Crop_Fungicide_Trials

Nicot PC, Bardin M (2012) Biological and integrated protection in the Mediterranean greenhouse: is disease management the weak link? IOBC WPRS Bull 80:11–17

Nicot PC, Decognet V, Fruit L et al (2002) Combined effect of microclimate and dose of application on the efficacy of biocontrol agents for the protection of pruning wounds on tomatoes against *Botrytis cinerea*. IOBC WPRS Bull 25(10):73–76

Nicot PC, Bardin M, Alabouvette C et al (2011) Potential of biological control based on published research. 1. Protection against plant pathogens of selected crops. In: Nicot PC (ed) Classical and augmentative biological control against diseases and pests: critical status analysis and review of factors influencing their success. IOBC/WPRS, Zurich, pp 1–11

Nicot PC, Bardin M, Debruyne F et al (2013) Effect of nitrogen fertilisation of strawberry plants on the efficacy of defence-stimulating biocontrol products against *Botrytis cinerea*. IOBC WPRS Bull 88:39–42

Nunes C, Usall J, Teixido N et al (2002a) Improved control of postharvest decay of pears by the combination of *Candida sake* (CPA-1) and ammonium molybdate. Phytopathology 92:281–287

Nunes C, Usall J, Teixido N et al (2002b) Control of Penicillium expansum and Botrytis cinerea on apples and pears with the combination of *Candida sake* and *Pantoea agglomerans*. J Food Prot 65:178–184

Okon Levy N, Katan J, Elad Y (2004) Integrated control of foliar infection by soil solarization and *Trichoderma*. IOBC WPRS Bull 27(8):65–70

Okon Levy N, Elad Y, Katan J et al (2005) *Trichoderma* and soil solarization induced microbial changes on plant surfaces. IOBC WPRS Bull 29(2):21–26

Okon Levy N, Meller Harel Y, Haile ZM et al (2015) Induced resistance to foliar diseases by soil solarization and *Trichoderma harzianum*. Plant Pathol 64:365–374

Ongena M, Henry G, Thonart P (2010) The role of cyclic lipopetides in the biocontrol activity of *Bacillus subtilis*. In: Gisi U, Chet I, Gullino ML (eds) Recent developments in the management of plant diseases. Plant Pathology in the 21st Century. Vol 1:59–69

Palmieri MC, Perazzolli M, Matafora V et al (2012) Proteomic analysis of grapevine resistance induced by *Trichoderma harzianum* T39 reveals specific defence pathways activated against downy mildew. J Exp Bot 63:6237–6251

Peng G, Sutton JC (1991) Evaluation of microorganisms for biocontrol of *Botrytis cinerea* in strawberry. Can J Plant Pathol 13:247–257

Perazzolli M, Dagostin S, Ferrari A et al (2008) Induction of systemic resistance against *Plasmopara viticola* in grapevine by *Trichoderma harzianum* T39 and benzothiadiazole. Biol Control 47:228–234

Perazzolli M, Roatti B, Bozza E, Pertot I (2011) *Trichoderma harzianum* T39 induces resistance against downy mildew by priming for defense without costs for grapevine. Biol Control 58:74–82

Perazzolli M, Moretto M, Fontana P et al (2012) Downy mildew resistance induced by *Trichoderma harzianum* T39 in susceptible grapevines partially mimics transcriptional changes of resistant genotypes. BMC Genomics 13:660

Quarles W (2009) Giant knotweed, plant disease protection and immortality. IPM Pract 22:1–6

Reeh KW, Cutler GC (2013) Laboratory efficacy and fungicide compatibility of *Clonostachys rosea* against *Botrytis* blight on lowbush blueberry. Can J Plant Sci 93:639–642

Ruocco M, Woo S, Vinale F et al (2011) Identified difficulties and conditions for field success of biocontrol. 2. Technical aspects: factors of efficacy. In: Nicot PC (ed) Classical and augmentative biological control against diseases and pests: critical status analysis and review of factors influencing their success. IOBC/WPRS, Zurich, pp 45–57

Schilder A (2013) Botector: a new biofungicide for control of Botrytis bunch rot of grapes http://msue.anr.msu.edu/news/botector_a_new_biofungicide_for_control_of_botrytis_bunch_rot_in_grapes

Schirmböck M, Lorito M, Wang Y et al (1994) Parallel formation and synergism of hydrolytic enzymes and peptaibol antibiotics, molecular mechanisms involved in the antagonistic activity of *Trichoderma harzianum* against phytopathogenic fungi. Appl Environ Microbiol 60:4364–4370

Schmitt A, Strathmann S, Emslie KA et al (1996) Use of *Regalia sachalinensis* extracts for induced resistance in integrated disease control: effects on *Botrytis cinerea*. XIth international Botrytis symp 23–27 June, Wageningen, 69

Schmitt A, Ibarra F, Francke W (2005) Resistance inducing constituents in extracts of *Reynoutria sachalinensis*. In: Dehne HW, Gisi U, Kuck KH, Russell PE, Lyr H (eds) Modern fungicides and antifungal compounds IV: proceedings of the 14th international Reinhardsbrunn symposium, April 25–29 2004, Friedrichroda, Germany. British Crop Production Council, Alton

Schoonbeek HJ, Jacquat-Bovet AC, Mascher F et al (2007) Oxalate-degrading bacteria can protect *Arabidopsis thaliana* and crop plants against *Botrytis cinerea*. Mol Plant Microbe Interact 20:1535–1544

Segarra G, Casanova E, Borrero C et al (2007) The suppressive effects of composts used as growth media against *Botrytis cinerea* in cucumber plants. Eur J Plant Pathol 117:393–402

Segarra G, Santpere G, Elena G et al (2013) Enhanced *Botrytis cinerea* resistance of *Arabidopsis* plants grown in compost may be explained by increased expression of defense-related genes, as revealed by microarray analysis. PLoS One 8(2):e56075. doi:10.1371/journal.pone.0056075

Shoresh M, Yedidia I, Chet I (2005) Involvement of jasmonic acid/ethylene signaling pathway in the systemic resistance induced in cucumber by *Trichoderma asperellum* T203. Phytopathology 95:76–84

Shoresh M, Mastouri F, Harman GE (2010) Induced systemic resistance and plant responses to fungal biocontrol agents. Annu Rev Phytopathol 48:21–43

Stapleton JJ, DeVay JE (1984) Thermal components of soil solarization as related to changes in soil and root microflora and increased plant growth response. Phytopathology 74:255–259

Stevens C, Khan VA, Rodriguez-Kabana R et al (2003) Integration of soil solarization with chemical, biological and cultural control for the management of soilborne disease of vegetables. Plant Soil 253:493–506

Su H (2012) Regalia bioprotectant in plant disease management. Outlook Pest Manag 23:30–34

Sylla J, Alsanius BW, Kruger E et al (2013) Leaf microbiota of strawberries as affected by biological control agents. Phytopathology 103:1001–1011

Tucci M, Ruocco M, De Masi L, De Palma M, Lorito M (2011) The beneficial effect of *Trichoderma* spp. on tomato is modulated by the plant genotype. Mol Plant Pathol 12:341–354

Vallad GE, Cooperband L, Goodman RM (2003) Plant foliar disease suppression mediated by composted forms of paper mill residuals exhibits molecular features of induced resistance. Physiol Mol Plant Pathol 63:65–77

Viterbo A, Haran S, Friesem D et al (2001) Antifungal activity of a novel endochitinase gene (chit36) from *Trichoderma harzianum* Rifai TM. FEMS Microbiol Lett 200:169–174

Whiteman SA, Stewart A (1998) Suppression of *Botrytis cinerea* sporulation on irradiated grape leaf tissue by the antagonistic bacterium *Serratia liquefaciens*. N Z J Crop Hortic Sci 26:325–330

Wood RKS (1951) The control of diseases of lettuce by use if antagonistic organisms. I. The control of *Botrytis cinerea* Pers. Ann Appl Biol 38:203–216

Yang HH, Yang SL, Peng KC et al (2009) Induced proteome of *Trichoderma harzianum* by *Botrytis cinerea*. Fungal Biol 113:924–932

Yogev A, Raviv M, Hadar Y et al (2010) Induced resistance as a putative component of compost suppressiveness. Biol Control 54:46–51

Yu T, Yu C, Lu HP et al (2012) Effect of *Cryptococcus laurentii* and calcium chloride on control of *Penicillium expansum* and *Botrytis cinerea* infections in pear fruit. Biol Control 61:169–175

Zamani-Zadeh M, Soleimanian-Zad S, Sheikh-Zeinoddin M (2013) Biocontrol of gray mold disease on strawberry fruit by integration of *Lactobacillus plantarum* A7 with ajwain and cinnamon essential oils. J Food Sci 78:M1582–M1588

Zhang W, Han DY, Dick WA et al (1998) Compost and compost water extract-induced systemic acquired resistance in cucumber and *Arabidopsis*. Phytopathology 88:450–455

Zhang H, Wang L, Dong Y et al (2007) Postharvest biological control of gray mold decay of strawberry with *Rhodotorula glutinis*. Biol Control 40:287–292

Zhang CL, Zheng BQ, Lao JP et al (2008) Clavatol and patulin formation as the antagonistic principle of *Aspergillus clavatonanicus*, an endophytic fungus of *Taxus mairei*. Appl Microbiol Biotechnol 78:833–840

Zhang H, Wang L, Ma L et al (2009) Biocontrol of major postharvest pathogens on apple using *Rhodotorula glutinis* and its effects on postharvest quality parameters. Biol Control 48:79–83

Zhang H, Ma L, Jiang S et al (2010) Enhancement of biocontrol efficacy of *Rhodotorula glutinis* by salicyclic acid against gray mold spoilage of strawberries. Int J Food Microbiol 141:122–125

Zhao LN, Zhang HY, Li J et al (2012) Enhancement of biocontrol efficacy of *Pichia carribbica* to postharvest diseases of strawberries by addition of trehalose to the growth medium. Int J Mol Sci 13:3916–3932

Zhao LN, Zhang HY, Lin HT et al (2013) Effect of trehalose on the biocontrol efficacy of *Pichia caribbica* against post-harvest grey mould and blue mould decay of apples. Pest Manag Sci 69:983–989

Zimand G, Elad Y, Chet I (1996) Effect of *Trichoderma harzianum* on *Botrytis cinerea* pathogenicity. Phytopathology 86:1255–1260

Chapter 10
Chemical Control and Resistance Management of *Botrytis* Diseases

Sabine Fillinger and Anne-Sophie Walker

Abstract Chemical control remains the easiest way to manage *Botrytis* epidemics on many crops. Nevertheless, actual concerns about the environment, human health and control sustainability invite to a smarter use of fungicides, aiming to delay resistance evolution in pathogen populations. This chapter deals with the mode of action of botryticides (including multi-site toxicants and molecules affecting specifically respiration, cytoskeleton, osmoregulation, sterol and amino-acid biosynthesis) and associated resistance cases, mostly due to target site modifications. We also present original resistance mechanisms for fungi such as detoxification and multidrug resistance. Finally, this chapter introduces strategies available to decrease selection pressure exerted by fungicides on *Botrytis* spp. populations with the long-term aim to improve resistance management in the field.

Keywords Fungicides • Mode of action • Resistance mechanism • Efficacy • Strategy

10.1 Introduction

Plant diseases due to infections by *Botrytis cinerea* and other *Botrytis* species, if uncontrolled, may account for important crop losses, pre- and postharvest, with potentially high economic impact as described in the previous chapters. Integrated pest-management, including resistant cultivars, prophylactic means or application of biocontrol agents, is necessary but not always sufficient or available to prevent these diseases (see Chaps. 8, 9, and 11). Chemical control based on the application of mostly synthetic fungicides, therefore constitutes the principal means of efficient and reliable crop protection against grey mould. Control of diseases due to *Botrytis* and related species (*e.g.*, *Sclerotinia*, *Monilinia*) represents roughly 8 % of the global fungicide market (Phillips and McDougall 2012). Fungicide investment may differ among crops according to their economic value, their sensitivity to *Botrytis*

S. Fillinger • A.-S. Walker (✉)
UMR 1290 BIOGER, INRA, AgroParisTech, BP01, Avenue Lucien Brétignières, F-78850 Thiverval-Grignon, France
e-mail: sabine.fillinger@versailles.inra.fr; walker@versailles.inra.fr

S. Fillinger, Y. Elad (eds.), *Botrytis – the Fungus, the Pathogen and its Management in Agricultural Systems*, DOI 10.1007/978-3-319-23371-0_10

infection and their storage time. Among them, grapes constitute high value crops for grey mould control.

During the last decades, restriction in fungicide application became necessary to reduce the impact on the environment (Fenner et al. 2013) and to limit fungicide residues (Verger and Boobis 2013) on harvest, requiring optimized protection strategies. At the same time, acquired resistance to most botryticides arose in many agronomical situations, sometimes impeding field efficacy and leading to additional sprays (Brent and Hollomon 2007). Reaching a compromise between fungicide durability, human health and environment protection and valuable crop production may imply to optimize spray timing and molecule choice and to promote agro-ecological practices combining prophylaxis measures, natural regulations (*e.g.*, by the means of biocontrol agents) and conventional fungicide treatments. In this chapter, we will describe the main chemicals used against grey mould with a focus on the latest modes of action introduced; resistance phenomena will be described, as well as their occurrence in the field, with a special highlight on new resistance mechanisms discovered since the last edition of the "Botrytis" book (Leroux 2004). Finally, we will propose rules for decision-makers, to help them adapting fungicide strategies according to the risk-situation. Most information available on these important topics concern *B. cinerea*, but data about other *Botrytis* spp. will be mentioned, when available.

10.2 Botryticides: Mode of Action and Resistance

In this chapter we consider only fungicides that inhibit or reduce disease development through direct activity on the pathogenic fungus. Their *in vitro* activities are either fungicidal or fungistatic (blocking the fungal development without killing the fungus itself). Mostly preventive, only few fungicides have curative activities once the disease is installed (for review see McGrath 2004). As will be presented in the following sections, fungicides target either specifically essential cellular functions (single-site activity), or display multi-site activity, interfering with more than one cellular function. Most modern fungicides, active at low dosage, are highly specific through their single site activity. However, concomitant with this strong activity, the risk of resistance selection after target site modifications is also high, for many modes of action. Historically, at least five groups of unisite botryticides were introduced into the fungicide market and target distinct cellular functions: (i) the cytoskeleton (microtubules); (ii) mitochondrial respiration and ATP-synthesis; (iii) ergosterol biosynthesis; (iv) biosynthesis of proteins or amino acids; (v) signal transduction. No elicitor activity on the plant's defense has been reported so far for fungicides registered against grey mould. Nevertheless, biocontrol agents or natural antifungal molecules, such as polyoxins or potassium bicarbonate, are of particular interest, especially for organic farming (Chap. 9). Polyoxins are fermentation products of *Streptomyces cacaoi* var. *asoensis* that interfere with the fungal cell wall biosynthesis (competitive inhibitor of chitin synthase). This kind of fungicide has been used on sweet basil in Israel since the early 1990s without the selection of high resistant strains (Mamiev et al. 2013).

The major active ingredients (a.i.) and the corresponding formulated trademarks registered for grey mould control are summarized in Table 10.1. Their modes of action according to the FRAC classification (Fungicide Resistance Action Committee; www.frac.info) are described in the following sections. As we will focus our detailed descriptions on the molecules introduced since 2004, we invite the reader to refer to the corresponding chapter from 2004 (Leroux 2004) for details on older fungicide categories. Resistance to fungicides may be preexisting in a fungal population at the species level (natural resistance) or it may arise in populations after fungicide selection (acquired resistance). The susceptibility of fungal isolates to fungicides is measured by growth assays on ranges of fungicide concentrations, in order to determine the concentration inhibiting fungal growth by 50 % – also called EC_{50} – or eventually the minimal inhibitory concentration (MIC). Comparing the EC_{50} values of a given strain to those of sensitive reference strains – generally those isolated before the introduction of the fungicide – allows determining its resistance factor or level (RF or RL). We propose here to consider low resistance (LR) levels for EC_{50} ratios between 2 and 20, moderate resistance (MR) for RFs between 20 and 100; high resistance (HR) would be considered for EC50 ratios >100.

When a fungal population is treated with a given fungicide, the proportion of resistant isolates – the only individuals adapted to survive – increases. The speed of this increase may be considered as a balance between the intensity of the positive selective pressure (frequency of the applications, nature of modes of action) and of the negative selection pressure, *i.e.* the resistance cost observed in the resistant isolates relative to that of the sensitive ones. The efficacy of a given fungicide may be threatened if the frequency of highly resistant isolates in the fungal population is above a critical level, specific to each situation but often estimated to 20 % (Hollomon and Brent 2009). Anti-resistance strategies aiming to reduce the incidence of resistance in fungal populations need to combine the biological risk (inherent to the fungus' life traits), the fungicide risk (inherent to the fungicide's mode of action) and the agronomic risk (reflecting cultural practices and the intensity of fungicide use) (Kuck and Russell 2006).

10.2.1 Multisite Botryticides

Multisite toxicants figure among the eldest fungicides used in agriculture with the inorganic sulfur and copper salts described already in the nineteenth century (reviewed in Russell 2005). Against *Botrytis* diseases, molecules belonging to the chemical families of chloronitriles, phtalimides, sulfamides and dithiocarbamates (*e.g.*, folpet, thiram, tolylfluanid and chlorothalonil) are still registered in many countries, as well as in mixture with unisite fungicides, targeting numerous fungal pests including oomycetes. Most of the compounds cited in Tables 10.1 and 10.2, have highly reactive electron-rich groups with a potentially strong action on thiol

Table 10.1 Major fungicides against grey mould: mode of action, intrinsic toxicity and application rates

Fungicide			EC_{50} value (mg/l)			
Mode of action	Common name	Selected trade names	Conidia germination	Germ tube elongation	Mycelium growth	Application rate (g/ha)
Multisite	Dichlofluanid	Elvaron, Euparen	0.20	0.05	3.00	2,000
	Folpet	Folpan	3.00	0.40	≥10.00	1,500
	Thiram	Pomarsol, Thiram	≈1.00	0.10	≥10.00	2,000–3,000
	Chlorothalonil	Fongistop	0.2	0.07	3.5	1,000
Respiration	Fluazinam	Frowncide, Shirlan, Sekoya	0.10	0.04	0.08	750
	Azoxystrobin[a]	Amistar, Heritage	0.20	0.10	0.20	200
	Pyraclostrobin[a]	Pristine[b]	0.015	0.006	0.02	100–200
	Boscalid	Endura, Cantus, Pristine[b]	0.50	0.05	0.40	500–600
	Fluopyram	Luna privilege	0.50	0.05	0.60	300
Cystosqueleton	Carbendazim	Bavistin, Derosal	>10.00	0.04	0.03	500
	Thiophanate-methyl	Topsin	>10.00	0.35	0.15	1,120
	Diethofencarb[e]	Sumico[c], Jonk[c]	>10.00	0.08	0.04	500
Osmoregulation	Iprodione	Kidan, Rovral	2.00	0.80	0.15	750
	Procymidone	Sumisclex, Sumilex	2.50	0.80	0.15	750
	Fludioxonil	Geoxe, Switch[d]	0.06	0.015	0.004	500

Amino-acid synthesis	Cyprodinil	Switch[d]	≈0.10	0.008	0.01	375
	Mepanipyrim	Japica, Cockpit	≈0.25	0.03	0.04	400–600
	Pyrimethanil	Scala	≈0.30	0.05	0.08	400–800
Sterol synthesis	Prochloraz	Octave, Sportak	>10.00	0.03	0.10	250
	Tebuconazole	Horizon	>10.00	0.20	0.30	250
	Fenhexamid	Teldor, Lazulie	>10.00	0.05	0.01	750
	Fenpyrazamine	Prolectus	>10.00	0.15	0.03	600

Adapted from Leroux (2004)

[a]EC_{50}s estimated with addition of SHAM to inhibit the alternative oxidase

[b]Only in the mixture of boscalid and pyraclostrobin

[c]Mixture of carbendazim and diethofencarb

[d]Mixture of fludioxonil and cyprodinil

[e]EC_{50} values of *benA*^{E198A} genotype (Table 10.2)

Table 10.2 Mode of action of botryticides and resistance phenomena recorded in *Botrytis* spp. field strains

Mode of action	Chemical family	Fungicides	Specific resistance			Selected references[f]
			Phenotype[a]	Target gene	Resistance allele	
Multisite	Chloronitriles	Chlorothalonil[c]	–	Unknown	–	1–3
	Phthalimides	Captan[c], folpet[c]	–	Unknown	–	
	Sulfamides	Dichlofluanid[c], tolylfluanid[c]	LR	Unknown	–	4–6
	Dithiocarbamates	Thiram[c], mancozeb[c], maneb[c]	LR	Unknown	–	2, 7, 8
Cyto-squeleton[e]	Thiophanates-benzimidazoles (Ben)	Thiophanate-methyl, carbendazim[c]	Ben^{HR} NPC^{S}	*benA* (*tubA*/*btuB*/*mbc1*)	E198A/V	9–18
	N-phenyl-carbamates (NPC)	Diethofencarb[c]	$Ben^{MR/HR}$ NPC^{R}	*benA* (*tubA*/*btuB*/*mbc1*)	F200Y	
					E198K	
Osmotic signal transduction	Dicarboximides (Imi)	Iprodione, procymidone[c], vinclozolin[c]	Imi^{LR} Phe^{S}	*bos1* (*daf1*)	I365N/R/S, V368F[b], Q369P[b], N373S[b], T447S[b], $S1040^{STOP}$	19–26
	Phenylpyrroles (Phe)	Fludioxonil	–	Unknown	–	
Methionine biosynthesis	Anilinopyrimidines	Cyprodinil, pyrimethanil, mepanipyrim	MR-HR	Unknown	–	27–29
Sterol C4-demethylation	Hydroxyanilides (Hyd)	Fenhexamid, fenpyrazamine[d]	HR ($HydR3^{+}$)	*erg27*	F412S/I/V/C,	30–36
			MR ($HydR3^{-}$)	*erg27*	F26S T63I, V192L, L195F, N196T, I232M, P250S, V309M, A314V, S336C, N369D, L400F/S, F427V A461S	
			LR-MR (HydR2)	Unknown, close to *bik*	–	37

Respiration: uncoupler	Pyridinamines	Fluazinam	LR/MR	Unknown	–	38, 39
Respiration: complex II	Pyridine carboxamides	Boscalid (Bos)	Bos^{HR} Fpm$^{S/HS}$	*sdhB*	H272Y/R	40–46
			Bos$^{MR/HR}$ Fpm$^{MR/HR}$	*sdhB*	P225T/F/L, N230I, H272L/V	
	Pyridinyl-ethyl-benzamides	Fluopyram[d] (Fpm)	Bos^{MR} Fpm$^{LR/MR}$	*sdhD*	H132R	
			Bos$^{LR/MR}$ Fpm$^{LR/MR}$	Unknown	–	
Respiration: complex III	Strobilurins	Azoxystrobin, pyraclostrobin	HR	*cytb*	G143A	43, 47–52

[a]The indicated phenotypes are *S* sensitive, *LR* low resistance, *MR* moderate resistance, *HR* high resistance according to the definition cited in Sect. 10.2

[b]Mutations found in combination with each other or with I365S

[c]Old fungicides which may not be registered any more in some countries or to protect some crops

[d]Recent fungicides which may not be already registered in all countries and on all crops

[e]*B. cinerea* wild-type phenotype is Ben^{S} NPC^{Rf}1= Corbett et al. 1984; 2= Tremblay et al. 2003; 3= Zhang et al. 2009; 4= Malathrakis 1989; 5= Pollastro et al. 1996; 6= Rewal et al. 1991; 7= Delen et al. 1984; 8= Roberts et al. 1999; 9= Bollen and Scholten 1971; 10= Leroux and Clerjeau 1985; 11= Leroux et al. 2002b; 12= Nakazawa and Yamada 1997; 13= Park et al. 1997; 14= Davidse and Ishii 1995; 15= Yarden and Katan 1993; 16= Zhao et al. 2010; 17= Banno et al. 2008; 18= Kim et al. 2009; 19= Faretra and Pollastro 1991; 20= Fujimura et al. 2000; 21= Cui et al. 2004; 22= Ma et al. 2007; 23= Carisse and Tremblay 2007; 24= Vignutelli et al. 2002; 25= Rosslenbroich and Stuebler 2000; 26= Oshima et al. 2006; 27= Fritz et al. 2003; 28= Forster and Staub 1996; 29= Bardas et al. 2008; 30= Albertini and Leroux 2004; 31= Billard et al. 2012; 32= Fillinger et al. 2008; 33= Leroux et al. 2002a; 34= Grabke et al. 2013; 35= Debieu et al. 2013; 36= Debieu and Leroux in press; 37= Schumacher et al. 2013; 38= Guo et al. 1991; 39= Tamura 2000; 40= Veloukas and Karaoglanidis 2012; 41= Kim and Xiao 2010; 42= Yin et al. 2011; 43= Leroux et al. 2010; 44= Laleve et al. 2014b; 45= Veloukas et al. 2014; 46= De Miccolis Angelini et al. 2014; 47= Ishii et al. 2009; 48= Ishii et al. 2007; 49= Banno et al. 2009; 50= Yin et al. 2010; 51= Yin et al. 2012; 52= Vallieres et al. 2011

(SH-) groups of fungal enzymes, inhibiting their reducing activity and/or the formation of disulfur-bonds (Corbett et al. 1984; Bernard and Gordon 2000). Resistance to multisite fungicides has been observed only in a few cases in *Botrytis* spp. and seems to involve detoxification (reviewed in Leroux 2004). Although less exposed to resistance development than unisite fungicides, some multisite toxicants might be withdrawn from certain countries or markets for toxicological reasons after their evaluation for re-registration, due to the high application rates necessary for these contact fungicides with solely preventive activity.

10.2.2 Unisite Fungicides

10.2.2.1 Cytoskeleton Inhibitors

The first systemic fungicides synthesized by the chemical companies were those affecting the cytoskeleton (*i.e.*, benzimidazoles, thiophanates, and N-phenyl-carbamates) through microtubular binding, with severe effects on cell division, mitosis and protein secretion (Gessler et al. 1981; Temperli et al. 1991; Jochova et al. 1993; Pedregosa et al. 1995; Davidse and Ishii 1995). The N-phenylcarbamate diethofencarb mainly used against grey mould and the benzamide zoxamide, an anti-oomycete, display a similar mode of action. They were the first fungicides with curative activity against many fungal diseases, but due to massive application, most fungi including *Botrytis* spp. became resistant to these unisite fungicides, especially towards benzimidazoles and thiophanates (Bollen and Scholten 1971; reviewed in Leroux 2004) (Table 10.2). Here and at later instances of this chapter, we will not use the nomenclature of the observed phenotypes, as these may be different among authors.

Two major phenotypes of resistance to cytoskeleton inhibitors have been described for *B. cinerea*. In the first one high resistance to benzimidazoles is associated with increased sensitivity to N-phenyl-carbamates and to zoxamide (Ben^{HR}, NPC^{S}). The second phenotype displays positive cross-resistance towards the three categories of cytoskeleton inhibitors ($Ben^{HR/MR}$, NPC^{R}). In both cases, point mutations in the β-tubulin encoding gene *benA* (synonymous of *tubA*/*btuB*/*mcb*) are responsible for theses phenotypes. The amino-acid changes E198A/V were observed in the Ben^{HR}, NPC^{S} strains and E198K/L or F200Y replacement in $Ben^{HR/MR}$, NPC^{R} strains (Yarden and Katan 1993; Park et al. 1997; Banno et al. 2009; Zhang et al. 2010; Ziogas et al. 2009; Kim et al. 2009). Probably with a low resistance cost, the E198A mutants are widely distributed among *B. cinerea* populations even in the absence of selection pressure. This contrasts with those harbouring the F200Y mutation whose frequency rapidly decreases when the application of the mixture between carbendazim and diethofencarb is stopped (Walker et al. 2013). At last, resistance to benzimidazoles was detected in *B. alii*, *B. elliptica*, and *B. tulipae* (Hsiang and Chastagner 1991, 1992). Due to high efficacy losses linked with resistance selection, in many situations, and the development of other botryticides with higher intrinsic activity, anti-microtubules have now little use.

10.2.2.2 Fungicides Affecting Signal-Transduction (Osmoregulation)

Two chemical categories, applied against *Botrytis* infections, interfere with the fungal signal transduction: the dicarboximides and the phenylpyrroles, which are structural analogs of the natural antifungal compound pyrrolnitrin (Chap. 9). The exact targets of dicarboximides (*e.g.* iprodione, vinclozolin, procymidone) and the phenylpyrrole fludioxonil are still unknown. Nevertheless, these botryticides induce physiological changes, characteristic of an over-stimulation of the stress response signal-transduction (for details, see Chaps. 13 and 14), namely glycerol-accumulation, lipid peroxidation, plasma membrane leakage (reviewed in Tanaka and Izumitsu 2010; Hayes et al. 2014). They inhibit conidial germination and mycelial growth of a variety of plant pathogenic fungi (Leroux 1996). Due to extensive use, dicarboximides rapidly lost their efficiency against grey mould after the selection and generalization of specific resistance among *B. cinerea* populations (refer to Leroux 2004). Only very restricted applications of these botryticides are allowed on some crops to limit the selection of dicarboximde resistant strains, which seem to exhibit high resistance cost in the field as well (Walker et al. 2013). Resistance to dicarboximides was also easily found in *B. squamosa* on onion and on *B. elliptica* on flower bulbs in Canada but the resistance mechanism was not explored (Hsiang and Chastagner 1992; Carisse and Tremblay 2007). On the opposite, the phenylpyrrole fludioxonil does not suffer real resistance problems since only rare cases of specific resistance have been reported in *Botrytis* isolates (Vignutelli et al. 2002; Zhao et al. 2010; Ma et al. 2007). This is probably not only due to resistance management, but rather to the strongly affected fitness of fludioxonil resistant mutants. The analysis of laboratory induced fludioxonil mutants revealed reduced conidiation rate and pathogenicity, increased sensitivity to osmotic and other stresses associated with high resistance levels to fludioxonil and cross-resistance to dicarboximides in a phenotype (Viaud et al. 2006; Ma et al. 2007; Fillinger et al. 2012).

The majority of *B. cinerea* field or laboratory mutants resistant to dicarboximides and/or fludioxonil harbor mutations in the histidine-kinase gene *bos1* (syn.: *daf1*; Table 10.2). The Bos1 protein probably senses the fungicides and transmits this signal to the downstream MAP-kinase BcSak1 (and potentially other pathways), thereby stimulating the cellular response leading to cell wall breakdown, cell swelling and burst (Liu et al. 2008; reviewed in Tanaka and Izumitsu 2010).

The modifications observed in the Bos1 protein either completely abolish its function (loss of function) leading to cross-resistance between fludioxonil and dicarboximides, mostly observed in laboratory mutants, or they interfere with the N-terminal, helical HAMP-domains of the protein involved in signal transduction. Indeed the replacements of hydrophobic residues in these domains (*e.g.*, I365S) are thought to abolish their helical structure and consequently signal transduction (Fillinger et al. 2012). If the histidine-kinase Bos1 constitutes the target of either the dicarboximides or the phenylpyrroles still remains unknown. Pillonel and Meyer (1997) showed differences in protein kinase inhibition profiles between phenylpyrrole and dicarboximide fungicides. It may be, as suggested by Hayes et al. (2014) that both fungicides induce cell death through over-stimulation of the BcSak1 MAP-kinase.

10.2.2.3 Inhibitors of Amino-Acid Biosynthesis

The anilinopyrimidines mepanipyrim, pyrimethanil and cyprodinil are registered against grey mould on various crops, *solo* or in mixture with fludioxonil. They are suspected to inhibit amino-acid biosynthesis, especially that of methionine (Fritz et al. 1997). However, enzyme assays could not prove any effect of pyrimethanil on cystathione-β-lyase activity (Fritz et al. 2003) and no specific mutations were recorded, in resistant strains, either in the sequence of the corresponding gene *BcmetC*, nor in those encoding cystathionine γ-synthase, cystathionine γ-lyase, or cystathionine β-synthase, also involved in methionine synthesis (Sierotzki et al. 2002; De Miccolis Angelini et al. 2012). Therefore, the direct target of anilinopyrimidines remains unknown.

Isolates displaying moderate or high resistance to anilinopyrimidines were found a few years after the introduction of these molecules (Leroux et al. 2002b). This specific resistance is conferred by a single gene and may be suspected as target site mutation (Chapeland et al. 1999). Additional genetic analyses conducted by De Miccolis Angelini et al. (2012) indicated strong instability of anilinopyrimidine resistance during vegetative growth without selective pressure, suggesting most resistant isolates to be heterokaryons. This was confirmed by the lethality of homokaryotic anilinopyrimidine resistant ascospores. Anilinopyrimidine resistance is detected in most grey mould populations. Resistance management, restricting their application allows maintaining acceptable resistance frequencies, and efficacy, while enabling negative selection pressure to operate (Walker et al. 2013).

10.2.2.4 Ergosterol Biosynthesis Inhibitors (SBIs)

Since ergosterol is specific to the fungal kingdom and the major sterol present in the membranes of most fungi, its biosynthesis constitutes an important target for general fungicides. Despite the number of active ingredients acting as SBIs, grey mould control relies on two to four molecules, the C4-demethylation inhibitors fenhexamid (late 1990s) and fenpyrazamine (2012) and, to a lesser extent, the 14α-demethylation inhibitors (DMI), tebuconazole and prochloraz. The C4-demethylation inhibitors have a spectrum of activity limited to *Botrytis* and closely related species (Rosslenbroich 1999; Debieu et al. 2013), but *Botrytis pseudocinerea* is naturally resistant to fenhexamid (Leroux et al. 2002a; Walker et al. 2011).

The hydroxyanilide fenhexamid and the amino-pyrazolinone fenpyrazamine inhibit the 3-ketoreductase of the C4-demethylation complex, stopping ergosterol synthesis and leading to the accumulation of toxic intermediates (Debieu et al. 2001; Tanaka, Botrytis Symposium 2013, oral comm.). The selectivity of these molecules can be explained by differential affinities of fenhexamid towards the 3-ketoreductase target enzyme of different fungal species (Debieu et al. 2013).

Genetic studies have shown that acquired resistance to fenhexamid (and also to fenpyrazamine prior to its introduction) in *B. cinerea* is due to target modifications in most strains (Fillinger et al. 2008; Billard et al. 2012). The principal highly resistant strains display a replacement of the phenylalanine at position 412 in the Erg27 protein, whereas 20 single modifications have been identified in moderately resistant strains (Albertini and Leroux 2004; Esterio et al. 2011; Fillinger et al. 2008; Grabke et al. 2013; Saito et al. 2014). These modifications decrease the affinity of fenhexamid for the 3-ketoreductase isoenzymes (Debieu et al. 2013), allowing the enzyme to be active even at high fenhexamid concentrations. Although specific resistance arose in populations a few years after fenhexamid registration, no or low efficacy losses are recorded for this molecule, possibly because the restrictions of use (*e.g.* one yearly on grapevine) keep resistant strains at an acceptable frequency and also because a low to moderate cost entails the fitness of resistant isolates (Billard et al. 2012).

10.2.2.5 Fungicides Affecting Fungal Respiration

Other essential cellular functions targeted by synthetic fungicides are fungal respiration and energy production. Eukaryotic cells use the mitochondrial electron transport chain (ETC) to oxidize the coenzyme NADH through electron exchange, but most importantly, the electrochemical proton gradient produced across the inner mitochondrial membrane allows the production of ATP, the cellular energy source necessary for any metabolic activity. As outlined in Fig. 10.1, the ETC is composed of four enzymatic complexes, involved in electron exchange through redox reactions, and the final enzyme, ATP synthase. Complex II, namely succinate dehydrogenase (SDH), also has an enzymatic function in the tricarboxylic acid (TCA) cycle. Five functional categories of respiration inhibitor fungicides have been developed, and three of them are registered as botryticides (Fig. 10.1): those inhibiting complex II (SDHIs), complex III (QoIs) and uncouplers of oxidative phosphorylation.

Uncouplers

Uncouplers reduce the proton-gradient across the mitochondrial membrane and therefore decrease or even inhibit ATP synthesis (Russell 2005). External uncouplers are generally hydrophobic compounds with a delocalized negative charge, which penetrate the mitochondrial membrane. The sole fungicide, classified as uncoupler acting on oxidative phosphorylation, is the dinitro-aniline – or pyridine – amine – fluazinam with a broad spectrum of preventive activity, used in particular against oomycetes and grey mould (reviewed in Terada 1981; Kadenbach 2003). It acts as uncoupler involving protonation/deprotonation reactions due to a protonophoric cycle (Brandt et al. 1992). Several authors suggested additional activities for fluazinam on mitochondrial respiration, *e.g.* inhibition of thiol groups (Brandt et al. 1992), release of cytochrome c into the cytosol and inhibition of complex I of the

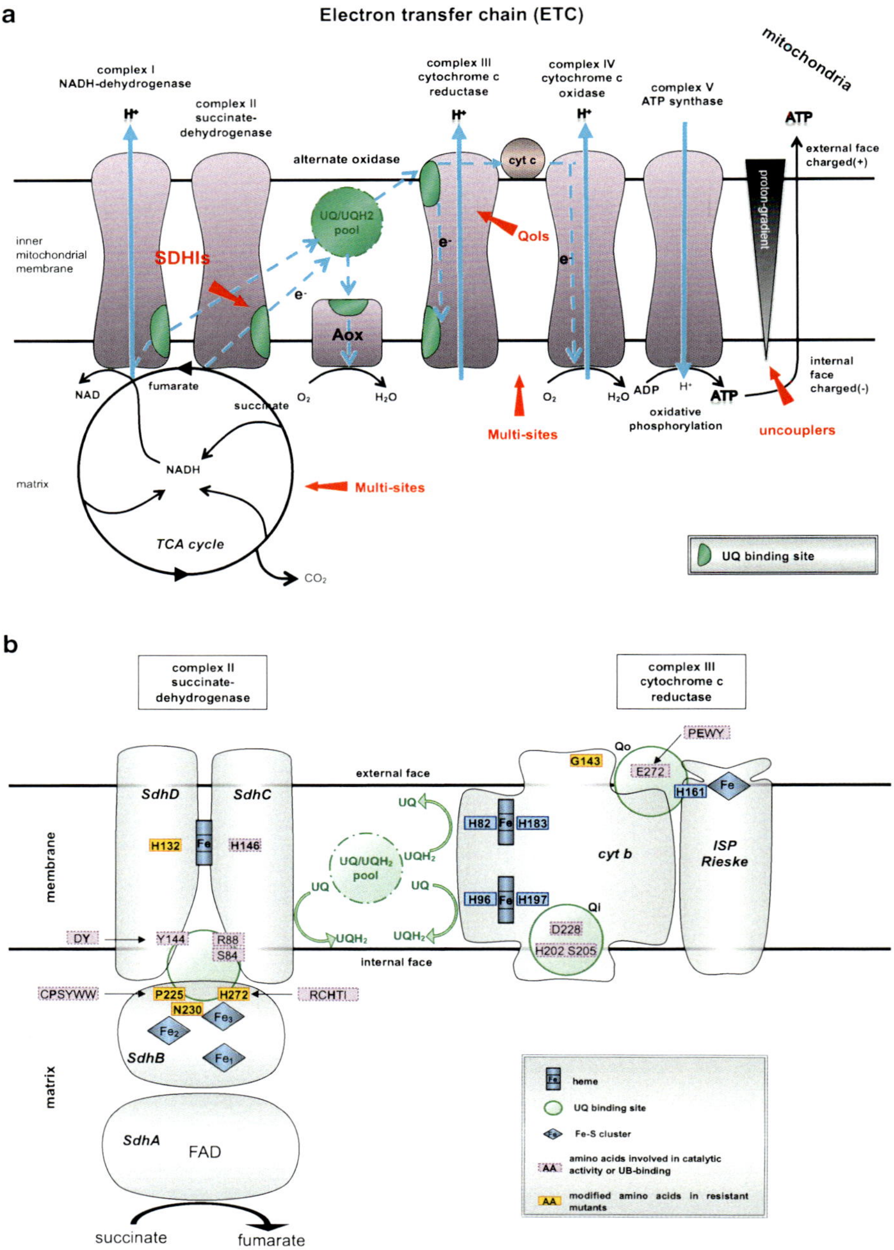

Fig. 10.1 Targets of respiration inhibitors and modified residues involved in resistance with focus on complex II inhibitors (SDHIs). (**a**) Enzyme complexes and target sites of respiration inhibitors on the electron transfer chain (Adapted from Leroux and Walker 2010); (**b**) Detailed view of complexes II and III of the ETC, with the conserved residues in the catalytic or ubiquinone (*UQ*) binding site (Adapted from Leroux and Walker 2010); (**c**) Chemical structure of some SDHIs (Adapted from Glättli et al. 2011); (**d**) Resistance profile of *B. cinerea* mutants to different SDHIs – field isolates (field) compared to *sdhB* site-directed mutants (lab). When particularly high SDHI concentrations are necessary to determine the EC_{50} value, only the minimum resistance factors are indicated over the columns (>RF) (Adapted from Lalève et al. 2014b)

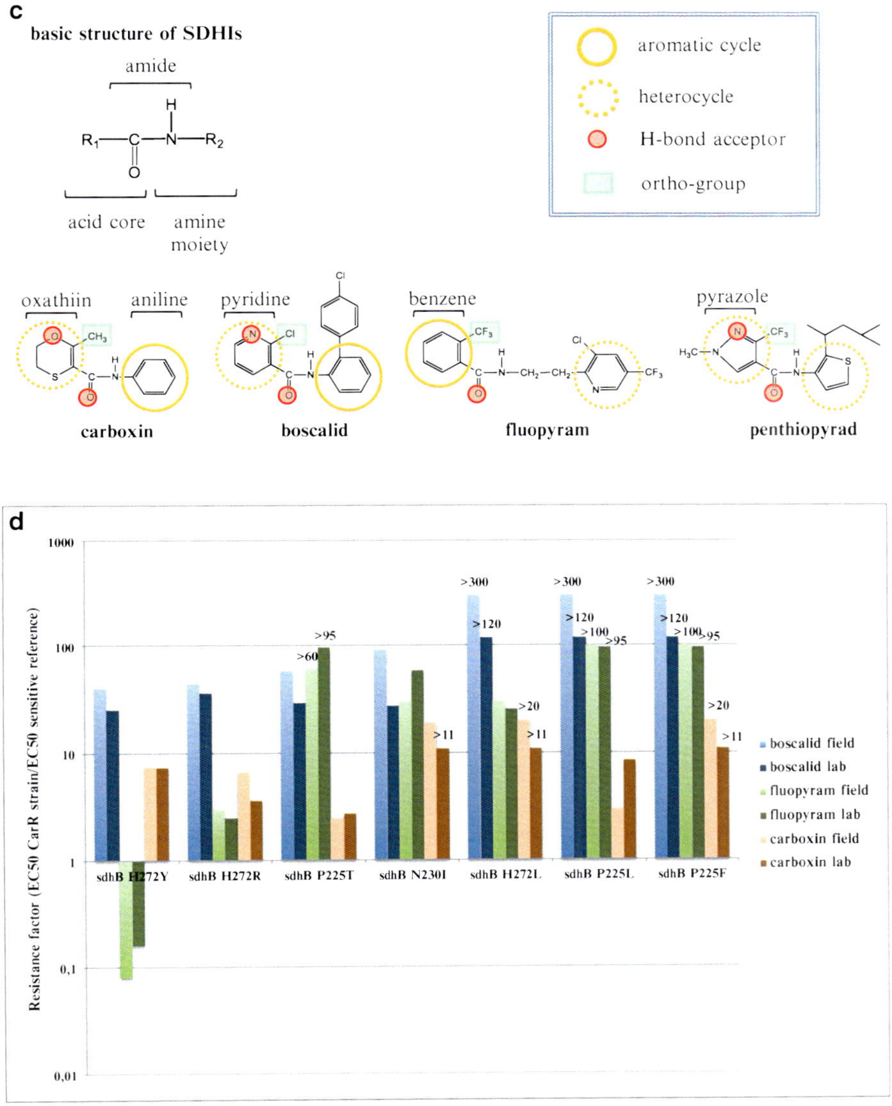

Fig. 10.1 (continued)

ETC (Akagi et al. 1996). These multiple activities on fungal respiration may explain the broad spectrum of fungitoxicity, but also why only few cases of fluazinam resistance have been reported so far for *B. cinerea* (Table 10.2). If resistance to fluazinam involves detoxication eventually through the action of GSTs (gluthathion-S-transferases), as suggested by Leroux (2004), remains to be investigated.

Inhibitors of Complex II: Succinate Dehydrogenase Inhibitors (SDHIs)

SDH couples the oxidation of succinate to fumarate in the mitochondrial matrix with ubiquinone reduction in the inner mitochondrial membrane. It is a complex of four proteins (*i.e.*, SdhA; SdhB; SdhC; SdhD) encoded by nuclear genes: the soluble entity, responsible for the succinate dehydrogenase activity of the enzyme, consists of subunits A and B; the SdhC and SdhD subunits form the integral membrane component, anchoring the enzyme complex to the inner mitochondrial membrane. The ubiquinone-binding site (Q-site) involves amino acids from SdhB, SdhC and SdhD (Cecchini et al. 2003; Hagerhall 1997). Fungicides of the carboxamide family inhibit ubiquinone reduction by binding to the Q-site of SDH. The intact carboxamide structure (R_1-CO-NH-R_2) seems to be required for full fungicidal activity. SDHIs build hydrogen-bonds (H-bonds) with the conserved residues of the Q-site through the heteroatoms highlighted in Fig. 10.1c and hydrophobic or π interactions through the aromatic cycle of the amine moiety (Glättli et al. 2011).

Six classes of SDHIs can be defined on the basis of the chemical structure of the acidic moiety of the molecule (reviewed by Leroux et al. 2010; Sierotzki and Scalliet 2013) (R_1): benzamides (*e.g.* fluopyram, flutolanil), furan carboxamides (*e.g.* fenfuram), oxathiin carboxamides (*e.g.* carboxin, oxycarboxin), pyrazole carboxamides (*e.g.* bixafen, isopyrazam, penthiopyrad), pyridine carboxamides (*e.g.* boscalid) and thiazole carboxamides (*e.g.* thifluzamide). The benzamides can also be subdivided into two groups on the basis of differences in the amine moiety (R_2): phenyl benzamides (*e.g.*, flutolanil) and pyridinyl ethylbenzamides (*e.g.* fluopyram). Actually, boscalid (2002), penthiopyrad (2009), isopyrazam (2010) and fluopyram (2012) figure among the latest registered fungicides against *Botrytis* spp., but similar molecules from other companies may be introduced (*e.g.*, benzovindiflupyr, isofetamid).

Once the baseline sensitivity to boscalid was established by different methods on *B. cinerea* isolates from different hosts and regions (Stammler and Speakman 2006; Zhang et al. 2007; Myresiotis et al. 2008), the first isolates resistant to boscalid were reported in 2007 (Stammler et al. 2007) and since then successively on many crops in several countries (Kim and Xiao 2010; Leroux et al. 2010; Yin et al. 2011; Veloukas et al. 2011; Fernandez-Ortuno et al. 2012; De Miccolis Angelini et al. 2014; Amiri et al. 2014 and others). The carbon source seems to be a critical issue in bioassays, as glucose may compensate SDHI toxicity and should be replaced by acetate, succinate or glycerol. As in other fungi, mutations were found in the genes encoding the subunits B and D of succinate-dehydrogenase, *sdhB* and *sdhD*, especially for the residues $SdhB^{P225}$ and $SdhB^{H272}$ of the ubiquinone binding site, or $SdhD^{H132}$ involved in heme-binding (Fig. 10.1b, d), but also the N230I modification in SdhB. Although modifications of SdhC have also been found in *B. cinerea* strawberry isolates, strict correlation with resistance to boscalid and fluopyram could not be found (Mosbach et al. 2014).

The highest levels of resistance have been recorded for the $SdhB^{P225F/L}$ and $SdhB^{H272L/V}$ substitutions. $SdhB^{H272R}$ and $SdhB^{H272Y}$ are the most frequently detected substitutions in boscalid-resistant strains. Genetic analyses and site directed mutagenesis showed that these modifications of SdhB confer boscalid resistance (De Miccolis Angelini et al. 2010; Laleve et al. 2014b). In fact, they are responsible for different levels of resistance to this pyridine carboxamide, but also for different spectra of cross-resistance to fluopyram, to the oxathiin carboxamide carboxin and to other SDHIs (Leroux et al. 2010; Veloukas et al. 2014) (Fig. 10.1d). Lalève and colleagues (2014b) showed for the *sdhB* mutations a strong correlation between the affinity of SDHIs for the SDH isoforms, SDH inhibition and *in vivo* growth inhibition confirming the key roles of H272, P225 and N230 in carboxamide binding (reviewed in Sierotzki and Scalliet 2013; Laleve et al. 2014b). The $sdhB^{H272Y}$ mutation, leading to fluopyram hypersensitivity, had no effect on SDH activity or respiration. This category of SDHI-resistant mutants, which is currently the most frequently isolated in many agronomic situations, may therefore be well controlled by alternating or mixed applications of boscalid and fluopyram, at least in the coming years. Resistance to SDHIs is associated with fitness cost, either on field mutants (Veloukas et al. 2014) or on isogenic laboratory mutants (Laleve et al. 2014a). Despite discrepancies between the results, both studies revealed more or less important fitness penalty on several life traits linked to the *sdhB* mutations. Veloukas and colleagues' competition assays (2014) on apple between SDHI-resistant and sensitive strains showed clear differences according to the selective pressure. In the presence of fluopyram, for example, the $sdhB^{P225F}$ isolates dominated the population.

The concurrent use of boscalid and fluopyram (and also of future SDHIs) could change the structure of resistant populations, favoring already known or new *sdh* alleles conferring strong positive cross-resistance between all molecules (Amiri et al. 2014). Continuous monitoring studies either with biological assays or combined with molecular tools (De Miccolis Angelini et al. 2014) are necessary to evaluate the impact of variations in SDHI selection pressure on resistance development.

Inhibitors of Complex III (QoIs)

The last two decades was the period of "raise and fall" of strobilurins on many crops. Synthetic molecules derived from the secondary metabolite strobilurin A, produced by basidiomycetes such as *Strobilurus tenacellus*, have been introduced on the fungicide market since 1992 (Russell 2005). They bind to the quinol oxidation (Qo) site of cytochrome b (complex III of the ETC) and thereby stop electron transfer between complex III and IV, inhibiting NADH oxidation and ATP synthesis in many fungal pathogens (reviewed in Balba 2007). Strobilurins are referred to as QoI fungicides, as they bind to the inner Qo-site (Fig. 10.1b). Two QoIs (azoxystrobin and pyraclostrobin), often associated to other modes of action, were used on several crops to control *Botrytis* disease and other fungi at the same time (Table 10.1). Indeed, QoIs have low intrinsic activity on *Botrytis* sp., due to the constitutive expression of the terminal alternative oxidase (AOX). AOX allows electrons to bypass the blockage of the cytochrome pathway caused by strobilurins (Ishii et al. 2009).

This category of fungicides bares a high risk of resistance development, as the target, cytochrome b, is encoded by a mitochondrial gene: *cyt b* mutations responsible for resistance (Table 10.2) confer resistance also under heteroplasmic conditions and are maternally transmitted, but probably also through hyphal fusion (reviewed in Gisi et al. 2002; Villani and Cox 2014; De Miccolis Angelini et al. 2012). Most phytopathogenic fungi treated with QoIs became resistant through the acquisition and dispersal of the G143A and two other minor mutations in the *cyt b* gene (Gisi et al. 2002; Russell 2005). Although in some *B. cinerea* strains the presence of an intron located precisely at codon 143 counter-selects the G143A mutation (Banno et al. 2009; Ishii et al. 2009; Leroux et al. 2010; Jiang et al. 2009; Asadollahi et al. 2013; Vallieres et al. 2011), QoI resistance is now generalized in *Botrytis* populations (due to the presence of resistance phenotype even in heteroplasmic cells), in agreement with the lack of fitness penalty associated with the *cyt* b^{G143A} allele (Veloukas et al. 2014). This resistance was also generalized on crops (*e.g.*, grapevine) that never received QoIs against *Botrytis* spp., suggesting that it can be unintentionally selected *via* sprays targeting other diseases. Considering the QoI resistance risk in *Botrytis* spp. and the limited intrinsic activities of these molecules, those should be replaced, whenever possible, by specific botryticides efficient on local populations. Finally, a novel QoI, the benzylcarbamate pyribencarb, with promising efficiency on QoI resistant strains, due to poor cross-resistance with strobilurins, is actually in the registration process (Takagaki et al. 2011). Indeed, it was suggested that pyribencarb might differ slightly in the binding sites within cytochrome b, compared to other QoIs (Kataoka et al. 2010).

10.2.3 Resistance Mechanisms Unlinked to the Target

Besides specific resistance to given fungicides due to target site modification, several other mechanisms have been extensively studied in *Botrytis* spp. in the last decade, eventually conferring cross resistance to unrelated chemical compounds, because the principal mechanism induces the reduction of the intracellular concentration of toxic compounds.

10.2.3.1 Multi-drug Resistance

Botrytis cinerea isolates displaying monogenic low-to-moderate resistance to several fungicides have been detected in French vineyards since the 1990s (Chapeland et al. 1999), probably due to the high concomitant selective pressure of various chemical families. Three patterns of cross-resistance were described (Leroux et al. 1999; Leroux and Walker 2013), respectively named MDR1, MDR2 and MDR3. All display cross resistance (low to medium RLs) to anilinopyrimidines, diethofencarb, iprodione, fludioxonil, some respiration inhibitors, but also to the clinical sterol-biosynthesis inhibitor tolnaftate (high RLs). Additional resistances allow distinguishing between MDR1 and MDR2 strains as shown in Fig. 10.2, while MDR3 strains combine resistance spectra of both phenotypes with additive resistance factors.

Multi-drug resistance (MDR) is a well-known phenomenon in the medical sector. Generally due to increased efflux of unrelated toxic compounds, it involves the upregulation of membrane transporters, either ABC (ATP-binding cassette) transporters or those of the major-facilitator superfamily (MFS) (for reviews see Moye-Rowley 2003; Morschhäuser 2010). Fungicide efflux is also at work in *B. cinerea* MDR strains correlated to membrane transporter overexpression. While MDR1 strains show overexpression of the ABC transporter gene *atrB* linked to single modifications in the transcription factor Mrr1, the MDR2 phenotype is due to *mfsM2* overexpression, itself originating from the insertion of a retrotransposon like element. MDR3 strains derive from recombination of both *mdr* mutations, probably after sexual crosses (Kretschmer et al. 2009; Mernke et al. 2011). In Champagne vineyards, more than 60 % of collected grey mould populations display an MDR phenotype (Walker et al. 2013), approximately 20 % of each phenotype. Despite this high frequency, no loss in field efficacy is observed with current fungicides at recommended application rates, as the resistance factors of MDR strains are low to moderate (Fig. 10.2).

In German strawberry fields the situation seems more severe with large proportions of MDR strains cumulating specific resistance(s) due to target site mutations, therefore leading to high resistance levels to many fungicides. In addition, the strawberry specific *B. cinerea* group S (described in Chap. 6) contains a new MDR1 phenotype named MDR1h, with two to three times higher resistance levels than previously identified MDR1 strains to cyprodinil and fludioxonil. A 3 bp deletion in the *mrr1* gene (Δ^{L497}) leads to 150–300 fold overexpression of *atrB*, three to six times higher than in MDR1 strains (Leroch et al. 2013). The combination of *mdr* mutations with specific resistance alleles may lead to serious crop losses, if the frequency of highly multi-resistant strains reaches a certain threshold and if their fitness is not too much affected.

10.2.3.2 Detoxification

Detoxification of chemical drugs through enzymatic metabolisation involving gluthathion-S-transferases (GSTs), cytochrome P450s, hydrolases or esterases constitutes a resistance mechanism widespread in insect pests and weeds (Delye et al. 2013; Ffrench-Constant 2013). Such as MDR, detoxification can confer cross-resistance to pesticides with different modes of action. In phytopathogenic fungi, this mechanism has been rarely involved in fungicide resistance. Some *B. cinerea* strains have been shown to be sensitive to the anti-oomycete fungicide cymoxanil through metabolic activation of this profungicide (Tellier et al. 2008, 2009). This phenomenon does not interfere with grey mould control, as this compound is not used against *Botrytis* spp. Detoxification was proposed as a possible resistance

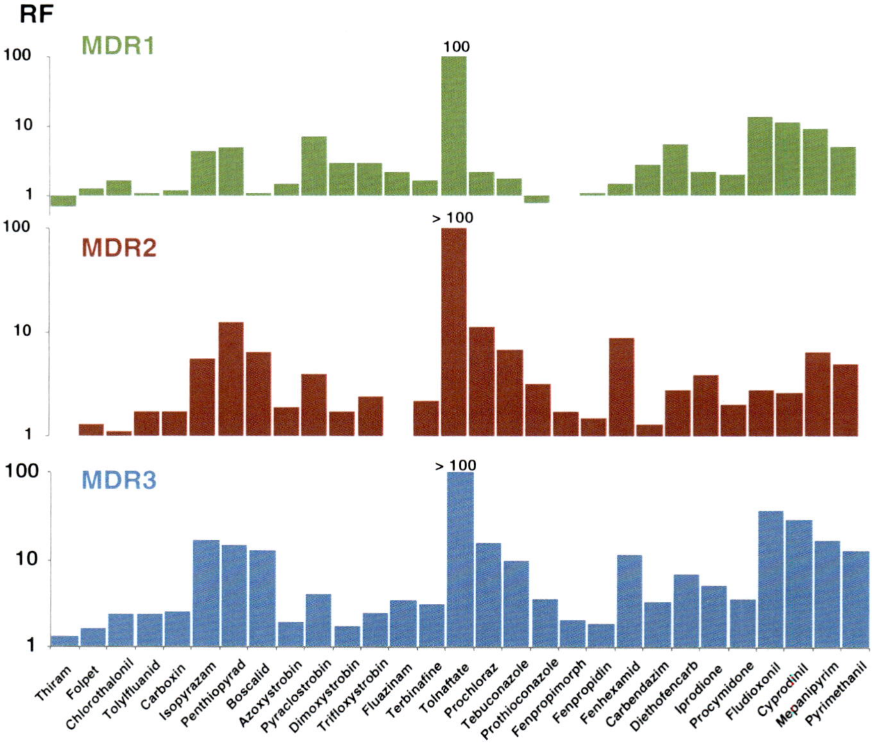

Fig. 10.2 Resistance factors of *B. cinerea* MDR strains on various fungicides (Adapted from Leroux and Walker 2013)

mechanism against multisite fungicides and fluazinam, as described in the review of Leroux (2004), but only few field isolates resistant to these compounds have been found so far.

The natural resistance of the new species *B. pseudocinerea* (Chap. 6) to the hydroxyanilide fenhexamid (formerly known as HydR1 phenotype) seems to involve detoxification. Besides the reduced affinity of fenhexamid for its target enzyme in *B. pseudocinerea* compared to the *B. cinerea* enzyme (Debieu et al. 2013), Leroux and colleagues observed synergy between fenhexamid and DMIs on *B. pseudocinerea*'s mycelial growth (Leroux et al. 2002a) and studies conducted by Bayer SAS showed metabolisation of fenhexamid by *B. pseudocinerea* strains (Suty et al. 1999). Later, Billard et al. identified a cytochrome P450 similar to the DMI target Cyp51, named Cyp684, whose inactivation nearly completely abolished fenhexamid resistance in *B. pseudocinerea* (Billard et al. 2011), indicating that Cyp684 is a major player in *B. pseudocinerea*'s natural resistance, potentially through fenhexamid metabolisation. The expression of *cyp684* displays higher induction levels in *B. pseudocinerea* strains after fenhexamid treatment, than in *B. cinerea* strains, but the metabolisation products still remain unknown (Billard et al. unpublished).

A rarely observed fenhexamid resistance phenotype in *B. cinerea*, named HydR2 (see Table 10.2), seems to be linked to fungicide detoxification as well. Synergy between DMIs and fenhexamid suggests the involvement of a cytochrome P450 (Leroux et al. 2002a) and no changes in *erg27* sequence, nor its expression, were observed with HydR2 isolates (Billard, unpublished). As the above mentioned *cyp684* was excluded from HydR2 phenotype (Billard, unpublished), the cytochrome P450 involved remains to be identified. Genetic analyses indicated its genomic location close to the *bik* gene cluster involved in bikaverin biosynthesis (Schumacher et al. 2013; see Chap. 13).

10.3 Fitness Cost of Fungicide Resistance

Resistance to fungicides may be associated with a cost, as generally reported for fungal populations subjected to fungicide-mediated selection pressure (Milgroom et al. 1989). Characterization of the cost of resistance in resistant isolates may make it possible to predict the rate of evolution of such isolates in the population. This characterization is therefore essential for estimation of the extent to which resistant isolates constitute a risk to disease control by fungicides and for the optimization of anti-resistance strategies. As an example, detecting a fitness cost may be of great interest in strategies alternating fungicidal modes of action since it may substantially delay resistance evolution between two applications (REX Consortium 2013).

Fitness is the ability of an individual to survive in its environment and to contribute successfully to the next generation (Orr 2009). Differences in fitness between individuals may arise from differences in performance at any stage of the life cycle, and any variation of these fitness components can contribute to differences in total fitness between individuals (Antonovics and Alexander 1989). Fitness can be measured using two approaches. More generally research groups measure several parameters on fungicide resistant field isolates in comparison to sensitive strains. The traits generally measured for phytopathogenic fungi are conidiation, conidia germination, hyphal growth and virulence (Antonovics and Alexander 1989) and should be chosen all along the life cycle. These analyses globally hint to fitness penalties (or not) of the phenotypic category considered (Bardas et al. 2008; Saito et al. 2014; Veloukas et al. 2014), but they need to be performed on a statistically significant set of representative strains, because the genetic and phenotypic polymorphism of natural *Botrytis* isolates may hide or exaggerate the phenotype linked to the resistance allele.

An alternative approach was developed these recent years by the construction of isogenic mutants using site-directed mutagenesis through homologous gene replacement. Briefly, all mutant strains are identical except for the resistance allele. Comparing their biological features allows precisely attributing a fitness cost to each allele by *in vitro* and *in planta* measurements (Billard et al. 2012; Laleve et al. 2014a).

However, both types of analyses do not necessarily give the same results. As in the case of SDHI resistant mutants, the study of isogenic laboratory mutants revealed the highest fitness penalty for the *sdhB*^{H272R} allele (Laleve et al. 2014a), whereas the very similar analysis of field-strains gave the lowest fitness penalty to this allele among all *sdhB* alleles tested (Veloukas et al. 2014). The genetic context of the respective field and laboratory strains may influence the biological parameters of the resistant mutants. Moreover, fitness is only estimated via a limited number of life traits, which may not be relevant. Therefore, conclusions about fitness penalties to predict the risk of a given resistance to persist and spread should be drawn with precautions when using this approach.

Another set of methods tends to evaluate fitness as a whole, *i.e.* trying to measure the survival of resistant strains and the evolution of their frequency in populations. This can be approached *in vitro* or *in planta* with competition experiments, measuring the frequency of each genotype after each subculture cycle (see examples in Veloukas et al. 2013; Laleve et al. 2014a) but cannot fully mimic biotic interactions, as they may happen in the field. Total fitness can also be estimated mathematically, by modeling changes in allele frequencies in populations subject to natural selection (Orr 2009). This can be achieved while detecting cline patterns, *i.e.* a gradient of resistant allele frequency over a geographical transect. Parameters of cline models may be direct indicators of selection, either negative or positive, and of migration, as demonstrated in resistance to insecticides and fungicides (Lenormand et al. 1999; Walker and Fournier 2014). Non-spatial models, modelling the evolution of resistance frequency all along the fungus life cycle may also help inferring these parameters (Walker and Fournier 2014).

10.4 Resistance Management

Anti-resistance strategies are based on the skillful deployment of tools (prophylaxis, plant resistance genes and antifungal compounds) to delay resistance. Prophylaxis against *Botrytis*, even of partial efficacy, can be deployed in many crops and mainly deals with decreasing the plant vigor (*via* fertilization management, host density), creating a dry climate around the susceptible organs (*via* pruning, green harvest, climate regulation in greenhouses) and preventing wounds on susceptible organs (control of insect vectors, adaptation of mechanical tools) (Chap. 8). Additionally, crops may have at least partial resistance to *Botrytis* in some cultivars.

Dealing with antifungals, either synthetic or natural, several strategies are available. Firstly, fungicides may be limited in their use, as early as registration. This may be of great efficacy in decreasing the selective pressure for a given mode of action but not adapted to crops that need a large number of sprays. As fungicides often have lower intrinsic activities against *Botrytis* than against other fungi, the mixture strategy, may not be the appropriate or should be restricted to the most powerful inhibitors, *i.e.* which suffer dose reduction (for economic, environmental or toxicological reasons). As mixture is based on the redundant killing of fungal species, both partners should be fully efficient against *Botrytis* local populations, not to expose one of the modes of action (*e.g.*, QoIs + SDHIs). Regular mixture applications may also select for generalist resistance mechanisms, such as MDR (*e.g.*, fludioxonil + cyprodinil).

Limiting the use of each botryticide at the multi-seasonal scale and alternating active ingredients at full dose with different modes of action seems to be a suitable approach in many situations, particularly in cases of emerging resistance (*e.g.* resistance to SDHIs). Indeed, this strategy allows the expression of resistance cost, as the same molecules may target distinct pathogen generations. This strategy has been shown to decrease the frequency (*e.g.*, dicarboximide or benzimidazole resistant strains) or to delay the emergence of resistant strains (*e.g.* to anilinopyrimidines or to fenhexamid), for example in French vineyards. The key elements of the actual management of fungicides resistance in *Botrytis* have been summarized as a decision tree based on the observation of mechanism, frequency and phenotype of field resistant mutants (Walker et al. 2013).

At last, as anti-resistance strategies delay but not fully prevent resistance, *Botrytis* control can only be efficient and durable if innovative modes of action are regularly released on the market. Keeping the diversity of modes of action, even with partial efficacy, is crucial in resistance management. During the last decades, important breakthroughs were achieved in the discovery of new resistance mechanisms, their genetic determinants (see also Chap. 3), the development of molecular tools to detect and quantify resistance phenomena in grey mould populations. Altogether, these achievements may help optimizing the chemical control of this threatening disease. In addition, resistance monitoring, with adapted technical procedures, relevant sampling sizes and observed areas, should accompany the fungicides' life, to identify and optimize the anti-resistance strategies to local situations.

Acknowledgements We are grateful to Pierre Leroux for critical reading and corrections of the manuscript.

References

Akagi T, Mitani S, Komyoji T et al (1996) Quantitative structure-activity relationships of fluazinam and related fungicidal N-phenylpyridinamines: preventive activity against *Sphaerotheca fuliginea*, *Pyricularia oryzae* and *Rhizoctonia solani*. J Pestic Sci 21(1):23–29

Albertini C, Leroux P (2004) A *Botrytis cinerea* putative 3-keto reductase gene (*ERG27*) that is homologous to the mammalian 17- beta-hydroxysteroid dehydrogenase type 7 gene. Eur J Plant Pathol 110(7):723–733

Amiri A, Heath SM, Peres NA (2014) Resistance to fluopyram, fluxapyroxad, and penthiopyrad in *Botrytis cinerea* from strawberry. Plant Dis 98(4):532–539

Antonovics J, Alexander HM (1989) The concept of fitness in plant-fungal pathogen systems. In: Plant disease epidemiology, vol 2: Genetics, resistance, and management. McGraw-Hill Publishing Company, New York, pp 185–214

Asadollahi M, Szojka A, Fekete E et al (2013) Resistance to QoI fungicide and cytochrome b diversity in the Hungarian *Botrytis cinerea* population. J Agric Sci Technol 15(2):397–407

Balba H (2007) Review of strobilurin fungicide chemicals. J Environ Sci Health B-Pestic Food Contam Agric Waste 42(4):441–451

Banno S, Fukumori F, Ichiishi A et al (2008) Genotyping of benzimidazole-resistant and dicarboximide-resistant mutations in *Botrytis cinerea* using real-time polymerase chain reaction assays. Phytopathology 98(4):397–404

Banno S, Yamashita K, Fukumori F et al (2009) Characterization of QoI resistance in *Botrytis cinerea* and identification of two types of mitochondrial cytochrome b gene. Plant Pathol 58(1):120–129

Bardas GA, Myresiotis CK, Karaoglanidis GS (2008) Stability and fitness of anilinopyrimidine-resistant strains of *Botrytis cinerea*. Phytopathology 98(4):443–450

Bernard BK, Gordon EB (2000) An evaluation of the common mechanism approach to the Food Quality Protection Act: captan and four related fungicides, a practical example. Int J Toxicol 19(1):43–61

Billard A, Fillinger S, Leroux P et al (2011) Fenhexamid resistance in the *Botrytis* species complex, responsible for grey mould disease. In: Thajuddin N (ed) Fungicides – beneficial and harmful aspects. InTech, Rijeka, Croatia. pp 61–78 http://www.intechopen.com/books/fungicides-beneficial-and-harmful-aspects

Billard A, Fillinger S, Leroux P et al (2012) Strong resistance to the fungicide fenhexamid entails a fitness cost in *Botrytis cinerea*, as shown by comparisons of isogenic strains. Pest Manag Sci 68(5):684–691

Bollen G, Scholten G (1971) Acquired resistance to benomyl and some other systemic fungicides in a strain of *Botrytis cinerea* in cyclamen. Neth J Plant Pathol 77:83–90

Brandt U, Schubert J, Geck P et al (1992) Uncoupling activity and physicochemical properties of derivatives of fluazinam. BBA 1101(1):41–47

Brent KJ, Hollomon DW (2007) Fungicide resistance: the assessment of risk, FRAC Monograph 2. Croplife International, Brussels

Carisse O, Tremblay DM (2007) Incidence and significance of iprodione-insensitive isolates of *Botrytis squamosa*. Plant Dis 91(1):41–46

Cecchini G, Maklashina E, Yankovskaya V et al (2003) Variation in proton donor/acceptor pathways in succinate:quinone oxidoreductases. FEBS Lett 545(1):31–38

Chapeland F, Fritz R, Lanen C et al (1999) Inheritance and mechanisms of resistance to anilinopyrimidine fungicides in *Bortytis cinerea* (*Botryotinia fuckeliana*). Pestic Biochem Physiol 64:85–100

Corbett J, Wright K, Baillie A (1984) The biochemical mode of action of pesticides, 2nd edn. Academic, London

Cui W, Beever RE, Parkes SL et al (2004) Evolution of an osmosensing histidine kinase in field strains of *Botryotinia fuckeliana* (*Botrytis cinerea*) in response to dicarboximide fungicide usage. Phytopathology 94(10):1129–1135

Davidse L, Ishii T (1995) Biochemical and molecular aspects of benzimidazoles, *N*-phenylcarbamates and *N*-phenylformamidoximes and the mechanisms of resistance to the compounds. In: Lyr H (ed) Modern selective fungicides. Gustav Fischer Verlag, Iena

De Miccolis Angelini RM, Habib W, Rotolo C et al (2010) Selection, characterization and genetic analysis of laboratory mutants of *Botryotinia fuckeliana* (*Botrytis cinerea*) resistant to the fungicide boscalid. Eur J Plant Pathol 128(2):185–199

De Miccolis Angelini RM, Pollastro S, Faretra F (2012) Genetics of fungcide resistance in *Botrytis cinerea*. In: Thind TS (ed) Fungicide resistance in crop protection: risk and management. CAB International, Oxfordshire, pp 237–250

De Miccolis Angelini RM, Masiello M, Rotolo C et al (2014) Molecular characterisation and detection of resistance to succinate dehydrogenase inhibitor fungicides in *Botryotinia fuckeliana* (*Botrytis cinerea*). Pest Manag Sci 70(12):1884–1893

Debieu D, Leroux P (in press) Mechanisms of resistance: sterol biosynthesis inhibitors – C-4 demethylation. In: Ishii H, Hollomon DW (eds) Fungicide resistance in plant pathogens: principles and a guide to practical management. Springer, Japan

Debieu D, Bach J, Hugon M et al (2001) The hydroxyanilide fenhexamid, a new sterol biosynthesis inhibitor fungicide efficient against the plant pathogenic fungus *Botryotinia fuckeliana* (*Botrytis cinerea*). Pest Manag Sci 57:1060–1067

Debieu D, Bach J, Montesinos E et al (2013) Role of sterol 3-ketoreductase sensitivity in susceptibility to the fungicide fenhexamid in *Botrytis cinerea* and other phytopathogenic fungi. Pest Manag Sci 69(5):642–651

Delen N, Yildiz M, Maraite H (1984) Benzimidazole and dithiocarbamate resistance of *Botrytis cinerea* on greenhouse crops in Turkey. Meded Faculteit Landbouwwetenschappen Rijksuniversiteit Gent 49(2a):153–161

Delye C, Jasieniuk M, Le Corre V (2013) Deciphering the evolution of herbicide resistance in weeds. Trends Genet 29(11):649–658

Esterio M, Ramos C, Walker AS et al (2011) Phenotypic and genetic characterization of Chilean isolates of Botrytis cinerea with different levels of sensitivity to fenhexamid. Phytopathol Mediterr 50(3):414–420

Faretra F, Pollastro S (1991) Genetic basis of resistance to benzimidazole and dicarboximide fungicides in *Botryotinia fuckeliana* (*Botrytis cinerea*). Mycol Res 95(8):943–951

Fenner K, Canonica S, Wackett LP et al (2013) Evaluating pesticide degradation in the environment: blind spots and emerging opportunities. Science 341(6147):752–758

Fernandez-Ortuno D, Chen FP, Schnabel G (2012) Resistance to pyraclostrobin and boscalid in *Botrytis cinerea* isolates from strawberry fields in the Carolinas. Plant Dis 96(8):1198–1203

Ffrench-Constant RH (2013) The molecular genetics of insecticide resistance. Genetics 194(4):807–815

Fillinger S, Leroux P, Auclair C et al (2008) Genetic analysis of fenhexamid resistant field isolates of the phytopathogenic fungus *Botrytis cinerea*. Antimicrob Agents Chemother 52(11):3933–3940

Fillinger S, Ajouz S, Nicot PC et al (2012) Functional and structural comparison of pyrrolnitrin- and iprodione-induced modifications in the class III histidine-kinase Bos1 of *Botrytis cinerea*. PLoS One 7(8):e42520

Forster B, Staub T (1996) Basis for use strategies of anilinopyrimidine and phenylpyrrole fungicides against *Botrytis cinerea*. Crop Prot 15(6):529–537

Fritz R, Lanen C, Colas V et al (1997) Inhibition of methionine biosynthesis in *Botrytis cinerea* by the anilinopyrimidine fungicide pyrimethanil. Pestic Sci 49:40–46

Fritz R, Lanen C, Chapeland-Leclerc F et al (2003) Effect of the anilinopyrimidine fungicide pyrimethanil on the cystathionine beta-lyase of *Botrytis cinerea*. Pestic Biochem Physiol 77(2):54–65

Fujimura M, Ochiai N, Ichiichi A et al (2000) Sensitivity to phenylpyrrole fungicides and abnormal glycerol accumulation in *os* and *cut* mutant strains of *Neurospora crassa*. J Pestic Sci 25:31–36

Gessler C, Sozzi D, Kern H (1981) Benzimidazole fungicides – mode of action and problems. Ber Schweizerischen Botanischen Ges 90(1–2):45–54

Gisi U, Sierotzki H, Cook A et al (2002) Mechanisms influencing the evolution of resistance to Qo inhibitor fungicides. Pest Manag Sci 58(9):859–867

Glättli A, Grote T, Stammler G (2011) SDH-inhibitors: history, biological performance and molecular mode of action. In: Dehne HW, Deising HB, Gisi U, Kuck KH, Russell PE, Lyr H (eds) Modern fungicides and antifungal compounds VI. DPG-Verlag, Braunschweig, pp 159–170

Grabke A, Fernandez-Ortuno D, Schnabel G (2013) Fenhexamid resistance in *Botrytis cinerea* from strawberry fields in the Carolinas is associated with four target gene mutations. Plant Dis 97(2):271–276

Guo ZJ, Miyoshi H, Komyoji T et al (1991) Uncoupling activity of a newly developed fungicide, fluazinam [3-chloro-N-(3-chloro-2,6-dinitro-4-trifluoromethylphenyl)-5-trifluoromethyl-2-pyridinamine]. BBA 1056(1):89–92

Hagerhall C (1997) Succinate: quinone oxidoreductases. Variations on a conserved theme. BBA 1320(2):107–141

Hayes BME, Anderson MA, Traven A et al (2014) Activation of stress signalling pathways enhances tolerance of fungi to chemical fungicides and antifungal proteins. Cell Mol Life Sci 71(14):2651–2666

Hollomon DW, Brent KJ (2009) Combating plant diseases—the Darwin connection. Pest Manag Sci 65(11):1156–1163

Hsiang T, Chastagner GA (1991) Growth and virulence of fungicide-resistant isolates of 3 species of *Botrytis*. Can J Plant Pathol 13(3):226–231

Hsiang T, Chastagner GA (1992) Production and viability of sclerotia from fungicide-resistant and fungicide-sensitive isolates of *Botrytis cinerea, B. elliptica* and *B. tulipae*. Plant Pathol 41(5):600–605

Ishii H, Yano K, Date H et al (2007) Molecular characterization and diagnosis of QoI resistance in cucumber and eggplant fungal pathogens. Phytopathology 97(11):1458–1466

Ishii H, Fountaine J, Chung WH et al (2009) Characterisation of Qol-resistant field isolates of *Botrytis cinerea* from citrus and strawberry. Pest Manag Sci 65(8):916–922

Jiang JH, Ding LS, Michailides TJ et al (2009) Molecular characterization of field azoxystrobin-resistant isolates of *Botrytis cinerea*. Pestic Biochem Physiol 93(2):72–76

Jochova J, Rupes I, Peberdy JF (1993) Effect of the microtubule inhibitor benomyl on protein secretion in *Aspergillus nidulans*. Mycol Res 97:23–27

Kadenbach B (2003) Intrinsic and extrinsic uncoupling of oxidative phosphorylation. BBA 1604(2):77–94

Kataoka S, Takagaki M, Kaku K et al (2010) Mechanism of action and selectivity of a novel fungicide, pyribencarb. J Pestic Sci 35(2):99–106

Kim YK, Xiao CL (2010) Resistance to pyraclostrobin and boscalid in populations of *Botrytis cinerea* from stored apples in Washington State. Plant Dis 94(5):604–612

Kim J, Min JY, Bae YS et al (2009) Molecular analysis of *Botrytis cinerea* causing ginseng grey mould resistant to carbendazim and the mixture of carbendazin plus diethofencarb. Plant Pathol J 25(4):322–327

Kretschmer M, Leroch M, Mosbach A et al (2009) Fungicide-driven evolution and molecular basis of multidrug resistance in field populations of the grey mould fungus *Botrytis cinerea*. PLoS Pathog 5(12):e1000696

Kuck K, Russell PE (2006) FRAC: combined resistance risk assessment. Asp Appl Biol 78:3–10

Lalève A, Fillinger S, Walker AS (2014a) Fitness measurement reveals contrasting costs in homologous recombinant mutants of *Botrytis cinerea* resistant to succinate dehydrogenase inhibitors. Fungal Genet Biol 67:24–36

Lalève A, Gamet S, Walker AS et al (2014b) Site-directed mutagenesis of the P225, N230 and H272 residues of succinate dehydrogenase subunit B from *Botrytis cinerea* highlights different roles in enzyme activity and inhibitor binding. Environ Microbiol 16(7):2253–2266

Lenormand T, Bourguet D, Guillemaud T et al (1999) Tracking the evolution of insecticide resistance in the mosquito *Culex pipiens*. Nature 400(6747):861–864

Leroch M, Plesken C, Weber RW et al (2013) Gray mould populations in German strawberry fields are resistant to multiple fungicides and dominated by a novel clade closely related to *Botrytis cinerea*. Appl Environ Microbiol 79(1):159–167

Leroux P (1996) Recent developments in the mode of action of fungicides. Pestic Sci 47(2):191–197

Leroux P (2004) Chemical control of *Botrytis cinerea* and its resistance to chemical fungicides. In: Elad Y, Williamson B, Tudzynski P, Delen N (eds) *Botrytis*: biology, pathology and control. Kluwer Academic Publishers, Dordrecht, pp 195–222

Leroux P, Clerjeau M (1985) Resistance of *Botrytis cinerea* Pers. and *Plasmopara viticola* (Berl. and de Toni) to fungicides in French vineyards. Crop Prot 4(2):137–160

Leroux P, Walker AS (2010) Les fongicides affectant les processus respiratoires. Episode 1: Modes d'action et phénomènes de résistance chez les anciennes substances (multisites et unisites affectant la biodisponibilité de l'ATP) et les nouvelles de type SDHI. Phytoma – La Défense des Végétaux 631:8–11

Leroux P, Walker AS (2013) Activity of fungicides and modulators of membrane drug transporters in field strains of *Botrytis cinerea* displaying multidrug resistance. Eur J Plant Pathol 135(4):683–693

Leroux P, Chapeland F, Desbrosses D et al (1999) Patterns of cross-resistance to fungicides in *Botryotinia fuckeliana* (*Botrytis cinerea*) isolates from French vineyards. Crop Prot 18(10):687–697

Leroux P, Debieu D, Albertini C et al (2002a) The hydroxyanilide botryticide fenhexamid/ mode of action and mechanism of resistance. In: Dehne H-W, Gisi U, Kuck KH, Russel PE, Lyr H (eds) Modern fungicides and antifungal compounds III. AgroConcept GmbH, Bonn, pp 29–40, Th. Mann Verlag, Gelsenkirchen, Germany, Andover, Hampshire

Leroux P, Fritz R, Debieu D et al (2002b) Mechanisms of resistance to fungicides in field strains of *Botrytis cinerea*. Pest Manag Sci 58(9):876–888

Leroux P, Gredt M, Leroch M et al (2010) Exploring mechanisms of resistance to respiratory inhibitors in field strains of *Botrytis cinerea*, the causal agent of gray mould. Appl Environ Microbiol 76(19):6615–6630

Liu W, Leroux P, Fillinger S (2008) The HOG1-like MAP kinase Sak1 of *Botrytis cinerea* is negatively regulated by the upstream histidine kinase Bos1 and is not involved in dicarboximide- and phenylpyrrole-resistance. Fungal Genet Biol 45(7):1062–1074

Ma Z, Yan L, Luo Y et al (2007) Sequence variation in the two-component histidine kinase gene of *Botrytis cinerea* associated with resistance to dicarboximide fungicides. Pestic Biochem Physiol 88(3):300–306

Malathrakis NE (1989) Resistance of *Botrytis cinerea* to dichlofluanid in greenhouse vegetables. Plant Dis 73(2):138–141

Mamiev M, Korolev N, Elad Y (2013) Resistance to polyoxin AL and other fungicides in *Botrytis cinerea* collected from sweet basil crops in Israel. Eur J Plant Pathol 137(1):79–91

McGrath MT (2004) What are fungicides? http://www.apsnet.org/edcenter/intropp/topics/Pages/Fungicides.aspx

Mernke D, Dahm S, Walker AS et al (2011) Two promoter rearrangements in a drug efflux transporter gene are responsible for the appearance and spread of multidrug resistance phenotype MDR2 in *Botrytis cinerea* isolates in French and German vineyards. Phytopathology 101(10):1176–1183

Milgroom M, Levin S, Fry W (1989) Population genetics theory and fungicide resistance. In: Plant disease epidemiology. Leonard KJ, Fry WE, McGraw-Hill, New York

Morschhäuser J (2010) Regulation of multidrug resistance in pathogenic fungi. Fungal Genet Biol 47(2):94–106

Mosbach A, Edel D, Kirchhofer L et al (2014) Mutagenesis studies and field resistance mechanisms to SDHIs in the grey mould pathogen *Botrytis cinerea*. In: Dehne H-W, Deising HB, Fraaije BA et al (eds) Modern fungicides and antifungal compounds. Deutsche Phytomedizinische Gesellschaft, Friedrichroda, pp 91–96

Moye-Rowley WS (2003) Transcriptional control of multidrug resistance in the yeast *Saccharomyces*. Prog Nucleic Acid Res Mol Biol 73:251–279

Myresiotis CK, Bardas GA, Karaoglanidis GS (2008) Baseline sensitivity of *Botrytis cinerea* to pyraclostrobin and boscalid and control of anilinopyrimidine- and benzimidazole-resistant strains by these fungicides. Plant Dis 92(10):1427–1431

Nakazawa Y, Yamada M (1997) Chemical control of gray mold in Japan – a history of combating fungicide resistance. Agrochem Jpn 71:2–6

Orr HA (2009) Fitness and its role in evolutionary genetics. Nat Rev Genet 10(8):531–539

Oshima M, Banno S, Okada K et al (2006) Survey of mutations of a histidine kinase gene BcOS1 in dicarboximide-resistant field isolates of *Botrytis cinerea*. J Gen Plant Pathol 72(1):65–73

Park SY, Jung OJ, Chung YR et al (1997) Isolation and characterization of a benomyl-resistant form of beta-tubulin-encoding gene from the phytopathogenic fungus *Botryotinia fuckeliana*. Mol Cell 7(1):104–109

Pedregosa AM, Rios S, Monistrol IF et al (1995) Effect of the microtubule inhibitor methyl benzimidazol-2-yl carbamate (mbc) on protein secretion and microtubule distribution in *Cladosporium cucumerinum*. Mycol Res 99:43–48

Phillips MWA, McDougall J (2012) Crop protection market trends and opportunities for new active ingredients. Abstracts of Papers of the American Chemical Society 244

Pillonel C, Meyer T (1997) Effect of phenylpyrroles on glycerol accumulation and protein kinase activity of *Neurospora crassa*. Pestic Sci 49:229–236

Pollastro S, Faretra F, DiCanio V et al (1996) Characterization and genetic analysis of field isolates of *Botryotinia fuckeliana* (*Botrytis cinerea*) resistant to dichlofluanid. Eur J Plant Pathol 102(7):607–613

Rewal N, Coley-Smith JR, Sealy-Lewis HM (1991) Studies on resistance to dichlofluanid and other fungicides in *Botrytis cinerea*. Plant Pathol 40(4):554–560

REX Consortium I (2013) Heterogeneity of selection and the evolution of resistance. Trends Ecol Evol 28(2):110–118

Roberts T, Hutson D, Jewess P et al (1999) Metabolic pathways of agrochemicals – Part 2: Insecticides and fungicides. Royal Society of Chemistry, Cambridge

Rosslenbroich H-J (1999) Efficacy of fenhexamid (KBR 2738) against *Botrytis cinerea* and related fungal pathogens. Pflanzenschutz-Nachr 52:127–144

Rosslenbroich HJ, Stuebler D (2000) *Botrytis cinerea* – history of chemical control and novel fungicides for its management. Crop Prot 19(8–10):557–561

Russell PE (2005) A century of fungicide evolution. J Agric Sci 143:11–25

Saito S, Cadle-Davidson L, Wilcox WF (2014) Selection, fitness, and control of grape isolates of *Botrytis cinerea* variably sensitive to fenhexamid. Plant Dis 98(2):233–240

Schumacher J, Gautier A, Morgant G et al (2013) A functional bikaverin biosynthesis gene cluster in rare strains of *Botrytis cinerea* is positively controlled by *VELVET*. PLoS One 8(1):e53729

Sierotzki H, Scalliet G (2013) A review of current knowledge of resistance aspects for the next-generation succinate dehydrogenase inhibitor fungicides. Phytopathology 103(9):880–887

Sierotzki H, Wullschleger J, Alt M et al (2002) Potential mode of resistance to anilinopyrimidine fungicides, in *Botrytis cinerea*. Paper presented at the 13th International Reinhardsbrunn Symposium, Friedrichroda, Germany, 14–18 May 2001

Stammler G, Speakman J (2006) Microtiter method to test the sensitivity of *Botrytis cinerea* to boscalid. J Phytopathol 154(7–8):508–510

Stammler G, Brix B, Nave B et al (2007) Studies on the biological performance of boscalid and its mode of action, Modern fungicides and antifungal compounds V. In: Dehne HW, Deising HB, Gisi U, Kuck KH, Russell PE, Lyr H (eds) vol 15th international reinhardsbrunn symposium, Friedrichroda

Suty A, Pontzen R, Stenzel K (1999) Fenhexamid – sensitivity of *Botrytis cinerea*: determination of baseline sensitivity and assessment of the resistance risk. Pflanzenschutz-Nachr Bayer 52(2):149–161

Takagaki M, Kataoka S, Kida K et al (2011) A method for monitoring the sensitivity of *Botrytis cinerea* to pyribencarb. J Pestic Sci 36(2):255–259

Tamura O (2000) Resistance development of grey mould on beans towards fluazinam and relevant countermeasures. Paper presented at the 10th symposium of research committee of fungicide resistance, Okayama, Japan, 5 Apr 2000

Tanaka C, Izumitsu K (2010) Two-component signaling system in filamentous fungi and the mode of action of dicarboximide and phenylpyrrol fungicides. In: Carisse O (ed) Fungicides, vol I. InTech, pp 523–538

Tellier F, Fritz R, Kerhoas L et al (2008) Characterization of metabolites of fungicidal cymoxanil in a sensitive strain of *Botrytis cinerea*. J Agric Food Chem 56(17):8050–8057

Tellier F, Fritz R, Kerhoas L et al (2009) Metabolism of fungicidal cyanooximes, cymoxanil and analogues in various strains of *Botrytis cinerea*. Pest Manag Sci 65(2):129–136

Temperli E, Roos UP, Hohl HR (1991) Germ tube growth and the microtubule cytoskeleton in *Phytophthora infestans* – effects of antagonists of hyphal growth, microtubule inhibitors, and ionophores. Mycol Res 95:611–617

Terada H (1981) The interaction of highly-active uncouplers with mitochondria. BBA 639(3–4):225–242

Tremblay DM, Talbot BG, Carisse O (2003) Sensitivity of *Botrytis squamosa* to different classes of fungicides. Plant Dis 87(5):573–578

Vallieres C, Trouillard M, Dujardin G et al (2011) Deleterious effect of the Qo inhibitor compound resistance-conferring mutation G143A in the intron-containing cytochrome b gene and mechanisms for bypassing it. Appl Environ Microbiol 77(6):2088–2093

Veloukas T, Karaoglanidis GS (2012) Biological activity of the succinate dehydrogenase inhibitor fluopyram against *Botrytis cinerea* and fungal baseline sensitivity. Pest Manag Sci 68(6):858–864

Veloukas T, Leroch M, Hahn M et al (2011) Detection and molecular characterization of boscalid-resistant *Botrytis cinerea* isolates from strawberry. Plant Dis 95(10):1302–1307

Veloukas T, Markoglou AN, Karaoglanidis GS (2013) Differential effect of *sdhB* gene mutations on the sensitivity to sdhi fungicides in *Botrytis cinerea*. Plant Dis 97(1):118–122

Veloukas T, Kalogeropoulou P, Markoglou AN et al (2014) Fitness and competitive ability of *Botrytis cinerea* field isolates with dual resistance to SDHI and QoI fungicides, associated with Several *sdhB* and the *cytb* G143A mutations. Phytopathology 104(4):347–356

Verger PJ, Boobis AR (2013) Global food supply. Reevaluate pesticides for food security and safety. Science 341(6147):717–718

Viaud M, Fillinger S, Liu W et al (2006) A class III histidine kinase acts as a novel virulence factor in *Botrytis cinerea*. Mol Plant Microbe Interact 19:1042

Vignutelli A, Hilber-Bodmer M, Hilber UW (2002) Genetic analysis of resistance to the phenylpyrrole fludioxonil and the dicarboximide vonclozolin in *Botryotinia fuckeliana* (*Botrytis cinerea*). Mycol Res 106(3):329–335

Villani SM, Cox KD (2014) Heteroplasmy of the cytochrome b gene in *Venturia inaequalis* and its involvement in quantitative and practical resistance to trifloxystrobin. Phytopathology 104(9):945–953

Walker A-S, Gautier A, Confais J et al (2011) *Botrytis pseudocinerea*, a new cryptic species causing gray mold in French vineyards in sympatry with *Botrytis cinerea*. Phytopathology 101(12):1433–1445

Walker A-S, Fournier E (2014) Habitat- and host-specific differentiation in the multihost pathogen *Botrytis cinerea* and evidence for fungicide selection in populations. In: Dehne DW DH, Fraaije B, Gisi U, Hermann D, Mehl A, Oerke EC, Russell PE, Stammler G, Kuck KH, Lyr H (eds) 17th international Reinhardsbrunn conference, Friedrichroda, 21–25 Apr 2013. DPG Spectrum Phytomedizin

Walker A-S, Micoud A, Rémuson F et al (2013) French vineyards provide information that opens ways for effective resistance management of Botrytis cinerea (grey mould). Pest Manag Sci 69(6):667–678

Yarden O, Katan T (1993) Mutations leading to substitutions at amino acids 198 and 200 of beta-tubulin that correlate with benomyl-resistance phenotypes of field strains of *Botrytis cinerea*. Phytopathology 83(12):1478–1483

Yin Y, Kim Y, Xiao C (2010) Characterization of pyraclostrobin resistance and detection of the Bcbi-143/144 intron in the cytochrome b gene in *Botrytis cinerea* isolates from apple. Phytopathology 100(6):S143–S143

Yin YN, Kim YK, Xiao CL (2011) Molecular characterization of boscalid resistance in field isolates of *Botrytis cinerea* from apple. Phytopathology 101(8):986–995

Yin YN, Kim YK, Xiao CL (2012) Molecular characterization of pyraclostrobin resistance and structural diversity of the cytochrome b gene in *Botrytis cinerea* from apple. Phytopathology 102(3):315–322

Zhang CQ, Yuan SK, Sun HY et al (2007) Sensitivity of *Botrytis cinerea* from vegetable greenhouses to boscalid. Plant Pathol 56(4):646–653

Zhang CQ, Hu JL, Wei FL et al (2009) Evolution of resistance to different classes of fungicides in *Botrytis cinerea* from greenhouse vegetables in eastern China. Phytoparasitica 37(4):351–359

Zhang CQ, Liu YH, Zhu GN (2010) Detection and characterization of benzimidazole resistance of *Botrytis cinerea* in greenhouse vegetables. Eur J Plant Pathol 126(4):509–515

Zhao H, Kim YK, Huang L et al (2010) Resistance to thiabendazole and baseline sensitivity to fludioxonil and pyrimethanil in *Botrytis cinerea* populations from apple and pear in Washington State. Postharvest Biol Technol 56(1):12–18

Ziogas BN, Nikou D, Markoglou AN et al (2009) Identification of a novel point mutation in the beta-tubulin gene of *Botrytis cinerea* and detection of benzimidazole resistance by a diagnostic PCR-RFLP assay. Eur J Plant Pathol 125(1):97–107

Chapter 11
Control Strategies for Postharvest Grey Mould on Fruit Crops

Gianfranco Romanazzi and Samir Droby

Abstract Grey mould, incited by *Botrytis cinerea* is considered as a major cause of postharvest losses in a wide range of crops including fruits, vegetables, cut flowers and flower bulbs. The infection can occur in the field, remain quiescent during the season, and develop after harvest. The pathogen is capable of infecting plant tissues through surface injuries inflicted during harvesting and subsequent handling and develop to mold during storage at 0 °C and above. The extent of postharvest decay is affected by preharvest disease management and by postharvest practices. To minimize postharvest grey mould, control programs rely mainly on applications of fungicides in the field. However, mounting concerns of consumers and health authorities about risks associated with chemical residues in food have led to imposing strict regulations and even banning the use of certain chemical groups. These developments have been strongly driving the search for alternative management strategies that are safe and effective. In this chapter, different control strategies are discussed while presenting their advantages and disadvantages. These strategies include the use of chemical fungicides, biocontrol agents, physical means, natural antimicrobials, and decontaminating agents. Based on examples to control grey mould on specific crops, it is concluded that an integrated management program in which adopting a holistic approach is the key for meeting the challenge of minimizing postharvest losses caused by *B. cinerea*. To optimize the efficacy of treatments, it is essential to understand their mechanism of action. Information about direct and indirect effects of each approach on the pathogen will be presented.

Keywords Cold storage • Decontaminating agents • Gray mold • Natural antimicrobials • Postharvest decay

G. Romanazzi (✉)
Department of Agricultural, Food and Environmental Sciences,
Marche Polytechnic University, Via Brecce Bianche, 60131 Ancona, Italy
e-mail: g.romanazzi@univpm.it

S. Droby (✉)
Department of Postharvest Science ARO, The Volcani Center, Bet Dagan, Israel
e-mail: samird@volcani.agri.gov.il

S. Fillinger, Y. Elad (eds.), *Botrytis – the Fungus, the Pathogen and its Management in Agricultural Systems*, DOI 10.1007/978-3-319-23371-0_11

11.1 Introduction

In a report by the United Nations Food and Agricultural Organization (FAO), it was estimated that one-third of the food produced worldwide for human consumption is lost after harvest (Gastavsson et al. 2011). Losses inflicted throughout the supply chain due to pathogen-induced diseases are a major component of food wastage. Pathogen attack may take place during harvesting and subsequent handling, storage, marketing, and post consumer purchase. Among these pathogens, *Botrytis cinerea*, the cause of grey mould, is considered the most important postharvest decay of fresh fruits and vegetables (Droby and Lichter 2004). According to a recent review, *B. cinerea* ranked second into the word Top 10 fungal plant pathogens list based on scientific and economic importance (Dean et al. 2012). *B. cinerea* is an important postharvest pathogen because of the conducive conditions prevailing throughout the postharvest handling chain, including injuries, high humidity, senescing plant tissue and high sugar content. Considerable postharvest losses, due to *B. cinerea* are inflicted in a long list of fresh fruits: blackberry, blueberry, currant, grape, kaki, kiwi, pomegranate, quince, raspberries, strawberry, grapes and many others (Droby and Lichter 2004; Romanazzi and Feliziani 2014). In other fruits (*e.g.*, apple, apricot, lemon, orange, peach, pear, plum, sweet cherry), although it is not the main pathogen, it is still capable of inducing considerable postharvest damage.

Harvested agricultural commodities are highly vulnerable to pathogen attack since they undergo accelerated senescence processes in which ethylene is playing a major role in enhancing susceptibility to grey mould as well as to other postharvest diseases (Lougheed et al. 1978). Manipulation of fruit ripening processes using various postharvest technologies (*e.g.*, inhibition of ethylene production or action, modified atmosphere (MA) and controlled atmosphere (CA), plant hormones) can greatly affect infection and development of postharvest grey mould (Crisosto et al. 2002).

B. cinerea can survive in the field under a wide range of conditions as a saprophyte, where it colonizes flower residues, fruit juice drops, dead leaves, or other non-living plant portions. This type of survival is well known in strawberry where the pathogen overwinters on dead leaves and starts its pathogenic activity at flowering, and then it can remain latent on stamens below sepals, and infect the fruit close to or soon after harvest (Powelson 1960). For this reason, the origin of most infections in strawberry fruit is located close to the sepals (Fig. 11.1a), which often covers under flower residues. In many cases it is possible to find grey mould developing on packed produce in the market, with the pathogen that started infection from infected petals. In grapes, colonization of flower residues by *B. cinerea* is considered to be an important mode of infection, that can remain into the cluster and starts additional infections of the berries when environmental conditions are favorable to the development of the disease (Pearson and Goheen 1988). In this case, treatment at pre-bunch closure is recommended in table grapes to avoid infections soon before and after harvest. This is due to the current lack of systemic active ingredients that

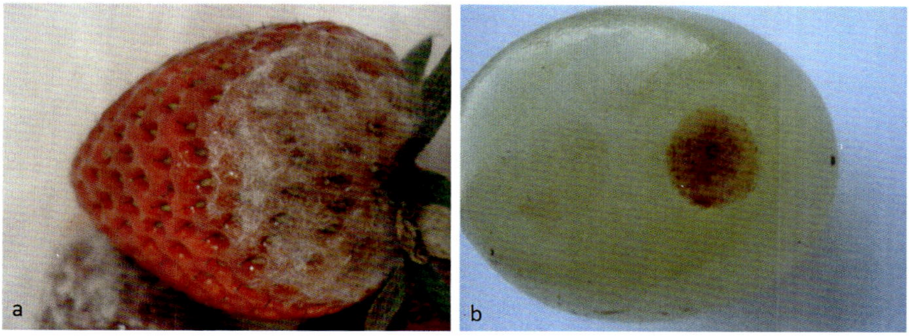

Fig. 11.1 (**a**) Infection of grey mould on strawberry fruit, starting close to the pedicel; (**b**) Initial infection of postharvest grey mould on table grape berry

target *B. cinerea* after bunch closure. These infections occur because the inoculum of *B. cinerea* surviving on flower residues is capable of initiating infections on tissue lesions due to biotic (grape moth, powdery mildew infections, fruit fly) or abiotic damage (striking among berries, hail, wind) (Fig. 11.1b).

After harvest, *B. cinerea* is capable of infecting fruits and vegetables through the damaged tissue of cut surface in the stem end, which is rich in nutrient exudates. Stem end infections can develop and spray to the entire fruit. This mode of infection is mostly known in Kiwifruit as the majority of fruits are affected through picking wounds (Michailides and Elmer 2000). In addition, there is substantial evidence indicating an important role of insects in mediating contamination of harvested agricultural commodities with *B. cinerea* inoculum. In this relation, *Thrips obscuratus* and honeybees were shown to facilitate deposition of conidia into fruit injuries and surface cracks (Michailides and Elmer 2000).

Efforts have been made to minimize grey mould infections and subsequent development of decay through better understanding of the biology and etiology of grey mould on harvested commodities as well as developing adequate pre- and postharvest handling technologies and control strategies for the pathogen. Among these approaches, the use of biocontrol agents (BCA) or natural compounds, when applied shortly before or soon after harvest, was found relatively successful (Calvo-Garrido et al. 2014). Overall, to reduce the infections of the fruit during storage is considered easier as compared to those inflicted in the field, and several appropriate disease management strategies have been suggested in this regard (Ippolito and Nigro 2000; Feliziani and Romanazzi 2013; Teles et al. 2014).

This chapter will provide a general overview of strategies and approached for management of postharvest rots caused by *B. cinerea*. It will touch on research done on this subject and provide details about current control strategies and alternative approaches to grey mould management.

11.2 Postharvest Control of Grey Mould in Conventional and Organic Agriculture

In conventional agriculture, we cannot avoid the use of chemical fungicides, and there is a long list of registered active ingredient on different crops for grey mould control pre- and post-harvest (Romanazzi and Feliziani 2014). However, growers are currently stimulated to adopt alternative approaches as stand-alone treatments or in conjunction with chemical fungicide. This development is taking place due to several reasons, including requirements from supermarket chains for commodities with low number of residual pesticides (e.g. max 4–5 residues) used during production and subsequent postharvest handling. In addition to the limited number of active ingredients on the fruit, the overall level of residues should not exceed 80 % of the total allowed maximum residue limits (MRLs). For example, if we have four residual active ingredients, each should be in an average at the level of 20 % of the allowed MRL. Unfortunately, these commercial policies do not take into consideration that presence of fungicide residues in the fruit below certain threshold will allow the pathogen to develop after harvest, resulting in significant losses along the entire postharvest handling chain. Furthermore, the presence of sub-lethal concentrations of fungicides in the fruit will increase the chances for development of resistant biotypes of the pathogen (Chap. 10).

In recent years, there have been registrations of several low risk active compounds used for the control of grey mould allowing reduction of preharvest application intervals (e.g. fenhexamid) to few days prior to harvest (e.g. strawberry). At the same time, more toxic older active ingredients, such as most benzymidazoles, are not available anymore in the market in the EU or are close to be banned (mostly dicarboximides) in other countries because of high frequency of resistant strains and lack of interest of companies to continue their marketing due to loss of profitability. In addition to chemicals used in conventional agriculture, there is increasing interest in using alternatives to chemical fungicides for the control of postharvest decay, based on the use of registered biocontrol agents in conjunction with chemicals for the purpose of managing resistance problems and fungicide residues in the fruit.

In recent years, we have been witnessing an increase in the number of products in the market place registered as adjuvants that are able to promote plant defense, but containing living organisms (biocontrol agents) or chemical plant stimulators such as glutathione, oligosaccharides, laminarin, chitosan that already proved effective in the control of postharvest decay. Those compounds usually have dual inhibitory effects on the disease due to reducing growth of the pathogens and induction of defense mechanisms in the host tissues. As an example by either fruit tissue and the antagonist through direct effect and induction of resistance mechanisms *Metschnikowia pulcherrima* induced iron depletion in apple wounds resulting in decreased infection by *B. cinerea* (Saravanakumar et al. 2008). Treatment with chitosan, benzothiadiazole and a mixture of calcium and organic acids reduced pathogen growth and increased the expression of enzymes linked to defense mechanisms in strawberry tissues (Landi et al. 2014). Reg. UE 2014/563 included chitosan chloride as first member of basic substance list of plant protection products (as planned with Reg. 2009/1107), so it can be used in plant disease management since July 1, 2014.

In organic agriculture, the use of synthetic chemical active ingredients is not allowed for controlling grey mould infections. This resides in the philosophy of growing organic crops, where negative effects can be prevented through a list of cultural and agronomic approaches (Elad and Shtienberg 1995) that include the reduction of fertilizers, the use of right cultivars, and where it is possible, the cultivation in areas not conducive to diseases development. However, the application of copper is allowed in organic agriculture, since it has broad spectrum of activity and has also a slight phytotoxicity effect that strengthens plant cuticle and make it more resistant to the infection by *B. cinerea* propagules. However, copper accumulates in the soil, and it is toxic to the resident microflora.

In a study aimed to characterize the effect of cropping system on epiphytic microbial community on grapes, Schmid et al. (2011) showed that in organically grown grapevine plants, the number of antagonists, such as *Aureobasidium pullulans* was enhanced. In this relation, *A. pullulans* was reported as the base of different biocontrol products against *Botrytis* (Botector; bio-ferm, Tulln, Austria). In recent years, major companies involved in crop protection (including Syngenta, Bayer, and BASF) have been investing in the field of biocontrol, natural compounds and resistance inducers, because of consumer demand for fruits free of pesticide residues along with increased restrictions imposed by legislation. Moreover, the market of organic agriculture is growing and these companies are targeting this niche.

The use of organic and inorganic salts is becoming increasingly popular in several organic crops (Nigro et al. 2006; Feliziani et al. 2013; Khamis and Sergio 2014). The application of calcium chloride is widely used in Southern Italy (Nigro et al. 2006) and it can be considered as one of the few examples of success of pre-harvest treatments alternative to synthetic fungicides to control postharvest decay on table grapes (Romanazzi et al. 2012). However, it still needs to take into consideration that those salts can leave a visible residue on the berry, and can delay the ripening process as demonstrated for calcium chloride (Nigro et al. 2006) or as the case with potassium sorbate, giving an earlier harvest (Feliziani et al. 2013).

11.3 Management Grey Mould on Stored Products

Once harvested, most fruits need to be cooled as quickly as possible to lower field temperature and to keep the produce as long as possible in its harvest quality. This practice is particularly important when air temperature at harvest is relatively high, and can lead to enhanced loss of water resulting in drying that starts from stems or pedicels. Loss of water from the commodity has great negative impact on overall quality of the final product (Kader and Rolle 2004).

Usually fresh fruit is cold stored at a temperature ranging between 0 and 10 °C, depending on the commodity, for a period from few days (in small berries) up to 2 months (for some table grape cultivars as Crimson seedless) or several months (for kiwifruit, apples or pears). Reduction of the temperature in a period as short as possible is indispensable for perishable fruits and vegetables. For example, wild strawberry (*Fragaria vesca*) fruits, that are highly perishable, are harvested directly into containers placed in a cold proof box with an ice pad on the bottom. In these conditions, the fruits can have a 3–4 days shelf life. For strawberry, packinghouses give a higher price to growers when fruits are harvested in early morning, and it is considered that harvesting fruits at each hour after 10 am results in 1 day less shelf life (G. Savini, personal communication).

Table grapes are usually packed directly in the field to avoid bruising and wounding, and are immediately pre-cooled using forced air ventilated rooms to reduce the temperature to about 0–1 °C. Rapid reduction of the temperature of the packed fruit usually leads to condensation of water on fruit surface, conditions that facilitate conidia germination and penetration through cracks or microlesions that can occur during harvest and subsequent handling. These conditions are ideal for infection because fruit tissues are less reactive due to weakening of defense mechanisms. Once decay is developed, it can progress rapidly by contact to nearby healthy fruits. This type of infection is known as nesting, because of clustering of infected fruit close to a source of inoculum. Low temperatures during storage slow down but do not stop the growth of *B. cinerea* since it is able to grow at temperatures, ranging from −0.5 to 32 °C (Coertze and Holz 1999).

Use of synthetic fungicides for controlling pathogens on most commodities is not allowed after harvest in most EU countries. In grapes and some other fruits, however, the use of sulfur dioxide during storage is permitted since it is considered as processing aid and not as a fungicide. In most cases, sulfur dioxide has negative effects on humans allergic to sulfur. For that reason, a tolerance limit of 10 $\mu L/L^{-1}$ for sulfite residues in table grapes was established by the U.S. Environmental Protection Agency (Anonymous 1989). In California, most of organic growers use ozone fumigation of grapes after harvest (Romanazzi et al. 2012), and this technology is becoming also popular among packinghouses working with conventionally grown grapes. An interesting side of ozone treatment resides in its oxidant activity that can reduce fungicide residues on the berries (Mlikota Gabler et al. 2010a, b).

While the use of sulfur dioxide is unappealing to the consumers, at the same time it can damage the fruit by causing surface cracks and bleaching. In addition, the treatment is non-selective in eliminating the vast majority of epiphytic microflora left on the fruit without natural protection allowing grey mould to develop more readily compared to non-fumigated fruit. To achieve good levels of control, usually SO_2 applied in storage room of grapes weekly, following a first treatment prior to cold storage (Luvisi et al. 1992), and/or by packing the grapes with a pad or generator containing sodium metabisulfite, which releases SO_2 when hydrated by water vapor inside the boxes (Droby and Lichter 2004). Due to the problematic use of SO_2, there are several reports about alternative methods, including application of ethanol after harvest (Karabulut et al. 2003), ethanol in conjunction with chitosan or calcium chloride (Romanazzi et al. 2007; Chervin et al. 2009), organic salts (Nigro et al. 2006), controlled atmosphere (Crisosto et al. 2002), or ozone (Palou et al. 2002; Feliziani et al. 2014). However, few of these methods are used at a commercial scale (Romanazzi et al. 2012). Recently, Teles et al. (2014) reported that a 40 % CO_2

for 48 h pre-storage treatment followed by controlled atmosphere markedly reduced grey mould incidence. High CO_2 pre-storage alone limited disease incidence in both naturally and artificially infected grapes, but was more effective when combined with CA storage.

11.4 Potential of Alternative Strategies for Controlling Postharvest Grey Mould

Synthetic chemical fungicide treatment has been the primary strategy for managing postharvest pathogens. However, there are many risks associated with these chemicals, including the development of fungicidal resistance (Chap. 10), mounting health concerns of consumers and health authorities leading to the demand to reduce human and environmental exposure to chemicals, increased restrictions imposed by regulatory agencies on specific agro-chemicals and/or their allowable residues, especially after harvest. These driving forces have been behind a significant research effort during the past 25 years to develop effective and useful alternative technologies to the synthetic fungicides to preserve quality and prolong the storage and shelf life of the produce. Innovations in this area can be grouped in four categories of treatments: (i) microbial biocontrol agents (BCAs); (ii) natural antimicrobials; (iii) disinfecting agents; and (iv) physical means. Among these, considerable work focused on the use of various microbial antagonists (yeasts and bacteria) that occur naturally on fruit surfaces and disrupt the ability of postharvest pathogens to establish infections in wounded fruits. Grey mould is one of the main targets of these antagonists.

Biocontrol at pre-harvest stage is described in Chap. 9. The research on BCAs for postharvest resulted in several commercial products able to control *B. cinerea* (Droby et al. 2009; Nunes 2012; Feliziani and Romanazzi 2013; Liu et al. 2013; Mari et al. 2014). These products (e.g. Shemer, Bayer CropScience, Germany) Candifruit and Pantovital (Sipcam-Inagra, Spain), Boniprotect (Bioferm, Germany), Yield Plus (Anchor Yeast, South Africa), Nexy (BioNext, Belgium), Biosave (JetHarvest, USA) have reached the market and their use has been promising (Feliziani and Romanazzi 2013; Mari et al. 2014). However, considering the costs for registration as plant protection products, the number of registered BCAs is very low as compared to the huge mass of work that has been conducted in this field because it is often particularly difficult to move from the discovery phase of an effective antagonist to the development until its commercial stage. Despite these possible limitations, the main companies working with synthetic fungicides are getting interest to show in their portfolio also a list of BCAs, then have acquired specialized companies in the field. Nowadays the research on discovery and characterization of old and new BCAs is very active. Studies on mechanisms of action of *M. pulcherrima* showed an increase of chitinase production in a strain that has a higher biocontrol activity on *B. cinerea* (Saravanakumar et al. 2009). *Pichia angusta* was effective in controlling *B. cinerea* on apples (Fiori et al. 2008), whereas the activity of *Wickerhamomyces anomalus* in the control of postharvest decay of sweet cherries was reported (Oro et al. 2014).

A large variety of volatile compounds along with other plant extracts and animal-derived materials with antifungal activity have been reported. Plant volatiles such as acetaldehyde, benzaldehyde, benzyl alcohol, ethanol, methyl salicylate, ethyl benzoate, ethyl formate, hexanal, (E)-2-hexenal, lipoxygenases, jasmonates, allicin, glucosinolates and isothiocyanates have been shown to have strong effect on *Botrytis* infection on various commodities when tested under laboratory and small scale conditions (Tripathi and Dubey 2004). Although proven effective under laboratory and small scale experimental levels, their efficacy still need to be confirmed under large scale and commercial conditions, and safety issues need to be addressed as most of these compounds when applied at concentrations effective in inhibiting the pathogens are harmful to the environment and human health. Treatment with chitosan was effective in the control of preharvest grey mould in wine grapes (Elmer and Reglinski 2006), and in the management of postharvest grey mould on different fruits (Romanazzi et al. 2015). The use of essential oils is becoming popular for the control of postharvest decay. These materials were reported to control grey mould of table grapes (Abdollahi et al. 2010, 2012), and were applied alone or together with other treatments (Sivakumar and Bautista-Baños 2014).

Disinfecting agents (ethanol, acetic acid, electrolyzed oxidizing water) have been used for fruit surface sterilization, mainly when the process of washing is included in postharvest fruit packing. Acetic acid was successfully used as fumigant to control postharvest decay of table grapes (Sholberg et al. 1996), as well as ethanol (Mlikota Gabler et al. 2005). The application of electrolyzed oxidizing water is effective in disinfection of water used in packinghouses operations and shown to decrease conidia contamination of different pathogens, including *B. cinerea* (Guentzel et al. 2010).

The use of physical means (UV-C irradiation, ozone, CA/MA, hypobaric or hyperbaric treatments) has been demonstrated to be effective in controlling grey mould on table grapes (Romanazzi et al. 2012). These control means have the advantage in that they avoid direct contact with the fruit (Sanzani et al. 2009), although often their effects last only as long as they are applied.

A strategy to further improve the effectiveness of alternatives control methods is the integration of different approaches. However, once a treatment is considered effective, it is necessary to carefully verify its potential introduction at a commercial scale in the packinghouse, transport and market chain (Romanazzi et al. 2012). To have effectiveness comparable to the synthetic fungicides, two or more alternative approaches need to be combined for the control of postharvest decay. Several combinations were applied in the case of grey mould. For example, application of hydroxypropyl methylcellulose and beeswax edible coatings reduced grey mould of stored tomatoes (Fagundes et al. 2014), the BCA *Muscodor albus* performed better in the control of grey mould of grapes when combined with ozone (Mlikota Gabler et al. 2010a) and the application of garlic extract and clove oil decreased infections of *B. cinerea* on apples (Daniel et al. 2015). However, an effectiveness in the lab need to be confirmed in large scale tests and possible side effects, such as organoleptic and quality changes, need to be addressed.

11.5 Concluding Remarks and Future Challenges

Postharvest decay caused by grey mould has great economic importance and in some cases can lead to complete loss of the product. Reducing these losses to a level that is acceptable still pose great challenge for producers, packers and marketing at the wholesale and retail levels. In this regard, grey mould is still a challenge to control in certain highly perishable crops such as small berries.

Extensive research has been done and will continue in the future to find effective management technologies and innovative approaches for the control of *Botrytis* decay on fresh fruit and vegetables after harvest. Most of the efforts, however, have been devoted to develop management programs at the pre-harvest level. These programs have been developed and applied specifically for each crop and largely dependent on epidemiological and etiological information. While lacking such information to manage development of postharvest grey mould, the availability and utilization of simple cultural methods previously discussed could greatly reduce postharvest losses due to grey mould on harvested commodities. Although chemical fungicides constitute still the common practice for controlling development of *Botrytis* in the field/orchard or in the packinghouse, their use after harvest on fruits is not allowed in many countries. Their continued use as preharvest treatment has come under scrutiny and their use as control strategy in long term is somewhat questionable. This is because of problems associated with (1) failure to effectively control pre and postharvest grey mould due to development of resistance; (2) consumers desire to reduce human and environmental exposure to chemicals and (3) increased restrictions imposed by marketing chains and governmental regulatory agencies on the use of agro-chemicals and/or allowable residues in fresh agricultural commodities. These have been the driving forces for the development of postharvest diseases measures that do not rely on chemical fungicides. Currently, the use of alternative methods as standalone treatments for the control of postharvest grey mould, however, does not provide the efficacy and consistency required for commercial situations.

Over the past 20 years, biological control of postharvest diseases has grown into a mature field of science and in a relatively short period of time seen successes in commercial application. Commercial development of postharvest biocontrol products has been rather limited and some products have had very short life spans because of their inability to capture a market large enough, or because of being developed and sold by small companies without a large market presence. However, the largest obstacle to their widespread use is providing a product that performs effectively and reliably under a wide array of conditions, and also adapts easily to a range of commercial processing systems. The reasons for the variability in performance are not clearly understood and may be due to the presence of pre-established infections, high levels of inoculum, poor storage of the biocontrol product prior to application, or improper application. Considerable effort has been made to integrate the use of postharvest biocontrol products into a production systems approach. The use of preharvest applications as well as postharvest applications, and the incorporation of various additives are two methods that have been used to increase the applicability, effectiveness, and reliability of postharvest BCAs.

B. cinerea attack fruits and vegetables before and after harvest using different modes of infection, depending on the type of the crop. To increase the chances to reduce the rate of such infections, it is important to influence the process of infection at different levels: the pathogen, the micro-environment, and the host. For example, application of a BCA or any other alternative method at a time that prevents establishment of the pathogen in the host tissue, given that the attachment of pathogen propagules to the host surfaces and the early stages of germination are critical to successful infection. The micro-environment (e.g. surface wounds) can also be altered to directly or indirectly affect the pathogen. The pH and nutritional composition of the infection site can be manipulated by the addition of salts, organic acids, or surfactants/adjuvants. Susceptibility of the commodity (host) may also be reduced by changing its physiology using various treatments to either slow down senescence or induce natural resistance.

It is anticipated that the continuing withdrawal of key postharvest fungicides from the market, due to exclusion by regulatory agencies or the high-cost of re-registration, will lead to an absence of effective tools for reducing postharvest losses due to grey mould. Hence, the use of alternative control methods is expected to gain momentum in the coming years and become more widely accepted as a component of an integrated strategy to managing postharvest diseases. Along with this approach, an effective alternative control strategy would rely on: (i) the classical (microbial antagonists), (ii) natural plant resistance, and (iii) natural antimicrobials which are the product of a biological process, (iv) combination between the above cited methods with physical means such as plant growth regulators, ethylene inhibitors, MA, CA, and heat treatments. Also, it is very important to reduce the inoculum load and conditions conducive to establishment of infections through well-established cultural and management practices.

References

Abdollahi A, Hassani A, Ghosta Y et al (2010) Study on the potential use of essential oils for decay control and quality preservation of Tabarzeh table grape. J Plant Prot Res 50:45–52

Abdollahi A, Hassani A, Ghosta Y et al (2012) Evaluation of essential oils for maintaining post-harvest quality of Thompson seedless table grape. Nat Prot Res 26:77–83

Anonymous (1989) Pesticide tolerance for sulfur dioxide. Fed Regist 40(20):125–126

Calvo-Garrido C, Vinas I, Elmer PAG et al (2014) Suppression of *Botrytis cinerea* on nectrotic grapevine tissues by early season applications of natural products and biocontrol agents. Pest Manag Sci 70:595–602

Chervin C, Lavigne D, Westercamp P (2009) Reduction of gray mold development in table grapes by preharvest sprays with ethanol and calcium chloride. Postharvest Biol Technol 54:115–117

Coertze S, Holz G (1999) Surface colonization, penetration, and lesion formation on grapes inoculated fresh or after cold storage with single airborne conidia of *Botrytis cinerea*. Plant Dis 83:917–924

Crisosto C, Garner D, Crisosto G (2002) Carbon dioxide-enriched atmospheres during cold storage limit losses from *Botrytis* but accelerate rachis browning of 'Redglobe' table grapes. Postharvest Biol Technol 26:181–189

Daniel CK, Lennox CL, Vries FA (2015) *In vivo* application of garlic extracts in combination with clove oil to prevent postharvest decay caused by *Botrytis cinerea*, *Penicillium expansum* and *Neofabraea alba* on apples. Postharvest Biol Technol 99:88–92

Dean R, Van Kan JAL, Pretorius ZA et al (2012) The Top 10 fungal pathogens in molecular plant pathology. Mol Plant Pathol 13:414–430

Droby S, Lichter A (2004) Post-harvest *Botrytis* infection: etiology, development and management. In: Elad Y, Williamson B, Tudzynski P, Delen N (eds) *Botrytis*: biology pathology and control. Kluwer Academic Publishers, Dordrecht, pp 349–367

Droby S, Wisniewski M, Macarisin D, Wilson C (2009) Twenty years of postharvest biocontrol research: is it time for a new paradigm? Postharvest Biol Technol 52:137–145

Elad Y, Shtienberg D (1995) *Botrytis cinerea* in greenhouse vegetables: chemical, cultural, physiological and biological controls and their integration. Int Pest Manag Rev 1:15–29

Elmer PAG, Reglinski T (2006) Biosuppression of *Botrytis cinerea* in grapes. Plant Pathol 55:155–177

Fagundes C, Palou L, Monteiro AR, Pérez-Gago MB (2014) Effect of antifungal hydroxypropyl methylcellulose-beeswax edible coatings on gray mold development and quality attributes of cold-stored cherry tomato fruit. Postharvest Biol Technol 92:1–8

Feliziani E, Romanazzi G (2013) Preharvest application of synthetic fungicides and alternative treatments to control postharvest decay of fruit. Stewart Postharvest Rev 3(4):1–6

Feliziani E, Smilanick JL, Margosan DA et al (2013) Preharvest fungicide, potassium sorbate, or chitosan use on quality and storage decay of table grapes. Plant Dis 97:307–314

Feliziani E, Romanazzi G, Smilanick J (2014) Application of low concentration of ozone during cold storage of table grapes. Postharvest Biol Technol 93:38–48

Fiori S, Fadda A, Giobbe S et al (2008) *Pichia angusta* is an effective biocontrol yeast against postharvest decay of apple fruit caused by *Botrytis cinerea* and *Monilia fructicola*. FEMS Yeast Res 8:961–963

Gastavsson J, Cederberg C, Sonesson U (2011) Global food losses and food waste. Food and Agriculture Organization (FAO) of the United Nations, Rome

Guentzel JL, Lam KL, Callan MA et al (2010) Postharvest management of gray mold and brown rot on surfaces of peaches and grapes using electrolyzed oxidizing water. Int J Food Microbiol 143:54–60

Ippolito A, Nigro F (2000) Impact of preharvest application of biological control agents on postharvest diseases of fresh fruits and vegetables. Crop Prot 19:715–723

Kader AA, Rolle RS (2004) The role of postharvest management in assuring the quality and safety of horticultural produce. Food and Agriculture Organization (FAO) of the United Nation. Bulletin No. 152, 125 pp

Karabulut OA, Smilanick JL, Mlikota Gabler F et al (2003) Near-harvest applications of *Metschnikowia fructicola*, ethanol, and sodium bicarbonate to control postharvest diseases of grape in central California. Plant Dis 87:1384–1389

Khamis Y, Sergio RR (2014) Applications of salt solutions before and after harvest affect the quality and incidence of postharvest gray mold of 'Italia' table grapes. Postharvest Biol Technol 87:95–102

Landi L, Feliziani E, Romanazzi G (2014) Expression of defense genes in strawberry fruit treated with different resistance inducers. J Agric Food Chem 62:3047–3056

Liu J, Sui Y, Wisniewski M, Droby S, Liu Y (2013) Review: utilization of antagonistic yeasts to manage postharvest fungal diseases of fruit. Int J Food Microbiol 167:153–160

Lougheed EC, Murr DP, Berard L (1978) Low pressure storage for horticultural crops. HortScience 13:21–27

Luvisi D, Shorey H, Smilanick JL et al (1992) Sulfur dioxide fumigation of table grapes. Bulletin 1932. University of California, Division of Agriculture and Natural Resources, Oakland, 21pp

Mari M, Di Francesco A, Bertolini P (2014) Control of fruit postharvest diseases: old issues and innovative approaches. Stewart Postharvest Rev 1(1):1–4

Michailides TJ, Elmer PAG (2000) Botrytis gray mold of kiwifruit caused by *Botrytis cinerea* in the United States and New Zealand. Plant Dis 84:208–223

Mlikota Gabler F, Smilanick JL, Ghosoph JM, Margosan DA (2005) Impact of postharvest hot water or ethanol treatment of table grapes on gray mold incidence, quality, and ethanol content. Plant Dis 89:309–316

Mlikota Gabler F, Smilanick JL, Mansour MF, Karaca H (2010a) Influence of fumigation with high concentrations of ozone gas on postharvest gray mold and fungicide residues on table grapes. Postharvest Biol Technol 55:85–90

Mlikota Gabler F, Mercier J, Jiménez JI, Smilanick JL (2010b) Integration of continuous biofumigation with *Muscodor albus* with pre-cooling fumigation with ozone or sulfur dioxide to control postharvest gray mold of table grapes. Postharvest Biol Technol 55:78–84

Nigro F, Schena L, Ligorio A et al (2006) Control of table grape storage rots by pre-harvest applications of salts. Postharvest Biol Technol 42:142–149

Nunes CA (2012) Biological control of postharvest diseases of fruit. Eur J Plant Pathol 133:181–196

Oro L, Feliziani E, Ciani M et al (2014) Biocontrol of postharvest brown rot of sweet cherries and population dynamic of *Saccharomyces cerevisiae* Disva599, *Metschnikowia pulcherrima* Disva267 and *Wickerhamomyces anomalus* Disva2 strains. Postharvest Biol Technol 96:64–68

Palou L, Crisosto CH, Smilanick JL et al (2002) Effects of continuous 0.3 ppm ozone exposure on decay development and physiological responses of peaches and table grapes in cold storage. Postharvest Biol Technol 24:39–48

Pearson RC, Goheen AC (1988) Compendium of grape diseases. APS Press, St. Paul, 96 pp

Powelson RL (1960) Initiation of strawberry fruit rot caused by *Botrytis cinerea*. Phytopathology 50:491–494

Romanazzi G, Feliziani E (2014) *Botrytis cinerea*. In: Bautista-Baños S (ed) Postharvest decay: control strategies. Elsevier, London, pp 131–146. ISBN 9780124115521

Romanazzi G, Karabulut OA, Smilanick JL (2007) Combination of chitosan and ethanol to control gray mold of table grapes. Postharvest Biol Technol 45:134–140

Romanazzi G, Lichter A, Mlikota Gabler F, Smilanick J (2012) Natural and safe alternatives to conventional methods to control postharvest gray mold of table grapes. Postharvest Biol Technol 63:141–147

Romanazzi G, Feliziani E, Bautista-Baños S, Sivakumar D (2015) Shelf life extension of fresh fruit and vegetables by chitosan treatment. Crit Rev Food Sci Nutr. doi:10.1080/10408398.2014.900474, in press

Sanzani SM, Nigro F, Mari M, Ippolito A (2009) Innovations in the control of postharvest diseases of fresh fruits and vegetables. Arab J Plant Prot 27:240–244

Saravanakumar D, Ciarovella A, Spadaro D et al (2008) *Metschnikowia pulcherrima* strain MACH1 outcompetes *Botrytis cinerea*, *Alternaria alternata* and *Penicillium expansum* in apples through iron depletion. Postharvest Biol Technol 49:121–128

Saravanakumar D, Spadaro D, Garibaldi A, Gullino ML (2009) Detection of enzymatic activity and partial sequence of a chitinase gene in *Metschnikowia pulcherrima* strain MACH1 used as post-harvest biocontrol agent. Eur J Plant Pathol 123:183–193

Schmid F, Moser G, Müller H, Berg G (2011) Functional and structural microbial diversity in organic and conventional viticulture: organic farming benefits natural biocontrol agents. Appl Environ Microbiol 77:2188–2191

Sholberg PL, Reynolds AG, Gaunce AP (1996) Fumigation of table grapes with acetic acid to prevent postharvest decay. Plant Dis 80:1425–1428

Sivakumar D, Bautista-Baños S (2014) A review on the use of essential oils for postharvest decay control and maintenance of fruit quality during storage. Crop Prot 64:27–37

Teles CS, Benedetti BC, Gubler WD, Crisosto CH (2014) Prestorage application of high carbon dioxide combined with controlled atmosphere storage as a dual approach to control *Botrytis cinerea* in organic 'Flame Seedless' and 'Crimson Seedless' table grapes. Postharvest Biol Technol 89:32–39

Tripathi P, Dubey NK (2004) Exploitation of natural products as an alternative strategy to control postharvest fungal rotting of fruit and vegetables. Postharvest Biol Technol 32:235–245

Chapter 12
Infection Process and Fungal Virulence Factors

Celedonio González, Nélida Brito, and Amir Sharon

Abstract The knowledge about the molecular mechanisms underlying *Botrytis cinerea* attack on susceptible hosts has expanded greatly in the last years. While many details are still missing, it is possible now to generate a higher resolution image of the molecular patterns that regulate pathogenic development of this broad host range necrotroph. Several decades of molecular research, including the complete genome sequencing and analyses of two *B. cinerea* strains, have culminated into a working model outlining the molecular strategy used by *B. cinerea* in the infection process. It is now apparent that a complex signalling network regulates secretion of a large set of proteins and phytotoxic secondary metabolites, which are necessary for progression of the infection from the early to late stages. Furthermore, manipulation of the plant hyper-sensitive response (HR), a form of programmed cell death (PCD), has been proposed to play a central role in the pathogenic strategy of *B. cinerea*. Although the molecular details of this aspect are largely uncharacterized, it is possible that some of the secreted proteins and metabolites function as effectors that target the PCD machinery. The virulence factors that have been revealed in these various categories and in others are reviewed in this chapter, with special emphasis on secreted proteins and PCD.

Keywords Pathogenicity factors • Secretome • Effectors • RNA interference • Programmed cell death • PCD

C. González • N. Brito (✉)
U.D. Bioquímica y Biología Molecular, Universidad de La Laguna, 38206 La Laguna, Tenerife, Spain
e-mail: cglez@ull.edu.es; nbrito@ull.edu.es

A. Sharon (✉)
Department of Molecular Biology and Ecology of Plants, Tel Aviv University, Tel Aviv 69978, Israel
e-mail: AmirSH@tauex.tau.ac.il

S. Fillinger, Y. Elad (eds.), *Botrytis – the Fungus, the Pathogen and its Management in Agricultural Systems*, DOI 10.1007/978-3-319-23371-0_12

12.1 Introduction

Molecules (proteins, RNAs, low molecular weight compounds) which contribute, in a significant manner, to any part of the infection cycle (adhesion, penetration, expansion of the lesions, dispersion of the disease, modulation of the plant immune system) are considered virulence factors. Traditionally, virulence factors have been postulated according to phenotype of mutant strains showing alteration in pathogenicity. Thus, genes associated with a range of processes, such as signalling, metabolism, catabolism, adherence, and cell cycle and architecture, have been proposed to encode virulence factors (Choquer et al. 2007; Nakajima and Akutsu 2014). Clearly, most of the so-called virulence factors have additional functions, which in many cases are the primary role of the molecule, while the effect on virulence may be a side effect due to e.g., reduced fitness, slow growth, etc. Hence, a pathogenicity phenotype is not sufficient for determination that a certain gene or molecule has a direct effect on virulence. Likewise, deletion of putative virulence factors might not show discernible pathogenicity phenotype, e.g. due to functional redundancy (Cuesta Arenas et al. 2010; Van Kan et al. 1997), or due to limited analytical capacities.

In recent years, the term "effector" was introduced, which refers to those virulence factors that are produced during pathogenic interaction with the purpose of inhibiting the plant immune system thus enhancing pathogen virulence (Bent and Mackey 2007; Thomma et al. 2011), while having limited or unknown side effect on other processes in the pathogen. Numerous effectors have been described in bacteria, Oomycetes (mainly *Phytophthora* sp.) and more lately also in biotrophic fungal pathogens (Ellis et al. 2009; Plett et al. 2011; Stergiopoulos and de Wit 2009). The paradigm of pathogenicity in necrotrophic pathogens, and specifically in *B. cinerea*, had suggested brutal damage of host tissues via massive secretion of toxic substances and hydrolytic enzymes (Williamson et al. 2007). However, recent works show that the interaction of *B. cinerea* with plants is more complicated than previously estimated. Two clear phases are distinguished – an early stage characterized by local necrosis, and a later stage that is characterized by spreading lesions (Van Kan 2006). Accordingly, a current model predicts that *B. cinerea* uses different types of virulence/effector molecules at each stage. For example, secretion of local necrosis-inducing factors is expected at the early stage, while production of spreading cell death-inducing factors are expected at the second phase (Shlezinger et al. 2011). In this chapter, we will describe the recent literature on virulence factors in *B. cinerea*. We will briefly mention the different types of virulence factors, and then focus on the role of secreted proteins, in particular proteins that promote necrosis and other types of host cell death.

12.2 Overview of *B. Cinerea* Virulence Factors

Virulence factors in *B. cinerea* have been found among different types of proteins, metabolites and even small RNAs. One important group is the one comprising genes/proteins related to signalling pathways including heterotrimeric G proteins,

MAP kinases, adenylate cyclases, protein phosphatases, etc. *B. cinerea* signalling modules affecting virulence of the fungus are covered in Chap. 13. Another group of virulence factors are proteins that contribute to the production or catabolism of reactive oxygen species (ROS). *B. cinerea* produces ROS during infection, which add to the oxidative burst generated by the plant. Several *B. cinerea* proteins that contribute to ROS production have been shown to also affect fungal virulence. Chapter 14 describes the mechanisms of ROS generation in *B. cinerea*, how it can cope with them, and the role of ROS in the virulence of the fungus.

Botrytis cinerea also produces several types of small molecules (secondary metabolites), some of which have been shown to contribute to virulence. These include well know molecules such as botrydial, a phytotoxic bicyclic sesquiterpene which contributes to virulence in some *B. cinerea* strains but not in others. Another example is oxalic acid, which *B. cinerea* produces and secretes in large amounts both in planta (Verhoeff et al. 1988) and in axenic culture (Gentile 1954). Oxalic acid can promote virulence in multiple ways: it lowers pH to values more appropriate for *B. cinerea* extracellular hydrolases, facilitates breakdown of the pectin layer in cell walls by sequestration of Ca^{2+} ions, and can promote programmed cell death (PCD) in the host (Kim et al. 2008). The diversity of virulence-related secondary metabolites and their role in the infection process are described in a separate chapter (Chap. 15).

The tetraspanin BcPLS1 has been shown to be required for full virulence in *B. cinerea* (Gourgues et al. 2004; Siegmund et al. 2013). *Bcpls1* mutants were generated in two *B. cinerea* strains, T4 and B05.10, and in both cases the resulting strains were less pathogenic than the corresponding wild-type strains, although to a different degree. Mutants in both strains were able to produce appressoria-like structures, but the mutants were defective in penetration, showing complete lack of penetration in the background of the T4 strain and 50 % reduced penetration in the background of the B05.10 strain. Once inside the plant, both mutants were able to colonize the plant tissue as efficiently as the wild-type. Consistent with a role in appressorium function, *Bcpls1* was selectively expressed at the hyphal tips. The exact role or mechanism of BcPLS1 is unclear, but it might be associated with regulation of ROS levels since the phenotype of the *Bcpls1* mutants closely resemble that of mutants in *Bcnoxb*, a gene coding for a catalytic subunit of NADPH oxidase (Siegmund et al. 2013).

12.3 Virulence Factors Among Secreted Proteins

One of the areas where one would expect to find virulence factors is, of course, among the proteins secreted by *B. cinerea* to the medium. According to the signal peptide prediction system signalP, the genome of *B. cinerea* codes for 1910 proteins predicted to enter the secretion pathway, which constitutes about 12 % of all the proteins coded by the genome (González et al. 2012). Not all these proteins would end in the extracellular medium as soluble proteins, and maybe not all of the putative genes coding for extracellular proteins are expressed in planta. Proteomic studies (reviewed in Chap. 16) have been carried out to identify the proteins secreted by

B. cinerea in different conditions (Espino et al. 2010; Fernández-Acero et al. 2010; Cherrad et al. 2012; Li et al. 2012; Shah et al. 2009a, b), including some supposed to mimick the "*in planta*" environment (Espino et al. 2010). Globally, these studies have identified 279 non-redundant proteins, that is, about 15 % of all proteins with a signal peptide. These proteins were classified in the categories previously reported in one the above papers (Fig. 12.1). Two features of these data are worth mention, in first place the highly redundant nature of the protein set secreted by *B. cinerea*. Even considering only the proteins which have been experimentally detected in the extracellular medium, and not the hypothetical secretome coded by the genome, it is quite common to find proteins represented by more than one isoform. For example, in the study about the early secretome (Espino et al. 2010), more than one isoform was found for cutinase, ß-glucosidase, ß-1,3-glucanosyltransferase, glucoamylase, and much more. Secondly, the number of proteases secreted by *B. cinerea* is surprisingly high compared with other fungi. Thirty-four types of proteases have been experimentally found in the secretome, and some of them are very abundant. The most extreme case is that of the aspartic protease BcAP8, which constituted about one fourth of the protein mass secreted by *B. cinerea* early after germination (Espino et al. 2010). These features imply a difficulty in identifying virulence factors in *B. cinerea* by the generation of mutants affected in a single gene, since often the single protein absent in the mutant is substituted by additional secretome members with overlapping enzymatic activities. Alternative ways to address this problem are the generation of gain-of-function (over-expression)

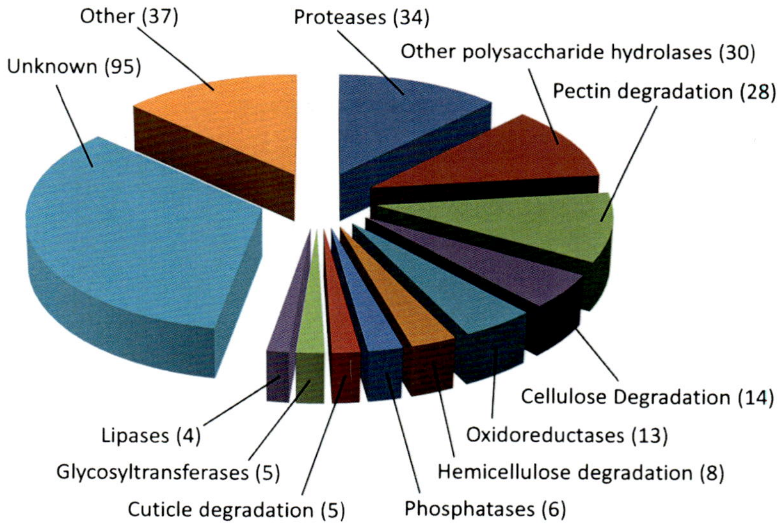

Fig. 12.1 Distribution of proteins experimentally found in the *B. cinerea* secretome. All the proteins which have been reported (Espino et al. 2010; Fernández-Acero et al. 2010; Cherrad et al. 2012; Li et al. 2012; Shah et al. 2009a, b) for the *B. cinerea* secretome (279 proteins in total) were classified according to function. Slice areas correspond to the number of proteins in each category, also displayed in brackets

strains, instead of loss-of-function (deletion), or the simultaneous silencing of all members of given protein families by RNAi technology, as has been carried out with in other fungi (Nguyen et al. 2011).

12.3.1 Secreted Proteases

The fact that the most abundant family among the secretome is that of proteases was unexpected, because proteases were originally not foreseen as candidate virulence factors. However, several reasons may help explaining their abundance. In first place proteases generate amino acids to sustain fungal growth since proteins are certainly an abundant component of cells and amino acids are a rich nutrient as carbon, nitrogen, and sulfur source. Secondly, proteases may contribute to the degradation of plant cell wall. Proteins are important components of plant cell walls (Chap. 18), too often overlooked. Proteomic studies with plant cell walls have revealed hundreds of proteins with diverse functions (Albenne et al. 2013), including structural proteins tightly bound to the cell wall. Degradation of these proteins with proteases may therefore help hyphal growth inside the plant tissue. Lastly, proteases may be secreted by the fungus as a counterattack measure against plant defense proteins. Despite this abundance, no virulence factor has been found yet among secreted proteases. Single and double mutants have been generated for six of the aspartic proteases coded by the *B. cinerea* genome (ten Have et al. 2010), which included two proteins with predicted extracellular localization, BcAp5 and BcAP8. From all the mutants generated, only the one affected in the *Bcap8* gene showed a reduction in the protease activity detected in the extracellular medium, which was around 70 % lower than in the wild-type. No difference in virulence was observed for any of the aspartic protease mutants, as compared to the wild-type. In light of the 34 proteases that have been found experimentally in the extracellular medium, it is not surprising that the absence of one or two of them goes unnoticed.

12.3.2 Plant Cuticle and Cell Wall Degrading Enzymes

Another abundant group of secretory proteins are those involved in the degradation of the plant cuticle, the first defensive barrier of plant cells. Five different cutinases have been found experimentally in the secretome, including one (CutA) which has also been purified form the extracellular fraction (Van der Vlugt-Bergmans et al. 1997). Only the gene for this last enzyme has been knocked-out in *B. cinerea* (Van Kan et al. 1997) resulting in a strain equally infective as the wild-type. Of course, this result does not imply that cutinase activity is dispensable for virulence as the rest of cutinases could substitute the absent one, but intriguingly, there are other reports indicating that *Arabidopsis thaliana* mutant plants with a weaker cuticle, contrary to initial expectations, are not more susceptible to *B. cinerea*, but more resistant (Bessire et al. 2007; Chassot et al. 2007). This is certainly counterintuitive, and is explained by the authors with the hypothesis that a weaker cuticle may allow the plant to better perceive elicitors and mount a faster and more intense defense response.

Lipases, also present in the secretome, could contribute to pathogenesis not only by the generation of nutrients assimilable by the fungal cells, but also by aiding in penetration, as these enzymes could potentially degrade components of the cuticle. Reports of a possible involvement of lipases in penetration came originally from experiments in which the addition of anti-lipase antibodies to *B. cinerea* conidia prior to inoculation caused inability to penetrate the plant cuticle (Commenil et al. 1998). However, the only lipase that has been studied by gene knock-out did not seem to contribute to virulence (Reis et al. 2005). In addition this particular lipase, Lip1, has never been found experimentally in the secretome.

The following defensive barrier that the fungus encounters is the cell wall, whose components have traditionally been divided into cellulose, hemicelluloses and pectin. Fourteen enzymes have been found in the secretome potentially involved in the degradation of cellulose. These enzymes catalyse the hydrolysis of the ß-1,4 glycosidic bonds in the cellulose polymer in several ways: endo-ß-1,4-glucanases cleave internal glycosidic bonds leaving shorter polysaccharide chains, cellobiohydrolases release the disaccharide cellobiose from the non-reducing ends, and exo-ß-1,4-glucanases liberate successive glucose units from the polymer ends. Only one mutant has been generated for one of the *B. cinerea* endo-ß-glucanases genes, *cel5A* (Espino et al. 2005), which was shown by Q-RT-PCR to be expressed (Espino et al. 2005) both in axenic culture and in planta. However, the corresponding protein has never been found experimentally in the secretome and deletion of *cel5A* had no effect on virulence.

Another important component of the cell wall are hemicelluloses, which include several polysaccharides such as xylan, arabinoxylan, and xyloglucan. Only eight enzymes that can be clearly classified as hemicellulases have been found in the secretome, but given the heterogeneous nature of this group of enzymes, it is possible that more hemicelluloses are included with other polysaccharides degrading enzymes. In principle, hemicellulases are not good candidates as virulence factors in *B. cinerea*, or at least not all of them, for instance when this fungus infects soft plant tissues poor in hemicelluloses. Two xylanase genes have been knocked-out in *B. cinerea*. One of them belongs to family 10 of glycosyl hydrolases and was identified in a screening for genes under the control of the *B. cinerea* Gα subunit of the heterotrimeric G protein Bcg1 (Gronover et al. 2004), and the other was the family-11 xylanase Xyn11A (Brito et al. 2006). Mutants in the gene for the first protein were equally virulent as the wild-type (J. Schumacher, pers. commun.) while mutants in *xyn11A* displayed a reduction in pathogenicity for tomato leaves and grape berries. Retransformation of the *xyn11A* mutants with the wild-type gene rescued the virulence defect, unequivocally showing that Xyn11A contributes to virulence (Brito et al. 2006). The most obvious explanation for this phenotype, *i.e.* xylanase activity is necessary for an efficient break down of the cell walls and therefore necessary for virulence, not only presented the problem of the low xylan content of the walls in some tissues of *B. cinerea* hosts, but also did not agree well with xylanase redundancy in the *B. cinerea* genome. There are five xylanase genes in the genome and, accordingly, the *xyn11A* mutant shows a modest reduction in the growth rate with xylan as the only carbon source as well as in the xylanase activity secreted to the culture medium. An alternative explanation for the reduced virulence of the *xyn11A* mutants is that this protein may contribute to the infection process not with the xylanase activity, but with the xylanase's ability to elicit defense responses in plants. This last activity has been shown for several different xylanases and requires the binding of these protein elicitors to a receptor of the plant immune system that has been characterized in tomato (Ron and Avni 2004). By working with purified Xyn11A, as well as with four site-directed *xyn11A* mutants lacking xylan hydrolytic activity, it was shown (Noda et al. 2010) that this protein is indeed able to induce necrosis in plants independently of its hydrolytic activity. Moreover, Xyn11A variants lacking xylanolytic activity were able to complement the lower virulence of the *xyn11A* mutant, showing that it was the necrotizing activity, and not the ability to hydrolyse xylan, what was contributing to virulence. This also means that we still do not know to what extent xylanase activity is required by *B. cinerea* to invade its hosts.

Pectin is the third group of plant cell wall compounds and is probably the most abundant one in the cells which are primordially attacked by *B. cinerea*, so it had been anticipated that their hydrolysis would be crucial in the invasion strategy of this fungus. Pectin is also composed of different polymers, with homogalacturonans and rhamnogalacturonans being the most prominent. Among the proteins experimentally found in the secretome, there are 28 enzymes potentially involved in the degradation of pectin which include endopolygalacturonases, pectin methylesterases, and rhamnogalacturonases. From this group, endopolygalacturonases and pectin methyl esterases have been analyzed for their contribution to virulence. Six endopolygalacturonases have been extensively studied and two of them, BcPG1 and BcPG2, are required for full virulence (Kars et al. 2005a; ten Have et al. 1998), while the rest seem to be dispensable for pathogenesis (Williamson et al. 2007). It is interesting to note that BcPG1 and BcPG2 were the only endopolygalacturonases found in the early secretome (Espino et al. 2010), and that they both were among the ten most abundant proteins.

Pectin methylesterases demethylate pectin and make it more prone to degradation by depolymerizing enzymes such as endopolygalacturonases. Four pectin methylesterases have been found in the proteomic studies of the secretome, out of which two genes have been knocked out, *Bcpme1* and *Bcpme2* (Kars et al. 2005b; Valette-Collet et al. 2003). The first gene has been mutated in two different *B. cinerea* strains, B05.10 and Bd90. *Bcpme1* knock-out mutants in the *B. cinerea* background B05.10 showed the same virulence as the wild-type strain, while deletion of *Bcpme1* in the background of Bd90 reduced the virulence of the deletion strain compared to the virulence of the corresponding wild-type strain. Thus, the search for virulence factors may be more complicated than previously anticipated since effect on virulence also depends on the genetic background of the strain. When assessing the contribution of one specific protein to the infectious process one needs to take into account that different results may be obtained when using different host plants, but also that different results may be obtained for different fungal strains. This is particularly important in the process of validating potential targets for new fungicides: a protein which is required for virulence in laboratory strains under controlled laboratory conditions may not be so for natural strains in field conditions.

12.3.3 Other Secreted Proteins

Another important group of proteins found in the secretome are oxidoreductases, most of them with unknown function. Some of them, for example, seem to be similar to esterases, but there is no clue about which ester they may hydrolyse. One of the enzymes found in this group is the Cu-Zn-superoxide dismutase BcSOD1, which has been actually shown to be a virulence factor (Rolke et al. 2004). Superoxide dismutases convert superoxide anion into oxygen and hydrogen peroxide, and it is not clear whether the detoxification of superoxide or the production of hydrogen peroxide, or both, contribute to the infection. *B. cinerea* is known to actively produce hydrogen peroxide early during infection and this compound is considered to contribute, itself, to virulence (Heller and Tudzynski 2011). More details about ROS are given in Chap. 14.

An interesting member of the secretome, which has been found in almost every *B. cinerea* secretome study, belongs to the cerato-platanin family of proteins (Pfam family PF07249). This family received its name from its founding member, the phytotoxic protein cerato-platanin, from the fungus *Ceratocystis platani* (formerly *C. fimbriata*). Proteins of this family are known to elicit a range of defense responses in plants including hypersensitive (HR)-like cell death (Frías et al. 2011; Pazzagli et al. 1999; Yang et al. 2009). BcSpl1, one of the two cerato-platanins coded by the *B. cinerea* genome, is one of the most abundant proteins found in the early secretome (Espino et al. 2010), although it seems that cerato-platanins are mainly cell wall proteins (Boddi et al. 2004; Frías et al. 2014) that just happen to leak into the extracellular medium. Knock-out mutants have been made for *Bcspl1*, and the resulting strains showed reduced virulence compared to the wild type (Frías et al. 2011). The recombinant BcSpl1 protein expressed in yeast caused fast and strong necrosis when infiltrated in plant leaves, and the treated area showed several symptoms of HR such as induction of reactive oxygen species, electrolyte leakage, cytoplasm shrinkage, and cell autofluorescence, as well as the induction of defense genes considered as HR markers (Frías et al. 2011). Moreover, existing evidences point to the recognition of cerato-platanins as Pathogen Associated Molecular Patterns (PAMPs) by the plant immune system as the molecular basis of the effects they cause on plants. The eliciting motif has been identified in BcSpl1 and shown to account by itself for the protein's contribution to virulence (Frías et al. 2014). However, no plant receptor for cerato-platanins has been identified yet. The fact that BcSpl1 elicits plant defenses and at the same time contributes to virulence is certainly unexpected but may be explained by the idea that necrotrophs such as *B. cinerea* could benefit from cell death resulting from an intense defense response (HR, paragraph 12.5). Cerato-platanins could therefore be just fungal cell-wall proteins with an eventual extensin-like role (Baccelli et al. 2013; de Oliveira et al. 2011), whose recognitions as PAMPs by plant cells might favour *B. cinerea* infection strategy.

BcNEP1 and BcNEP2 are two members of the NEP1-like family of proteins (Pfam PF05630) of the *B. cinerea* genome. The proteins of this family also induce plant defense responses (callose apposition, accumulation of reactive oxygen species, induction of defense genes, etc.) and, at high doses, the death of the treated cells (Gijzen and Nürnberger 2006). Although both genes are expressed in the fungal cells (Cuesta Arenas et al. 2010), with different features, only BcNEP2 has been found experimentally in the secretome. Both proteins, expressed in yeast, are able to induce the synthesis of ethylene and necrosis in plant leaves (Cuesta Arenas et al. 2010). However, the mutants lacking either of the two NEPs are not affected in virulence, indicating that they might not contribute individually to the infection process, while attempts to generate the double mutant were unsuccessful (Cuesta Arenas et al. 2010).

Finally, one key feature of the proteins in the *B. cinerea* secretome is the high frequency of *O*-glycosylation among them. *O*-glycosylation usually occurs along the secretory pathway for protein regions rich in Ser and Thr. An in-silico search for such regions, among the proteins predicted to be secreted by *B. cinerea*, found that 41 % display at least one region with at least a 40 % Ser/Thr content in a minimum length of 20 residues (González et al. 2012). Additionally, about one fourth of secretory proteins were actually predicted to have hyper-*O*-glycosylated regions. Protein *O*-glycosylation is initiated by protein *O*-mannosyltransferases (PMTs), which add the first mannose residue to the protein substrates in the ER. The three genes coding PMTs in the *B. cinerea* genome, *Bcpmt1*, *Bcpmt2*, and *Bcpmt4*, have been individually deleted (González et al. 2013), and the corresponding Δ*Bcpmt* mutants have a wide array of defects in growth, morphology, and virulence. Especially significant are the defects in the adherence and the penetration of intact plant surfaces, as well as a low production of extracellular matrix. These defects, along with the fact that plants do not possess PMTs, point to these enzymes as possible targets in the design of novel control strategies against *B. cinerea*.

12.4 Small RNAs as Virulence Factors

One interesting feature of *B. cinerea*'s infection strategy is its capacity to secrete small RNA molecules (sRNAs) able to selectively silence certain plant genes putatively involved in plant defense, by hijacking the plant RNA interference system. In a recent paper Weiberg et al. (2013) described, for the first time, that *B. cinerea* secretes sRNAs that are taken up by plant cells and, once there, make use of the plant silencing machinery to promote degradation of defense-specific plant mRNAs, thus enhancing susceptibility to the fungus. These sRNAs were identified in a search, by deep sequencing, for *B. cinerea* sRNAs that were more abundant in plant infections compared to axenic culture. This approach resulted in the identification of 73 sRNAs with the potential to silence genes both in *A. thaliana* and tomato. Three of these were further characterized by Weiberg et al. (2013). These authors showed that they are pre-processed in the fungal cells by *B. cinerea*'s own Dicer enzymes and, once secreted and taken up by *A. thaliana* cells, they bind to one specific plant Argonaute protein to carry out gene silencing, resulting in diminished mRNA levels of the target plant genes. Silencing of these targets, which are suspected to play a role in the plant immune system, results in enhanced susceptibility to *B. cinerea*, both in *A. thaliana* and tomato. Silencing of the *A. thaliana* targets for the three characterized sRNAs was shown to result in lower expression of the *BOTRYTIS-INDUCED KINASE1* gene *BIK1*, whose mutation is known to increase susceptibility to necrotrophic plant pathogens including *B. cinerea* and *Alternaria brassicicola* (Veronese et al. 2006) (Chap. 19).

The number of sRNAs that are potentially involved in silencing plant genes (Weiberg et al. 2013) was surprisingly high; expression of 832 sRNAs was enhanced during plant infection, and potential plant targets were identified for 73 of them in *A. thaliana* and tobacco. Since the strategy used by these researchers aimed at the identification of sRNAs fulfilling certain stringent requirements (small size, higher expression in planta, non-complete identity with plant genes), other sRNAs not fulfilling these criteria may also be important in pathogenesis. Therefore we can expect more sRNAs and corresponding plant targets to be discovered in the future. Overall, these numbers point to a very prominent attack with sRNAs carried out by the fungus during infection.

12.5 Targeting of Host HR and PCD Machinery

In the broadest sense, all pathogens can be divided into killers and non-killers (Sharon and Shlezinger 2013). The ultimate goal of killer pathogens is the conversion of living tissue into dead organic matter. Therefore, killer pathogens are programmed to efficiently kill host cells, while non-killer pathogens aim to keep their host alive, and are therefore programmed to prevent host cells from dying. Accordingly, host organisms use opposite defense strategies against killer and non-killer pathogens: they are programmed to activate self-cell death when facing non-killer pathogens, and block cell death when attacked by killer pathogens. Control of cell death is therefore a central element in all types of pathogen–host interactions. Broadly, two main types of cell death are distinguished – uncontrolled death processes, collectively referred to as necrotic cell death, and controlled death, collectively referred to as PCD.

When examining the virulence strategy of *B. cinerea*, we should consider these basic principles: as a killer pathogen, *B. cinerea*'s main strategy would be fast killing of the host tissue, and it could use either or both of necrosis- and PCD-inducing virulence factors to achieve this goal. According to the model proposed by Shlezinger et al. (2011), necrosis-inducing factors are essential during the early infection phase, whereas PCD-inducing factors are probably more abundant at the later phase.

12.6 Local Necrosis-Inducing Virulence Factors

Similar to other necrotrophic pathogens, *B. cinerea* is unable to extract nutrients from living tissues. This entails instant killing of host tissues upon the first encounter. One alternative to destroy the plant tissues would be secretion of massive amounts of toxins and hydrolytic enzymes. However, considering that conidia are

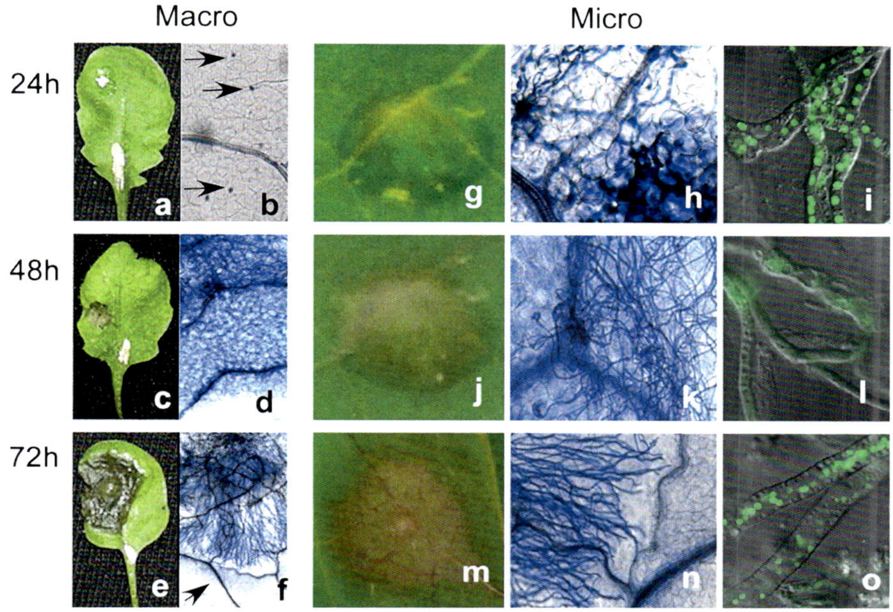

Fig. 12.2 Stages and developmental changes during *B. cinerea* infection. *Left panel* shows macro images of infected *A. thaliana* leaves at 24, 48 and 72 h post-inoculation (PI). Trypan blue (**b**, **d**, **f**) stains dead plant cells and living fungal hyphae. *Arrows* show appearance of hypersensitive-like local necroses 24 h PI (**b**) and formation of a ring of dead plant cells at 48 and 72 h PI (**d**, **f**). The formation of local necroses 24 h PI is visible on leaves at higher magnification (**g**). Organization and biomass of fungal hyphae and mycelium change from thin, isolated hyphae at 24 h PI (**h**) to a bulk of mycelium composed of thin hyphae at 48 h (**k**), to a well-organized mycelium composed of bundles of thick hyphae that grows from the centre of infection outward at 72 h PI (**n**). At 24 h PI, hyphae are viable, as can be seen by strong nuclear signal in a strain expressing a histone-GFP construct (**i**). Massive cell death is evidenced by disappearance of the nuclei at 48 h PI (**l**). At 72 h PI the fungus recovers and new hyphae are formed (**o**) (Modified from Shlezinger et al. 2011)

the main source of *B. cinerea* infections, it is clear that at the early infection phase, production and secretion of proteins is limited to very small amounts. Hence, massive secretion of substances at the early stages of infection is unlikely. Furthermore, germinating conidia are exposed to toxic host metabolites, which induce cell death in the fungus and destroy a large part of the new hyphae. To counter the host-induced cell death, the fungus needs an active anti-PCD machinery, which enables rescue of a few cells during this critical stage (Shlezinger et al. 2011). Continuation of infection depends on the ability of the fungus to form a fully necrosed region, before being eliminated by the plant toxic compounds. Thus, even in successful infections, the active fungal biomass remains small until the end of the first infection phase.

Close examination of *A. thaliana* leaf tissues 12–24 h after inoculation with *B. cinerea* shows development of HR-like, small necroses (Fig. 12.2). HR, while highly effective in restricting biotrophic pathogens, is ineffective against *B. cinerea*. In fact, this fungus needs the host HR for full virulence; suppression of HR, either by chemicals or through mutations, restricts *B. cinerea* infection, whereas induced HR enhances the disease (El Oirdi and Bouarab 2007; Govrin et al. 2006; Govrin and Levine 2000; Hoeberichts et al. 2003; Rossi et al. 2011). Hence, it is reasonable to assume that certain factors, such as the necrosis-inducing BcNEPs and the cerato-platanin BcSpl1, would activate the HR response to induce fast necrosis. In this event, a very small amount of protein will be necessary, just enough to activate the plant's self-destroy machinery. Furthermore, it can be expected that the effect of such molecules will be highly restricted to the site of protein application/secretion, similar to the nature of HR response. Indeed, both BcNEPs and the cerato-platanin BcSpl1 are active at super low (picomole) amounts, and induce local necrosis (Cuesta Arenas et al. 2010; Frías et al. 2011; Qutob et al. 2006). It is intuitive to predict that *B. cinerea* produces additional necrosis-inducing factors (proteins as well as other types of molecules) for the same purpose. One class of such molecules are the earlier mentioned toxins – oxalic acid, botrydial and alike.

12.7 Induction of Spreading Cell Death

Once established within a fully necrosed tissue, the fungus is no longer exposed to the plant defense substances, and fungal cells that had survived the first encounter with living plant tissue can nourish on the dead tissue and produce new biomass in the protected zone. A clear change is visible in *A. thaliana* leaves at this stage in disease development, characterized by fast development of wet rotting (Fig. 12.2). This change in disease progression is accompanied by dramatic changes in development of the fungus: massive amount of fungal mycelium develops, which is characterized by bundles of thick hyphae organized as a radial colony that grows outward from the necrotic centre. While at the centre of the colony the plant tissue is fully macerated, a ring of intact, but dead plant cells is visible around the infection court (Fig. 12.2 and, e.g., Fig. 7 in Rowe et al. 2010). These changes in disease spreading and fungal development imply deployment of a new set of virulence factors, which are different in the nature and mode of action from the necrosis-inducing factors that are necessary for establishment of infection at the early phase. Instead, one can expect to find factors of a more diffusible nature, which promote spreading cell death in tissues that are not in direct contact with the fungus.

The notion that *B. cinerea* uses specific virulence factors (effectors) to promote spreading cell death is relatively new, and such molecules have not (yet) been reported in *B. cinerea*. Nevertheless, it is possible that such putative effectors would also target the HR and PCD machinery, although with different effects than those of the early infection phase effectors. A putative PCD-inducing virulence factor has been described in *Sclerotinia sclerotiorum*, another broad host range necrotrophic pathogen that is closely related to *B. cinerea*. *S. sclerotiorum* mutants that do not produce oxalic acid (OA) trigger HR-like response, but are unable to initiate disease. OA has multiple effects on the plant, one of which is induction of apoptotic-like cell death (Kim et al. 2008). Transgenic tobacco plants expressing anti-apoptotic genes did not develop apoptosis when treated with OA, and were highly resistant to infection by *S. sclerotiorum* as well as by *B. cinerea* (Dickman et al. 2001). Recent studies showed that OA suppresses autophagy, a process that is tightly connected with PCD and essential for survival of the tissue. *A. thaliana* mutants lacking autophagy genes showed full sensitivity to the *S. sclerotiorum* OA-deficient mutant, as opposed to wild-type plants that are completely resistant to the mutant (Kabbage et al. 2013). These and other findings led to the conclusion that OA induces apoptosis and suppresses autophagy during *S. sclerotiorum* infection, thereby initiating a death process while blocking autophagy, the pro-survival element of PCD (Zhou et al. 2014). Without OA, *S. sclerotiorum* is unable neither to trigger PCD nor to block autophagy, and hence the plant defense can be activated and block pathogen spreading. It should be noted that *B. cinerea* also produces large quantities of OA, but its effect on virulence was not shown in this system.

The possibility that *B. cinerea* effectors target the plant PCD and HR machineries receives additional support from work on the regulation of HR-associated PCD in plants (Chaps. 17, 18, and 19). PCD is part of the normal plant development and life cycle. It is associated with various developmental processes such as xylogenesis, reproduction and senescence (Greenberg 1996; Pennell and Lamb 1997). External stimuli, both biotic and abiotic, can also lead to PCD. Perhaps the most familiar form of induced plant PCD is associated with HR following pathogen attack. The development of apoptotic markers during HR response has been known for a long time. In a normal HR response, PCD is limited to a small number of cells around the pathogen invasion site. Studies in *A. thaliana* have shown that a series of negative and positive regulators of PCD control the onset and termination of PCD during HR. The zinc-finger transcription factor LSD1 negatively regulates spreading of cell death to cells surrounding infection sites (Dietrich et al. 1997). In *lsd1* mutants, cell death control is impaired resulting in spreading of cell death to the entire leaf, a phenomenon known as 'runway cell death' (RCD). An *A. thaliana* double mutant affected in both the NADPH oxidase AtrbohD and LSD1, shows enhanced RCD and increased sensitivity to *B. cinerea* (Torres et al. 2005). Mutation in *LOL1*, a *LSD1* paralogue gene, has opposite effects to the *lsd1* mutation. An *lsd1/lol1* double mutant shows reduced RCD, whereas over-expression of *LOL1* induces HR in wild-type as well as in *lsd1* plants (Epple et al. 2003). In wild type plants, LOL1 is antagonized by LSD1, which restricts the activity of LOL1 to a limited number of cells. In *lsd1* plants, LOL1 is not blocked, resulting in the RCD phenomenon. These results suggest that LSD1 might act as an antagonist of PCD by retaining LOL1 and possibly additional transcription factors that are positive regulators of PCD, such as AtbZIP10, in the cytoplasm, thereby preventing activation of apoptosis by this group of pro-apoptotic transcription elements. This model resembles the control of cell death in metazoan by the IAP family of zinc-finger proteins, which maintain a threshold for cell death by modulation of caspases. If this analogy is correct, caspase activity should be affected in the corresponding *A. thaliana* mutant plants (Sharon and Finkelshtein 2009).

Indeed, recent work by Dangl and co-workers further support this model. AtMC1, one of three metacaspases found in *A. thaliana*, showed strong interaction with LSD1 and the *lsd1*-associated RCD phenotype was abolished in *atmc1* mutant background (Coll et al. 2010). Furthermore, there was no interaction between LSD1 and the second metacaspase AtMC2, although the RCD phenotype was intensified in *atmc2* mutant background. These and other data show that HR-associated PCD is executed by AtMC1 and restricted in adjunct cells by the antagonistic activity of AtMC2.

Although there is no information on *B. cinerea* effectors that target these or similar regulators of HR/PCD, the RCD phenotype provides a mechanism that can explain the sharp shift from local death to fast-spreading lesion observed during *B. cinerea* infection. While early phase effectors act locally at super low levels and induce local necrosis, effectors at the second phase could possibly target regulators of the PCD machinery to generate the phenomenon of RCD. In view of the *B. cinerea* ability to manipulate plant cells via RNA interference mechanisms (Weiberg et al. 2013), already discussed above, these effectors could be secreted in the form of sRNAs with the capacity to silence key regulators of PCD.

Further evidences connecting PCD with the plant defense machinery come from work with anti-apoptotic genes. p35, an IAP-like protein from baculovirus suppresses apoptosis in mammalian cells by inhibiting caspases (Clem et al. 1996). Expression of the anti-apoptotic baculovirus protein p35 in plants significantly affected disease caused by various pathogens. Transgenic tomato plants expressing the anti-apoptotic p35 baculovirus protein were protected from AAL toxin-induced cell death and showed enhanced resistance to infection by the necrotrophic fungal pathogen *A. alternata* (Lincoln et al. 2002). A p35 binding site mutant clone that is inactive against human caspase 3 did not protect plants form the pathogens or the toxin, indicating the involvement of plant caspases in the HR response. The anti-apoptotic protein BI-1 blocks PCD in plants. BI-1 is necessary for successful infection of barley by the powdery mildew fungus *Blumeria graminis*, a non-killer (biotrophic) pathogen (Eichmann et al. 2004), whereas over-expression of BI-1 reduces plants' susceptibility to killer pathogens such as *Fusarium graminearum* and *B. cinerea* (Babaeizad et al. 2009; Imani et al. 2006). AtBI-1, one of several homologs of human BI-1 found in *A. thaliana*, blocked Bax- H_2O_2- and salicylic acid-induced PCD and reduced sensitivity to *B. cinerea* when overexpressed in *A. thaliana* and tobacco (Kawai-Yamada et al. 2004). Similarly, expression of Barley BI-1 in carrots limited necrosis and restricted spreading of *B. cinerea* (Imani et al. 2006). Finally, tobacco plants expressing anti-apoptotic members of the Bcl-2 family of proteins, provided protection against *B. cinerea* and *S. Sclerotiorum* (Dickman et al. 2001). DNA laddering, a hallmark of apoptosis that occurred after infection with pathogens in wild type tobacco plants did not occur in the transgenic plants, suggesting that infection by these necrotrophic pathogens is promoted by induced-PCD, which enhances plant sensitivity to the pathogens.

The above examples show that plant cell death machinery, both necrosis and PCD, is target for *B. cinerea* effectors, which act to induce abrupt cell death by affecting essential processes (necrosis-inducing factors) and modify expression of these self-destroy mechanisms (PCD) to their own benefit.

12.8 Concluding Remarks

The paradigm of *B. cinerea* infection has been revisited during the past years. It is now clear that the pathogenic development of *B. cinerea* is far more complex than previously appreciated. Disease progression is tightly regulated during the entire infection process, and the fungus makes developmental adaptations that match the different infection stages. Such sophisticated interaction between the fungus and the plant entails involvement of molecules with internal and external effects. One class of such molecules are highly specific virulence factors (effectors), similar to the situation in other patho-systems. Recent works on *B. cinerea* proteome and transcriptome, along with advanced studies on the regulation of the plant HR defense response, suggest that *B. cinerea* virulence factors target and manipulate the plants' HR and the associated self-cell death (PCD) machinery. Discovery of the fungal factors/effectors and identification of their plant targets, hold potential for improvement of plant tolerance to this devastating pathogen.

References

Albenne C, Canut H, Jamet E (2013) Plant cell wall proteomics: the leadership of *Arabidopsis thaliana*. Front Plant Sci 4:e111

Babaeizad V, Imani J, Kogel KH et al (2009) Over-expression of the cell death regulator BAX inhibitor-1 in barley confers reduced or enhanced susceptibility to distinct fungal pathogens. Theor Appl Genet 118:455–463

Baccelli I, Luti S, Bernardi R et al (2013) Cerato-platanin shows expansin-like activity on cellulosic materials. Appl Microbiol Biotechnol 98:175–184

Bent AF, Mackey D (2007) Elicitors, effectors, and R genes: the new paradigm and a lifetime supply of questions. Annu Rev Phytopathol 45:399–436

Bessire M, Chassot C, Jacquat A-C et al (2007) A permeable cuticle in *Arabidopsis* leads to a strong resistance to *Botrytis cinerea*. EMBO J 26:2158–2168

Boddi S, Comparini C, Calamassi R et al (2004) Cerato-platanin protein is located in the cell walls of ascospores, conidia and hyphae of *Ceratocystis fimbriata* f. sp. *platani*. FEMS Microbiol Lett 233:341–346

Brito N, Espino JJ, González C (2006) The endo-ß-1,4-xylanase Xyn11A is required for virulence in *Botrytis cinerea*. Mol Plant-Microbe Interact 19:25–32

Chassot C, Nawrath C, Metraux JP (2007) Cuticular defects lead to full immunity to a major plant pathogen. Plant J 49:972–980

Cherrad S, Girard V, Dieryckx C et al (2012) Proteomic analysis of proteins secreted by *Botrytis cinerea* in response to heavy metal toxicity. Metallomics 4:835–846

Choquer M, Fournier E, Kunz C et al (2007) *Botrytis cinerea* virulence factors: new insights into a necrotrophic and polyphageous pathogen. FEMS Microbiol Lett 277:1–10

Clem RJ, Hardwick JM, Miller LK (1996) Anti-apoptotic genes of baculoviruses. Cell Death Differ 3:9–16

Coll NS, Vercammen D, Smidler A et al (2010) *Arabidopsis* type I metacaspases control cell death. Science 330:1393–1397

Commenil P, Belingheri L, Dehorter B (1998) Antilipase antibodies prevent infection of tomato leaves by *Botrytis cinerea*. Physiol Mol Plant Pathol 52:1–14

Cuesta Arenas Y, Kalkman ERIC, Schouten A et al (2010) Functional analysis and mode of action of phytotoxic Nep1-like proteins of *Botrytis cinerea*. Physiol Mol Plant Pathol 74:376–386
de Oliveira AL, Gallo M, Pazzagli L et al (2011) The structure of the elicitor Cerato-platanin (CP), the first member of the CP fungal protein family, reveals a double psibeta-barrel fold and carbohydrate binding. J Biol Chem 286:17560–17568
Dickman MB, Park YK, Oltersdorf T et al (2001) Abrogation of disease development in plants expressing animal antiapoptotic genes. Proc Natl Acad Sci U S A 98:6957–6962
Dietrich RA, Richberg MH, Schmidt R et al (1997) A novel zinc finger protein is encoded by the *Arabidopsis LSD1* gene and functions as a negative regulator of plant cell death. Cell 88:685–694
Eichmann R, Schultheiss H, Kogel KH et al (2004) The barley apoptosis suppressor homologue BAX inhibitor-1 compromises nonhost penetration resistance of barley to the inappropriate pathogen *Blumeria graminis* f. sp. *tritici*. Mol Plant-Microbe Interact 17:484–490
El Oirdi M, Bouarab K (2007) Plant signalling components EDS1 and SGT1 enhance disease caused by the necrotrophic pathogen *Botrytis cinerea*. New Phytol 175:131–139
Ellis JG, Rafiqi M, Gan P et al (2009) Recent progress in discovery and functional analysis of effector proteins of fungal and oomycete plant pathogens. Curr Opin Plant Biol 12:399–405
Epple P, Mack AA, Morris VR et al (2003) Antagonistic control of oxidative stress-induced cell death in *Arabidopsis* by two related, plant-specific zinc finger proteins. Proc Natl Acad Sci U S A 100:6831–6836
Espino JJ, Brito N, Noda J et al (2005) *Botrytis cinerea* endo-ß-1,4-glucanase Cel5A is expressed during infection but is not required for pathogenesis. Physiol Mol Plant Pathol 66:213–221
Espino JJ, Gutiérrez-Sánchez G, Brito N et al (2010) The *Botrytis cinerea* early secretome. Proteomics 10:3020–3034
Fernández-Acero FJ, Colby T, Harzen A et al (2010) 2-DE proteomic approach to the *Botrytis cinerea* secretome induced with different carbon sources and plant-based elicitors. Proteomics 10:2270–2280
Frías M, González C, Brito N (2011) BcSpl1, a cerato-platanin family protein, contributes to *Botrytis cinerea* virulence and elicits the hypersensitive response in the host. New Phytol 192:483–495
Frías M, Brito N, González M et al (2014) The phytotoxic activity of the cerato-platanin BcSpl1 resides in a two-peptide motif in the protein surface. Mol Plant Pathol 15:342–351
Gentile AC (1954) Carbohydrate metabolism and oxalic acid synthesis by *Botrytis cinerea*. Plant Physiol 29:257–261
Gijzen M, Nürnberger T (2006) Nep1-like proteins from plant pathogens: recruitment and diversification of the NPP1 domain across taxa. Phytochemistry 67:1800–1807
González M, Brito N, González C (2012) High abundance of serine/threonine-rich regions predicted to be hyper-*O*-glycosylated in the extracellular proteins coded by eight fungal genomes. BMC Microbiol 12:213
González M, Brito N, Frías M et al (2013) *Botrytis cinerea* protein *O*-mannosyltransferases play critical roles in morphogenesis, growth, and virulence. PLoS ONE 8:e65924
Gourgues M, Brunet-Simon A, Lebrun MH et al (2004) The tetraspanin BcPls1 is required for appressorium-mediated penetration of *Botrytis cinerea* into host plant leaves. Mol Microbiol 51:619–629
Govrin EM, Levine A (2000) The hypersensitive response facilitates plant infection by the necrotrophic pathogen *Botrytis cinerea*. Curr Biol 10:751–757
Govrin EM, Rachmilevitch S, Tiwari BS et al (2006) An elicitor from *Botrytis cinerea* induces the hypersensitive response in *Arabidopsis thaliana* and other plants and promotes the gray mold disease. Phytopathology 96:299–307
Greenberg JT (1996) Programmed cell death: a way of life for plants. Proc Natl Acad Sci U S A 93:12094–12097
Gronover CS, Schorn C, Tudzynski B (2004) Identification of *Botrytis cinerea* genes up-regulated during infection and controlled by the G alpha subunit BCG1 using suppression subtractive hybridization (SSH). Mol Plant-Microbe Interact 17:537–546

Heller J, Tudzynski P (2011) Reactive oxygen species in phytopathogenic fungi: signaling, development, and disease. Annu Rev Phytopathol 49:369–390

Hoeberichts FA, Ten HA, Woltering EJ (2003) A tomato metacaspase gene is upregulated during programmed cell death in *Botrytis cinerea*-infected leaves. Planta 217:517–522

Imani J, Baltruschat H, Stein E et al (2006) Expression of barley BAX inhibitor-1 in carrots confers resistance to *Botrytis cinerea*. Mol Plant Pathol 7:279–284

Kabbage M, Williams B, Dickman MB (2013) Cell death control: the interplay of apoptosis and autophagy in the pathogenicity of *Sclerotinia sclerotiorum*. PLoS Pathog 9:e1003287

Kars I, Krooshof GH, Wagemakers L et al (2005a) Necrotizing activity of five *Botrytis cinerea* endopolygalacturonases produced in *Pichia pastoris*. Plant J 43:213–225

Kars I, McCalman M, Wagemakers L et al (2005b) Functional analysis of *Botrytis cinerea* pectin methylesterase genes by PCR-based targeted mutagenesis: *Bcpme1* and *Bcpme2* are dispensable for virulence of strain B05.10. Mol Plant Pathol 6:641–652

Kawai-Yamada M, Ohori Y, Uchimiya H (2004) Dissection of *Arabidopsis* Bax inhibitor-1 suppressing Bax-, hydrogen peroxide-, and salicylic acid-induced cell death. Plant Cell 16:21–32

Kim KS, Min JY, Dickman MB (2008) Oxalic acid is an elicitor of plant programmed cell death during *Sclerotinia sclerotiorum* disease development. Mol Plant-Microbe Interact 21:605–612

Li B, Wang W, Zong Y et al (2012) Exploring pathogenic mechanisms of *Botrytis cinerea* secretome under different ambient pH based on comparative proteomic analysis. J Proteome Res 11:4249–4260

Lincoln JE, Richael C, Overduin B et al (2002) Expression of the antiapoptotic baculovirus p35 gene in tomato blocks programmed cell death and provides broad-spectrum resistance to disease. Proc Natl Acad Sci U S A 99:15217–15221

Nakajima M, Akutsu K (2014) Virulence factors of *Botrytis cinerea*. J Gen Plant Pathol 80:15–23

Nguyen QB, Itoh K, Van Vu B et al (2011) Simultaneous silencing of endo-β-1,4 xylanase genes reveals their roles in the virulence of *Magnaporthe oryzae*. Mol Microbiol 81:1008–1019

Noda J, Brito N, Gonzalez C (2010) The *Botrytis cinerea* xylanase Xyn11A contributes to virulence with its necrotizing activity, not with its catalytic activity. BMC Plant Biol 10:38

Pazzagli L, Cappugi G, Manao G et al (1999) Purification, characterization, and amino acid sequence of Cerato-platanin, a new phytotoxic protein from *Ceratocystis fimbriata* f. sp. *platani*. J Biol Chem 274:24959–24964

Pennell RI, Lamb C (1997) Programmed cell death in plants. Plant Cell 9:1157–1168

Plett JM, Kemppainen M, Kale SD et al (2011) A secreted effector protein of *Laccaria bicolor* is required for symbiosis development. Curr Biol 21:1197–1203

Qutob D, Kemmerling B, Brunner F et al (2006) Phytotoxicity and innate immune responses induced by Nep1-like proteins. Plant Cell 18:3721–3744

Reis H, Pfiffi S, Hahn M (2005) Molecular and functional characterization of a secreted lipase from *Botrytis cinerea*. Mol Plant Pathol 6:257–267

Rolke Y, Liu SJ, Quidde T et al (2004) Functional analysis of H_2O_2-generating systems in *Botrytis cinerea*: the major Cu-Zn-superoxide dismutase (BCSOD1) contributes to virulence on French bean, whereas a glucose oxidase (BCGOD1) is dispensable. Mol Plant Pathol 5:17–27

Ron M, Avni A (2004) The receptor for the fungal elicitor ethylene-inducing xylanase is a member of a resistance-like gene family in tomato. Plant Cell 16:1604–1615

Rossi FR, Gárriz A, Marina M et al (2011) The sesquiterpene botrydial produced by *Botrytis cinerea* induces the hypersensitive response on plant tissues and its action is modulated by salicylic acid and jasmonic acid signaling. Mol Plant-Microbe Interact 24:888–896

Rowe HC, Walley JW, Corwin J et al (2010) Deficiencies in jasmonate-mediated plant defense reveal quantitative variation in *Botrytis cinerea* pathogenesis. PLoS Pathog 6:e1000861

Shah P, Atwood JA, Orlando R et al (2009a) Comparative proteomic analysis of *Botrytis cinerea* secretome. J Proteome Res 8:1123–1130

Shah P, Gutiérrez-Sánchez G, Orlando R et al (2009b) A proteomic study of pectin-degrading enzymes secreted by *Botrytis cinerea* grown in liquid culture. Proteomics 9:3126–3135

Sharon A, Finkelshtein A (2009) Programmed cell death in fungal-plant interactions. In: Deising HB (ed) The mycota XXII, vol V, 2nd edn, Plant relationships. Springer, Heidelberg, p 219

Sharon A, Shlezinger N (2013) Fungi infecting plants and animals: killers, non-killers, and cell death. PLoS Pathog 9:e1003517

Shlezinger N, Minz A, Gur Y et al (2011) Anti-apoptotic machinery protects the necrotrophic fungus *Botrytis cinerea* from host-induced apoptotic-like cell death during plant infection. PLoS Pathog 7:e1002185

Siegmund U, Heller J, Van Kan JA et al (2013) The NADPH oxidase complexes in *Botrytis cinerea*: evidence for a close association with the ER and the tetraspanin Pls1. PLoS ONE 8:e55879

Stergiopoulos I, de Wit PJ (2009) Fungal effector proteins. Annu Rev Phytopathol 47:233–263

ten Have A, Mulder W, Visser J et al (1998) The endopolygalacturonase gene *Bcpg1* is required for full virulence of *Botrytis cinerea*. Mol Plant-Microbe Interact 11:1009–1016

ten Have A, Espino JJ, Dekkers E et al (2010) The *Botrytis cinerea* aspartic proteinase family. Fungal Genet Biol 47:53–65

Thomma BP, Nurnberger T, Joosten MH (2011) Of PAMPs and effectors: the blurred PTI-ETI dichotomy. Plant Cell 23:4–15

Torres MA, Jones JD, Dangl JL (2005) Pathogen-induced, NADPH oxidase-derived reactive oxygen intermediates suppress spread of cell death in *Arabidopsis thaliana*. Nat Genet 37:1130–1134

Valette-Collet O, Cimerman A, Reignault P et al (2003) Disruption of *Botrytis cinerea* pectin methylesterase gene *Bcpme1* reduces virulence on several host plants. Mol Plant-Microbe Interact 16:360–367

Van der Vlugt-Bergmans CJB, Wagemakers CAM, Van Kan JAL (1997) Cloning and expression of the cutinase A gene of *Botrytis cinerea*. Mol Plant-Microbe Interact 10:21–29

Van Kan JAL (2006) Licensed to kill: the lifestyle of a necrotrophic plant pathogen. Trends Plant Sci 11:247–253

Van Kan JAL, Van't Klooster JW, Wagemakers CAM et al (1997) Cutinase A of *Botrytis cinerea* is expressed, but not essential, during penetration of gerbera and tomato. Mol Plant-Microbe Interact 10:30–38

Verhoeff K, Leeman M, Van Peer R et al (1988) Changes in pH and the production of organic acids during colonization of tomato petioles by *Botrytis cinerea*. J Phytopathol 122:327–336

Veronese P, Nakagami H, Bluhm B et al (2006) The membrane-anchored *BOTRYTIS-INDUCED KINASE1* plays distinct roles in *Arabidopsis* resistance to necrotrophic and biotrophic pathogens. Plant Cell 18:257–273

Weiberg A, Wang M, Lin FM et al (2013) Fungal small RNAs suppress plant immunity by hijacking host RNA interference pathways. Science 342:118–123

Williamson B, Tudzynski B, Tudzynski P et al (2007) *Botrytis cinerea*: the cause of grey mould disease. Mol Plant Pathol 8:561–580

Yang Y, Zhang H, Li G et al (2009) Ectopic expression of MgSM1, a Cerato-platanin family protein from *Magnaporthe grisea*, confers broad-spectrum disease resistance in *Arabidopsis*. Plant Biotechnol J 7:763–777

Zhou J, Yu J-Q, Chen Z (2014) The perplexing role of autophagy in plant innate immune responses. Mol Plant Pathol 15:637–645

Chapter 13
Signal Transduction Cascades Regulating Differentiation and Virulence in *Botrytis cinerea*

Julia Schumacher

Abstract Fungi are eukaryotic organisms that are more closely related to animals than to plants and algae, and therefore fungal cells share important features with mammalian cells. This is perhaps most apparent in the signaling cascades that mediate the communication between environmental signals and the cellular machinery regulating developmental programs. Conserved signaling systems present in all eukaryotes, from yeast to mammals, comprise cAMP and Ca^{2+} as second messengers, and further include Ras superfamily proteins and mitogen-activated (MAP) kinases. Apart from that, filamentous fungi evolved a light response system allowing for use of light and its absence, respectively, as environmental cues to control development and secondary metabolism. Recent progress in understanding the regulatory networks in *B. cinerea* by functional characterization of key components of the signal transduction pathways is summarized with emphasis on their impact on differentiation and virulence.

Keywords Signal transduction • cAMP • Calcium • Light • Transcription factor

13.1 Introduction

Signaling cascades transmit a signal (input) across membranes to the cytosol and/or the nucleus, where the respective cellular response is arranged (output). The signals (primary messengers) are perceived by receptors, transduced into intracellular signals involving phosphorylation events or secondary messenger systems, and finally passed through to the ultimate effector proteins: Regulation of transcription factors (TFs) allow for the reprogramming of the cell while the modification of enzyme activities allow for the quick adaptation of the cell metabolism to environmental

---

J. Schumacher (✉)
Institut für Biologie und Biotechnologie der Pflanzen, Westfälische Wilhelms-Universität Münster, Schlossplatz 8, 48143 Münster, Germany
e-mail: julia.schumacher@uni-muenster.de

S. Fillinger, Y. Elad (eds.), *Botrytis – the Fungus, the Pathogen and its Management in Agricultural Systems*, DOI 10.1007/978-3-319-23371-0_13

changes. Some core components of signal transduction pathways are present in yeast, filamentous fungi, plants and animals; some of them have been duplicated in the higher eukaryotes; others got lost or have been acquired by certain species. The repertoire of TFs and their target genes (signaling outputs) are species-specific and can be considered as the result of adaptive evolution. Hence, signaling cascades may trigger different responses even in closely related species addressing the need to study their functions in the particular organism. Due to the availability of the genome sequence and the optimization of the molecular techniques, the number of functionally characterized genes in *B. cinerea* has risen in recent years.

13.2 Monomeric G Proteins (Small GTPases)

Monomeric G proteins (GTPases of the Ras superfamily) function as binary molecular switches between "on" and "off" states and regulate a variety of biochemical reactions inside the cell. When bound to GDP, they are in an inactive configuration but the exchange of GDP to GTP, catalyzed by guanine nucleotide exchange factors (GEFs), results in a conformational change to an activated state ("on"). The GTPase interacts then with effector proteins transmitting the signal to downstream pathways. GTPase-activating proteins (GAPs) stimulate the otherwise low GTPase activity thereby restoring the inactive GDP-bound form ("off"). Furthermore, guanine nucleotide dissociation inhibitors (GDIs) may maintain small GTPases in the inactive state by preventing their activations by GEFs (reviewed in Weeks and Spiegelman 2003). The Ras superfamily is divided into five main families with conserved functions: RAS (cell proliferation), RHO (cell morphology/cytoskeletal dynamics), RAB (membrane trafficking), RAN (nuclear transport), and ARF (vesicular transport) (reviewed in Wennerberg et al. 2005).

Like the budding yeast *Saccharomyces cerevisiae*, *B. cinerea* possesses three RasGTPases. BcRas1, BcRas2 and B0510_6434 (Broad Database) represent the orthologs of yeast Ras1p, Ras2p, and Rsr1p/Bud1p, respectively. Mutants of *bcras1* and *bcras2* have been described: Δ*bcras1* mutants are severely affected in growth by forming hyper-branched and deformed hyphae (Minz Dub et al. 2013); Δ*bcras2* mutants are impaired in radial growth rates as well, but unlike Δ*bcras1*, they are able to differentiate conidia and sclerotia and to infect the plant tissue though with a considerable delay (Schumacher et al. 2008b).

S. cerevisiae has six RhoGTPases (Cdc42p, Rho1-5p); whereby Cdc42p and Rho1p are essential for viability. Cdc42p is required for polarized bud growth by regulating the organization of the actin cytoskeleton and the exocytic machinery; Rho1p regulates the activity of the (1,3)-β-D-glucan synthase and the cell wall integrity pathway (reviewed in Perez and Rincon 2010). *B. cinerea* as other filamentous ascomycetes lacks an ortholog of Rho5p, but contains instead another RhoGTPase named BcRac. The phenotype of Δ*bcrac* mutants resembles those caused by Δ*bcras1*, suggesting that both monomeric GTPases act in the same signaling pathway (Minz Dub et al. 2013). The phenotype of Δ*bccdc42* mutants is milder than that of *Δbcras1*/*Δbcrac* mutants: they exhibit slightly reduced growth rates, do not form sclerotia and produce fewer malformed conidia that possess altered germination and penetration properties (Kokkelink et al. 2011). Whether, in conformity with the yeast system, the *B. cinerea* orthologs of the p21-activated kinases (PAKs) Cla4p and Ste20p act as effector proteins of BcRac and BcCdc42 is currently under investigation (Minz Dub et al. unpublished). However, the RhoGEF BcCdc24 appears to be essential, possibly because of its GEF function for both, BcRac and BcCdc42. The formin BcSep1 which is involved in the assembly of F-actin cables, was recently identified as effector protein of RhoGTPase signaling; it is associated with BcCdc24 via the scaffold protein BcBem1 and is required for polarized growth and septum formation (Giesbert et al. 2014).

13.3 G Protein-Coupled Receptor (GPCR) Systems

Heterotrimeric G proteins are the most common signal transducers in eukaryotic cells and mediate the effects of various external signals. Prominent examples are the pheromone response in *S. cerevisiae* and the rhodopsin/transducin visual system in mammals. These G proteins act at the plasma membrane and are composed of three subunits: Gα contains a GTPase domain that is structurally identical to that of Ras superfamily proteins and interacts with a membrane-spanning G protein-coupled receptor (GPCR); Gβ interacts with Gα and Gγ. In the inactive state, the G protein subunits form a heterotrimer that is associated with the GPCR. Ligand binding to the receptor leads to the exchange of GTP for GDP in Gα, and Gα-GTP and the Gβγ dimer dissociate due to conformational changes. Gα and Gβγ can bind to and regulate membrane-localized effectors including adenylate cyclase, ion channels, phosphodiesterases, and phospholipases. Hydrolysis of the bound GTP by the intrinsic GTPase activity of Gα leads to the re-association of Gα-GDP with Gβγ and the GPCR (reviewed in Cabrera-Vera et al. 2003) (Fig. 13.1).

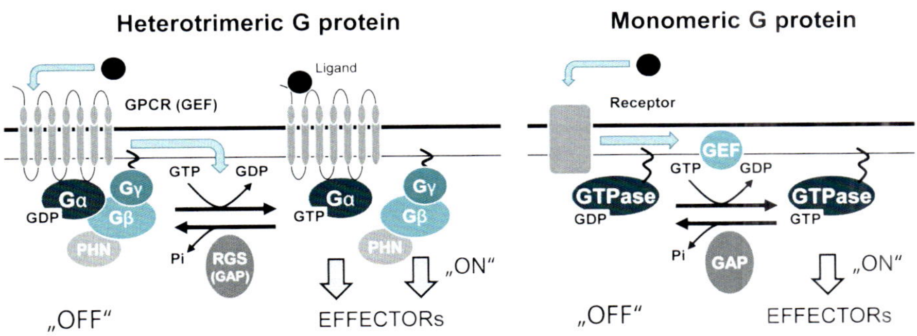

Fig. 13.1 G proteins – the mode of action. See text for explanation

In contrast to yeasts, which possess two Gα (Gpa1p, Gpa2p), *B. cinerea* and other filamentous fungi contain three Gα subunits. Respective genes named *bcg1*-*bcg3* in *B. cinerea* have been deleted revealing that all three Gα subunits are implicated in the infection process. Deletion of *bcg1* causes aberrant colony morphology and severely reduced virulence (spreading) that is accompanied by altered expression levels of genes involved in secondary metabolism (toxin biosynthesis) and plant cell wall degradation (Schulze Gronover et al. 2001, 2004). In contrast, the functions of the second Gα are less evident; Δ*bcg2* mutants do not exhibit altered growth characteristics and lesions provoked by Δ*bcg2* conidia on bean leaves spread with slightly reduced speed (Schulze Gronover et al. 2001). Gα3 (Bcg3) is required for conidial germination induced by carbon sources; its absence furthermore results in reduced conidiation, excessive sclerotia formation and delayed infection of tomato leaves (Döhlemann et al. 2006). Mutants lacking *bcg1* or *bcg3* contain lower cAMP levels compared to the wild type and their phenotypes are at least partially restored by exogenous cAMP indicating that Gα1 and Gα3 trigger cAMP production via activation of the adenylate cyclase (Schulze Gronover et al. 2001; Klimpel et al. 2002; Döhlemann et al. 2006).

B. cinerea possesses single genes (*bcgb1*, *bcgg1*) encoding Gβ and Gγ, respectively. As expected due to their function as subunits of membrane-bound G proteins, Gβ- and Gγ-GFP fusion proteins localize at the plasma membrane. Deletions of *bcgb1* and *bcgg1* result in identical phenotypes *i.e.*, reduced radial growth rates accompanied by increased aerial hyphae formation, loss of sclerotia formation, impaired conidiation, reduced virulence and increased intracellular cAMP levels (unpublished), corroborating the hypothesis that Gβ and Gγ act as a functional unit in *B. cinerea* to keep the Gα subunits in their inactive state. Notably, the latter function is not conserved in fungi as the disturbance of the Gβγ dimer in *Neurospora crassa* results in the posttranslational degradation of the Gα subunits and subsequently in lower intracellular cAMP contents (reviewed in Li et al. 2007).

Regulators of G protein signaling (RGSs) bind to activated Gα subunits and function as their GAPs (reviewed in Cabrera-Vera et al. 2003). The number and domain organization of RGSs differ among fungi; hence, five, seven and eight RGSs were described for *Aspergillus nidulans*, *Fusarium graminearum* and *Magnaporthe oryzae*, respectively (Wang et al. 2013). The search for RGSs in *B. cinerea* revealed seven candidates (Amselem et al. 2011) whereby BcRgs1 and BcRgs3 represent the orthologs of *A. nidulans* FlbA and RgsA which form specific RGS–Gα pairs with Gα1 (FadA) and Gα3 (GanB), respectively (reviewed in Yu 2006). Moreover, phosducin-like proteins (PhLPs) are required for proper G protein signaling by acting as chaperones during Gβγ assembly (Willardson and Howlett 2007). *B. cinerea* as *A. nidulans* possesses three PhLPs while *N. crassa* and *M. oryzae* just contain two (Amselem et al. 2011). BcPhnA, one of the three PhLPs, is involved in conidia and sclerotia production and is essential for plant infection (Kilani et al. unpublished).

The search for potential GPCRs in *B. cinerea* revealed 14 candidates including Ste2p/ Ste3p-like (pheromone) receptors (BcPre1, 2), the ortholog of the carbon sensor in yeast (BcGpr1), the cAMP receptor-like class which is absent from yeast

(BcGpr2-6), the group of Stm1-related proteins (BcGpr7-10), and the class of microbial opsins (Bop1, 2) (Li et al. 2007; Amselem et al. 2011). Deletion approaches have been undertaken for *bcgpr1-5* (*bcpgr6* is a pseudogene). Respective mutants do not exhibit altered growth behaviors or virulence (unpublished). However, Δ*bcgpr3* mutants are resistant to the toxic grapevine defensin VvAMP2 (Nanni et al. 2014). Another class of receptors in filamentous fungi is typified by *M. oryzae* PTH11, an integral membrane protein required for pathogenicity. Sixty-one PTH11-related proteins were identified in *M. oryzae* (Kulkarni et al. 2005), and BlastP analyses in *B. cinerea* databases using PTH11 as query revealed 80 PTH11-like proteins. Two of them, named BTP1 and BTP2, were shown to lack obvious functions in virulence (Schulze Gronover et al. 2005; unpublished). Until today, no GPCR could be assigned to any heterotrimeric G protein by mutant analysis. Possibly, G proteins are primarily regulated by other factors, such as the ortholog of Ric8 that was recently demonstrated to function as GEF for Gα subunits in *N. crassa* (Wright et al. 2011) and/or by PhLP and RGS family proteins.

13.4 The Cyclic AMP (cAMP) Cascade

A prominent effector protein of heterotrimeric G proteins is the membrane-associated adenylate cyclase (AC). This enzyme is activated by interaction with stimulatory Gα subunits after ligand-binding to the GPCR, and forms the second messenger cAMP. The primary receptor protein for cAMP in eukaryotes is the cAMP-dependent protein kinase (PKA). In its inactive state, the PKA is a tetramer composed of two regulatory and two catalytic subunits. After binding of cAMP to the regulatory subunits, the tetramer dissociates into a dimer of regulatory subunits and two active monomeric catalytic subunits that phosphorylate effector proteins such as TFs and metabolic enzymes (reviewed in Sassone-Corsi 2012).

In common with other filamentous fungi, *B. cinerea* contains single genes encoding the adenylate cyclase (*Bac*) and the regulatory PKA subunit (*BcPkaR*), and two genes encoding catalytic PKA subunits (*BcPka1*, *BcPka2*). Deletion of *bac* results in reduced intracellular cAMP levels, and affects colony growth, conidial germination and virulence (Klimpel et al. 2002; Döhlemann et al. 2006). Active Gα1, Gα3 and the monomeric G protein BcRas2 elevate intracellular cAMP levels by interacting with Bac (Klimpel et al. 2002; Schumacher et al. 2008b). BcPka1 represents the major catalytic PKA subunit as its deletion leads to significant growth retardation, reduced virulence and increased cAMP levels while the deletion of the second subunit does not result in obvious phenotypes (Schumacher et al. 2008b). However, both catalytic PKA subunits may substitute each other to maintain a basal level of PKA activity. Both Δ*bac* and Δ*bcpka1* mutants exhibit similarly reduced virulence on bean leaves, which is in line with the hypothesis that AC and PKA act in the same signaling pathway. Besides, both mutants have specific phenotypes; the defects in sugar-induced germination and sclerotia formation are unique to Δ*bac* suggesting additional effector proteins of Bac and cAMP. Indeed, two additional proteins with cAMP-binding domains have been identified that consequently may function downstream of Bac and in parallel to the PKA (Amselem et al. 2011). *B. cinerea* mutants deleted for *bcpkaR* share identical phenotypes with Δ*bcpka1* mutants and lack PKA activity. This situation differs from that found in other fungal systems in which the deletion of the regulatory subunit results in constitutively active PKA signaling (Lee et al. 2003). So far, it is not clear by which mechanism BcPka1 activity is abolished in the absence of *BcPkaR*.

Amplitude and duration of cAMP signaling is modulated by phosphodiesterases (PDEs) that degrade the secondary messenger. Yeast and filamentous fungi contain two PDEs with unrelated primary sequences and different affinities for the substrate. BcPde1 and BcPde2 represent the low- and high-affinity forms, respectively, and exhibit different subcellular localizations (Harren et al. 2013). BcPde2, the main PDE, is required for proper regulation of colony growth, conidiation, sclerotia formation and virulence. Absence of BcPde1 has no effect; however, the overexpression of this enzyme in the Δ*bcpde2*-background and the more pronounced phenotype of ΔΔ*bcpde1/2* mutants suggest that both enzymes have redundant functions and that the low-affinity PDE may compensate for the loss of the high-affinity PDE. Unlike the situation in other fungal systems in which the loss of PDE activity leads to elevated cAMP levels and in turn to constitutively active PKA signaling, *B. cinerea* PDE loss-of-function mutants contain slightly reduced cAMP levels and reduced PKA activities (Harren et al. 2013). In sum, there are several evidences that the regulation of cAMP signaling in *B. cinerea* differs widely from that found in other fungal species.

13.5 Ca^{2+}-Mediated Signal Transduction

Phospholipase C (PLC) isoforms are further effector proteins of GPCR systems. In mammalian cells, Gα, Gβγ, or both bind to membrane-localized PLCs after the dissociation of the G protein has taken place. PLCs catalyze the hydrolysis of phosphatidylinositol-4,5-bisphosphate (PIP_2) to inositol-1,4,5-trisphosphate (IP_3) and diacylglycerol (DAG), which are important secondary messengers. At this step, the signaling pathway splits into two branches. IP_3 binds to and thereby opens Ca^{2+} channels in the ER membrane leading to a quick rise in intracellular Ca^{2+} that affects the activity of Ca^{2+}-dependent proteins (see below). DAG, the other cleavage product of PIP_2 remains embedded in the membrane and promotes there the activation of the Ca^{2+}-dependent protein kinase (PKC) after Ca^{2+}-induced translocation of the inactive PKC to the plasma membrane has taken place (reviewed in Rhee 2001). In common with mammalian systems, IP_3 is produced by PLCs in fungi resulting in increased cytosolic Ca^{2+} levels; however, the characteristic IP_3-binding Ca^{2+} channels are missing and the molecular targets of IP_3 are still unknown. Moreover, it turned out that in yeast and possibly in filamentous fungi as well, the vacuole and not the ER is the main storage organelle for Ca^{2+}. In yeast, Ca^{2+} is released from this cell compartment *via* the vacuolar Ca^{2+} channel Yvc1p in response to particular stimuli (reviewed in Cunningham 2011).

B. cinerea possesses two PLCs (BcPlc1, BcPlc2) while both *N. crassa* and *M. oryzae* contain four enzymes of this type. Deletion of *bcplc1* results in a pleiotropic phenotype including defects in conidial germination, hyphal branching, sclerotia formation and virulence. BcPlc1, Gα1 and the Ca^{2+}-dependent calcineurin phosphatase (CN) coordinately regulate gene expression indicating a crosstalk between heterotrimeric G protein- and Ca^{2+}-dependent signal transduction (Schumacher et al. 2008c). The function of BcPlc2 is still unknown.

It is crucial for the cell to maintain Ca^{2+} levels within optimal ranges in the cytosol and the organelles. In yeast, Ca^{2+} enters the cytosol and is in turn transported into the vacuoles through the Ca^{2+} pump Pmc1p and the Ca^{2+}/H^{+} exchanger Vcx1p, or into the Golgi/ER through the Ca^{2+} pump Pmr1p. Finally, excessive Ca^{2+} in the secretory pathway is extruded from the cell *via* exocytosis (reviewed in Cui et al. 2009). In contrast to yeast, that possesses single copies of the Ca^{2+}-transporting proteins, genomes of filamentous fungi contain multiple genes encoding putative orthologs of Pmr1p and Vcx1p (Amselem et al. 2011). Import of Ca^{2+} is mediated *via* high- (HACS) and low-affinity (LACS) Ca^{2+} influx systems. Two putative HACS components *i.e.*, the Ca^{2+} channel proteins BcMid1 and BcCch1 have been studied in *B. cinerea*: as in yeast, the proteins are required for facilitating growth in low environmental Ca^{2+} conditions (Harren and Tudzynski 2013).

Free Ca^{2+} ions act as cofactor for several enzymes including PLCs and PKCs, and as secondary messenger by binding to Ca^{2+}-sensitive (EF-hand containing) proteins, such as calmodulin (CaM) and neuronal Ca^{2+} sensor (NCS) proteins. Upon Ca^{2+} binding, CaM undergoes a conformational change enabling its interaction with effector proteins such as Ca^{2+}/CaM-dependent kinases (CMKs) and the calcineurin (CN) phosphatase. Inactive CN represents a heterodimer composed of a catalytic (CNA) and a regulatory subunit (CNB). When Ca^{2+}/CaM binds to CNA, the autoinhibitory domain is released from the active site enabling binding and dephosphorylation of CN substrates (Stie and Fox 2008).

FK506 and cyclosporine A (CsA) are naturally derived compounds and are used to prevent transplant rejection by blocking T-cell activation (immunosuppression) in patients. Both compounds act *via* interaction with peptidyl-prolyl isomerases (FKBP12 and cyclophilin A, respectively); subsequent binding of the drug–protein complexes to CN prevents its catalytic activity (reviewed in Cyert 2003). Deletions of the *B. cinerea* orthologs, named BcPic5/BcFkbp12 and Bcp1, result as expected, in FK506- or CsA-resistant mutants (Gioti et al. 2006; Meléndez et al. 2009; Viaud et al. 2003). Treatment of wild-type conidia with CsA does not affect conidial germination *per se* but results in hyper-branched hyphae and subsequent growth arrest. Comparative gene expression studies using CsA for inhibiting CN activity revealed 18 CN-dependent (*CND*) genes (Viaud et al. 2003). Mutants deleted for *BccnA* and therefore missing CN activity are "viable" but are severely affected in hyphal growth and related differentiation processes such as the formation of a hyphal network and reproductive structures (Harren et al. 2012). The phosphoprotein BcRcn1 has been identified as a positive regulator of CN activity: it interacts with BcCnA *via* a CN-docking motif, and is required for the expression of *bccnA* and CN-dependent genes (Harren et al. 2012). Phosphorylation of BcRcn1 is likely carried out by the glycogen synthase kinase 3 (BcGsk3), the ortholog of yeast Mck1p (Harren et al. unpublished data).

Even though CN subunits and regulatory proteins are highly conserved in yeast, fungi and mammals – they do not exist in plants – striking differences exist with regard to the CN effectors. In yeast and mammals, CN acts on TFs, which are localized in the cytosol when they are inactive. The dephosphorylated proteins are translocated to the nucleus where they induce gene expression. The PxIxIT motif for CN-binding is conserved in yeast and mammalian TFs and is present in almost all CN-regulated proteins. However, yeast Crz1 and human NFAT (nuclear factor of activated T cell) TFs are evolutionary not related (reviewed in Aramburu et al. 2004). BcCrz1 is the first CN-regulated TF that has been described in a filamentous fungus (Schumacher et al. 2008a). *Bccrz1* partially complements the yeast Δ*crz1* mutation, and the subcellular localization of the GFP-BcCrz1 fusion protein depends on Ca^{2+} in a CN-dependent fashion in both, *S. cerevisiae* and *B. cinerea*. BcCrz1 is involved in distinct processes including growth on poor media, hyphal morphology, stress response, cell wall and membrane integrity, and virulence. As expected for a component of the Ca^{2+}/CN signaling pathway, high environmental Ca^{2+} becomes toxic for Δ*bccrz1* as the transcriptional response, usually enabling the adaptation to this stress condition is blocked (Schumacher et al. 2008a; Schumacher 2012). Recently, genome-wide expression studies have been initiated to identify Ca^{2+}-induced genes (CIGs), and to study their fate in the absence of BcRcn1 or BcCrz1. Preliminary data support the hypothesis that more CN effector proteins apart from BcCrz1 exist: a number of Ca^{2+}-responsive TFs (CTFs) was identified whose induction does not require BcCrz1 (Harren et al. unpublished data).

13.6 Mitogen-Activated Protein Kinase (MAPK) Cascades

Mitogen-activated protein kinase (MAPK) cascades function as key signal transducers that use protein phosphorylation/dephosphorylation cycles to channel information. MAPK modules in all eukaryotes consist of three interlinked protein kinases that are sequentially activated: Activated MAP kinase kinase kinase (MAPKKK) phosphorylates two Ser and/or Thr residues in the activation loop of MAP kinase kinases (MAPKK). Activated MAPKK in turn triggers the activation of MAPKs *via* dual phosphorylation of conserved Thr and Tyr residues (Fig. 13.2).

Finally, activated MAPKs phosphorylate Ser/Thr residues of effector proteins in the cytosol and/or the nuclei to achieve specific output responses. Four MAPK modules are present in *S. cerevisiae* controlling mating response (Ste11p–Ste7p–Fus3p/Kss1p), filamentous growth (Ste11p–Ste7p–Kss1p), cell wall integrity and cell cycle (Bck1p–Mkk1p/Mkk2p–Slt2p), or osmoregulation and stress response (Ssk2p/Ssk22p–Pbs2p–Hog1p) (reviewed in Zhao et al. 2007).

Based on sequence comparisons, three MAPKs and their upstream kinases have been identified in *B. cinerea*. The Fus3p/Kss1p ortholog represents Bmp1, whose deletion causes defects in hydrophobicity-induced germination, delayed vegetative growth, reduced size of conidia, lack of sclerotia formation, and loss of pathogenic-

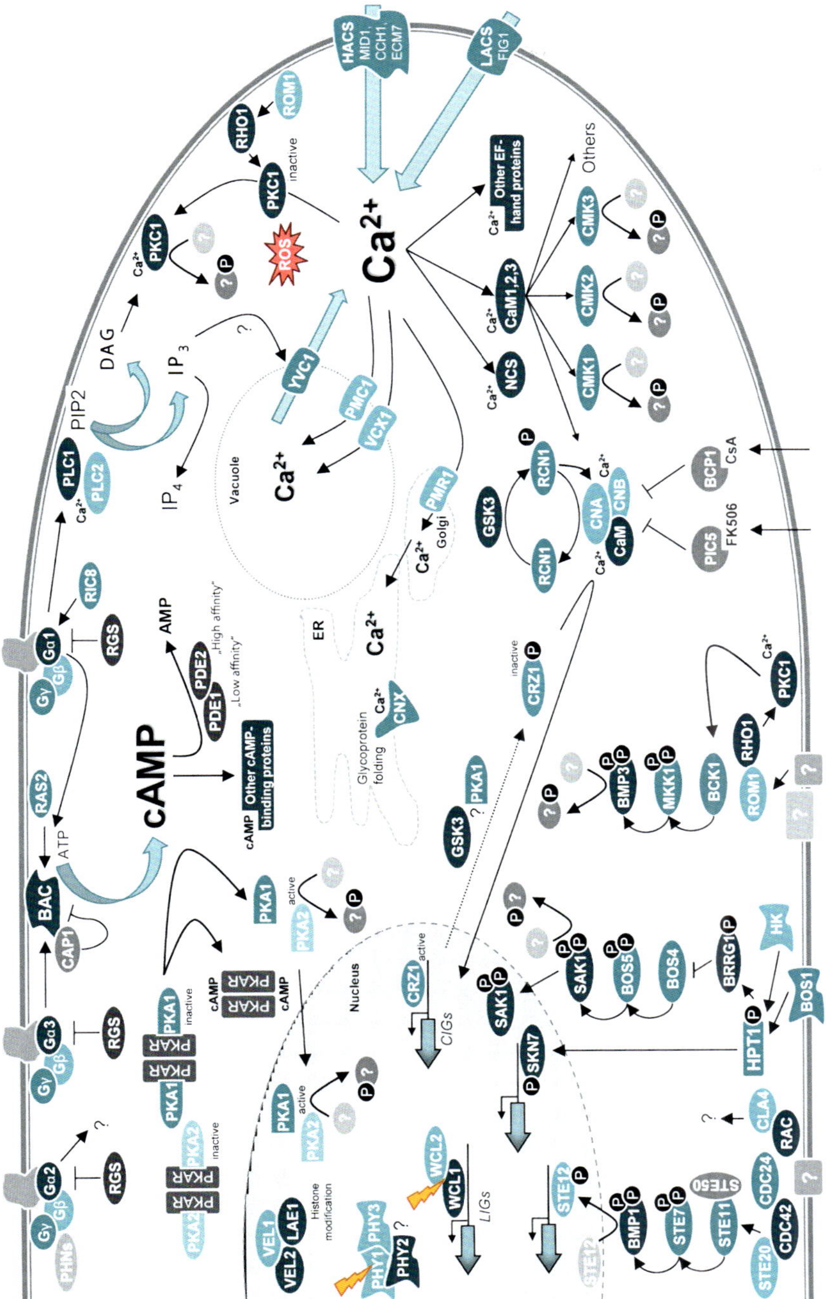

Fig. 13.2 The signal transduction network in *B. cinerea*. See text and Table 13.1 for explanation

ity. In line with the assumption that the components belong to the same MAPK module, the deletion mutants of the MAPKKK BcSte11, the MAPKK BcSte7 and the adaptor protein BcSte50 exhibit similar phenotypes as the Δ*bmp1* mutant. The homeobox/ Cys2His2-TF BcSte12 is considered as one of several effector proteins of the BcSte11–BcSte7–Bmp1 module; its deletion results in delayed infection due to low penetration efficiencies (in contrast to apathogenicity of the other mutants), loss of sclerotia formation and increased melanization (Schamber et al. 2010).

Bmp3 is orthologous to yeast Slt2p. Δ*bmp3* mutants exhibit a pleiotropic phenotype including a general growth defect that becomes more obvious under hypoosmotic conditions, increased susceptibility to oxidizing agents and fungicides, impaired conidiation and the lack of sclerotia formation. Moreover, the capacity of Δ*bmp3* to penetrate and to colonize the host tissue is significantly affected (Rui and Hahn 2007). To date, no further components of this MAPK module have been studied, but based on the high conservation of the MAPK components in fungi, BcBck1 and BcMkk1 are likely the upstream kinases of Bmp3.

The MAPK orthologous to yeast Hog1p is called Sak1 in *B. cinerea*, and is phosphorylated in response to oxidative and hyperosmotic stress as well as in response to treatment with fungicides and calcofluor white (CFW). Along with the observed action spectrum, Δ*bcsak1* mutants are hypersensitive to oxidative stress, hyperosmotic conditions and CFW treatment; however, the mutants exhibit wild-type-like sensitivity to fungicides (Segmüller et al. 2007; Liu et al. 2008, 2011). Interestingly, the subcellular localization of BcSak1 in response to the stimuli differs: while BcSak1 localizes to the nuclei rapidly after exposure to osmotic stress, it remains in the cytosol after exposure to oxidative stress suggesting that BcSak1 effector proteins in the cytosol/nuclei are selectively affected either in response to osmotic or oxidative stress (Heller et al. 2012). However, the mechanisms permitting the selective localization of BcSak1 are still unknown. The absence of BcSak1 affects furthermore the composition and the integrity of the cell wall (increased contents of β-1-3 glucan and chitin accompanied by decreased susceptibility to cell wall-degrading enzymes and cell wall-interfering agents) (Liu et al. 2011), and the accumulation of glycerol in response to fungicide and NaCl treatment (Liu et al. 2008). Unlike the other two MAPKs, BcSak1 is required for conidiation and dispensable for the formation of sclerotia. Δ*bcsak1* hyphae are unable to penetrate the intact plant tissue; however, after wounding the mutant colonizes the host tissue almost in a wild-type-like fashion (Segmüller et al. 2007). Bos4 (MAPKKK) and Bos5 (MAPKK) have been confirmed to act upstream of BcSak1 in the same module: along with Δ*bcsak1*, Δ*bos4* and Δ*bos5* mutants do not form conidia, and are affected in virulence and sensitivity to hyperosmotic stress. In addition, phosphorylation of BcSak1 does not occur in the mutant backgrounds (Yang et al. 2012; Yan et al. 2010).

Signaling cascades that involve phosphorylation *via* kinases require phosphatases to turn off signaling. As dephosphorylation of one of the two residues in MAPKs is already sufficient for their inactivation, MAPK activity may be modulated by Ser/Thr (PTCs), Tyr (PTPs) and dual-specificity phosphatases (DSPs) (reviewed in Martin et al. 2005). To date, four PTC- and two PTP-encoding genes have been characterized in *B. cinerea*. Unexpectedly, though BcPtpA, BcPtpB, and BcPtc1 are able to complement respective yeast deletion mutants by preventing the accumulation of phosphorylated Hog1p, the phosphorylation of Bmp3 and BcSak1 is not detectable or unchanged from wild type in respective *B. cinerea* deletion mutants (Yang et al. 2013b, c). This might be explained by the presence of other MAPK phosphatases that over-compensate the loss of these phosphatases in order to prevent the inappropriate accumulation of phosphorylated MAPKs.

13.7 Two-Component Signal Transduction

Two-component histidine kinase (HK) phosphorelay systems regulate many aspects of bacterial and fungal life including the responses to environmental stresses. The HK serves a sensory function and responds to a signal by catalyzing the autophosporylation of a conserved His residue. In a series of phosphotransfers, the phosphoryl group is ultimately transferred to a conserved Asp residue in the receiver domain of a response regulator (RR). Eukaryotic HKs are hybrids combining HK and RR in single proteins, and require additional phosphotransfer through a His phosphotransfer (HPT) protein and a second RR (reviewed in Fassler and West 2013).

B. cinerea, like yeast and other filamentous fungi, contains two RRs, named Brrg1 and BcSkn7, and a single HPT. However, *B. cinerea* possesses more HK-encoding genes (at least 20) than *S. cerevisiae* (1), *N. crassa* (11) and *Fusarium verticillioides* (16) (Catlett et al. 2003). The HKs share similar C-terminal regions with the essential domains for phosphotransfer, while the N-terminal regions are variable by housing protein domains for sensing different signals. Typical are GAF and PAS/PAC domains that allow for binding of cyclic GMP or chromophores or to sense the redox potential, cellular oxygen or light (reviewed in Catlett et al. 2003). Transmembrane domains are only present in a small number of HKs suggesting that HKs may also localize to the cytosol or the nuclei. Taken together, the presence of a high number of HKs with different sensing properties and a single phosphorelay step via HPT – turning *bchpt1* likely into an essential gene – allows the integration of multiple input signals into a single response. In *B. cinerea* as well as in other filamentous fungi the knowledge of HK functions is limited. To date, three HK-encoding genes have been studied in *B. cinerea*: deletion mutants of *bchk1* and *bchk5* lack obvious phenotypes (Table 13.1), while it turned out that Bos1 fulfills a similar function as Sln1p in yeast. Deletion of *bos1* results in the constitutive activation of the BcSak1 module accompanied by increased sensitivity to hyperosmotic conditions, oxidants, decreased sensitivity to fungicides, impaired virulence and loss of conidiation (Viaud et al. 2006; Liu et al. 2008).

Brrg1, the ortholog of Ssk1p, is required for MAPK activation, as BcSak1 is not phosphorylated in Δ*brrg1* mutants (Yan et al. 2011). Based on the fact that Bos1 regulates certain phenotypes such as fungicide susceptibility, superoxide tolerance, adaptation and conidiation under high osmolarity independently of BcSak1, an additional effector pathway of the HK is assumed (Liu et al. 2008). Recently, it was shown, as suggested by Tanaka and Izumitsu (2010), that the second RR BcSkn7 is involved in the osmotic stress response pathway downstream of Bos1 (Viefhues et al. 2015; Yang et al. 2015).

Table 13.1 List of functionally characterized signaling components in *B. cinerea*

Name	Description	D	V	References
BcRas1	RasGTPase	+	+	Minz Dub et al. (2013)
BcRas2	RasGTPase	+	+	Schumacher et al. (2008b)
BcCdc42	RhoGTPase	+	+	Kokkelink et al. (2011)
BcRac	RhoGTPase	+	+	Minz Dub et al. (2013)
BcRho3	RhoGTPase	+	+	An et al. (2015)
BcCdc24	RhoGEF	+	?	Giesbert et al. (2014)
BcBem1	Scaffold protein	+	+	Giesbert et al. (2014)
BcFar1	Scaffold protein	–	–	Giesbert et al. (2014)
BcSte20	p21-activated kinase	+	–	Giesbert and Tudzynski unpublished
BcCla4	p21-activated kinase	?	?	Minz Dub and Sharon unpublished
Bcg1	Gα subunit	+	+	Schulze Gronover et al. (2001)
Bcg2	Gα subunit	–	–	Schulze Gronover et al. (2001)
Bcg3	Gα subunit	+	+	Döhlemann et al. (2006)
BcGb1	Gβ subunit	+	+	Schumacher and Tudzynski unpublished
BcGg1	Gγ subunit	+	+	Strotbaum and Schumacher unpublished
BcPhnA	Phosducin	+	+	Kilani et al. unpublished
BcRgs1	Regulator of G protein	+	+	Choquer et al. unpub.
BcGpr1	GPCR, Git3-like	–	–	Strotbaum and Schumacher unpublished
BcGpr2	GPCR, Git3-like	–	–	Strotbaum and Schumacher unpublished
BcGpr3	GPCR, secretin-like	–	–	Strotbaum and. Schumacher unpublished
BcGpr4	GPCR, secretin-like	–	–	Strotbaum and Schumacher unpublished
BcGpr5	GPCR, family 2-like	–	–	Strotbaum and Schumacher unpublished
Bop1	GPCR, opsin	–	–	Heller et al. (2012)
Btp1	7-TM protein	–	–	Schulze Gronover et al. (2005)
Btp2	7-TM protein	–	–	Schumacher and Tudzynski unpublished
Bac	Adenylate cyclase	+	+	Klimpel et al. (2002)
BcPka1	Catalytic PKA subunit	+	+	Schumacher et al. (2008b)
BcPka2	Catalytic PKA subunit	–	–	Schumacher et al. (2008b)
BcPkaR	Regulatory PKA subunit	+	+	Schumacher et al. (2008b)
BcPde1	cAMP phosphodiesterase	–	–	Harren et al. (2013)
BcPde2	cAMP phosphodiesterase	+	+	Harren et al. (2013)
Bc1g01623	cAMP-binding protein	–	–	Harren and Tudzynski unpublished
BcPlc1	Phospholipase C	+	+	Schumacher et al. (2008c)
BcCnA	Catalytic CN subunit PP2B	+	+	Harren et al. (2012)
BcRcn1	Regulator of CN	+	+	Harren et al. (2012)

(continued)

Table 13.1 (continued)

Name	Description	D	V	References
BcGsk3	Glycogen synthase kinase 3	+	+	Harren and Tudzynski unpublished
Bcp1[a]	CsA-binding protein	+	+	Viaud et al. (2003)
BcFkbp12	FK506-binding protein	–	–	Gioti et al. (2008) and Meléndez et al. (2009)
BcMid1	Ca^{2+} channel protein	–	–	Harren and Tudzynski (2013)
BcCch1	Ca^{2+} channel protein	–	–	Harren and Tudzynski (2013)
Bmp1	MAPK (Fus3p/Kss1p-like)	+	+	Zheng et al. (2000) and Döhlemann et al. (2006)
BcSte7	MAPKK [*ad* BMP1]	+	+	Schamber et al. (2010)
BcSte11	MAPKKK [*ad* BMP1]	+	+	Schamber et al. (2010)
BcSte50	Adaptor protein [*ad* BMP1]	+	+	Schamber et al. (2010)
BcBic5	BMP1-interacting protein	–	–	Müller et al. unpublished
BcBic8	BMP1-interacting protein	–	–	Müller et al. unpublished
BcMsb2	Sensor protein [*ad* BMP1]	–	+	Leroch et al. (2015)
BcSho1	Sensor protein	–	+	Leroch and Hahn unpublished
BcSak1	MAPK (Hog1p-like)	+	+	Segmüller et al. (2007) and Heller et al. (2012)
Bos5[a]	MAPKK [*ad* BcSAK1]	+	+	Yan et al. (2010)
Bos4[a]	MAPKKK [*ad* BcSAK1]	+	+	Yang et al. (2012)
Bmp3	MAPK (Slt2p-like)	+	+	Rui and Hahn (2007)
BcPtc1[a]	Protein phosphatase 2C	+	+	Yang et al. (2013b)
BcPtc3[a]	Protein phosphatase 2C	+	+	Yang et al. (2013b)
BcPtc5[a]	Protein phosphatase 2C	–	–	Yang et al. (2013b)
BcPtc6[a]	Protein phosphatase 2C	–	–	Yang et al. (2013b)
BcPtpA[a]	Protein phosphatase PTP	+	+	Yang et al. (2013c)
BcPtpB[a]	Protein phosphatase PTP	+	+	Yang et al. (2013c)
Bos1	Histidine kinase	+	+	Viaud et al. (2006) and Liu et al. (2008)
BcHk1	Histidine kinase	–	–	Temme and Tudzynski unpublished
BcHk5	Histidine kinase	–	–	Cuesta Arenas and van Kan unpublished
Brrg1[a]	Response regulator	+	+	Yan et al. (2011)
BcSkn7	Response regulator	+	+	Viefhues et al. (2015)
BcSnf1	Snf1p-like Ser/Thr kinase	+	+	Schumacher and Tudzynski unpublished
Bpk2	Sch9p-like Ser/Thr kinase	–	–	Schulze Gronover and Tudzynski unpub.
Bpk3	Atg1p-like Ser/Thr kinase	–	–	Schulze Gronover and Tudzynski unpub.
Bpk4	Sks1p-like Ser/Thr kinase	–	–	Schulze Gronover and Tudzynski unpub.
BcPom1	Protein kinase	+	+	Kilani et al. unpublished
BcPp2A	Protein phosphatase 2A	+	+	Giesbert et al. (2012)

(continued)

Table 13.1 (continued)

Name	Description	D	V	References
BcNoxA	Catalytic NOX subunit A	+	+	Segmüller et al. (2008) and Roca et al. (2012)
BcNoxB	Catalytic NOX subunit B	+	+	Segmüller et al. (2008) and Roca et al. (2012)
BcNoxR	Regulatory NOX subunit	+	+	Segmüller et al. (2008) and Roca et al. (2012)
BcPls1	Tetraspanin	+	+	Gourgues et al. (2004) and Siegmund et al. (2013)
BcNoxD	ER protein	+	+	Siegmund et al. (2014)
BcVel1	Velvet protein (VeA-like)	+	+	Schumacher et al. (2012)
BcVel2[a]	Velvet protein (VelB-like)	+	+	Yang et al. (2013a)
BcLae1	Methyltransferase	+	+	Schumacher et al. (2015)
BcVvd1	Blue light sensor	+	?	Cohrs and Schumacher unpublished
BcWcl1	GATA-TF	+	+	Canessa et al. (2013)
BcLtf1	GATA-TF	+	+	Schumacher et al. (2014)
BcLtf2	Cys2His2-TF	+	?	Cohrs and Schumacher unpublished
BcLtf3	Cys2His2-TF	+	?	Cohrs and Schumacher unpublished
BcCrz1	Cys2His2-TF [*ad* BcCnA]	+	+	Schumacher et al. (2008a)
BcReg1	Transcriptional regulator	+	+	Michielse et al. (2011)
BcAtf1	bZIP-TF	+	+	Temme et al. (2012)
Bap1	bZIP-TF	–	–	Temme and Tudzynski (2009)
BcBip1	bZIP-TF	–	–	Choquer unpublished
BcYoh1	Cys2His2-TF	–	–	Simon et al. (2013)
BcSpt3	THIID-TF	+	+	Giesbert et al. (2012)
BcPro1	Zn2Cys6-TF	+	+	Marschall and Tudzynski unpublished
BcRum1	Zn2Cys6-TF	–	–	Leroch et al. (2013)
BcMrr1	Zn2Cys6-TF	?	?	Kretschmer et al. (2009)
BcPacC	Zn2Cys6-TF	+	+	Poussereau and Rascle unpublished
BcNaf1	Zn2Cys6-TF	–	–	Poussereau et al. unpub.
BcStf1	Zn2Cys6-TF	–	–	Poussereau and Kaiser unpublished
B0510_6651	Zn2Cys6-TF [*ad* BMP1]	–	–	Leroch et al. (2013)
BcCTF1α	Zn2Cys6-TF	+	–	Leroch and Hahn unpublished
B0510_1535	Zn2Cys6-TF	–	–	Müller and Hahn unpublished
B0510_8578	Zn2Cys6-TF	–	–	Müller and Hahn unpublished
B0510_602	Zn2Cys6-TF	–	–	Müller and Hahn unpublished
B0510_8863	Zn2Cys6-TF	–	–	Müller and Hahn, unpublished
BcXlr1	Zn2Cys6-TF	–	–	Schumacher unpublished
BcSte12	Homeobox-TF [*ad* BMP1]	+	+	Schamber et al. (2010)
BcHox8	Homeobox-TF	+	+	Antal et al. (2012)

D differentiation (includes all tested growth characteristics), *V* virulence altered in respective deletion mutant (+), not altered (–)

[a]Deletion done in other recipient strain than B05.10

13.8 Light Regulation

Filamentous fungi, unlike yeasts, possess several proteins (photoreceptors) that use chromophores to detect different wavelengths. Phytochromes are HKs using bilin to absorb red/far-red light; opsins are transmembrane proteins (GPCRs) that may use retinal to detect green light, and binding of flavin (*via* LOV or FAD-binding domains) allows for absorption of near-UV and blue light. Cryptochromes are FAD-binding proteins that may additionally exhibit photolyase activity (reviewed in Rodriguez-Romero et al. 2010).

Since 100 years it is known that light represents an important environmental cue for *B. cinerea*. Studies described its impact on morphogenesis by favoring conidiation and suppressing sclerotial development, and reported on tropic responses of conidiophores, conidial germ tubes and apothecia (reviewed in Jarvis 1977). However, confusions have been caused by the isolation and investigation of *B. cinerea* strains exhibiting altered or no responses to light: those strains produced either conidia or sclerotia under the same light conditions (constant darkness or light), or even failed to produce any reproductive structures (reviewed in Jarvis 1977; Lorbeer 1980). In the 1970s it was demonstrated that *B. cinerea* responds to near-UV, blue, red and far-red light. According to these observations, a "Two-receptor-model" was postulated in which near-UV/blue- and red/far-red-reversible photoreceptors ("mycochromes") are closely interacting to regulate asexual reproduction (reviewed in Epton and Richmond 1980). Along with the responses to different wavelengths, *B. cinerea* possesses near-UV-sensing cryptochromes (BcCry1, 2), potential blue light sensors (BcWcl1, BcVvd1, BcLov3, BcRgs5), opsins (Bop1, 2) as well as red/far-red light-sensing phytochromes (BcPhy1-3) (Schumacher and Tudzynski 2012). Strain B05.10 forms predominantly conidia in short-wave light (near-UV, blue) and sclerotia in long-wave light (yellow, red). However, during exposure to blue light fewer conidia and instead more aerial hyphae are formed which is in accordance with previous reports describing the capacity of blue light to inhibit conidiation and to cause the de-differentiation of already developing conidiophore and sclerotial initials (reviewed in Epton and Richmond 1980). The exposure of undifferentiated surface-grown mycelia of strain B05.10 to white light (60 min) affected the expression levels of 293 genes, separated into 249 light-induced (LIGs) and 49 light-repressed genes (LRGs). To the group of LIGs belong genes involved in photoperception, oxidative stress response, and transcription (six light-responsive TFs). Members of the latter group (BcLtf1, BcLtf2, BcLtf3) regulate the differentiation of conidiophores and conidia in the light (Schumacher et al. 2014; unpublished).

To get first access to the light signaling machinery in *B. cinerea*, the study of the *N. crassa* WHITE COLLAR (WC) orthologs has been initiated. BcWcl1 and BcWcl2 are GATA-type TFs that physically interact in the nuclei forming the WC complex (WCC) (Schumacher 2012). Deletion of *bcwcl1* results in mutants exhibiting persistent conidiation while the overexpression of the WCC prevents conidiation accompanied by increased aerial hyphae formation in the light. Additionally, the WCC is required for achieving full virulence in the light. As in other fungi, gene expression in response to white light involves the WCC; however, induction either depends totally or only partially on the WCC in *B. cinerea*. Shared targets of the WCC in *N. crassa* and *B. cinerea* include genes encoding photoreceptors, carotenogenic enzymes, and proteins putatively involved in running a circadian clock. Striking differences are observed for TFs reported to act downstream of the WCC in *N. crassa*, suggesting that signaling events downstream of and/or in parallel with the WCC differ in the two fungi (Canessa et al. 2013). Given the properties of other wavelengths to affect morphogenesis in *B. cinerea*, further photoreceptors may contribute to drive white light-responsive gene expression.

However, it is evident that the VELVET complex – not present in yeast – consisting of the VELVET-domain proteins BcVel1, BcVel2 and the putative methyltransferase BcLae1, is required for linking light signals with differentiation programs and secondary metabolism in *B. cinerea*, analogous to the situation in other filamentous fungi (Bayram and Braus 2012). Deletion of BcVel1 that acts as a scaffold by interacting with both, BcVel2 and BcLae1 (Schumacher et al. 2015), results in light-independent conidiation accompanied by the lack of sclerotial development, loss of oxalic acid production and reduced virulence. These phenotypes correspond to those observed for the wild strains T4 and 1750; and indeed, these strains carry single nucleotide polymorphisms (SNPs) in *bcvel1* leading to nonfunctional BcVel1 proteins (Schumacher et al. 2012, 2013).

13.9 Transcription Factors (TFs) as Ultimate Effector Proteins

TFs are crucial players in signal transduction pathways, being the link between signal input/flow and signaling output (alteration of the gene expression). The functionality of TFs e.g. whether they induce or repress transcription depends on many parameters, and their involvement in a particular signaling pathway is sometimes difficult to predict. As effector proteins of more than one signaling cascade, TFs are furthermore prominent candidates to interconnect different pathways.

The genome of *B. cinerea* contains 406 genes encoding TFs of distinct families (Simon et al. 2013), and only a small number of them have been functionally characterized to date. Generally, TFs are activated/inactivated by posttranslational modifications. For instance, kinases/phosphatases may define the phosphorylation status of the TF and thereby its subcellular localization and/or transcriptional activity as it is the case for BcCrz1 (Schumacher 2012). Alternatively, the activity of a TF may be modulated by direct oxidation of Cys residues; an example is Bap1, which induces gene expression in response to oxidative stress (Temme and Tudzynski 2009). TFs may also work as nuclear receptors; here, the binding of a ligand results in a conformational change that allows the TF to bind to respective promoters. An example is BcMrr1, which triggers gene expression in response to fungicides (Kretschmer et al. 2009). Another example represents BcWcl1, which associates with BcWcl2 in response to light turning on light-dependent gene expression (Canessa et al. 2013). However, TFs themselves may (additionally) underlie transcriptional regulation; prominent examples in *B. cinerea* are the light-responsive and the recently identified Ca^{2+}-responsive TFs (Schumacher et al. 2014; Harren et al. unpublished).

13.10 Conclusion

In the last decade, much new knowledge regarding signal transduction in *B. cinerea* has been gained from targeted mutation approaches. However, this is still the beginning; the identification of crosstalks between the conserved pathways, the fungal-specific light signaling machinery and ROS (reactive oxygen species)-regulated signaling will be the challenge for the future. The latter pathway, that exploits ROS as additional second messengers, is introduced and discussed in detail by U. Siegmund and A. Viefhues in Chap. 14.

Acknowledgments I am grateful to B. Tudzynski, P. Tudzynski, U. Siegmund, A. Viefhues and K. Beckervordersandfort for fruitful discussions, and M. Leroch, M. Hahn, A. Sharon, N. Poussereau, M. Choquer, C. Bruel and S. Fillinger for sharing unpublished data. Financial support from the German Research Foundation (DFG, grant SCHU 2833/2-1) is gratefully acknowledged.

References

Amselem J, Cuomo CA, Van Kan JA et al (2011) Genomic analysis of the necrotrophic fungal pathogens *Sclerotinia sclerotiorum* and *Botrytis cinerea*. PLoS Genet 7:e1002230

An B, Li B, Qin G et al (2015) Function of small GTPase Rho3 in regulating growth, conidiation and virulence of *Botrytis cinerea*. Fungal Genet Biol 75:46–55

Antal Z, Rascle C, Cimerman A et al (2012) The homeobox BcHOX8 gene in *Botrytis cinerea* regulates vegetative growth and morphology. PLoS One 7:e48134

Aramburu J, Heitman J, Crabtree GR (2004) Calcineurin: a central controller of signalling in eukaryotes. EMBO Rep 5:343–348

Bayram O, Braus GH (2012) Coordination of secondary metabolism and development in fungi: the velvet family of regulatory proteins. FEMS Microbiol Rev 36:1–24

Cabrera-Vera TM, Vanhauwe J, Thomas TO et al (2003) Insights into G protein structure, function and regulation. Endocr Rev 24:765–781

Canessa P, Schumacher J, Hevia MA et al (2013) Assessing the effects of light on differentiation and virulence of the plant pathogen *Botrytis cinerea*: characterization of the white collar complex. PLoS One 8:e84223

Catlett NL, Yoder OC, Turgeon BG (2003) Whole-genome analysis of two-component signal transduction genes in fungal pathogens. Eukaryot Cell 2:1151–1161

Cui J, Kaandorp JA, Sloot PM et al (2009) Calcium homeostasis and signaling in yeast cells and cardiac myocytes. FEMS Yeast Res 9:1137–1147

Cunningham KW (2011) Acidic calcium stores of *Saccharomyces cerevisiae*. Cell Calcium 50:129–138

Cyert MS (2003) Calcineurin signaling in *Saccharomyces cerevisiae*: how yeast go crazy in response to stress. Biochem Biophys Res Commun 311:1143–1150

Döhlemann G, Berndt P, Hahn M (2006) Different signalling pathways involving a Galpha protein, cAMP and a MAP kinase control germination of *Botrytis cinerea* conidia. Mol Microbiol 59:821–836

Epton HAS, Richmond DV (1980) Formation, structure and germination of conidia. In: Coley-Smith JR, Verhoeff K, Jarvis WR (eds) The biology of *Botrytis*. Academic, London, pp 41–83

Fassler JS, West AH (2013) Histidine phosphotransfer proteins in fungal two-component signal transduction pathways. Eukaryot Cell 12:1052–1060

Giesbert S, Schumacher J, Kupas V et al (2012) Identification of pathogenesis-associated genes by T-DNA-mediated insertional mutagenesis in *Botrytis cinerea*: a type 2A protein phosphatase and a SPT3 transcription factor have significant impact on virulence. Mol Plant Microbe Interact 25:481–495

Giesbert S, Siegmund U, Schumacher J et al (2014) Functional analysis of BcBem1 and its interaction partners in *B. cinerea*: impact on differentiation and virulence. Plos ONE 9:e95172

Gioti A, Simon A, Le Pêcheur P et al (2006) Expression profiling of *Botrytis cinerea* genes identifies three patterns of up-regulation *in planta* and an FKBP12 protein affecting pathogenicity. J Mol Biol 358:372–386

Gioti A, Pradier JM, Fournier E et al (2008) A *Botrytis cinerea* emopamil binding domain protein, required for full virulence, belongs to a eukaryotic superfamily which has expanded in euascomycetes. Eukaryot Cell 7:368–378

Gourgues M, Brunet-Simon A, Lebrun MH et al (2004) The tetraspanin BcPls1 is required for appressorium-mediated penetration of *Botrytis cinerea* into host plant leaves. Mol Microbiol 51:619–629

Harren K, Tudzynski B (2013) Cch1 and Mid1 are functionally required for vegetative growth under low-calcium conditions in the phytopathogenic ascomycete *Botrytis cinerea*. Eukaryot Cell 12:712–724

Harren K, Schumacher J, Tudzynski B (2012) The Ca^{2+}/calcineurin-dependent signaling pathway in the gray mold *Botrytis cinerea*: the role of calcipressin in modulating calcineurin activity. PLoS One 7:e41761

Harren K, Brandhoff B, Knödler M et al (2013) The high-affinity phosphodiesterase BcPde2 has impact on growth, differentiation and virulence of the phytopathogenic ascomycete *Botrytis cinerea*. PLoS One 8:e78525

Heller J, Ruhnke N, Espino J et al (2012) The MAP kinase BcSak1 of *Botrytis cinerea* is required for pathogenic development and has broad regulatory functions beyond stress response. Mol Plant Microbe Interact 25:802–816

Jarvis WR (1977) *Botryotinia* and *Botrytis* species: taxonomy, physiology and pathogenicity. Research Branch, Canada Department of Agriculture, Ottawa

Klimpel A, Schulze Gronover C, Williamson B et al (2002) The adenylate cyclase (BAC) in *Botrytis cinerea* is required for pathogenicity. Mol Plant Pathol 3:439–450

Kokkelink L, Minz A, Al-Masri M et al (2011) The small GTPase BcCdc42 affects nuclear division, germination and virulence of the gray mold fungus *Botrytis cinerea*. Fungal Genet Biol 48:1012–1019

Kretschmer M, Leroch M, Mosbach A et al (2009) Fungicide-driven evolution and molecular basis of multidrug resistance in field populations of the grey mould fungus *Botrytis cinerea*. PLoS Pathog 5:e1000696

Kulkarni RD, Thon MR, Pan H et al (2005) Novel G-protein-coupled receptor-like proteins in the plant pathogenic fungus *Magnaporthe grisea*. Genome Biol 6:R24

Lee N, D'Souza CA, Kronstad JW (2003) Of smuts, blasts, mildews, and blights: cAMP signaling in phytopathogenic fungi. Annu Rev Phytopathol 41:399–427

Leroch M, Kleber A, Silva E et al (2013) Transcriptome profiling of *Botrytis cinerea* conidial germination reveals upregulation of infection-related genes during the prepenetration stage. Eukaryot Cell 12:614–626

Leroch M, Mueller N, Hinsenkamp I et al (2015) The signaling mucin Msb2 regulates surface sensing and host penetration via BMP1-MAP kinase signaling in *Botrytis cinerea*. Mol Plant Pathol 16:787–798, in press. doi:10.1111/mpp.12234

Li L, Wright SJ, Krystofova S et al (2007) Heterotrimeric G protein signaling in filamentous fungi. Annu Rev Microbiol 61:423–452

Liu W, Leroux P, Fillinger S (2008) The HOG1-like MAP kinase Sak1 of *Botrytis cinerea* is negatively regulated by the upstream histidine kinase Bos1 and is not involved in dicar-boximide- and phenylpyrrole-resistance. Fungal Genet Biol 45:1062–1074

Liu W, Soulié MC, Perrino C et al (2011) The osmosensing signal transduction pathway from *Botrytis cinerea* regulates cell wall integrity and MAP kinase pathways control melanin biosynthesis with influence of light. Fungal Genet Biol 48:377–387

Lorbeer JW (1980) Variation in *Botrytis* and *Botryotinia*. In: Coley-Smith JR, Verhoeff K, Jarvis WR (eds) The biology of *Botrytis*. Academic, London, pp 19–39

Martín H, Flández M, Nombela C et al (2005) Protein phosphatases in MAPK signalling: we keep learning from yeast. Mol Microbiol 58:6–16

Meléndez HG, Billon-Grand G, Fèvre M et al (2009) Role of the *Botrytis cinerea* FKBP12 ortholog in pathogenic development and in sulfur regulation. Fungal Genet Biol 46:308–320

Michielse CB, Becker M, Heller J et al (2011) The *Botrytis cinerea* Reg1, a putative transcriptional regulator, is required for pathogenicity, conidiogenesis, and the production of secondary metabolites. Mol Plant Microbe Interact 224:1074–1085

Minz Dub A, Kokkelink L, Tudzynski B et al (2013) Involvement of *Botrytis cinerea* small GTPases BcRAS1 and BcRAC in differentiation, virulence, and the cell cycle. Eukaryot Cell 12:1609–1618

Nanni V, Schumacher J, Giacomelli L et al (2014) Vv-AMP2, a grapevine flower specific defensin capable of *Botrytis cinerea* growth inhibition: insights into its mode of action. Plant Pathol 63:899–910

Perez P, Rincón SA (2010) Rho GTPases: regulation of cell polarity and growth in yeasts. Biochem J 426:243–253

Rhee SG (2001) Regulation of phosphoinositide-specific phospholipase C. Annu Rev Biochem 70:281–312

Roca MG, Weichert M, Siegmund U et al (2012) Germling fusion via conidial anastomosis tubes in the grey mould *B. cinerea* requires NADPH oxidase activity. Fungal Biol 116:379–387

Rodriguez-Romero J, Hedtke M, Kastner C et al (2010) Fungi, hidden in soil or up in the air: light makes a difference. Annu Rev Microbiol 64:585–610

Rui O, Hahn M (2007) The Slt2-type MAP kinase Bmp3 of *Botrytis cinerea* is required for normal saprotrophic growth, conidiation, plant surface sensing and host tissue colonization. Mol Plant Pathol 8:173–184

Sassone-Corsi P (2012) The cyclic AMP pathway. Cold Spring Harb Perspect Biol 4:a011148

Schamber A, Leroch M, Diwo J et al (2010) The role of mitogen-activated protein (MAP) kinase signalling components and the Ste12 transcription factor in germination and pathogenicity of *Botrytis cinerea*. Mol Plant Pathol 11:105–119

Schulze Gronover C, Kasulke D, Tudzynski P et al (2001) The role of G protein alpha subunits in the infection process of *Botrytis cinerea*. Mol Plant Microbe Interact 14:1293–1302

Schulze Gronover C, Schorn C, Tudzynski B (2004) Identification of *Botrytis cinerea* genes up-regulated during infection and controlled by the Galpha subunit BCG1 using suppression subtractive hybridization (SSH). Mol Plant Microbe Interact 17:537–546

Schulze Gronover C, Schumacher J, Hantsch P et al (2005) A novel seven-helix transmembrane protein BTP1 of *Botrytis cinerea* controls the expression of GST-encoding genes, but is not essential for pathogenicity. Mol Plant Pathol 6:243–256

Schumacher J (2012) Tools for *Botrytis cinerea*: new expression vectors make the gray mold fungus more accessible to cell biology approaches. Fungal Genet Biol 49:483–497

Schumacher J, Tudzynski P (2012) Morphogenesis and infection in *Botrytis cinerea*. In: Pérez-Martín J, Di Pietro A (eds) Topics in current genetics, vol. Morphogenesis and pathogenicity in fungi. Springer, Berlin, pp 225–241

Schumacher J, de Larrinoa IF, Tudzynski B (2008a) Calcineurin-responsive zinc finger transcription factor CRZ1 of *Botrytis cinerea* is required for growth, development, and full virulence on bean plants. Eukaryot Cell 7:584–601
Schumacher J, Kokkelink L, Huesmann C et al (2008b) The cAMP-dependent signalling pathway and its role in conidial germination, growth and virulence of the grey mould fungus *Botrytis cinerea*. Mol Plant Microbe Interact 21:1443–1459
Schumacher J, Viaud M, Simon A et al (2008c) The Gα subunit BCG1, the phospholipase C (BcPLC1) and the calcineurin phosphatase co-ordinately regulate gene expression in the grey mould fungus *Botrytis cinerea*. Mol Microbiol 67:1027–1250
Schumacher J, Pradier JM, Simon A et al (2012) Natural variation in the VELVET gene *bcvel1* affects virulence and light-dependent differentiation in *Botrytis cinerea*. PLoS One 7:e47840
Schumacher J, Gautier A, Morgant G et al (2013) A functional bikaverin biosynthesis gene cluster in rare strains of *B. cinerea* is positively controlled by VELVET. PLoS One 8:e53729
Schumacher J, Simon A, Cohrs KC et al (2014) The transcription factor BcLTF1 regulates virulence and light responses in the necrotrophic plant pathogen *Botrytis cinerea*. PLoS Genet 10:e1004040
Schumacher J, Simon A, Cohrs KC et al (2015) The VELVET complex in the gray mold fungus *Botrytis cinerea*: impact of BcLAE1 on differentiation, secondary metabolism and virulence. Mol Plant Microbe Interact 28:659–674
Segmüller N, Ellendorf U, Tudzynski B et al (2007) BcSAK1, a stress-activated mitogen-activated protein kinase, is involved in vegetative differentiation and pathogenicity in *Botrytis cinerea*. Eukaryot Cell 6:211–221
Segmüller N, Kokkelink L, Giesbert S et al (2008) NADPH oxidases are involved in differentiation and pathogenicity in *Botrytis cinerea*. Mol Plant Microbe Interact 21:808–819
Siegmund U, Heller J, Van Kan JA et al (2013) The NADPH oxidase complexes in *B. cinerea*: evidence for a close association with the ER and the tetraspanin Pls1. PLoS One 8:e55879
Siegmund U, Marschall R, Tudzynski P (2014) BcNoxD, a putative ER protein, is a new component of the NADPH oxidase complex in *Botrytis cinerea*. Mol Microbiol 95:988–1005
Simon A, Dalmais B, Morgant G et al (2013) Screening of a *Botrytis cinerea* one-hybrid library reveals a Cys2His2 transcription factor involved in the regulation of secondary metabolism gene clusters. Fungal Genet Biol 52:9–19
Stie J, Fox D (2008) Calcineurin regulation in fungi and beyond. Eukaryot Cell 7:177–186
Tanaka C, Izumitsu K (2010) Two-component signaling system in filamentous fungi and the mode of action of dicarboximide and phenylpyrrol fungicides. In: Carisse O (ed) Fungicides, vol I. InTech, Rijeka, pp 523–538
Temme N, Tudzynski P (2009) Does *Botrytis cinerea* ignore H_2O_2-induced oxidative stress during infection? Characterization of *Botrytis* activator protein 1. Mol Plant Microbe Interact 22:987–998
Temme N, Oeser B, Massaroli M et al (2012) BcAtf1, a global regulator, controls various differentiation processes and toxin production in *Botrytis cinerea*. Mol Plant Pathol 13:704–718
Viaud M, Brunet-Simon A, Brygoo Y et al (2003) Cyclophilin A and calcineurin functions investigated by gene inactivation, cyclosporin A inhibition and cDNA arrays approaches in the phytopathogenic fungus *Botrytis cinerea*. Mol Microbiol 50:1451–1465
Viaud M, Fillinger S, Liu W et al (2006) A class III histidine kinase acts as a novel virulence factor in *Botrytis cinerea*. Mol Plant Microbe Interact 19:1042–1050
Viefhues A, Schlathoelter I, Simon A et al (2015) Unravelling the function of the response regulator BcSkn7 in the stress signaling network of *Botrytis cinerea*. Eukaryot Cell 14:636–651, in press. doi:10.1128/EC.00043-15
Wang Y, Geng Z, Jiang D et al (2013) Characterizations and functions of regulator of G protein signaling (RGS) in fungi. Appl Microbiol Biotechnol 97:7977–7987
Weeks G, Spiegelman GB (2003) Roles played by Ras subfamily proteins in the cell and developmental biology of microorganisms. Cell Signal 15:901–909
Wennerberg K, Rossman KL, Der CJ (2005) The Ras superfamily at a glance. J Cell Sci 118:843–846

Willardson BM, Howlett AC (2007) Function of phosducin-like proteins in G protein signaling and chaperone-assisted protein folding. Cell Signal 19:2417–2427

Wright SJ, Inchausti R, Eaton CJ et al (2011) RIC8 is a guanine-nucleotide exchange factor for Gα that regulates growth and development in *Neurospora crassa*. Genetics 189:165–176

Yan L, Yang Q, Sundin GW et al (2010) The mitogen-activated protein kinase kinase BOS5 is involved in regulating vegetative differentiation and virulence in *Botrytis cinerea*. Fungal Genet Biol 47:753–760

Yan L, Yang Q, Jiang J et al (2011) Involvement of a putative response regulator Brrg-1 in the regulation of sporulation, sensitivity to fungicides, and osmotic stress in *Botrytis cinerea*. Appl Microbiol Biotechnol 90:215–226

Yang Q, Yan L, Gu Q et al (2012) The mitogen-activated protein kinase kinase kinase BcOs4 is required for vegetative differentiation and pathogenicity in *Botrytis cinerea*. Appl Microbiol Biotechnol 96:481–492

Yang Q, Chen Y, Ma Z (2013a) Involvement of BcVeA and BcVelB in regulating conidiation, pigmentation and virulence in *Botrytis cinerea*. Fungal Genet Biol 50:63–71

Yang Q, Jiang J, Mayr C et al (2013b) Involvement of two type 2C protein phosphatases BcPtc1 and BcPtc3 in the regulation of multiple stress tolerance and virulence of *Botrytis cinerea*. Environ Microbiol 15:2696–2711

Yang Q, Yu F, Yin Y et al (2013c) Involvement of protein tyrosine phosphatases BcPtpA and BcPtpB in regulation of vegetative development, virulence and multi-stress tolerance in *Botrytis cinerea*. PLoS One 8:e61307

Yang Q, Yin D, Yin Y et al (2015) The response regulator BcSkn7 is required for vegetative differentiation and adaptation to oxidative and osmotic stresses in *Botrytis cinerea*. Mol Plant Pathol 16:276–287

Yu JH (2006) Heterotrimeric G protein signaling and RGSs in *Aspergillus nidulans*. J Microbiol 44:145–154

Zhao X, Mehrabi R, Xu JR (2007) Mitogen-activated protein kinase pathways and fungal pathogenesis. Eukaryot Cell 6:1701–1714

Zheng L, Campbell M, Murphy J et al (2000) The BMP1 gene is essential for pathogenicity in the gray mold fungus *Botrytis cinerea*. Mol Plant Microbe Interact 13:724–732

Chapter 14
Reactive Oxygen Species in the *Botrytis* – Host Interaction

Ulrike Siegmund and Anne Viefhues

Abstract Reactive oxygen species (ROS) are unavoidable byproducts of several metabolic processes. Due to their high reactivity they can cause molecular damages such as protein oxidations or DNA mutations, but they also serve as important signalling molecules within cells. Intracellular ROS primarily originate in the mitochondria; however enzymatic ROS generating systems such as the membrane associated NADPH oxidase complex (Nox) contribute to their production. In particular, during host-pathogen interactions ROS are of key importance for plant defence but also for fungal attack. As an early response to pathogen infestation the plant releases high amounts of reactive oxygen species to counteract the pathogen, known as the oxidative burst. It was shown that *Botrytis* exploits this plant defence reaction and even contributes to this oxidative burst by forming its' own ROS. Thus, the fungus needs a robust oxidative stress responsive (OSR) system to cope with ROS. In order to balance the intracellular redox state, effective antioxidant systems, including the thioredoxin and the glutathione system, are indispensable. Furthermore, catalases, superoxide dismutases and several peroxidases support ROS scavenging by enzymatic inactivation. Transcription factors such as the *Botrytis* activator protein (Bap1) and the response regulator Skn7 were shown to be involved in OSR. In this chapter we discuss the role of ROS in *Botrytis* – host interaction and both ROS generating and detoxifying systems and their importance for *Botrytis* pathogenicity.

Keywords ROS • NADPH oxidases • Thioredoxin • Glutathione • Oxidative stress

14.1 Reactive Oxygen Species (ROS)

Reactive oxygen species (ROS) are generated in all aerobic environments and play a major role for all organisms dependent on oxygen. The oxygen in our atmosphere facilitated the evolution of multi-cellular organisms, but is also the source for several

U. Siegmund (✉) • A. Viefhues
Institut fuer Biologie und Biotechnologie der Pflanzen, Westfaelische Wilhelms, Universitaet Muenster, Schlossplatz 8, 48143 Muenster, Germany
e-mail: ulrike.siegmund@uni-muenster.de

S. Fillinger, Y. Elad (eds.), *Botrytis – the Fungus, the Pathogen and its Management in Agricultural Systems*, DOI 10.1007/978-3-319-23371-0_14

challenges for all organisms. The weak reactant oxygen tends to form radicals, either by energy transfer reactions ($^1O_2$), or by electron transfer reactions ($O_2^{\cdot-}$, H_2O_2, $OH^{\cdot}$), which are highly reactive oxidants (Klotz 2002). ROS – like superoxide ($O_2^{\cdot-}$) and hydrogen peroxide (H_2O_2) – are on the one hand byproducts of aerobic metabolism or partial reductions of oxygen during respiration, especially in the mitochondria; on the other hand they are actively produced enzymatically (Heller and Tudzynski 2011).

ROS can be harmful causing DNA damage, protein inactivation or lipid peroxidations; thus it has been necessary for all organisms to evolve detoxification mechanisms to protect their own cells from these toxic effects and to maintain low intracellular ROS levels. Due to their potentially damaging nature, organisms can also use ROS as defence molecules, e.g. during the plant's oxidative burst. Besides their function as defence molecules, ROS also function as signalling molecules during cellular differentiation and development (Takemoto et al. 2007; Wojtaszek 1997). Functions of ROS as well as their production and detoxification systems with particular emphasis on *Botrytis cinerea* will be discussed in this chapter.

14.1.1 ROS during Cellular Differentiation and as Signaling Molecules

Over the past years it became obvious that ROS are involved in various differentiation processes, but also act as signalling molecules. Hansberg and Aguirre (1990) previously suggested differentiation as a response to oxidative stress and during fungal differentiation processes changes in ROS level were detected (Aguirre et al. 2005). ROS were also associated with development of fungal infection structures, polarized growth and fruiting body formation (summarized in Scott and Eaton 2008). Regarding roles of ROS as signalling molecules, data suggest that ROS are involved in signalling cascades, therewith regulating cellular functions and H_2O_2 seems to be the major messenger as it is diffusible and more stable than $O_2^{\cdot-}$ (Afanas'ev 2007; Giorgio et al. 2007). In microvessels (mice) and coronary arteries (human) H_2O_2 was shown to play an essential role as an endothelium-derived hyperpolarizing factor modulating the vascular tone (Matoba et al. 2000; Zhang et al. 2012). In tomato leaves it was suggested that ROS are generated near the cell walls of vascular bundle cells and serve as second messengers for the activation of defense genes in mesophyll cells (Orozco-Cardenas et al. 2001). Acting as second messengers is not the only role of ROS in plant defense.

14.1.2 ROS During Plant Infection

When attacked by a pathogen the plant reacts with an oxidative burst, which is the rapid and transient production of ROS; mainly produced by the plant's Rboh's (respiratory burst oxidase homolog) – homologs of mammalian NADPH oxidase gp91^{phox} (Torres et al. 2002; Wojtaszek 1997). ROS from the oxidative burst can drive the crosslinking of cell wall structural proteins and orchestrate the so-called hypersensitive response (HR; mechanism of plants to prevent spread of infections; Chaps. 12, 17, 18, and 19): they trigger programmed cell death in attacked cells, and are used as signals to induce cellular protectants like glutathione peroxidases (Brisson et al. 1994; Levine et al. 1994). In *Arabidopsis thaliana* it was shown that the inoculation of the plant with *B. cinerea* and also *Sclerotinia sclerotiorum* induces ROS production during the oxidative burst (Govrin and Levine 2000). Furthermore, virulence of *B. cinerea* was attenuated in HR-deficient *Arabidopsis* mutants and indicates suppressed growth of *B. cinerea* in these mutants compared to WT plants; similar results were obtained on transgenic plants expressing animal anti-apoptotic genes (Dickman et al. 2001). Additionally, growth was enhanced on plants whose HR was previously triggered by avirulent pathogens (Govrin and Levine 2000). Comparable results were obtained on *Arabidopsis* strains with delayed or reduced cell death response; here infection with different *Botrytis* strains was slower, while plants with accelerated response were more susceptible (Van Baarlen et al. 2007). Staining of ROS during infection showed an accumulation of H_2O_2 in the infection zones in tomato and bean leaves (chloronaphtol staining; Schouten et al. 2002). A literally closer look at ROS production indicated ROS accumulation around penetrated cell walls, at the plasma membrane and inside cells around the inoculation site during early stages of infection (Schouten et al. 2002; Tenberge 2004; Van Baarlen et al. 2007), while during later stages ROS accumulation proceeded inside infected tissue leading to death of the infected cells and colonization by *B. cinerea* (Muckenschnabel et al. 2002; Van Baarlen et al. 2007). High resolution imaging of infected *A. thaliana* showed a maximum of H_2O_2 accumulation at the inoculation site after 48 h. At this point it has also accumulated in cell walls, mitochondria and chloroplasts (Simon et al. 2013).

14.2 Fungal ROS Producing Systems

14.2.1 NADPH Oxidases

As mentioned, NADPH oxidases (Nox) constitute the most common enzymatic systems to produce ROS. They are multi-enzyme complexes which transport electrons through membranes reducing oxygen to superoxide using NADPH as an electron donor with $FADH_2$ and two hemes as cofactors (Bedard and Krause 2007). The mechanism of activation is best characterized in the phagocytic oxidase gp91^{phox} (also known as Nox2), which is responsible for the oxidative burst of neutrophils. The catalytical subunit gp91^{phox} is a transmembrane protein stabilized by another transmembrane protein p22^{phox}; together they form the integral transmembrane protein flavocytochrome b_{558}. Assembly of the active complex requires the translocation of the cytosolic subunits p40^{phox}, p47^{phox}, p67^{phox} and the small GTPase Rac to the membrane (Babior et al. 2002). Generally, Nox produce ROS in a regulated manner in different cell types as response to growth factors, cytokinesis and calcium signals (Lambeth 2004). Generation of ROS by Nox plays a role during defence responses as mentioned, but also participates in the signalling system in multicellular organisms (Takemoto et al. 2007). In mammals five Nox and two Duox (Dual oxidases = longer homologs of Nox2) are present, but occurrence is not restricted to mammals. Nox have been identified in most multicellular organisms, and recently even in the unicellular yeast, where a putative Nox ortholog has been described (Aguirre and Lambeth 2010; Rinnerthaler et al. 2012).

In fungi two homologs of gp91^{phox} have been identified, named NoxA and NoxB. In some fungi a third isoform was found (Aguirre et al. 2005), which contains an additional EF-hand (Ca^{2+} binding motif) at the N-terminus as described for mammalian Nox5 or plant Rboh's, but so far a function for this isoform is not known. Furthermore, homologs for the cytosolic components p67^{phox} and Rac were identified in fungi named NoxR and Rac. NoxR was shown to be a regulator of the two catalytical subunits NoxA and NoxB in several fungi (Segmüller et al. 2008; Takemoto et al. 2006), which were shown to be involved in different processes. Thus, NoxA (Nox1) is involved in fruiting body formation in *Aspergillus nidulans* (Lara-Ortiz et al. 2003), *Podospora anserina* (Malagnac et al. 2004) and *Neurospora crassa* (Cano-Dominguez et al. 2008), sclerotia formation in *Claviceps purpurea* (Giesbert et al. 2008) and virulence in *S. sclerotiorum* (Kim et al. 2011), *Alternaria alternata* (Yang and Chung 2012) and *C. purpurea* (Giesbert et al. 2008). On the other hand NoxB (Nox2) is necessary for penetration of host tissue in *Magnaporthe grisea* (Egan et al. 2007), sclerotia production in *C. purpurea* (Schürmann et al. 2013) and ascospore germination in *P. anserina* (Malagnac et al. 2004) as well as *N. crassa* (Cano-Dominguez et al. 2008). In the following we will concentrate on Nox occurrence, composition, function and involvement in pathogenicity in *B. cinerea*.

14.2.1.1 Components of *B. cinerea* Nox Complexes, Function and Organization

As most other fungi, *B. cinerea* possesses two catalytical subunits, BcNoxA and BcNoxB, the regulatory component NoxR and the GTPase Rac (see Fig. 14.1; Segmüller et al. 2008). Analyses of the respective mutants indicate that the two catalytical subunits exhibit distinct functions, as described above for the majority of other fungi. NoxA is involved in colonization of plant tissue, sclerotia formation and mutants were more sensitive to oxidative stress (Segmüller et al. 2008). Furthermore, it was recently shown that NoxA is necessary for CAT (conidial anastomosis tubes) fusions (Roca et al. 2012) and is localised primarily in the ER (Siegmund et al. 2013). In addition, BcNoxA is involved in infection cushion formation on glass surface (Siegmund et al. 2014), while in contrast BcNoxB is necessary for appressoria-mediated penetration leading to delayed infection. Furthermore, BcNoxB is localised in the ER and its deletion also led to increased sensitivity to oxidative stress (Segmüller et al. 2008; Siegmund et al. 2013). In contrast to what was previously published (smaller and fewer sclerotia) a more detailed functional

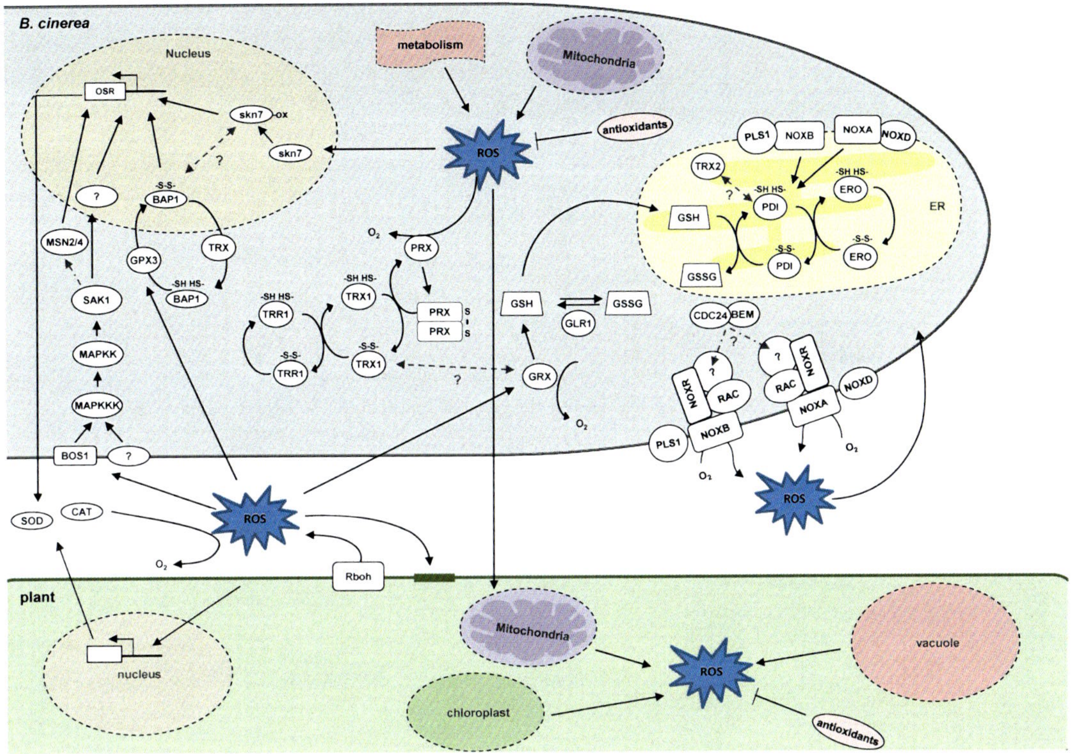

Fig. 14.1 Schematic overview of components involved in ROS production, detoxification or signaling in *B. cinerea* hyphae and the connection of ROS during fungus-plant interactions. *OSR* oxidative stress response, *Skn7* response regulator, *Bap1 Botrytis* activator protein, *GPX3* Glutathione peroxidase, *TRX* Thioredoxin, *TRR* Thioredoxin reductase, *SAK1* stress activated kinase, *BOS1* class III histidine-kinase, upstream of BcSak1 (Liu et al. 2008), *MSN2/4* multi-copy suppressor of Snf1, *SOD* superoxide dismutase, *CAT* catalase, *PRX* peroxiredoxin, *GSH* reduced form of glutathione, *GSSG* glutathione disulfide, *GLR* Glutathione reductase, *GRX* Glutaredoxin, *PDI* protein disulfide isomerase, *ERO* ER oxidoreductase, *NOX* (*A*/*B*) NADPH oxidase, *NOXD* Nox component, *PLS1* tetraspanin, *BEM1* scaffold protein, *CDC24* guanine nucleotide exchange factor, *NOXR* Nox regulator, *RAC* small GTPase

characterization of *BcnoxB* including new generated deletion mutants, showed that such mutants still formed normal sclerotia (Siegmund U, unpublished data). Analysis of the Δ*BcnoxR* mutant showed an additive phenotype of *BcnoxA* and *B* deletion mutants. Accordingly the mutant was defective in primary lesion formation and host tissue colonization, more sensitive to oxidative stress and it was impaired in sclerotia and CAT fusion formation. GFP-fusions showed a localisation in small aggregates throughout the hyphae and at the hyphal tip (Segmüller et al. 2008; Roca et al. 2012; Siegmund et al. 2013). Recently, analysis of the small GTPase BcRac suggested an involvement of the protein in polar growth, pathogenicity (deletion mutants were non-pathogenic) and cell-cycle progression through mitosis. Furthermore, deletion mutants had reduced growth rate with hyper-branching and deformed hyphae and were vegetatively sterile (Minz Dub et al. 2013).

So far analyses of known homologs of Nox components from mammals indicate that NoxR regulates NoxA and B in *B. cinerea*, but attempts to show a direct interaction of these components using yeast two-hybrid assays were not successful. Investigations from other fungi, e.g. *E. festucae*, suggest that the scaffold protein Bem1 and the GEF (guanine nucleotide exchange factor) Cdc24 are part of the Nox complex. In *E. festucae* and *C. purpurea* a direct interaction via the PB1 domains (Phox and Bem1p – mediates protein-protein interactions) was shown between NoxR, Bem1 and Cdc24 (Herrmann et al. 2014; Schürmann et al. 2013; Takemoto et al. 2011). Comparable studies in *B. cinerea* did not show a similar interaction of Bem1 and Cdc24 with NoxR. Only BcCdc24 and BcBem1 strongly interact with each other. Functional analysis of these proteins indicated an involvement of BcBem1 in conidia formation and germination as well as in penetration and host colonization. Nevertheless, BcCdc24 appears to be an essential protein as deletion seems to be lethal. A heterokaryotic mutant of the *Bccdc24* deletion showed a similar germination defect as Δ*Bcbem1* and was retarded in host penetration (Giesbert et al. 2014).

Besides these more obvious connections, recently two other proteins were suggested to be connected to fungal Nox complexes as potential stabilizing proteins. On one hand the tetraspanin Pls1 and Nox2 (homolog of BcNoxB), are similarly distributed in fungal genomes, meaning they are either both present or absent, hinting to similar functions. Furthermore, Pls1 was also shown to have a similar function in ascospore germination as Nox2 in *P. anserina* (Lambou et al. 2008). On the other hand, also in *P. anserina*, a connection of Nox1 (homolog of BcNoxA) to the putative ER protein PaNoxD (homolog of Pro41 from *Sordaria macrospora*) was suggested, as both deletion mutants exhibit similar phenotypes (Lacaze et al. 2014). The tetraspanin BcPls1 was analysed in the *B. cinerea* strains T4 and B05.10; in T4 it is involved in appressoria-mediated penetration (Gourgues et al. 2004), in B05.10 it is also necessary for appressoria-mediated penetration, for female fertility and it localises in the ER of the apical segment of the hyphae (Siegmund et al. 2013). The similar appressorium-mediated penetration defect and localisation of BcNoxB and BcPls1 hint to a close association of these two proteins and a concerted function during pathogenicity (Segmüller et al. 2008; Siegmund et al. 2013).

The *B. cinerea* ER protein BcNoxD (Pro41 homolog in *B. cinerea*) was functionally characterized and indeed showed an identical phenotype as the *BcnoxA* deletion mutant. BcNoxD is similarly involved in plant colonization, sclerotia formation and CAT fusions as shown for BcNoxA. Besides the phenotypic similarities, localisation studies revealed the presence of both proteins in the ER, BcNoxD being additionally present in vesicles (Siegmund et al. 2014). As both catalytical Nox subunits are located in the ER/nuclear envelope, a putative function of these proteins in this compartment needs to be considered. From mammals it is known that several Nox isoforms have functions in the ER, amongst others *via* a connection to the protein disulfide isomerase (Pdi), a protein involved in oxidative protein folding and degradation of mis-folded proteins (Laurindo et al. 2012, 2014). Reducing equivalents for Pdi are provided by glutathione (GSH, see below; Appenzeller-Herzog et al. 2010; Toledano et al. 2013), but the ER oxidoreductase Ero1 is the main antagonist of Pdi (Sevier and Kaiser 2008; see Fig. 14.1). We recently investigated a role of BcNoxA in the ER by complementing the deletion mutant with a construct retaining BcNoxA in the ER. The construct complemented the pathogenicity defect of the mutant, but not CAT fusions or sclerotia development, suggesting distinct roles of BcNoxA in the ER and after translocation (Siegmund et al. 2014).

Taken together NoxA and B and interaction partners have been analysed in *B. cinerea*, but their mode of activation and regulation remains to be elucidated. All of the mentioned proteins are involved in pathogenicity with varying severity. Their involvement in ROS production and its impact on pathogenicity are discussed in the next paragraph.

14.2.1.2 Involvement of Nox Derived ROS in Pathogenicity?

As described, components of the Nox complexes in *B. cinerea* are involved in various differentiation processes as well as pathogenicity, but surprisingly, no changes in the ROS level were detected in the corresponding mutants. In axenic cultures neither 3,3′-diaminobenzidine (DAB) staining of extracellular ROS (Siegmund et al. 2013), nor nitro blue tetrazolium (NBT) staining of superoxide in the hyphae detected any differences between wild type and *BcnoxA*, *B* or *R* mutants (Segmüller et al. 2008). Interestingly, flavoenzyme inhibition through diphenyleneiodonium chloride (DPI) led to loss of superoxide production in hyphae, while the Δ*BcnoxAB* double deletion mutant also produced superoxide as the WT (NBT staining, Viefhues A, unpublished data). Thus, another flavoenzyme seems to produce ROS as an alternative source. As all *bcnox* deletion mutants were impaired in some step of pathogenicity, ROS levels were also analysed during infection. NBT staining of hyphae after infection of onion epidermis did not show any difference in superoxide level, neither did the cytochemical detection of cerium perhydroxide formed by the reaction of cerium ions with H_2O_2 (Segmüller et al. 2008). Nox are not engaged in major ROS production in *B. cinerea* (at least not detectable with the applied methods), but rather ROS derived from Nox are necessary for signalling and differentiation. This would explain the various differentiation processes disturbed in the mutants (sclerotia formation, CAT fusions) and also that the mutants are impaired in pathogenicity, because they fail to properly form appressoria or infection cushions (on glass). Lastly, if ROS involved in pathogenicity are not produced by the enzymatic system of the NADPH oxidases, what alternative systems, other flavoenzymes for example, are responsible for ROS production and how is this process regulated?

14.2.2 Alternative ROS Producing Systems

Investigations regarding Nox in various fungi revealed the existence of alternative ROS sources, since deletion of Nox isoforms did not necessarily lead to reduced ROS production. Mitochondria produce predictably more ROS in quantitative terms than Nox in most cell types (Balaban et al. 2005; Laurindo et al. 2014). The mitochondria are thought to generate continuous levels of $O_2^{\cdot-}$ at complex I and III of the electron transport chain (Starkov 2008), which is then converted to H_2O_2 by mitochondrial superoxide dismutases (SOD). Mitochondrial ROS are involved in regulation of apoptosis in animal cells and it was suggested that they also participate in cell signalling (Werner and Werb 2002) and development (Aguirre et al. 2005; Severin and Hyman 2002). Fungal data regarding mitochondria as a major source for ROS are thus far limited (reviewed by Tudzynski et al. 2012).

Besides the mitochondria as an alternative ROS source, there are also other enzymatic systems producing ROS. Several ROS producing systems including laccases, GMC oxidoreductases, galactose oxidases and quinone reductases were implicated in plant biomass degradation (Grissa et al. 2010; Martinez et al. 2005). Laccases are copper-containing phenol oxidases with several functions. In *B. cinerea* laccase activity oxidizes tannins in grape berries and plays a role in degradation of the plant's secondary metabolite resveratrol (Adrian et al. 1998), suggesting a role in pathogenicity. Nevertheless, analysis of two laccases in *B. cinerea* BcLcc1 and BcLcc2 indicated expression only for *Bclcc2* during infection, but no altered virulence in the respective deletion mutant (Schouten et al. 2002). Another class of enzymes generating ROS is that of glucose oxidases (GOD); these enzymes generate H_2O_2 as a by-product during the oxidation of glucose. Two GODs have been identified in *B. cinerea* so far; one intracellular GOD, whose involvement in pathogenicity is unlikely, the second GOD found is a secreted enzyme, BcGod1, which was shown not to be involved in pathogenicity (Liu et al. 1998; Rolke et al. 2004).

14.3 ROS Detoxification and Scavenging

To counteract ROS either originating from normal physiological processes or from environmental changes, the fungus needs an efficient detoxification system. The intracellular redox status depends on the balance between oxidizing and reducing equivalents. To ensure this balance, different ROS detoxification systems are present in the cell. One can distinguish between enzymatic and non-enzymatic ROS degradation. While the glutathione system and antioxidants are non-enzymatic, the thioredoxin system belongs to the enzymatic systems as do detoxifying enzymes including catalases, peroxidases and superoxide dismutases. Moreover, transcription factors are known to be important for the regulation of these components. In the following we give insight into the different detoxification systems, important OSR (oxidative stress response) transcription factors and their role in pathogenicity and development.

14.3.1 Non-enzymatic ROS Scavenging Systems

14.3.1.1 The Glutathione System

In order to detoxify ROS, low molecular weight compounds interact with radicals and oxidants. These antioxidants differ in chemical structure and usually involve small soluble molecules that are oxidized by ROS and in doing so, eliminate oxidants from the surrounding. One key antioxidant is the cellular redox buffer glutathione, a tripeptide γ-L-glutamyl-L-cysteinyl-glycine, which is part of one of the major redox systems, the glutathione system. Glutathione is synthesized based on glutamate, cysteine, and glycine and found in most prokaryotes and eukaryotes (Meister and Anderson 1983; Foyer and Noctor 2005). The central cysteine allows the reversible oxidation and reduction of glutathione; accordingly it either exists as a reduced form (GSH) or oxidized as glutathione disulfide (GSSG). Besides the requirement for glutathione to buffer and maintain the general redox status, it can reduce glutaredoxins (Carmel-Harel and Storz 2000). Glutaredoxin (Grx) belongs to the thioredoxin family and contains a conserved hallmark sequence, consisting of two vicinal cysteine residues CxxC giving these proteins a high redox potential (Holmgren 1989). Grx is reduced by GSH *via* electrons from NADPH, whereupon GSH is oxidized to GSSG. Glutaredoxins can directly reduce oxidized substrates or hydrogen peroxide in a catalytic manner (Collinson et al. 2002). The phenotype of *grx* deletion, depending if either one or several *grx* are deleted, resulted in enhanced susceptibility towards oxidative stress and loss of oxidoreductase activity in yeasts, including *Candida glabrata* (Gutierrez-Escobedo et al. 2013), *C. albicans* (Chaves and da Silva 2012), *Saccaromyces cerevisiae* (Collinson et al. 2002; Pujol-Carrion et al. 2006; Tan et al. 2010) and *Schizosaccharomyces pombe* (Chung et al. 2004).

In order to reinstate the GSH pool in the cell GSSG can be recycled by the glutathione reductase (Glr) under consumption of electrons from NADPH (Grant et al. 1996). Glrs belong to the oxidoreductase family, which is featured by a NADPH binding site, an FAD prosthetic group and an active site containing redox-active disulfide (Mustacich and Powis 2000). Deletion of glutathione reductase in

Fusarium oxysporum (Sato et al. 2011), *A. nidulans* (Sato et al. 2009, 2011), *S. cerevisiae* (Muller 1996) *or S. pombe* (Lee et al. 1997) resulted in increased accumulation of GSSG and reduced growth rates in the presence of oxidants, ROS or elemental sulfur. In *B. cinerea* two glutathione reductases (BcGlr1, BcGlr2) were identified. Investigation of the deletion phenotype of *Bcglr2* revealed no changes in development, resistance or virulence. The effect of *Bcglr1* deletion was more pronounced. It could be shown that infection development on primary leaves of *Phaseolous vulgaris* was retarded for Δ*Bcglr1* (Viefhues et al. 2014). This defect could be attributed to delayed conidia germination, as infections with mycelium spread in a WT manner. Moreover, reduced growth on minimal medium was observed, indicating partial amino acid auxotrophy of the mutant. Restoration of growth by the addition of either cysteine or methionine suggests an involvement of the glutathione system in sulfur metabolism. BcGlr1 was shown to be essential for the reduction of oxidized glutathione. Analysis of the glutathione pool by redox-sensitive GFP confirmed that the loss of *Bcglr1* results in increased GSSG levels due to the loss of reductase function. Thus, *Bcglr1* is essential for the reversibility of GSH/GSSG reactions (Viefhues et al. 2014).

14.3.1.2 Antioxidants

Besides GSH several other compounds are involved in scavenging ROS. Small molecules like ascorbic acid, polyamines, polychelatins, carotenoids, flavonoids and alkaloids were described to take part in the detoxification of ROS in yeast (Jamieson 1998). The data available for *Botrytis* on this topic are limited. However, it was shown that the addition of antioxidants like ascorbic acid can rescue pathogenicity defects of mutants disrupted in the tolerance of oxidative stress, as was postulated for the deletion of the light dependent transcription factor *Bcltf1* (Schumacher et al. 2014). On the other hand, the addition of antioxidants to a conidial suspension of *Botrytis* was shown to suppress infection, depending on the type and the concentration of the antioxidants (Elad 1992). In contrast to fungi, plants clearly depend on antioxidants in response to pathogen attack, and as mentioned above their antioxidants alter pathogen infections. Upon H_2O_2 application plants already accumulate antioxidants like GSH (Levine et al. 1994). Also, under pathogen attack antioxidants and detoxifying enzymes could be detected at the infection site of *P. vulgaris* (Elad and Lapsker 2001). In *Botrytis*-infected tomato leaves a clear shift in the antioxidant pool could be observed.

While the content of ascorbic acid did not significantly change, a strong accumulation of dehydroascorbate and a prooxidative change in the ascorbate redox status in chloroplasts could be seen (Kuzniak and Sklodowska 1999). Furthermore Kuzniak and Sklodowska (1999) confirmed a decrease in glutathione content and an increase in Glr activity after infection. In *A. thaliana* a decrease of ascorbic acid was confirmed (Muckenschnabel et al. 2002). Depending on the time, a decrease in the antioxidant level makes sense when considering the oxidative stress present in the biological tissues of the plant (Noctor and Foyer 1998). Von Tiedemann (1997) demonstrated that the aggressiveness of *Botrytis* correlates with the intensity of the oxidative burst it induces. Thus, it is even possible that the fungus uses plant derived ascorbic acid for the production of oxalate and H_2O_2 (Loewus 1999; Muckenschnabel et al. 2002).

14.3.2 Enzymatic Systems

14.3.2.1 The Thioredoxin System

Besides non-enzymatic ROS detoxification one of the main enzymatic systems is the thioredoxin system. In diverse organisms this complex, consisting of thioredoxin (Trx) and thioredoxin reductase (Trr), was shown to regulate several metabolic enzymes, which form disulfide bonds during their catalytic cycle and function as scavengers of ROS (Malagnac et al. 2007). Thioredoxins, similar to Glutaredoxins, belong to the family of oxidoreductases that allow reversible thiol exchange reactions to reduce oxidized substrates (Arner and Holmgren 2000; Holmgren 1989). The thioredoxin reductase functionally belongs to the flavoprotein family of nucleotide-disulphide oxidoreductases, including the glutathione reductase described previously. The thioredoxin system was shown to be involved in numerous central cellular processes such as synthesis of desoxyribonucleotides, sulfur metabolism, regulation of gene expression and oxidative stress defence (Yoshida et al. 2003). The data available on the thioredoxin system in phytopathogenic fungi is still limited so far. However, it was shown that this system is of major importance for development and stress response in several fungi. Investigations on the thioredoxin system in *S. cerevisiae* (Gan 1991; Izawa et al. 1999; Garrido and Grant 2002; Oliveira et al. 2010), *C. albicans* (da Silva Dantas et al. 2010), *A. nidulans* (Thon et al. 2007), *S. pombe* (Casso and Beach 1996), *P. anserina* (Malagnac et al. 2007), *N. crassa* (Onai and Nakashima 1997) and *Cryphonectria parasitica* (Kim and Kim 2006) showed effects on vegetative growth, ROS tolerance, amino acid synthesis, sexual reproduction and even viability.

In *B. cinerea* one thioredoxin reductase (BcTrr1) and several thioredoxin-like proteins were identified. As BcTrr1 is a component of an important redox system, it is not surprising that the oxidative stress resistance was shown to be affected. The mutant was highly sensitive towards H_2O_2 and showed constitutive expression of oxidative stress responsive (OSR) genes, such as glutathione reductase, glutaredoxin, glutathione peroxidase and peroxiredoxin. Moreover, the mutant itself showed enhanced ROS production (Viefhues et al. 2014). Between numerous Trx-like proteins only one candidate was annotated as a *bona fide* Trx (BcTrx1). Sequence analysis revealed a second putative thioredoxin (BcTrx2), which shows the conserved Trx-site. Deletion of *Bctrx1* resulted in reduced tolerance towards H_2O_2. Analysis of the redox status by examining changes in the intracellular glutathione pool using a redox sensitive GFP (Heller et al. 2012a; Meyer et al. 2007) revealed no influence of *Bctrx1* deletion on GSH/GSSG ratio. The phenotype of Δ*Bctrx2* did not differ from the WT and no influence on oxidative stress tolerance could be observed (Viefhues et al. 2014). As BcTrx2 could be localized in the ER, the missing response to oxidative stress might hint towards a role in ER processes including protein folding rather than direct a role in redox control. In mammals it was shown that Trx-like proteins may have reductase activity *in vitro* and are able to participate in protein folding processes to support proteins of the folding complex (Sugiura et al. 2010). Although interplay between Trx, Trr and putative substrates is likely, yeast two-hybrid analyses could not confirm interactions, yet.

Thus far controversial results regarding the importance of ROS detoxification during pathogenesis were obtained (Rolke et al. 2004; Schouten et al. 2002; Temme and Tudzynski 2009). *Bctrx1* and *Bctrr1* knock-outs had a severe influence on pathogenicity as they exhibited strongly retarded lesion formation. Tissue maceration or spreading lesions could hardly be observed and the plant remained viable (Viefhues et al. 2014). *B. cinerea* was shown to exploit the plants oxidative burst and even contributes to it; however, as Δ*Bctrx1* and Δ*Bctrr1* were highly sensitive toward H_2O_2, this appears to be too much ROS to handle for the other detoxification systems. Redox balance cannot be fully restored without BcTrx1 and BcTrr1 and presumably results in a highly oxidized surrounding. The fact that Δ*Bctrr1* produced higher amounts of ROS and was greatly impaired in vegetative growth suggests not only a total disruption of ROS regulation, but shows the thioredoxin system to be a key player for several metabolic processes. Moreover, the data obtained for the thioredoxin system indicate that redox processes and regulation are important for host-pathogen interactions in *B. cinerea* but also for its own development.

14.3.2.2 Other ROS Detoxifying Enzymes

In addition to the thioredoxin system several intra- and extra-cellular ROS scavenging enzymes are present, including catalases (CAT), superoxide dismutases (SOD), peroxiredoxins (PRX) and several peroxidases. Catalases are enzymes with high turnover rates that convert H_2O_2 into O_2 and H_2O. Schouten et al. (2002) investigated the extracellular catalase gene *Bccat2*, one of seven catalases present in *B. cinerea. Bccat2* expression was rapidly induced upon exposure to H_2O_2*in vitro* and mRNA was already detected in very early stages of plant infection. Disruption of *Bccat2* resulted in an enhanced sensitivity to H_2O_2 and up-regulation of *Bcgst1* (glutathione S-transferase) and *Bcubi4* (polyubiquitin). Induction of stress responsive genes indicated that the Δ*Bccat2* mutant was generally more stressed than the WT; however, an influence on pathogenicity was not observed (Schouten et al. 2002).

The intracellular peroxisomal catalase BcCatA described by Van der Vlugt-Bergmans et al. (1997) noticeably differs structurally from BcCat2. BcCatA was expressed under H_2O_2 conditions; however, *BccatA* was not expressed *in planta* and does not have any influence on virulence. The tomato catalase encoding gene *cat1* was expressed under infection conditions, indicating that oxidative stress is occurring during infection (Van der Vlugt-Bergmans et al. 1997). Not only catalases but also superoxide dismutases (SODs) detoxify ROS. SODs convert $O_2^{\cdot-}$ to H_2O_2, which can be further reduced by catalases or peroxidases to H_2O. BcSod1 is involved in oxidative stress response and pathogenicity, as deletion mutants were significantly retarded in lesion development and showed sensitivity to paraquat. Thus, BcSod1 appears to be a virulence factor; however, it remains to be elucidated if superoxide from the host, fungus or even from both are substrates for BcSod1 (Rolke et al. 2004). Peroxidases, like guaiacol peroxidase, glutathione peroxidase or ascorbic peroxidase are capable of H_2O_2 detoxification (Gil-ad et al. 2000). Glutathione peroxidase for example can convert H_2O_2 to H_2O upon oxidation of glutathione. However, investigations of the glutathione peroxidase BcGpx3 showed neither involvement in pathogenicity nor in response to oxidative stress (Viefhues et al. 2015). Because of the numerous ROS detoxifying enzymes, the effect of a single deletion can be masked in most cases. Interplay of all enzymes is necessary to provide a stable redox environment and to allow a successful infection process.

14.3.3 Transcriptional Regulation of OSR Genes

All previously mentioned enzymes are regulated transcriptionally by a diversity of stress responsive factors. One important component for evoking stress response is the mitogen-activated protein kinase (MAPK) Hog1, whose homologs were shown to be involved in the reaction to different types of stress in various fungi (Dixon et al. 1999; Du et al. 2006; Moriwaki et al. 2006; Zhang et al. 2002). When Hog1 is activated, it can either phosphorylate transcription factors or proteins to execute their functions by structural changes or gene expression or, depending on stress-mediating transcriptional activators, it translocates to the nucleus and associates to stress responsive promoters to recruit components of the transcriptional machinery (Alepuz et al. 2001; Chap. 13). Depending on the stress applied the *B. cinerea* Hog1 homolog BcSak1 either localized to the nucleus (osmotic stress) or remained cytosolic (oxidative stress). Interestingly, during early stages of infection a translocation of the MAPK to the nucleus was verified, indicating either that *B. cinerea* suffers from osmotic stress during infection or other factors might induce the translocation, like fungal programmed cell death–inducing factors secreted by plants (Heller et al. 2012b). Analysis of a *Bcsak1* disruption mutant revealed a function of this gene in conidiospore formation and plant penetration. Accordingly, the Δ*Bcsak1* mutant was not able to colonize unwounded tissue (Segmüller et al. 2007). Moreover, reduced secretion of major phytotoxins could be observed (Heller et al. 2012b). Expression analyses confirmed that BcSak1 is necessary for the expression of several stress responsive genes under oxidative stress (Heller et al. 2012b).

BcSak1 shares some target genes with the bZIP transcription factor Bap1 (*Botrytis* activator protein), which mainly regulates OSR genes. Macroarray data revealed a group of common target genes, amongst others *BccatA*, *BccatB*, *BccatC*, *Bcgrx1*, and *Bcgpx3*. Nonetheless, Bap1 seems to be regulated independently of the BcSak1 pathway, as knock-out phenotypes differed; still crosstalk between both pathways is likely due to their shared role in oxidative stress response. The mode of action of Bap1 upon oxidative stress is assumed to be regulated as described for the yeast activator protein Yap1. Under oxidative stress glutathione peroxidases or peroxiredoxins sense H_2O_2 and direct the oxidation of two cysteines in the C- and N-terminal cysteine rich domain region of Yap1. This oxidation results in disulfide bond formation and a subsequent conformational change of the transcription factor. A nuclear export signal located in the C-terminal region gets masked and therewith inhibits binding of the nuclear export regulator (NER). Accumulation in the nucleus occurs and thus Yap1 can regulate gene expression. The reversion of the oxidized condition is achieved by the expression of thioredoxin, which results in a negative feedback regulation, enabling Yap1 to translocate back to the cytoplasm (Delaunay et al. 2000). The results for the deletion of *bap1* were surprising; no effect on pathogenicity was observed, although several OSR genes were not expressed in the knock-out mutant and tolerance towards H_2O_2 and menadione was greatly decreased. However, expression analysis *in planta* (2 dpi) revealed that Bap1 target genes were not even expressed in the WT, although H_2O_2 was measurable (Temme and Tudzynski 2009). So does the fungus sense oxidative stress during infection?

The yeast response regulator (RR) Skn7 was described to be involved in the OSR and to work in concert with Yap1 in gene regulation (Calvo et al. 2012; Mulford and Fassler 2011). Depending on the stress being applied Skn7 signalling pathways differ. Under osmotic or cell wall stress conditions, a phosphotransferase changes the phosphorylation status of Skn7 and thereby evokes the transcription of stress responsive genes. However, if oxidative stress is applied, Skn7 is directly oxidized and regulates gene expression. A homolog of this RR was identified in *B. cinerea* (BcSkn7). High sensitivity to oxidative stress and localisation in the nucleus supports a probable function of BcSkn7 in stress responses. Moreover, Bap1 and BcSkn7 share several target genes, including *Bctrx1*, *Bctrr1*, *Bcgpx3*, and *Bcgrx1* and thus a coordinated regulation of Bap1 and BcSkn7 is possible. Nevertheless, the role of BcSkn7 in pathogenicity is only minor (Yang et al. 2014; Viefhues et al. 2015). Besides the previously mentioned transcription factors, the Cys2His2 zinc finger protein Msn2/4 could play a role in ROS response in concert with Bap1 and Skn7. In yeast a noteworthy sensitivity to different types of stress and the importance for the expression of stress responsive genes was established (Hasan et al. 2002; Martinez-Pastor et al. 1996).

All the factors are important for the regulation of numerous OSR genes. However, the impact on pathogenicity when compared to their essential function in gene regulation is only minor. Next to the deletion of single OSR genes also the deletion of stress response regulators appears to be partially redundant. The components might form a complex regulatory network (see Fig. 14.1) and work as back-up systems when other components fail. Still the question remains: Is enzymatic ROS detoxification in the *Botrytis*-host interaction important?

14.4 Future Perspectives

The plant's defence response, especially the oxidative burst, seems to be essential for successful infection of *B. cinerea*. As *Botrytis* also contributes to the process by triggering its own ROS, the question emerges as to how the fungus produces and copes with such large amounts of ROS. As described above a variety of ROS producing (focus on Nox) and detoxifying systems have been analysed in detail in *B. cinerea* and some of them are involved in pathogenicity at various modes. ROS play a dual role, during infection: on the one hand causing oxidative stress, which needs to be handled by the fungus and on the other hand ROS are needed as signalling molecules for differentiation processes. Regarding the role of ROS during differentiation processes investigations of BcNox complexes revealed that these enzymes in *B. cinerea* produce only minor amounts of ROS. Nevertheless, these are essential for differentiation processes such as appressoria and sclerotia formation or CAT fusions and the functional interaction between the fungus and its host. The discovered function of Nox in the ER of *B. cinerea* was unexpected. However, it provides another hint that Nox are rather not involved in ROS synthesis leading to oxidative stress for the plant, but in complex regulation mechanisms. Regulation through Nox derived ROS takes, amongst others, place in the ER, where Pdi might serve as a link to couple ER-stress to Nox dependent ROS generation (Laurindo et al. 2012). The locations of Nox and their putative complex partners, as well as the connection to the components in charge of the redox status in the ER (Pdi, Ero1, GSH) are depicted in Fig. 14.1. In addition this scheme illustrates the complex connections and locations of the proteins involved in various parts of ROS scavenging, and the connection to ROS during infection of the plant.

Although many single components have been investigated, by now their role in ROS scavenging and detoxification during infection still needs to be elucidated. Additional factors including antioxidants and complex interplay of the different systems provide a challenge to define the contribution of single factors in the (de-) toxification process during infection. A concentration on interactions in these networks rather than a continuative analysis of single genes might be beneficial. The diverse ROS scavenging/redox systems have different influences on virulence and their detailed role in detoxification of ROS, especially derived from the plant's oxidative burst, will have to be addressed in the future.

Acknowledgements We thank Paul Tudzynski, Marty Dickman and Julia Schumacher for critical reading and/or discussion, Bettina Richter for excellent technical assistance and the DFG (Tu 50/19) and the GRK (Graduate Research School) 1409 for funding.

References

Adrian M, Rajaei H, Jeandet P et al (1998) Resveratrol oxidation in *Botrytis cinerea* conidia. Phytopathology 88:472–476

Afanas'ev IB (2007) Signaling functions of free radicals superoxide & nitric oxide under physiological & pathological conditions. Mol Biotechnol 37:2–4

Aguirre J, Lambeth JD (2010) Nox enzymes from fungus to fly to fish and what they tell us about Nox function in mammals. Free Radic Biol Med 49:1342–1353

Aguirre J, Ríos-Momberg M, Hewitt D et al (2005) Reactive oxygen species and development in microbial eukaryotes. Trends Microbiol 13:111–118

Alepuz PM, Jovanovic A, Reiser V et al (2001) Stress-induced map kinase Hog1 is part of transcription activation complexes. Mol Cell 7:767–777

Appenzeller-Herzog C, Riemer J, Zito E et al (2010) Disulphide production by Ero1 alpha-PDI relay is rapid and effectively regulated. EMBO J 29:3318–3329

Arner ES, Holmgren A (2000) Physiological functions of thioredoxin and thioredoxin reductase. Eur J Biochem 267:6102–6109

Babior BM, Lambeth JD, Nauseef W (2002) The neutrophil NADPH oxidase. Arch Biochem Biophys 397:342–344

Balaban RS, Nemoto S, Finkel T (2005) Mitochondria, oxidants, and aging. Cell 120:483–495

Bedard K, Krause KH (2007) The NOX family of ROS-generating NADPH oxidases: physiology and pathophysiology. Physiol Rev 87:245–313

Brisson LF, Tenhaken R, Lamb C (1994) Function of oxidative cross-linking of cell wall structural proteins in plant disease resistance. Plant Cell 6:1703–1712

Calvo IA, Garcia P, Ayte J et al (2012) The transcription factors Pap1 and Prr1 collaborate to activate antioxidant, but not drug tolerance, genes in response to H_2O_2. Nucleic Acids Res 40:4816–4824

Cano-Dominguez N, Alvarez-Delfin K, Hansberg W et al (2008) NADPH oxidases NOX-1 and NOX-2 require the regulatory subunit NOR-1 to control cell differentiation and growth in *Neurospora crassa*. Eukaryot Cell 7:1352–1361

Carmel-Harel O, Storz G (2000) Roles of the glutathione- and thioredoxin-dependent reduction systems in the *Escherichia coli* and *Saccharomyces cerevisiae* responses to oxidative stress. Annu Rev Microbiol 54:439–461

Casso D, Beach D (1996) A mutation in a thioredoxin reductase homolog suppresses p53-induced growth inhibition in the fission yeast *Schizosaccharomyces pombe*. Mol Gen Genet 252:518–529

Chaves GM, da Silva WP (2012) Superoxide dismutases and glutaredoxins have a distinct role in the response of *Candida albicans* to oxidative stress generated by the chemical compounds menadione and diamide. Mem Inst Oswaldo Cruz 107:998–1005

Chung WH, Kim KD, Cho YJ et al (2004) Differential expression and role of two dithiol glutaredoxins Grx1 and Grx2 in *Schizosaccharomyces pombe*. Biochem Biophys Res Commun 321:922–929

Collinson EJ, Wheeler GL, Garrido EO et al (2002) The yeast glutaredoxins are active as glutathione peroxidases. J Biol Chem 277:16712–16717

da Silva Dantas A, Patterson MJ, Smith DA et al (2010) Thioredoxin regulates multiple hydrogen peroxide-induced signaling pathways in *Candida albicans*. Mol Cell Biol 30:4550–4563

Delaunay A, Isnard AD, Toledano MB (2000) H_2O_2 sensing through oxidation of the Yap1 transcription factor. EMBO J 19:5157–5166

Dickman MB, Park YK, Oltersdorf T et al (2001) Abrogation of disease development in plants expressing animal antiapoptotic genes. Proc Natl Acad Sci U S A 98:6957–6962

Dixon KP, Xu JR, Smirnoff N et al (1999) Independent signaling pathways regulate cellular turgor during hyperosmotic stress and appressorium-mediated plant infection by *Magnaporthe grisea*. Plant Cell 11:2045–2058

Du C, Sarfati J, Latge JP et al (2006) The role of the sakA (Hog1) and tcsB (sln1) genes in the oxidant adaptation of *Aspergillus fumigatus*. Med Mycol 44:211–218

Egan MJ, Wang ZY, Jones MA et al (2007) Generation of reactive oxygen species by fungal NADPH oxidases is required for rice blast disease. Proc Natl Acad Sci U S A 104:11772–11777

Elad Y (1992) The use of antioxidants (free-radical scavengers) to control grey mould (*Botrytis cinerea*) and white mould (*Sclerotinia sclerotiorum*) in various crops. Plant Pathol 41:417–426

Elad Y, Lapsker Z (2001) Involvement of reactive oxygen species and antioxidative processes in the disease caused by *B. cinerea* on bean leaves and in its biological control by means of *Trichoderma harzianum* T39. IOBC Bull 24(3):21–25

Foyer CH, Noctor G (2005) Redox homeostasis and antioxidant signaling: a metabolic interface between stress perception and physiological responses. Plant Cell 17:1866–1875

Gan ZR (1991) Yeast thioredoxin genes. J Biol Chem 266:1692–1696

Garrido EO, Grant CM (2002) Role of thioredoxins in the response of *Saccharomyces cerevisiae* to oxidative stress induced by hydroperoxides. Mol Microbiol 43:993–1003

Giesbert S, Schürg T, Scheele S et al (2008) The NADPH oxidase Cpnox1 is required for full pathogenicity of the ergot fungus *Claviceps purpurea*. Mol Plant Pathol 9:317–327

Giesbert S, Siegmund U, Schumacher J et al (2014) Functional analysis of BcBem1 and its interaction partners in *Botrytis cinerea*: impact on differentiation and virulence. PLoS One 9(5):e95172. doi:10.1371/journal.pone.0095172

Gil-ad NL, Bar-Nun N, Noy T et al (2000) Enzymes of *Botrytis cinerea* capable of breaking down hydrogen peroxide. FEMS Microbiol Lett 190:121–126

Giorgio M, Trinei M, Migliaccio E et al (2007) Hydrogen peroxide: a metabolic by-product or a common mediator of ageing signals? Nat Rev Mol Cell Biol 8:722–728

Gourgues M, Brunet-Simon A, Lebrun MH et al (2004) The tetraspanin BcPls1 is required for appressorium-mediated penetration of *Botrytis cinerea* into host plant leaves. Mol Microbiol 51:619–629

Govrin EM, Levine A (2000) The hypersensitive response facilitates plant infection by the necrotrophic pathogen *Botrytis cinerea*. Curr Biol 10:751–757

Grant CM, Collinson LP, Roe JH et al (1996) Yeast glutathione reductase is required for protection against oxidative stress and is a target gene for yAP-1 transcriptional regulation. Mol Microbiol 21:171–179

Grissa I, Bidard F, Grognet P et al (2010) The Nox/Ferric reductase/Ferric reductase-like families of Eumycetes. Fungal Biol 114:766–777

Gutierrez-Escobedo G, Orta-Zavalza E, Castano I et al (2013) Role of glutathione in the oxidative stress response in the fungal pathogen *Candida glabrata*. Curr Genet 59:91–106

Hansberg W, Aguirre J (1990) Hyperoxidant states cause microbial cell differentiation by cell isolation from dioxygen. J Theor Biol 142:201–221

Hasan R, Leroy C, Isnard AD et al (2002) The control of the yeast H_2O_2 response by the Msn2/4 transcription factors. Mol Microbiol 45:233–241

Heller J, Tudzynski P (2011) Reactive oxygen species in phytopathogenic fungi: signaling, development, and disease. Annu Rev Phytopathol 49:369–390

Heller J, Meyer AJ, Tudzynski P (2012a) Redox-sensitive GFP2: use of the genetically encoded biosensor of the redox status in the filamentous fungus *Botrytis cinerea*. Mol Plant Pathol 13:935–947

Heller J, Ruhnke N, Espino JJ et al (2012b) The mitogen-activated protein kinase BcSak1 of *Botrytis cinerea* is required for pathogenic development and has broad regulatory functions beyond stress response. Mol Plant-Microbe Interact 25:802–816

Herrmann A, Tillmann B, Schürmann J et al (2014) Small GTPase-associated signaling by the guanine nucleotide exchange factors CpDock180 and CpCdc24, the GTPase effector CpSte20 and the scaffold CpBem1 in *Claviceps purpurea*. Eukaryot Cell 13(4):470–482

Holmgren A (1989) Thioredoxin and glutaredoxin systems. J Biol Chem 264:13963–13966

Izawa S, Maeda K, Sugiyama K et al (1999) Thioredoxin deficiency causes the constitutive activation of Yap1, an AP-1-like transcription factor in *Saccharomyces cerevisiae*. J Biol Chem 274:28459–28465

Jamieson DJ (1998) Oxidative stress responses of the yeast *Saccharomyces cerevisiae*. Yeast 14:1511–1527

Kim JH, Kim DH (2006) Cloning and characterization of a thioredoxin gene, CpTrx1, from the chestnut blight fungus *Cryphonectria parasitica*. J Microbiol 44:556–561

Kim H, Chen C, Kabbage M et al (2011) Identification and characterization of *Sclerotinia sclerotiorum* NADPH oxidases. Appl Environ Microbiol 77:7721–7729

Klotz LO (2002) Oxidant-induced signaling: effects of peroxynitrite and singlet oxygen. Biol Chem 383:443–456

Kuzniak E, Sklodowska M (1999) The effect of *Botrytis cinerea* infection on ascorbate-glutathione cycle in tomato leaves. Plant Sci 148:69–76

Lacaze I, Lalucque H, Silar P et al (2014) Identification of NoxD/Pro41 as the homologue of the p22phox NADPH oxidase subunit in fungi. Mol Microbiol. doi:10.1111/mmi.12876

Lambeth JD (2004) NOX enzymes and the biology of reactive oxygen. Nat Rev Immunol 4:181–189

Lambou K, Malagnac F, Barbisan C et al (2008) The crucial role of the Pls1 tetraspanin during ascospore germination in *Podospora anserina* provides an example of the convergent evolution of morphogenetic processes in fungal plant pathogens and saprobes. Eukaryot Cell 7:1809–1818

Lara-Ortiz T, Riveros-Rosas H, Aguirre J (2003) Reactive oxygen species generated by microbial NADPH oxidase NoxA regulate sexual development in *Aspergillus nidulans*. Mol Microbiol 50:1241–1255

Laurindo FR, Pescatore LA, Fernandes Dde C (2012) Protein disulfide isomerase in redox cell signaling and homeostasis. Free Radic Biol Med 52:1954–1969

Laurindo FR, Araujo TL, Abrahao TB (2014) Nox NADPH oxidases and the endoplasmic reticulum. Antioxid Redox Signal. doi:10.1089/ars.2013.5605

Lee J, Dawes IW, Roe JH (1997) Isolation, expression, and regulation of the pgr1(+) gene encoding glutathione reductase absolutely required for the growth of *Schizosaccharomyces pombe*. J Biol Chem 272:23042–23049

Levine A, Tenhaken R, Dixon R et al (1994) H_2O_2 from the oxidative burst orchestrates the plant hypersensitive disease resistance response. Cell 79:583–593

Liu S, Oeljeklaus S, Gerhardt B et al (1998) Purification and characterization of glucose oxidase of *Botrytis cinerea*. Physiol Mol Plant Pathol 53:123–132

Liu W, Leroux P, Fillinger S (2008) The HOG1-like MAP kinase Sak1 of *Botrytis cinerea* is negatively regulated by the upstream histidine kinase Bos1 and is not involved in dicarboximide- and phenylpyrrole-resistance. Fungal Genet Biol 45:1062–1074

Loewus FA (1999) Biosynthesis and metabolism of ascorbic acid in plants and of analogs of ascorbic acid in fungi. Phytochemistry 52:193–210

Malagnac F, Lalucque H, Lepere G et al (2004) Two NADPH oxidase isoforms are required for sexual reproduction and ascospore germination in the filamentous fungus *Podospora anserina*. Fungal Genet Biol 41:982–997

Malagnac F, Klapholz B, Silar P (2007) PaTrx1 and PaTrx3, two cytosolic thioredoxins of the filamentous ascomycete *Podospora anserina* involved in sexual development and cell degeneration. Eukaryot Cell 6:2323–2331

Martinez AT, Speranza M, Ruiz-Duenas FJ et al (2005) Biodegradation of lignocellulosics: microbial, chemical, and enzymatic aspects of the fungal attack of lignin. Int Microbiol 8:195–204

Martinez-Pastor MT, Marchler G, Schuller C et al (1996) The *Saccharomyces cerevisiae* zinc finger proteins Msn2p and Msn4p are required for transcriptional induction through the stress-response element (STRE). EMBO J 15:2227–2235

Matoba T, Shimokawa H, Nakashima M et al (2000) Hydrogen peroxide is an endothelium-derived hyperpolarizing factor in mice. J Clin Invest 106:1521–1530

Meister A, Anderson ME (1983) Glutathione. Annu Rev Biochem 52:711–760

Meyer AJ, Brach T, Marty L et al (2007) Redox-sensitive GFP in *Arabidopsis thaliana* is a quantitative biosensor for the redox potential of the cellular glutathione redox buffer. Plant J 52:973–986

Minz Dub A, Kokkelink L, Tudzynski B et al (2013) Involvement of *Botrytis cinerea* small GTPases BcRAS1 and BcRAC in differentiation, virulence, and the cell cycle. Eukaryot Cell 12:1609–1618

Moriwaki A, Kubo E, Arase S et al (2006) Disruption of SRM1, a mitogen-activated protein kinase gene, affects sensitivity to osmotic and ultraviolet stressors in the phytopathogenic fungus *Bipolaris oryzae*. FEMS Microbiol Lett 257:253–261

Muckenschnabel I, Goodman BA, Williamson B et al (2002) Infection of leaves of *Arabidopsis thaliana* by *Botrytis cinerea*: changes in ascorbic acid, free radicals and lipid peroxidation products. J Exp Bot 53:207–214

Mulford KE, Fassler JS (2011) Association of the Skn7 and Yap1 transcription factors in the *Saccharomyces cerevisiae* oxidative stress response. Eukaryot Cell 10:761–769

Muller EG (1996) A glutathione reductase mutant of yeast accumulates high levels of oxidized glutathione and requires thioredoxin for growth. Mol Biol Cell 7:1805–1813

Mustacich D, Powis G (2000) Thioredoxin reductase. Biochem J 346(Pt 1):1–8

Noctor G, Foyer CH (1998) Ascorbate and glutathione: keeping active oxygen under control. Annu Rev Plant Physiol Plant Mol Biol 49:249–279

Oliveira MA, Discola KF, Alves SV et al (2010) Insights into the specificity of thioredoxin reductase-thioredoxin interactions. A structural and functional investigation of the yeast thioredoxin system. Biochemistry 49:3317–3326

Onai K, Nakashima H (1997) Mutation of the cys-9 gene, which encodes thioredoxin reductase, affects the circadian conidiation rhythm in *Neurospora crassa*. Genetics 146:101–110

Orozco-Cardenas ML, Narvaez-Vasquez J, Ryan CA (2001) Hydrogen peroxide acts as a second messenger for the induction of defense genes in tomato plants in response to wounding, systemin, and methyl jasmonate. Plant Cell 13:179–191

Pujol-Carrion N, Belli G, Herrero E et al (2006) Glutaredoxins Grx3 and Grx4 regulate nuclear localisation of Aft1 and the oxidative stress response in *Saccharomyces cerevisiae*. J Cell Sci 119:4554–4564

Rinnerthaler M, Buttner S, Laun P et al (2012) Yno1p/Aim14p, a NADPH-oxidase ortholog, controls extramitochondrial reactive oxygen species generation, apoptosis, and actin cable formation in yeast. Proc Natl Acad Sci U S A 109:8658–8663

Roca MG, Weichert M, Siegmund U et al (2012) Germling fusion via conidial anastomosis tubes in the grey mould *Botrytis cinerea* requires NADPH oxidase activity. Fungal Biol 116:379–387

Rolke Y, Liu SJ, Quidde T et al (2004) Functional analysis of H_2O_2-generating systems in *Botrytis cinerea*: the major Cu-Zn-superoxide dismutase (BCSOD1) contributes to virulence on French bean, whereas a glucose oxidase (BCGOD1) is dispensable. Mol Plant Pathol 5:17–27

Sato I, Shimizu M, Hoshino T et al (2009) The glutathione system of *Aspergillus nidulans* involves a fungus-specific glutathione S-transferase. J Biol Chem 284:8042–8053

Sato I, Shimatani K, Fujita K et al (2011) Glutathione reductase/glutathione is responsible for cytotoxic elemental sulfur tolerance via polysulfide shuttle in fungi. J Biol Chem 286:20283–20291

Schouten A, Tenberge KB, Vermeer J et al (2002) Functional analysis of an extracellular catalase of *Botrytis cinerea*. Mol Plant Pathol 3:227–238

Schumacher J, Simon A, Cohrs KC et al (2014) The transcription factor BcLTF1 regulates virulence and light responses in the necrotrophic plant pathogen *Botrytis cinerea*. PLoS Genet 10:e1004040

Schürmann J, Buttermann D, Herrmann A et al (2013) Molecular characterization of the NADPH oxidase complex in the ergot fungus *Claviceps purpurea*: CpNox2 and CpPls1 are important for a balanced host-pathogen interaction. Mol Plant-Microbe Interact 26:1151–1164

Scott B, Eaton CJ (2008) Role of reactive oxygen species in fungal cellular differentiations. Curr Opin Microbiol 11:488–493

Segmüller N, Ellendorf U, Tudzynski B et al (2007) BcSAK1, a stress-activated mitogen-activated protein kinase, is involved in vegetative differentiation and pathogenicity in *Botrytis cinerea*. Eukaryot Cell 6:211–221

Segmüller N, Kokkelink L, Giesbert S et al (2008) NADPH oxidases are involved in differentiation and pathogenicity in *Botrytis cinerea*. Mol Plant-Microbe Interact 21:808–819

Severin FF, Hyman AA (2002) Pheromone induces programmed cell death in *S. cerevisiae*. Curr Biol 12:R233–R235

Sevier CS, Kaiser CA (2008) Ero1 and redox homeostasis in the endoplasmic reticulum. Biochim Biophys Acta 1783:549–556

Siegmund U, Heller J, Van Kan JAL et al (2013) The NADPH Oxidase Complexes in *Botrytis cinerea*: evidence for a close association with the ER and the tetraspanin Pls1. PLoS One 8:e55879

Siegmund U, Marschall R, Tudzynski P (2014) BcNoxD, a putative ER protein, is a new component of the NADPH oxidase complex in *Botrytis cinerea*. Mol Microbiol. doi:10.1111/mmi.12869

Simon UK, Polanschutz LM, Koffler BE et al (2013) High resolution imaging of temporal and spatial changes of subcellular ascorbate, glutathione and H_2O_2 distribution during *Botrytis cinerea* infection in Arabidopsis. PLoS One 8:e65811

Starkov AA (2008) The role of mitochondria in reactive oxygen species metabolism and signaling. Ann N Y Acad Sci 1147:37–52

Sugiura Y, Araki K, Iemura S et al (2010) Novel thioredoxin-related transmembrane protein TMX4 has reductase activity. J Biol Chem 285:7135–7142

Takemoto D, Tanaka A, Scott B (2006) A p67Phox-like regulator is recruited to control hyphal branching in a fungal-grass mutualistic symbiosis. Plant Cell 18:2807–2821

Takemoto D, Tanaka A, Scott B (2007) NADPH oxidases in fungi: diverse roles of reactive oxygen species in fungal cellular differentiation. Fungal Genet Biol 44:1065–1076

Takemoto D, Kamakura S, Saikia S et al (2011) Polarity proteins Bem1 and Cdc24 are components of the filamentous fungal NADPH oxidase complex. Proc Natl Acad Sci U S A 108:2861–2866

Tan SX, Greetham D, Raeth S et al (2010) The thioredoxin-thioredoxin reductase system can function *in vivo* as an alternative system to reduce oxidized glutathione in *Saccharomyces cerevisiae*. J Biol Chem 285:6118–6126

Temme N, Tudzynski P (2009) Does *Botrytis cinerea* ignore H_2O_2-induced oxidative stress during infection? Characterization of *Botrytis* activator protein 1. Mol Plant Microbe Interact 22:987–998

Tenberge KB (2004) Morphology and cellular organization in *Botrytis* interaction with plants. In: Elad Y, Williamson P, Tudzinski P, Delen N (eds) *Botrytis*: biology, pathology and control. Kluwer Academic, Dordrecht, pp 67–84

Thon M, Al-Abdallah Q, Hortschansky P et al (2007) The thioredoxin system of the filamentous fungus *Aspergillus nidulans*: impact on development and oxidative stress response. J Biol Chem 282:27259–27269

Toledano MB, Delaunay-Moisan A, Outten CE et al (2013) Functions and cellular compartmentation of the thioredoxin and glutathione pathways in yeast. Antioxid Redox Signal 18:1699–1711

Torres MA, Dangl JL, Jones JD (2002) *Arabidopsis* gp91phox homologues AtrbohD and AtrbohF are required for accumulation of reactive oxygen intermediates in the plant defense response. Proc Natl Acad Sci U S A 99:517–522

Tudzynski P, Heller J, Siegmund U (2012) Reactive oxygen species generation in fungal development and pathogenesis. Curr Opin Microbiol 15:653–659

Van Baarlen P, Woltering EJ, Staats M et al (2007) Histochemical and genetic analysis of host and non-host interactions of *Arabidopsis* with three *Botrytis* species: an important role for cell death control. Mol Plant Pathol 8:41–54

Van der Vlugt-Bergmans CJB, Wagemakers CAM, Dees DCT et al (1997) Catalase A from *Botrytis cinerea* is not expressed during infection on tomato leaves. Physiol Mol Plant Pathol 50:1–15

Viefhues A, Heller J, Temme N (2014) Redox systems in *Botrytis cinerea*: impact on development and virulence. Mol Plant Microbe Interact 27:858–874

Viefhues A, Schlathoelter I, Simon A, et al (2015) Unravelling the function of the response regulator BcSkn7 in the stress signaling network of *Botrytis cinerea*. Eukaryot Cell pii: EC.00043-15. [Epub ahead of print] von Tiedemann A (1997) Evidence for a primary role of active oxygen species in induction of host cell death during infection of bean leaves with *Botrytis cinerea*. Physiol Mol Plant Pathol 50:151–166

Werner E, Werb Z (2002) Integrins engage mitochondrial function for signal transduction by a mechanism dependent on Rho GTPases. J Cell Biol 158:357–368

Wojtaszek P (1997) Oxidative burst: an early plant response to pathogen infection. Biochem J 322(Pt 3):681–692

Yang SL, Chung KR (2012) The NADPH oxidase-mediated production of hydrogen peroxide (H_2O_2) and resistance to oxidative stress in the necrotrophic pathogen *Alternaria alternata* of citrus. Mol Plant Pathol 13:900–914

Yang Q, Yin D, Yin Y et al (2014) The response regulator BcSkn7 is required for vegetative differentiation and adaptation to oxidative and osmotic stresses in *Botrytis cinerea*. Mol Plant Pathol. doi:10.1111/mpp.12181

Yoshida T, Oka S, Masutani H et al (2003) The role of thioredoxin in the aging process: involvement of oxidative stress. Antioxid Redox Signal 5:563–570

Zhang Y, Lamm R, Pillonel C et al (2002) Osmoregulation and fungicide resistance: the *Neurospora crassa* os-2 gene encodes a HOG1 mitogen-activated protein kinase homologue. Appl Environ Microbiol 68:532–538

Zhang DX, Borbouse L, Gebremedhin D et al (2012) H_2O_2-induced dilation in human coronary arterioles: role of protein kinase G dimerization and large-conductance Ca^{2+}-activated K^+ channel activation. Circ Res 110:471–480

Chapter 15
Secondary Metabolism in *Botrytis cinerea*: Combining Genomic and Metabolomic Approaches

Isidro G. Collado and Muriel Viaud

Abstract Filamentous fungi produce a wide variety of bioactive secondary metabolites that play roles in development and fitness. In the pre-genomic era of *Botrytis cinerea* research, already about eight families of secondary metabolites have been isolated from *in vitro* mycelium. In particular, the predominant metabolites botrydial and botcinic acid were identified as two unspecific phytotoxins contributing to the necrotrophic and polyphagous lifestyle of the fungus. Sequencing and annotation of the complete genome revealed more than 40 clusters of genes dedicated to the synthesis of polyketides, terpenes, non-ribosomal peptides and alkaloids which indicates that *B. cinerea* has the potential to produce many metabolites that have not been described so far. By a combination of transcriptomic, mutagenesis and metabolomic approaches the genes responsible for the production of botcinic acid and botrydial were identified and significant progress was made in the elucidation of the corresponding polyketidic and terpenic biosynthesis pathways. Mutagenesis also revealed that, although these two toxins play together a significant role in plant tissues colonization, some of the other secondary metabolites seem to be crucial for the necrotrophic processes as well. A major bottleneck in the identification of these compounds is that they are not produced in sufficient amounts during standard *in vitro* conditions. However, progress in understanding the regulation of fungal secondary metabolism, through transcription factors and epigenetic mechanisms, provide new strategies to "wake up" biosynthetic genes during *in vitro* growth. This will pave the way for the characterization of these metabolites that play an important if not essential role in *B. cinerea* virulence.

Keywords Genomics • Metabolomic • Sesquiterpenes • Polyketides • Pigments • Detoxification

---

I.G. Collado (✉)
Departamento de Química Orgánica, Facultad de Ciencias, Universidad de Cádiz, Campus Rio San Pedro, 11510, Puerto Real, Cádiz, Spain
e-mail: isidro.gonzalez@uca.es

M. Viaud (✉)
INRA UMR1290 BIOGER, Avenue Lucien Brétignières, 78850 Thiverval-Grignon, France
e-mail: viaud@versailles.inra.fr

S. Fillinger, Y. Elad (eds.), *Botrytis – the Fungus, the Pathogen and its Management in Agricultural Systems*, DOI 10.1007/978-3-319-23371-0_15

15.1 Introduction

Filamentous fungi produce a wide variety of medically (e.g. antibiotics) and agriculturally important (e.g. mycotoxins) secondary metabolites (SMs). These compounds can be classified as terpenes, polyketides, non-ribosomal peptides, prenylated tryptophan derivatives, or as hybrids between these four main classes (Wiemann and Keller 2014). Although, the native biological role of many of them remains unclear, they are believed to allow the fungi to survive in their different ecological niches. One well-known example is the production of pigments that act as UV-protectants. Fungi also produce bioactive compounds that help them to compete with other organisms. In addition, SMs could act as virulence factors both on animal and plant hosts (Scharf et al. 2014).

The grey mould fungus *Botrytis cinerea* infects more than 200 plant species (see Chaps. 2, 7, and 21) thereby encountering very different abiotic and biotic stresses where SMs can be a selective advantage. The most obvious SM produced by this pathogen is melanin that gives the typical colour of grey mould and protects survival structures i.e. conidia and sclerotia from UV (Doss et al. 2003). But *B. cinerea* also produces several structurally unique terpenes and polyketides that were isolated from *in vitro* grown mycelium. The terpenes include abscisic acid (ABA), botrydial and other botryane derivatives while the polyketides include botcinic acid, botcinins and botrylactone (Collado et al. 2000, 2007; Shiina and Fukui 2009). Purified botrydial and botcinic acid showed nonspecific phytotoxic activities (Rebordinos et al. 1996; Cutler et al. 1993, 1996) and were therefore considered as putative weapons allowing *B. cinerea* to kill the cells of a large range of plant species and to feed on dead tissues.

With the availability of the genome sequence (Amselem et al. 2011; Staats and Van Kan 2012), the genetic determinants of SMs production were more easily investigated. This chapter shows how recent studies provided new insights into the arsenal of SMs that *B. cinerea* can produce, their biosynthesis routes and their functions in fungal development and virulence.

15.2 Genome Sequencing Revealed a Repertoire of Secondary Metabolism Genes

Although, the identification of genes involved in the biosynthesis of botrydial and ABA was initiated already a decade ago (Siewers et al. 2004, 2005, 2006), sequencing of *B. cinerea* genome provided a complete view of the repertoire of genes involved in the synthesis of SMs. A total of 44 genes encoding enzymes responsible for the committed biosynthetic step (called "Key" Enzyme, KE) i.e. terpene synthases, PolyKetide Synthases (PKSs), Non-Ribosomal Peptide Synthetases (NRPSs), PKS-NRPS hybrids and DiMethylAllylTryptophan Synthases (DMATS) were predicted from the sequences of the B05.10 and T4 strains (Table 15.1).

Table 15.1 Repertoire of secondary metabolism key enzymes (KEs) in *Botrytis cinerea*

KE name(s)	Gene ID in B05.10 and T4 strains (BC1G/BofuT4)	Metabolite when isolated or predicted
BcStc1/Bot2	16381.1/not predicted	Botrydial (Pinedo et al. 2008)
BcStc2	09560.1/P130680.1	Unknown sesquiterpenes
BcStc3	06357.1/P062370.1	
BcStc4	14308.1/P096230.1	
BcStc5	10537.1/P162160.1	
BcStc6	Absent/P020710.1	
BcDtc1	13295.1/P044420.1	Unknown diterpenes
BcDtc2	06148.1/P017060.1	
BcDtc3	06751.1/P132450-60.1	
BcPax1	01823.1/P135790.1	Unknown indole-diterpene
BcPhs1	13908.1/P163490.1	Retinal (Schumacher et al. 2014)
BcPks6/Boa6	16086-7.1/not predicted	Botcinic acid (Dalmais et al. 2011; Massaroli et al. 2013)
BcPks9/Boa9	15837-39.1/not predicted	
BcPks12	06876.1/P088370.1	Melanin (Schumacher et al. 2014)
BcPks13	14497.1/P077210.1	
BcBik1	Absent/absent	Bikaverin (Schumacher et al.2013)
BcPks1	13366.1/P088990.1	Unknown polyketides
BcPks2	02586.1/P108860.1	
BcPks4	04786.1/P037780.1	
BcPks8	02704.1/P119030.1	
BcPks10	04310.1/P125860.1	
BcPks11	09042.1/P059830.1	
BcPks14	08227.1/P034170-90.1	
BcPks15	01752.1/P135120.1	
BcPks16	10686-7/P152140.1	
BcPks17	01953.1/P052990.1	
BcPks18	06884-5.1/P088290.1	
BcPks19	16074.1/P119530.1	
BcPks20	07029-30.1/P085340.1	
BcPks21	13114.1/P103050.1	
BcChs1/Bpks	06032.1/P125060.1	Pyrones, resorcylic acids and resorcinols (Jeya et al. 2012)
BcNps2	03511.1/P120010-20.1	Ferrichrome siderophores (Bushley and Turgeon 2010)
BcNps3	10927-8.1/P150070.1	
BcNps7	15494.1/Not predicted	
BcNps6	10566-7.1/P162470.1	Coprogene siderophore (Bushley and Turgeon 2010)
BcNps1	07441-2.1/P002220.1	Unknown peptides
BcNps4	02495.1/P109840.1	
BcNps5	10622.1/P115630.1	
BcNps8	04782.1/P037730.1	
BcNps9	09040-1.1/P059820.1	

(continued)

Table 15.1 (continued)

KE name(s)	Gene ID in B05.10 and T4 strains (BC1G/BofuT4)	Metabolite when isolated or predicted
BcPks3	00695.1/P020270.1	Unknown amino-acid containing-polyketides (PKS-NRPS hybrids)
BcPks5	15478-79.1/P085990.1	
BcPks7	15702-4.1/P071080.1	
BcDms1	08209.1/P033980.1	Unknown alkaloids
BcDms2	Not predicted/P098370.1	

Indicated gene IDs refer to Amselem et al. (2011). Other gene IDs and inactivated genes are available at http://botbioger.versailles.inra.fr/botmut/. Corresponding metabolites are indicated when identified or predicted

Stc sesquiterpene cyclase, *Dtc* diterpene cyclase, *Pks* polyketide synthase, *Nps* non-ribosomal peptide synthetases, *Dms* dimethylallyltryptophan synthase

As in other fungal genomes, most of these KE encoding-genes are co-localized with other genes encoding enzymes (e.g. P450 monooxygenases) that likely contribute to the biosynthesis of the same SM by modifications of the initial skeleton. The number of these gene clusters suggests that *B. cinerea*, as several other fungal pathogens, may be able to produce approximately 40 different families of compounds. Notably, the repertoire of the produced SMs must be specific to *B. cinerea* since less than half of the KEs are shared with the closely related fungus *Sclerotinia sclerotiorum* that is also necrotrophic and polyphageous (Amselem et al. 2011).

Functional annotation of *B. cinerea* genes by itself did not allow the identification of those responsible for the biosynthesis of the compounds previously isolated from *in vitro* cultures. Nevertheless, fungal PKS and NRPS enzymes contain highly conserved functional domains that have been used for phylogenetic studies. This revealed that only a couple of these KEs are widely distributed among fungi and therefore present in *B. cinerea*. The first case is the PKS involved in melanin synthesis (Kroken et al. 2003; see paragraph 4.2). The second case concerns the NRPSs responsible for the biosynthesis of siderophores (Bushley and Turgeon 2010; Table 15.1; Fig. 15.10). Additionally, the genome of *B. cinerea* contains a cluster of genes that are orthologous to those responsible for the production of paxilline suggesting that an indole-diterpene could be synthetized. The biological activity most commonly associated with indole diterpenes is entomotoxicity which may contribute to protect fungal reproductive structures from insects (Saikia et al. 2008).

Below we present the set of phytotoxins and other SMs that have been experimentally investigated in *B. cinerea* so far. When available, we include the structure of the compounds, their biosynthesic routes, the corresponding required genes and hypotheses about their role in the *B. cinerea* life cycle.

15.3 Terpenes

15.3.1 Botrydial and Other Botryanes

The first SMs isolated from *B. cinerea* culture broth were the sesquiterpenes (15 carbone-terpenes) botrydial and related compounds. Since the isolation and characterization of botrydial (**1**) and dihydrobotrydial (**2**), an important number of derivatives have been characterized (Collado et al. 2000, 2007). Although the botryanes have an apparently non-isoprenoid structure, their sesquiterpenoid origin was established in a series of papers by Hanson and co-workers who focussed on defining the manner in which farnesyl pyrophosphate (FPP) is folded in fungal sesquiterpenoids (Bradshaw et al. 1981, 1982; Hanson 1981). The NMR spectra of these compounds have been assigned and have facilitated the elucidation of the structures of many additional botryane metabolites that have subsequently been isolated from the fungus (Fig. 15.1).

The formation of the various metabolites is dependent on the growth and culture conditions, such as agitation and pH. Compounds differing in the oxidation level of C-10 and C-15 are common. A number of the carboxylic acids were isolated as their methyl esters (compounds **3** and **4**) or lactones (**5** and **6**) whilst the 10:15-hemiacetals have also been obtained either free (compound **2**) or as their methyl ethers (**7** and **8**) (Collado et al. 1995, 2007; Fig. 15.1). A detailed examination of the fermentation broth led to the isolation of 1-epibotrydial (**1a**), 8,9-epibotrydial (**1b**) and 1,8,9-epibotrydial (**1c**; Durán-Patrón et al. 1999). Further series of botryane metabolites are dehydration products of botrydial (**1**) (botryendial (**9**), botryenalol (**10**), and botrydiendial (**11**)). The diene may undergo dehydrogenation to form aromatic metabolites (10-oxodehydrodihydrobotrydial (**12**) and dehydrobotrydienol (**13**); Collado et al. 1996; Durán-Patrón et al. 2000).

Additional biochemical studies have revealed detailed information about the stereochemistry of the biosynthetic steps to generate botrydial and its derivatives from FPP (Fig. 15.2; Bradshaw et al. 1981, 1982; Durán-Patrón et al. 2001, 2003). The folding of FPP from which botryane carbon skeleton is formed was distinguished by investigating the labelling pattern arising from feeding of [1,2-$^{13}C_2$]-acetate, [4,5-$^{13}C_2$]-mevalonate and the induced couplings arising from a pulse-feeding experiment of [1-^{13}C]-acetate to intact culture of *B. cinerea* (Hanson 1981). The results were consistent with the folding of FPP shown in Fig. 15.2, which was reminiscent of the biosynthetic scheme leading to the formation of caryophyllene.

The stereochemical fate of the mevalonoid hydrogen atoms shed further light on the mechanism and stereochemistry of the enzymatic formation of presilphiperfolan-8-ol (PSP; Hanson 1981). Incorporation experiments of labelled acetate and mevalonate established, in addition to the origin of the botrydial (**1**) carbon skeleton, the operation of a backbone rearrangement as well as that a 1,3-hydrogen shift

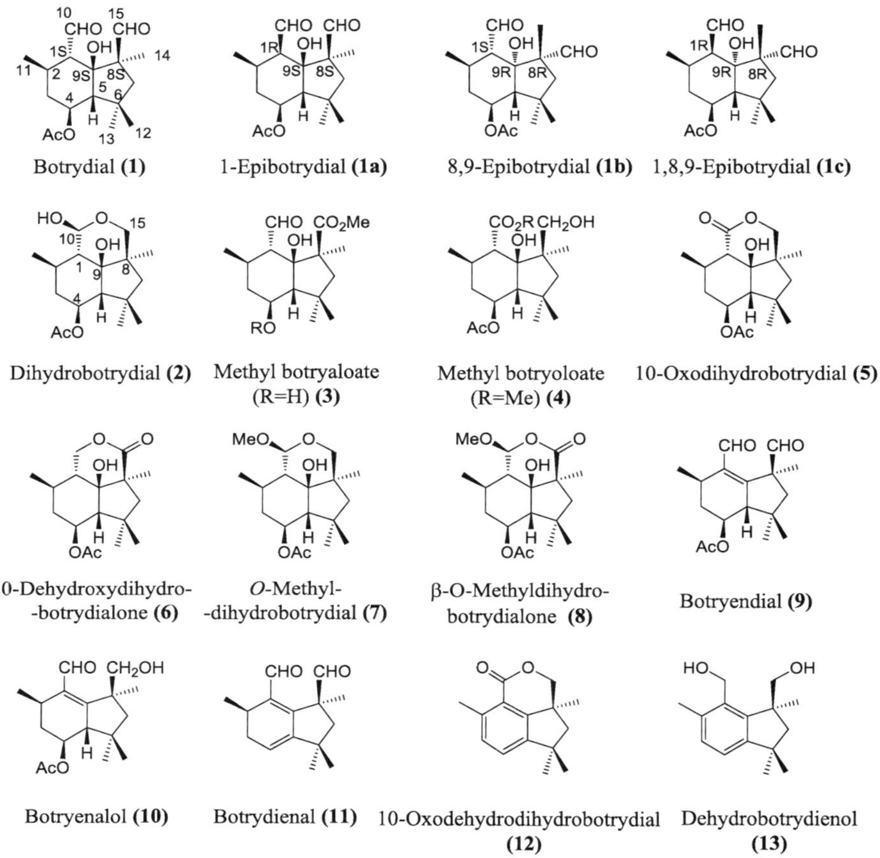

Fig. 15.1 Some of the most relevant botryane metabolites isolated from *B. cinerea*

had occurred during the course of FPP cyclization (Bradshaw et al. 1982). Furthermore, studies with $H_2{}^{18}O$ revealed that the hydroxyl group at C-9 originated from water consistent with the discharge of a presilphiperpholane C-9 cation (Hanson 1981; Fig. 15.2).

The first gene shown to be involved in botrydial synthesis (*Bcbot1*) was identified already a decade ago. It encodes a P450 monooxygenase responsible for one of the final steps of botrydial biosynthesis (Fig. 15.2; Siewers et al. 2005). Indeed, *ΔBcbot1* deletion mutants are unable to produce either 10β,15α-dihydroxyprobotryane (**2e**), botrydial (**1**), or dihydrobotrydial (**2**) (Fig. 15.2). Genome sequencing revealed that *Bcbot1* is part of a cluster of five co-expressed genes all encoding biosynthetic enzymes (Fig. 15.3; Pinedo et al. 2008).

The *Bcstc1* gene, also named *Bcbot2*, was shown to encode the SesquiTerpene Cyclase (STC) that is responsible for the corresponding step in the biosynthesis of botrydial. Indeed targeted deletion of the *Bcbot2* gene abolished production of

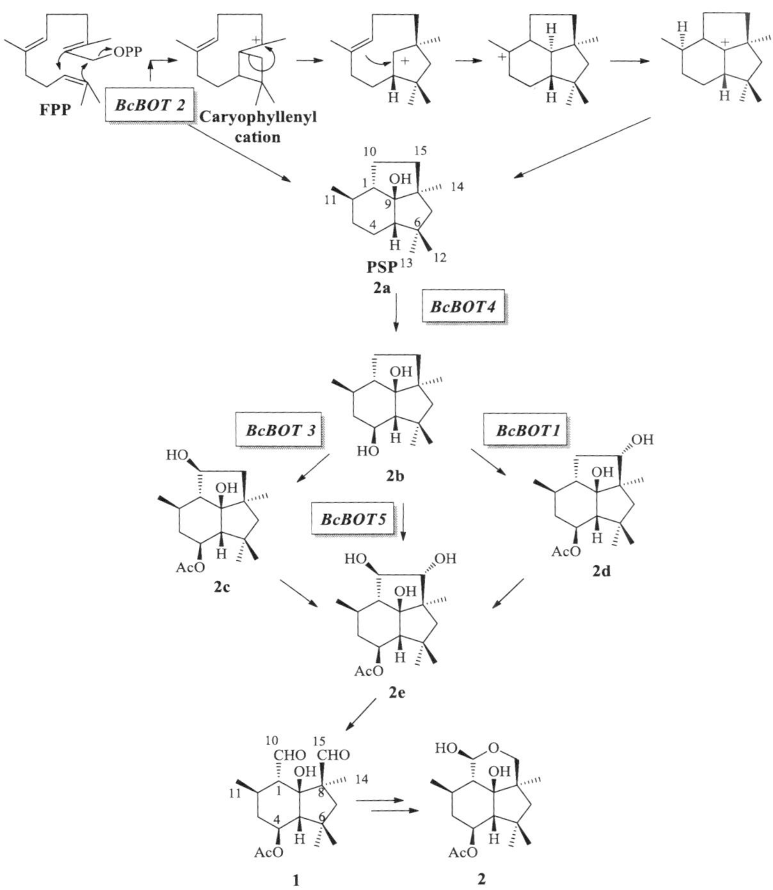

Fig. 15.2 Biosynthetic pathway from FPP to Botrydial (**1**)

botrydial and all related probotryane metabolites. Direct evidence for the biochemical function of *Bcbot2* came from the demonstration that recombinant BcBot2 protein converted FPP to the parent tricyclic alcohol PSP (**2a**; Fig. 15.2; Pinedo et al. 2008). Moreover, preparative scale incubation of labelled FPP with the BcBot2 enzyme confirmed the cyclisation processes suggested in earlier studies (Wang et al. 2009).

The cluster also includes two other P450 monooxygenase-encoding genes, *Bcbot3*, and *Bcbot4* as well as a fifth gene which showed high homology to an acetyl transferase (*Bcbot5*), suggesting that it is likely responsible for the introduction of the characteristic acetyl group of botrydial (**1**) and related late-stage biosynthetic intermediates (**2c**, **2d**; Fig. 15.2). Gene inactivation of *Bcbot3* and *Bcbot4* confirmed that they are required for botrydial synthesis (Collado et al. *unpublished*). In summary, PSP **2a** is converted to botrydial by the action of three

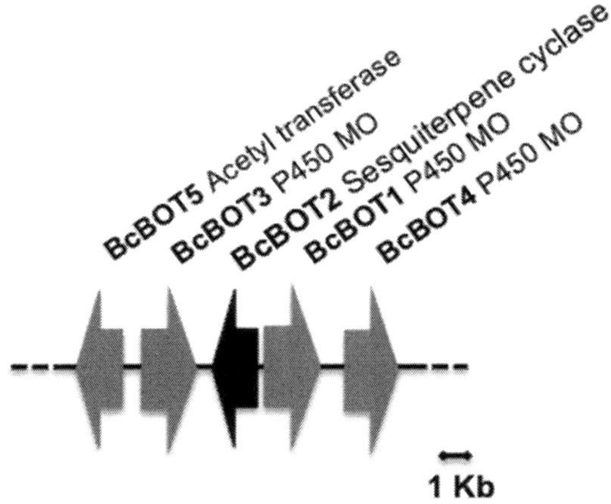

Fig. 15.3 Schematic view of the botrydial genes cluster. The *dotted line* indicates AT-rich DNA. *MO* MonoOxygenase

cytochrome P450s encoded by the genes *Bcbot1*, *Bcbot3*, and *Bcbot4* as well as by an acetyl transferase (probably BcBot5), acting in as yet an undetermined sequence.

An intriguing biosynthetic step which is not yet resolved is the cleavage of the C10–C15 bond on the trihydroxypresilphiperfolane intermediate **2e**. Incorporation experiments with isotopically labelled precursors have established the stereochemistry of this process. These results and the isolation of the probotryane-presilphiperfolane metabolites **2a**–**2e** confirmed that the dialdehyde is formed by the cleavage of a *trans* probotryane 10β:15α glycol. The structures of these biosynthetic intermediates showed that the probotryane skeleton is hydroxylated, by the three cytochrome P450s BcBot1, BcBot3, and BcBot4, in two steps, with concomitant acetylation of hydroxyl group on C4, to give the glycol **2e** with the above proposed stereochemistry (Fig. 15.2). Although the cleavage of the C-10:C-15 bond may take place by a direct enzymatic process, there is the possibility of neighbouring group participation from the C-9 hydroxyl group. Dehydrogenation of the cis 9β:10β-diol may afford a transient 9:10 epidioxide which could then undergo an acid catalyzed fragmentation of the O-O bond, leading in turn to fission of the C-10:C-15 bond and the formation of botrydial (**1**) (Collado et al. 2007).

Botrydial was first shown to be phytotoxic by Rebordinos et al. (1996) who inoculated the compound in tobacco leaves and observed typical chlorotic symptoms of grey mould. This powerful phytotoxic activity is also visible on bean leaves at concentration as low as 1 ppm (Colmenares et al. 2002a). Dihydrobotrydial and other botryane compounds are also phytotoxic but to a lesser extent (Collado et al. 1996, 1999; Durán-Patrón et al. 1999; Colmenares et al. 2002a). The toxicity of botrydial and its derivatives is not specific to plants as cytotoxic activity was also observed on human cell lines (Reino et al. 2003).

The isolation of epibotrydials (**1a–1c**), as well as the high number of botrydial derivatives isolated from *B. cinerea* broth, has allowed to obtain important conclusion about the structure activity relationship for this family of toxins, revealing that botrydial (**1**) and 8,9-epimer (**1b**), which showed the highest phytotoxic activity, presented a similar spatial disposition of carbonyl groups, with parallel aldehyde groups where the aldehydic hydrogens and the oxygen of tertiary hydroxyl groups must be coplanar, as shown in Fig. 15.4. The phytotoxic activity correlated with the aldehydic oxidation state of the substituents on C-1 and C-8 carbon atoms. Furthermore, differences in the biological activities of compounds **1** and **1c** support the idea that bioactivity is strongly correlated with the stereochemistry of substituents on C-1, C-8 and C-9 carbon atoms. Especially, the relative configuration (*S*) of the C-1 substituent seems to play a critical role in the binding of the substrate to the chemoreceptor (Durán-Patrón et al. 1999; Reino et al. 2003).

Botrydial was detected in a range of hosts infected by *B. cinerea*. On infected bean leaves, the concentration of the toxin reaches approximately 50 ppm (Deighton et al. 2001). Altogether, these data suggested that botrydial could act as a weapon to kill the cells of a large range of hosts to allow the pathogen to feed on dead tissues. The availability of *ΔBcbot1* and *ΔBcbot2* null mutants later revealed that the toxin was a strain-dependant virulence factor. Indeed, gene inactivation in the T4 strain resulted in an impaired plant tissues colonisation while the gene inactivation in the B05.10 or SAS56 backgrounds gave mutants that were as virulent as the wild-type strains (Siewers et al. 2005; Pinedo et al. 2008). This absence of virulence defect could be due to a functional redundancy with botcinic acid that is produced by the strains B05.10 and SAS56 but not by the T4 strain. Further evidence of redundancy

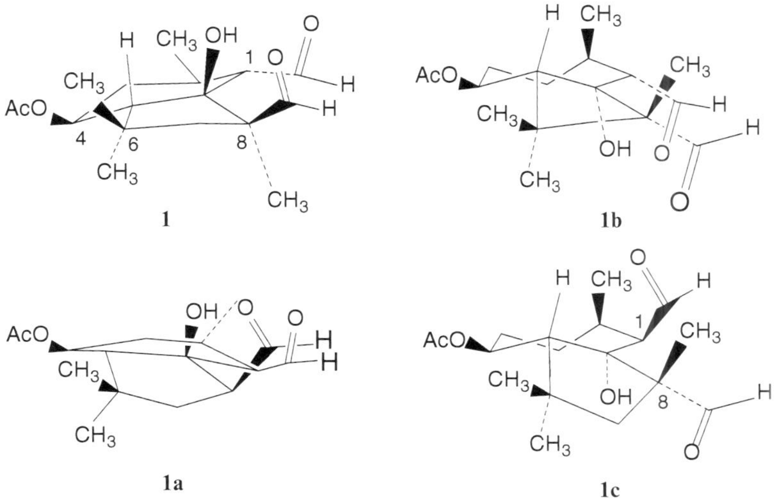

Fig. 15.4 Comparison of the spatial arrangement of carbonyl and hydroxyl groups in the botrydial and derivatives (compounds 1–1c)

was provided by the construction of double mutants that were impaired both in botrydial and in botcinic acid production (See part 4.1).

The cellular target of botrydial remains unknown but the recent work of Rossi et al. (2011) provided clues about its mode of action during plant infection. Indeed, application of botrydial on several lines of *Arabidopsis thaliana* indicated that it induces the hypersensitive response through the salicylic acid and jasmonic acid signaling pathways. Therefore, botrydial may act as effector of plant cell death.

When botrydial is present at high concentrations (50–250 ppm), it inhibits the growth of *B. cinerea* itself (Durán-Patrón et al. 2004). Interestingly, the fungus seems to possess a detoxification mechanism. Evidently, when the concentration of botrydial (**1**) reaches a particular level (approx. 100 ppm), the fungus transforms it into less active phytotoxins including dihydrobotrydial (**2**) and secobotrytrienediol (**14**; Fig. 15.5). Labelling studies with [1-^{13}C]- and [1,2-$^{13}C_2$]-acetate confirmed that secobotrytrienediol (**14**) is biosynthesized from FPP by the botryane pathway (Durán-Patrón et al. 2003). Further investigations of these degradative reactions confirmed that they are two main detoxification pathways which depend on the ambient pH as shown in Fig. 15.5 (Durán-Patrón et al. 2004; Daoubi et al. 2006).

15.3.2 Abscisic Acid

The sesquiterpene abscisic acid (ABA) (**15**; Fig. 15.6) is a well-known plant hormone, but its production by fungi was only reported for some of species including *B. cinerea* (Marumo et al. 1982). Other authors have detected 1′-deoxy-ABA in *B. cinerea* culture filtrate and presented evidences that 1′,4′-trans-ABA-diol (**16**) is an endogenous precursor of ABA. Furthermore, four new metabolites, structurally related to ABA have been reported (**17–20**; Wang et al. 2008; Fig. 15.6). Early steps of ABA biosynthesis in *B. cinerea* suggested a route from FPP via allofarnesene and ionylideneethane (Innomata et al. 2004).

Using a *B. cinerea* isolate that over-produces ABA, Siewers et al. (2004) identified a first gene involved in the biosynthesis of this sesquiterpene. The *Bcaba1* gene

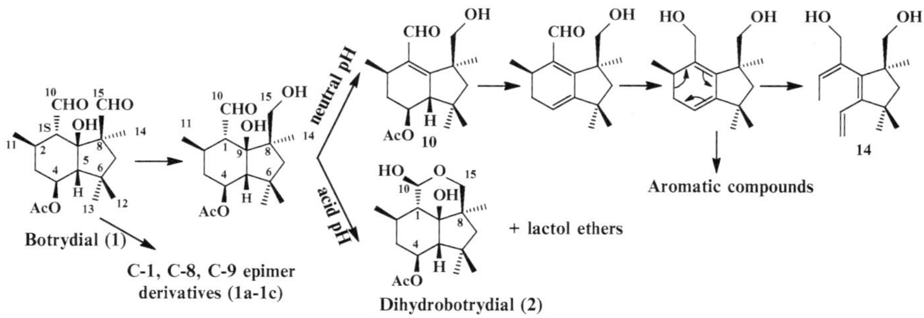

Fig. 15.5 Detoxification of botrydial and relationship between toxins produced by *B. cinerea*

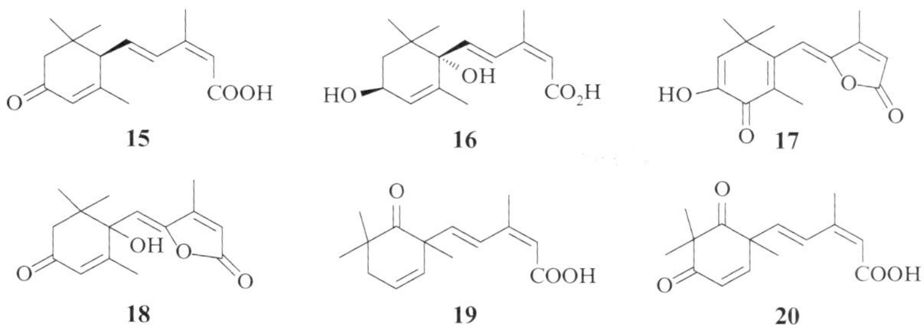

Fig. 15.6 Abscisic acid and some isolated derivatives

encoding a P450 monooxygenase was shown to be over-expressed in the presence of the ABA precursor mevalonic acid and gene inactivation resulted in the absence of ABA. The genomic locus includes three other genes (*Bcaba2-4*) encoding an additional P450 monooxygenase, a putative dehydrogenase and an unknown protein. Gene inactivation demonstrated that they are, as BcAba1, involved in the later stages of ABA biosynthesis (Siewers et al. 2006). Analysis of the intermediate compounds that are still produced by the mutants supported the biosynthesis route suggested by Innomata et al. (2004). Strikingly, the *Bcaba1-4* locus and the neighbouring genomic region do not contain a gene encoding a STC that would be expected for the corresponding biosynthesis step from FPP. The current hypothesis is that the *Bcstc* gene responsible for ABA production is localized elsewhere in the genome thus suggesting that not all genes involved in the synthesis route of one SM (here ABA) are systematically clustered in the *B. cinerea* genome. Identification of the BcStc involved in ABA biosynthesis will allow the study the whole chemical pathway.

The production of the plant hormone ABA by a pathogen is intriguing and raises questions about its possible role in manipulating the host. ABA produced by plants has often been considered a negative regulator of disease resistance through down-regulation of defense responses (reviewed in Mauch-Mani and Mauch 2005). Notably, tomato ABA-deficient mutants display enhanced resistance to *B. cinerea* which is characterized by an rapid epidermal cell wall fortification and higher induction of defense-related genes (Asselbergh et al. 2007; Chap. 18). On the other hand, the phenotype of *B. cinerea ΔBcaba1* null mutant showed that ABA production by the pathogen is not essential for virulence (Tudzynski et al. unpublished).

15.3.3 Carotenoid/Retinal

During stress conditions such as intensive illumination, *B. cinerea* mycelium sometimes shows an orange coloration that could be due to the production of carotenoids, a group of pigments that are widely produced among plants and fungi to protect them against oxidative stress. Genome annotation and transcriptomic analysis revealed a cluster of genes encoding enzymes putatively involved in the biosynthesis of carotenoid pigments i.e. a phytoene dehydrogenase (BcPhd1) and a phytoene synthase (BcPhs1). Additionally, as first described in *Fusarium fujikuroi* (Prado-Cabrero et al. 2007), the cluster also contains genes that are predicted to encode an opsin (Bop2) and a carotenoid oxygenase (BcCao1) involved in the synthesis of retinal, the chromophore for opsin. The four *B. cinerea* clustered genes are up-regulated by light which fits with the expected protective role of carotenoids and the function of opsin as photoreceptor (Schumacher et al. 2014; Chap. 13).

15.4 Polyketides

15.4.1 Botcinic Acid, Botcinins and Botrylactone

The second well-known family of toxins described in *B. cinerea* is constituted of highly hydroxylated nonanolactone polyketides called botcinolides (Cutler et al. 1993; Collado et al. 2000). In 2005, Nakajima's group reported the isolation of a group of antifungal metabolites, which they designated as botcinins (**21–30**; Fig. 15.7). The absolute configuration of botcinin A (**24**) was determined by applying the modified Mosher's method and a careful reinvestigation of the spectroscopic data reported for botcinolide analogues allowed to revise the structures of botcinolide derivatives to botcinic (**21**) and botcineric acids (**22**) and to their ring-closing derivatives, botcinins (**24–30**) (Tani et al. 2005, 2006; Fig. 15.7).

The revised structures of this group of natural products have been unequivocally determined through their total synthesis (Fukui and Shiina 2008; Fukui et al. 2009) and a revision comparing the botcinolides with their corresponding botcinin structures has been reported (Shiina and Fukui 2009). In addition, new polyketides called botrylactones (**31–33**) were described as bearing an interesting lactone skeleton featuring two oxirane bridges. The structure of the compound **31** originally identified by Welmar et al. 1979 was recently revised (Moraga et al. 2011). Botcinins (**21–30**) and botrylactones (**31–33**) (Fig. 15.7) are related polyketides sharing an identical C1–C8 fragment (Massaroli et al. 2013). Preliminary isotopic labelling experiments, in which labelled sodium acetates, propionate and L-methylmethionine, were fed to cultures of *B. cinerea* clearly demonstrated that botcinins skeleton arise from an acetate-derived polyketide, being assembled from carbon C-8 to C-1, with its four pendent methyl groups originating from l-methylmethionine (Fig. 15.7).

The iterative nature of fungal PKSs means that, in the majority of cases, there is only one PKS involved in the synthesis of one particular polyketide. However, botcinic acid as few other fungal polyketides requires the action of two PKS, BcBoa6 and BcBoa9 (Table 15.1; Fig. 15.8). These two PKS encoding genes were identified as highly co-expressed during tomato leaf infection. Deletion of either *Bcboa6* or *Bcboa9* gene indicated that they act in concert to synthetize botcinic acid (**21**): *ΔBcboa6* and *ΔBcboa9* null mutants did not produce botcinic acid or its derivatives,

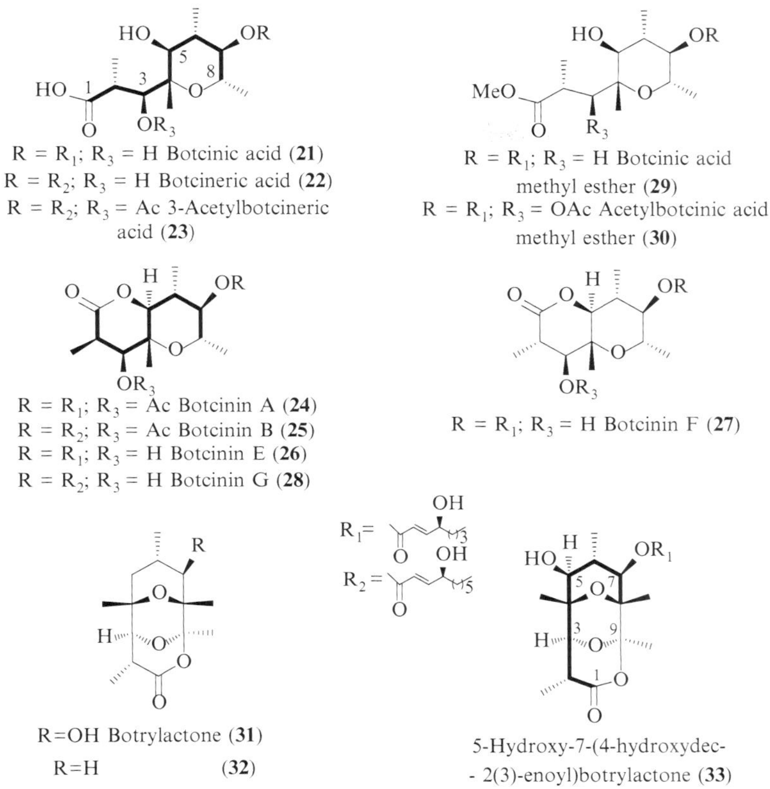

Fig. 15.7 Some of the polyketidic metabolites isolated from *B. cinerea*

but interestingly *ΔBcboa9* overproduced botrylactone (**31**) (Dalmais et al. 2011; Massarolli et al. 2013).

Further isotopic labelling experiments showed that BcBoa6 mediates the formation of the per-methylated pentaketide core, common to both botcinins and botrylactones, whereas BcBoa9 would be responsible for the 7-*O*-acyltetraketide side chain of botcinic acid and its derivatives (**21–30**) (Massaroli et al. 2013). These results enable to propose and confirm the biosynthetic route (Fig. 15.9), where the pathways to botrylactone (**31**) and botcinins (**21–30**) diverge in the botrylactone precursor (**31a**) opening up through a retro-Claisen mechanism to produce the corresponding botcinin and botrylactone derivatives. Furthermore, the biosynthetic pathway shed light on the origin of the acetate starter unit solving a longstanding mystery in the biosynthesis of botcinins.

The *Bcboa6* and *Bcboa9* genes are both clustered with other co-regulated *Bcboa* genes (*Bcboa1-6* and *Bcboa7-17* respectively, Fig. 15.8; Dalmais et al. 2011) that are expected to contribute to the biosynthesis of botcinins and botrylactones. The update

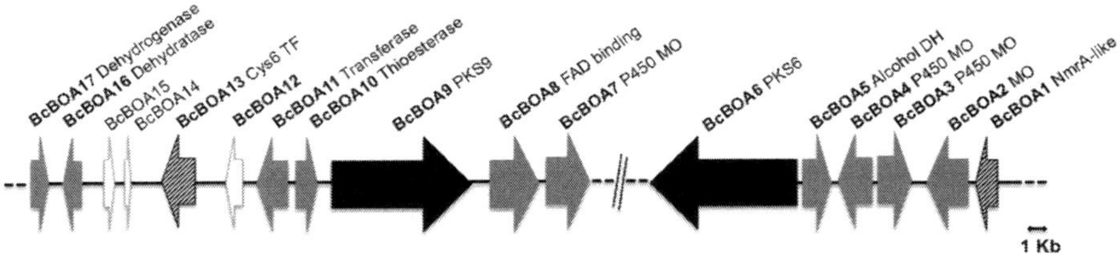

Fig. 15.8 Schematic view of the botcinic acid genes cluster. The *dotted line* indicates AT-rich DNA. *TF* Transcription factor, *MO* Monooxygenase, *DH* DeHydrogenase

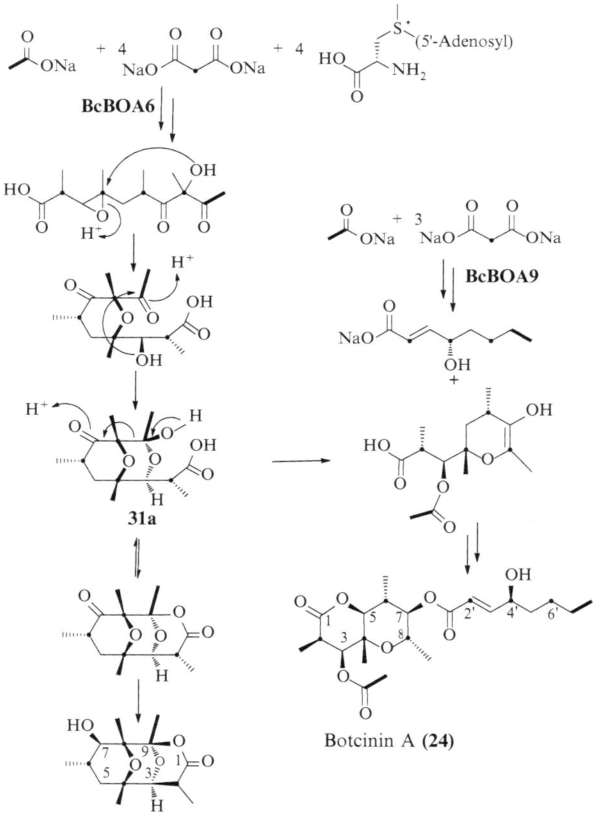

Fig. 15.9 Common biosynthetic pathway to botcinins and botrylactones

of the T4 strain genome sequence (Staats and Van Kan 2012) revealed that *Bcboa1-6* and *Bcboa7-17* loci are physically linked by a 8 kb region of AT-rich DNA.

Botcinic acid and derivatives have been described as phytotoxins provoking chlorosis and necrosis both in dicotyledonous (bean, tobacco) and monocotyledonous (wheat, corn) plants (Cutler et al. 1993, 1996). Deletion of the key enzyme encoding genes (*Bcboa6* or *Bcboa9*) did not modify significantly the virulence of B05.10 in standard pathotests. In contrast, double mutants (*ΔBcboa6 ΔBcbot2*) that do not produce botcinic acid nor botrydial were severely impaired in necrotrophic development suggesting that the two compounds have a redundant function in plant tissue colonization (Dalmais et al. 2011).

In addition to their phytotoxic activity, metabolites of the botcinin family are of particular interest because of their antifungal activity against the fungus *Magnaporthe grisea*, the causal agent of the rice blast disease (Tani et al. 2005, 2006).

15.4.2 Polyketidic Pigments: Melanin and Bikaverin

B. cinerea produces melanin of the dihydroxynaphthalene (DHN) type (Doss et al. 2003) which contributes to the typical colour of the symptoms of grey mould disease. DHN-melanin is probably the polyketide that is most widely produced by fungi (Scharf et al. 2014) and the required key enzyme, a reducing PKS, is conserved (Kroken et al. 2003). In contrast to other fungi, *B. cinerea* possesses two PKSs, BcPks12 and BcPks13, possibly involved in DHN-melanin synthesis. The *Bcpks13* gene is clustered with two genes encoding the other enzymes required for melanin biosynthesis, scytalone dehydratase (BcScd1) and hydroxynaphthalene reductase (BcBrn2). On the basis of gene expression patterns during the life cycle of *B. cinerea*, it was suggested that BcPks12 and BcPks13 are responsible for the characteristic grey colour of sclerotia and conidia, respectively (Schumacher et al. 2014). In addition to its role as a protectant against UV and other exogenous stresses (oxidizing agents, desiccation), fungal melanin is often associated with the cell wall and thereby also contribute to fungal structures i.e. conidia, sclerotia and appressoria (Scharf et al. 2014).

Recently, it was shown that some rare strains of *B. cinerea* produce an additional polyketidic red pigment, bikaverin (Schumacher et al. 2013) that was previously only described in *F. fujikuroi* and closely related species. Phylogenetic studies suggested that an ancestor of the genus *Botrytis* acquired the six clustered genes required for bikaverin synthesis (*bik1-6*) by horizontal gene transfer (HGT) from *Fusarium* (Campbell et al. 2012; Schumacher et al. 2013). The bikaverin cluster would have then subsequently decayed to different degrees in different *Botrytis* species as a result of independent mutation and deletion events (Campbell et al. 2013). Notably, even in *B. cinerea*, most of the strains (including B05.10 and T4) are impaired in bikaverin production due to the loss of the PKS encoding gene (*bik1*) and deleterious mutations in two other biosynthesis genes. The reason why only few strains of *B. cinerea* retain the ability to produce bikaverin remains unknown, but two hypotheses could be raised: the production of bikaverin could be an additional way to protect the fungus from UV damage. This is supported by the fact that, bikaverin-producing *Fusarium* species lack the PKS-encoding gene responsible for DHN-melanin biosynthesis. Bikaverin was also previously described as toxic against protozoa, nematodes, ascomycetes and oomycetes (Limon et al. 2010).

Therefore, it may confer to *B. cinerea* a selective advantage to survive as a saprotrophe in soil in the presence of fungivore nematodes and competing microbes (Campbell et al. 2013).

15.5 "Orphan" Secondary Metabolites from *Botrytis cinerea*

It is worth noting that several orphan SMs, those whose genes have not been identified yet, have been reported in *B. cinerea*. They are presented in Fig. 15.10 together with a couple of compounds that have been isolated from other *Botrytis* species.

B. cinerea seems to produce several trihydroxamate siderophores under low-iron stress conditions. However, only ferrirhodin (**34**) was characterised as the predominant secreted siderophore (Konetschny-Rapp et al. 1988). The biosynthesis of this cyclic peptide could be dependant of one of the putative ferrichrome synthases (BcNps2, 3 or 7; Table 15.1). As iron is an indispensable nutrient for all eukaryotic organisms, the ability to overcome iron limitation is crucial for fungal fitness and in some cases for full virulence (Scharf et al. 2014).

The polyketide botrallin (**35**; Overeem and Van Dijkman 1968; Kameda et al. 1974) was isolated from *B. allii*, while the antimicrobial polyketides ramulosin (**36**)

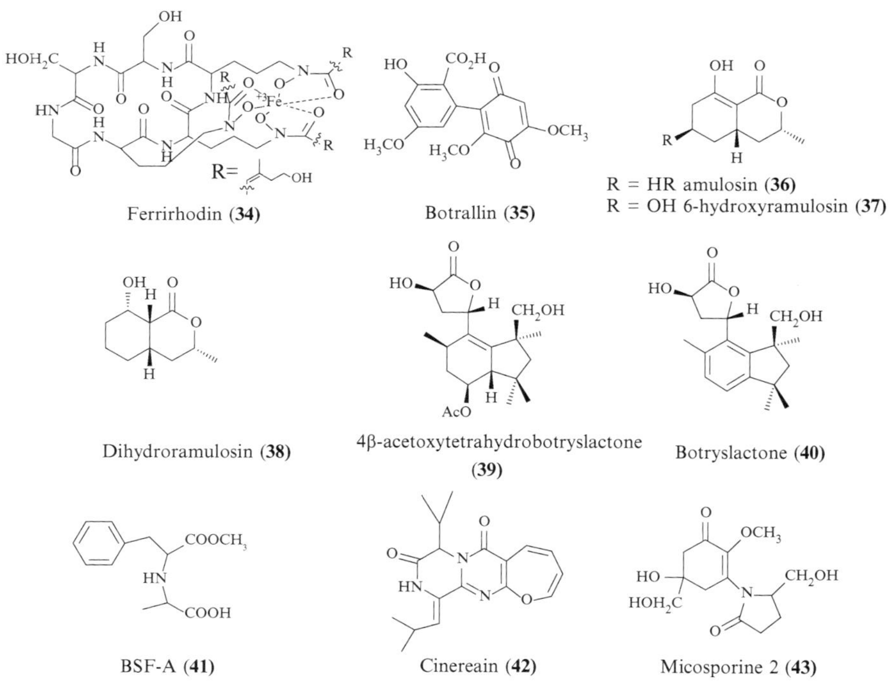

Fig. 15.10 Orphan metabolites isolated from *B. cinerea* and other *Botrytis* spp

and its derivatives (**37** and **38**), have been isolated from an undetermined *Botrytis* species (Stierle et al. 1998).

Several isolated metabolites from *Botrytis* spp. display chemical structures arising from mixed biosynthetic pathways. *B. cinerea* and *B. squamosa* produce compounds with a botryane skeleton and a five-membered lactone ring moiety at C-10: 4-acetoxytetrahydrobotryslactone (**39**; Colmenares et al. 2002b) and botryslactone (**40**; Kimura et al. 1995). The biosynthetic origin of carbons C-16, C-17, and C-18 in compounds **39** and **40** may arise from the condensation of a unit of acetate, a methyl unit and the corresponding botryane. Interestingly, compound (**40**) showed activity as plant growth inhibitor and root-promoting factor. Additional reported SMs (**41**–**42**) could be due to mixed biosynthetic pathways (possibly involving both a PKS and a NRPS): the plant growth regulator BSF-A produced by *B. squamosa* (**41**; Kimura et al. 1993) and the plant growth inhibitor cinereain produced by *B. cinerea* (**42**; Cutler et al. 1988).

Finally, the first described mycosporines have been isolated from *B. cinerea* (Fabre-Bonvin et al. 1976; Arpin et al. 1977). These SMs that are widespread in fungi absorb UV radiation and act as sunscreen. In addition, they are closely associated with light-induced conidiation and are suspected to have a sporogenic activity. In accordance with these general observations, *B. cinerea* mycosporine 2 (**43**) is produced by conidiating mycelium and absorbs UV light in the 260–340 nm range.

15.6 Strategies to Identify New (Cryptic) Metabolites

Over the past years, the field of genome mining has emerged and an important number of SMs has been characterized from sequenced fungi by genomics-guided approaches (Challis 2008). In *B. cinerea*, a range of interesting bioactive compounds have been characterized but still a high number of SMs remain completely unknown. Indeed, the majority (approx. 80 %) of the KEs listed in Table 15.1 contribute to the synthesis of undetermined compounds. Some of them must be involved in the synthesis of the "orphan" SMs presented in Fig. 15.10, but the remaining ones correspond to SMs that have not been isolated from cultures so far. One of the main reasons that hamper the identification of fungal SMs is the fact that some of the biosynthesis gene clusters are silent under standard cultivation conditions. Several global transcriptomic analyses of *B. cinerea* in different physiological stages, both during *in vitro* growth and during the infection process, have been conducted these last years (Amselem et al. 2011; Schumacher et al. 2012, 2014). From these different data sets, it appears that each SM biosynthesis cluster has its own expression pattern. Some clusters, like those involved in botrydial and botcinic acid biosynthesis, are expressed both *in vitro* and *in planta*, while some other clusters are more specifically induced at one specific developmental stage such as appressorium formation. Finally, some clusters of genes are very weakly expressed in all conditions tested so far. Therefore, one challenge for the coming years is to overcome this technical bottleneck to produce new fungal SMs from *in vitro* cultures. In this aspect, both the improvement of culture conditions and a better knowledge of the regulation of secondary metabolism would be helpful (reviewed in Sherlach and Hertweck 2009; Brakhage 2012; Wiemann and Keller 2014).

Production of SMs depends on environmental signals that can be either abiotic or biotic. Therefore, modulating the medium composition and incubation conditions could modify the patterns of compounds isolated from fungal cultures. For example, illumination and oxidative stress induce the expression of a range of SM biosynthesis genes in *B. cinerea* (Schumacher et al. 2014). Since some SMs play a role in interspecies interactions (e.g. antibiotics), co-cultivation with other microorganisms can also activate silent gene clusters as demonstrated in *Aspergillus* (Brakhage 2012).

An alternative approach to the improvement of culture conditions is genetic engineering. One possibility is to express the fungal KE gene in *Escherichia coli*, to purify the protein and to study its activity *in vitro*. As mentioned above, this allowed the identification of BcBot2 as a PSP synthase (Pinedo et al. 2008). Similarly, purified BcChs1/Bpks that belongs to the chalcone synthase superfamily (type III PKS) was shown to accept C4–C18 aliphatic acyl-CoAs and benzoyl-CoA as the starters to form pyrones, resorcylic acids and resorcinols through sequential condensation with malonyl-CoA (Joya et al. 2012). Although the biosynthetic activity of single KE can successfully be studied *in vitro*, it does not provide the identity of the final bioactive SM(s) produced by the fungus. In this respect, activation of the whole enzymatic pathway is required. Important advances in the knowledge of the control of fungal SM gene expression have been recently obtained in the model fungi *Aspergillus* spp. and *Fusarium* spp. It was demonstrated that SM gene expression is subject to regulatory controls involving both transcription factors (TFs) and chromatin structure remodelling. Secondary metabolism gene clusters often contain a gene encoding a TF required for the expression of its co-clustered genes (approx. 30 % of the clusters in *B. cinerea*). These pathway-specific TFs could act jointly with global TFs responsive to environmental factors to regulate SM gene expression. Nevertheless, the binding of TFs to the promoters of SM biosynthesis genes could be prevented by the formation of localized heterochromatin. This repression involves a fungal specific protein complex called *Velvet* and several enzymes involved in the modification of chromatin structure (Brakhage 2012).

Investigation of the signalling pathways in *B. cinerea* (fully described in Chap. 13) showed that botrydial and botcinic acid genes clusters are dependant of several TFs including BcCrz1 (Schumacher et al. 2008a), BcAtf1 (Temme et al. 2012), BcYoh1 (Simon et al. 2013), the light-responsive TF BcLtf1 (Schumacher et al. 2014) and the putative transcriptional regulator Reg1 (Michielse et al. 2011). BcCrz1 is acting downstream the calcineurin signalling pathway (Schumacher et al. 2008b) while BcAtf1 and Reg1 acts downstream the stress-induced mitogen-activated protein kinase cascade (Heller et al. 2012). Although the importance of these TFs in regulating toxin gene clusters, and other SM biosynthesis genes has been demonstrated, the evidence of a direct regulation by binding to the promoter was only determined for the global regulator BcYoh1 through a yeast-one-hybrid approach (Simon et al. 2013). Further knowledge of the specific and global TFs will provide new strategies (such as overexpression of TFs) to activate silent biosynthetic SM genes that will hopefully lead to the production of large quantities of SMs from culture broths.

As in other filamentous fungi, the *Velvet* complex was shown to link light-dependant development and secondary metabolism in *B. cinerea* (Schumacher et al. 2012, 2013, 2015; Chap. 13). The probable control of SMs production by chromatin-linked mechanisms is currently under investigation. This will pave the way for elaborating global approaches to inhibit heterochromatin formation either by epigenetic modifiers (e.g. inhibitors of histone deacetylases and of DNA methyltransferases) or by using mutants as already done in other fungi (Sherlach and Hertweck 2009; Brakhage 2012).

Once more SMs will have been characterized from *B. cinerea*, the remaining challenge for phytopathologists will be the elucidation of their role in the development of grey mould disease. From recent gene knock-out studies, three additional unknown terpenes and polyketides appear to be involved in the ability to invade plant tissues (Viaud et al. unpublished). This suggests that, beside botcinic acid and botrydial, *B. cinerea* produces several other SMs that are required for necrotrophy. When these compounds will be identified and purified, their biological activities could be investigated on the model plant *A. thaliana* as previously reported for botrydial (Rossi et al. 2011) and on other host plants.

In addition to their activities toward plants, new SMs isolated from *B. cinerea* could have interesting toxic activities toward other organisms (fungi, oomycetes, insects, bacteria, nematodes...) that can be exploited in biotechnologies. Indeed, as more fungal genome sequences are becoming available and more SM biosynthetic genes are being identified, fungi re-emerge as important resources for new therapeutic and agrochemical agents (Sherlach and Hertweck 2009).

Finally, studying *B. cinerea* secondary metabolism may provide useful start points for the identification of new biological targets to control this crop-devastating fungus. This wealth of knowledge may enable to design selective and rational structure-based fungicides. One promising example is the use of non-phytotoxic analogues of botrydial biosynthetic intermediates, that can either block the production of the toxin or generate fungitoxic compounds, both resulting in a reduced fitness of *B. cinerea* (Collado et al. 2007).

References

Amselem J, Cuomo CA, Van Kan JA et al (2011) Genomic analysis of the necrotrophic fungal pathogens *Sclerotinia sclerotiorum* and *Botrytis cinerea*. PLoS Genet 7:e1002230

Arpin N, Favre-Bonvin J, Thivend S (1977) Structure de la mycosporine 2, nouvelle molécule isolée de *Botrytis cinerea*. Tetrahedron Lett 18:819–820

Asselbergh B, Curvers K, França SC et al (2007) Resistance to *Botrytis cinerea* in *sitiens*, an abscisic acid-deficient tomato mutant, involves timely production of hydrogen peroxide and cell wall modifications in the epidermis. Plant Physiol 144:1863–1877

Bradshaw APW, Hanson JR, Nyfeler R (1981) Studies in terpenoid biosynthesis. Part 24. The formation of the carbon skeleton of the sesquiterpenoid dihydrobotrydial. J Chem Soc Perkin 1:1469–1472

Bradshaw APW, Hanson JR, Nyfeler R (1982) Studies in terpenoid biosynthesis. Part 25. The fate of the mevalonoid hydrogen atoms in the biosynthesis of the sesquiterpenoid, dihydrobotrydial. J Chem Soc Perkin Trans 1:2187–2192

Brakhage AA (2012) Regulation of fungal secondary metabolism. Nat Rev Microbiol 11:21–32

Bushley KE, Turgeon BG (2010) Phylogenomics reveals subfamilies of fungal nonribosomal peptide synthetases and their evolutionary relationships. BMC Evol Biol 10:26

Campbell MA, Rokas A, Slot J (2012) Horizontal transfer and death of a fungal secondary metabolic gene cluster. Genome Biol Evol 4:289–293

Campbell MA, Staats M, Van Kan JA et al (2013) Repeated loss of an anciently horizontally transferred gene cluster in *Botrytis*. Mycologia 105:1126–1134

Challis GL (2008) Mining microbial genomes for the new natural products and biosynthetic pathways. Microbiology 154:1555–1569

Collado IG, Hernández-Galán R, Durán-Patrón R et al (1995) Metabolites from a shake cultive of *Botrytis cinerea*. Phytochemistry 38:647–650

Collado IG, Hernández-Galán R, Prieto V et al (1996) New biologically active metabolites from the fungus *Botrytis cinerea*. Phytochemistry 41:513–517

Collado IG, Durán-Patrón R, Hernández-Galán R (1999) Some evidence on the role of exudated toxins in the pathogenesis of *Botrytis cinerea*. Rec Adv Allelopathy 1:479–484

Collado IG, Aleu J, Hernández-Galán R et al (2000) *Botrytis* species: an intriguing source of metabolites with a wide range of biological activities. Structure, chemistry and bioactivity of metabolites isolated from *Botrytis* species. Curr Org Chem 4:1261–1286

Collado IG, Macias-Sánchez AJ, Hanson JR (2007) Fungal terpene metabolites: biosynthetic relationships and the control of the phytopathogenic fungus *Botrytis cinerea*. Nat Prod Rep 24:674–686

Colmenares AJ, Aleu J, Duran-Patron R, Collado IG, Hernandez-Galan R (2002a) The putative role of botrydial and related metabolites in the infection mechanism of *Botrytis cinerea*. J Chem Ecol 28:997–1005

Colmenares AJ, Durán-Patrón RM, Hernández-Galán R et al (2002b) Four new lactones from *Botrytis cinerea*. J Nat Prod 65:1724–1726

Cutler HG, Springer JP, Arrendale RF et al (1988) Cinereain: a novel metabolite with plant growth regulating properties from *Botrytis cinerea*. Agric Biol Chem 52:1725–1733

Cutler HG, Jacyno JM, Harwood JS et al (1993) Botcinolide: a biologically active natural product from *Botrytis cinerea*. Biosci Biotechnol Biochem 57:1980–1982

Cutler HG, Parker SR, Ross SA et al (1996) Homobotcinolide: a biologically active natural homolog of botcinolide from *Botrytis cinerea*. Biosci Biotechnol Biochem 60:656–658

Dalmais B, Schumacher J, Moraga J et al (2011) The *Botrytis cinerea* phytotoxin botcinic acid requires two polyketide synthases for production and has a redundant role in virulence. Mol Plant Pathol 12:564–579

Daoubi M, Duran-Patron R, Hernandez-Galan R et al (2006) The role of botrydienediol in the biodegradation of the sesquiterpenoid phytotoxin botrydial by *Botrytis cinerea*. Tetrahedron 62:8256–8261

Deighton N, Muckenschnabel I, Colmenares AJ et al (2001) Botrydial is produced in plant tissues infected by *Botrytis cinerea*. Phytochemistry 57:689–692

Doss RP, Deisenhofer J, Krug von Nidda HA et al (2003) Melanin in the extracellular matrix of germlings of *Botrytis cinerea*. Phytochemistry 63:687–691

Durán-Patrón R, Hernández-Galán R, Rebordinos L et al (1999) Structure-activity relationships of new phytotoxic metabolites with the botryane skeleton from *Botrytis cinerea*. Tetrahedron 55:2389–2400

Durán-Patrón R, Hernández-Galán R, Collado IG (2000) Secobotrytriendiol and related sesquiterpenoids: new phytotoxic metabolites from *Botrytis cinerea*. J Nat Prod 63:182–184

Durán-Patrón R, Colmenares AJ, Hernández-Galán R et al (2001) Some key metabolic intermediates in the biosynthesis of botrydial and related compounds. Tetrahedron 57:1929–1933
Durán-Patrón R, Colmenares AJ, Montes A et al (2003) Studies on the biosynthesis of secobotryane skeleton. Tetrahedron 59:6267–6271
Durán-Patrón R, Cantoral JM, Hernández-Galán R et al (2004) The biodegradation pathways of the self-produced phytotoxic metabolite botrydial by *Botrytis cinerea*. J Chem Res 6:441–443
Favre-Bonvin J, Arpin N, Brevard C (1976) Structure de la mycosporine (P 310). Can J Chem 54:1105–1113
Fukui H, Shiina I (2008) Asymmetric total synthesis of botcinins C, D, and F. Org Lett 10:3153–3156
Fukui H, Tsuji K, Umezaki Y et al (2009) Total synthesis and the confirmation of the revised structures of botcinins A and B. Heterocycles 79:403–410
Hanson JR (1981) The biosynthesis of some sesquiterpenoids. Pure Appl Chem 53:1155–1162
Heller J, Ruhnke N, Espino JJ et al (2012) The mitogen-activated protein kinase BcSak1 of *Botrytis cinerea* is required for pathogenic development and has broad regulatory functions beyond stress response. Mol Plant Microbe Interact 25:802–816
Inomata M, Hirai N, Yoshida R, Ohigashi H (2004) The biosynthetic pathway to abscisic acid via ionylideneethane in the fungus *Botrytis cinerea*. Phytochemistry 65:2667–2678
Jeya M, Kim TS, Tiwari MK et al (2012) The *Botrytis cinerea* type III polyketide synthase shows unprecedented high catalytic efficiency toward long chain acyl-CoAs. Mol BioSyst 8:2864–2867
Kameda K, Aoki H, Namiki M (1974) An alternative structure for botrallin a metabolite of *Botrytis allii*. Tetrahedron Lett 1:103–106
Kimura Y, Fujioka H, Nakajima H et al (1993) BSF-A, a new plant growth regulator produced by the fungus *Botrytis squamosa*. Biosci Biotechnol Biochem 57:1584–1585
Kimura Y, Fujioka H, Hamasaki T et al (1995) Botryslactone, a new plant growth regulator produced by *Botrytis squamosa*. Tetrahedron Lett 36:7673–7676
Konetschny-Rapp S, Jung G, Huschka HG et al (1988) Isolation and identification of the principal siderophore of the plant pathogenic fungus *Botrytis cinerea*. Biol Met 1:90–98
Kroken S, Glass NL, Taylor JW et al (2003) Phylogenomic analysis of type I polyketide synthase genes in pathogenic and saprobic ascomycetes. Proc Natl Acad Sci U S A 100:15670–15675
Limon MC, Rodriguez-Ortiz R, Avalos J (2010) Bikaverin production and applications. App Microb Biotechnol 87:21–29
Marumo S, Katayama M, Komori E et al (1982) Microbial production of abscisic acid by *Botrytis cinerea*. Agric Biol Chem 46:1967–1968
Massaroli M, Moraga J, Bastos Borges K et al (2013) A shared biosynthetic pathway for botcinins and botrylactones revealed through gene deletion. ChemBioChem 14:132–136
Mauch-Mani B, Mauch F (2005) The role of abscisic acid in plant-pathogen interactions. Curr Opin Plant Biol 8:409–414
Michielse CB, Becker M, Heller J et al (2011) The *Botrytis cinerea* Reg1 protein, a putative transcriptional regulator, is required for pathogenicity, conidiogenesis, and the production of secondary metabolites. Mol Plant Microbe Interact 24:1074–1085
Moraga J, Pinedo C, Duran-Patron R et al (2011) Botrylactone: new interest in an old molecule, review of its absolute configuration and related compounds. Tetrahedron 67:417–420
Overeem JC, Van Dijkman A (1968) Botrallin, a novel quinone produced by *Botrytis allii*. Rec Trav Chim NL 87:940–944
Pinedo C, Wang CM, Pradier JM et al (2008) Sesquiterpene synthase from the botrydial biosynthetic gene cluster of the phytopathogen *Botrytis cinerea*. ACS Chem Biol 3:791–801
Prado-Cabrero A, Scherzinger D, Avalos J et al (2007) Retinal biosynthesis in fungi: characterization of the carotenoid oxygenase CarX from *Fusarium fujikuroi*. Eukaryot Cell 6:650–657
Rebordinos L, Cantoral JM, Prieto MV et al (1996) The phytotoxic activity of some metabolites of *Botrytis cinerea*. Phytochemistry 42:383–387

Reino JL, Durán-Patrón R, Segura I et al (2003) Chemical transformations on botryane skeleton. Effect on the cytotoxic activity. J Nat Prod 66:344–349

Rossi FR, Garriz A, Marina M et al (2011) The sesquiterpene botrydial produced by *Botrytis cinerea* induces the hypersensitive response on plant tissues and its action is modulated by salicylic acid and jasmonic acid signalling. Mol Plant Microbe Interact 24:888–896

Saikia S, Nicholson MJ, Young C et al (2008) The genetic basis for indole-diterpene chemical diversity in filamentous fungi. Mycol Res 112:184–199

Scharf DH, Heinekamp T, Brakhage AA (2014) Human and plant fungal pathogens: the role of secondary metabolites. PLoS Pathog 10:e1003859

Scherlach K, Hertweck C (2009) Triggering cryptic natural product biosynthesis in microorganisms. Org Biomol Chem 7:1753–1760

Schumacher J, de Larrinoa IF, Tudzynski B (2008a) Calcineurin-responsive zinc finger transcription factor CRZ1 of *Botrytis cinerea* is required for growth, development, and full virulence on bean plants. Eukaryot Cell 7:584–601

Schumacher J, Viaud M, Simon A et al (2008b) The Galpha subunit BCG1, the phospholipase C (BcPLC1) and the calcineurin phosphatase co-ordinately regulate gene expression in the grey mould fungus *Botrytis cinerea*. Mol Microbiol 67:1027–1050

Schumacher J, Pradier JM, Simon A et al (2012) Natural variation in the VELVET gene bcvel1 affects virulence and light-dependent differentiation in *Botrytis cinerea*. PLoS One 7:e47840

Schumacher J, Gautier A, Morgant G et al (2013) A functional bikaverin biosynthesis gene cluster in rare strains of *Botrytis cinerea* is positively controlled by VELVET. PLoS One 8:e53729

Schumacher J, Simon A, Cohrs KC et al (2014) The transcription factor BcLTF1 regulates virulence and light responses in the necrotrophic plant pathogen *Botrytis cinerea*. PLoS Genet 10:e1004040

Schumacher J, Simon A, Cohrs KC et al (2015) The VELVET complex in the gray mold fungus *Botrytis cinerea*: impact of BcLAE1 on differentiation, secondary metabolism and virulence. Mol Plant Microbe Interact 28:659–674. doi:http://dx.doi.org/10.1094/MPMI-12-14-0411-R

Shiina I, Fukui H (2009) Chemistry and structural determination of botcinolides, botcinins, and botcinic acids. Chem Commun 4:385–400

Siewers V, Smedsgaard J, Tudzynski P (2004) The P450 monooxygenase BcABA1 is essential for abscisic acid biosynthesis in *Botrytis cinerea*. Appl Environ Microbiol 70:3868–3876

Siewers V, Viaud M, Jiménez-Teja D et al (2005) Functional analysis of the cytochrome P450 monooxygenase gene *bcbot1* of *Botrytis cinerea* indicates that botrydial is a strain-specific virulence factor. Mol Plant Microbe Interact 18:602–612

Siewers V, Kokkelink L, Smedsgaard J et al (2006) Identification of an abscisic acid gene cluster in the grey mold *Botrytis cinerea*. Appl Environ Microbiol 72:4619–4626

Simon A, Dalmais B, Morgant G et al (2013) Screening of a *Botrytis cinerea* one-hybrid library reveals a Cys2His2 transcription factor involved in the regulation of secondary metabolism gene clusters. Fungal Genet Biol 52:9–19

Staats M, Van Kan JA (2012) Genome update of *Botrytis cinerea* strains B05.10 and T4. Eukaryot Cell 11:1413–1414

Stierle DB, Stierle AA, Kunz A (1998) Dihydroramulosin from *Botrytis* sp. J Nat Prod 61:1277–1278

Tani H, Koshino H, Sakuno E et al (2005) Botcinins A, B, C, and D, metabolites produced by Botrytis cinerea, and their antifungal activity against *Magnaporthe grisea*, a pathogen of rice blast disease. J Nat Prod 68:1768–1772

Tani H, Koshino H, Sakuno E et al (2006) Botcinins E and F and botcinolide *from Botrytis cinerea* and structural revision of botcinolides. J Nat Prod 69:722–725

Temme N, Oeser B, Massaroli M et al (2012) BcAtf1, a global regulator, controls various differentiation processes and phytotoxin production in *Botrytis cinerea*. Mol Plant Pathol 13:704–718

Wang TS, Zhou JY, Tan H (2008) Three new metabolites from *Botrytis cinerea*. J Asian Nat Prod Res 10:919–924

Wang CM, Hopson R, Lin X et al (2009) Biosynthesis of the sesquiterpene botrydial in *Botrytis cinerea*. Mechanism and stereochemistry of the enzymatic formation of presilphiperfolan-8beta-ol. J Am Chem Soc 131:8360–8361

Welmar K, Tschesche R, Breitmaier E (1979) Botrylacton, ein neuer Wirkstoff aus der Nährlösung des Pilzes *Botrytis cinerea* 2. Chem Ber 112:3598–3602

Wiemann P, Keller NP (2014) Strategies for mining fungal natural products. J Ind Microbiol Biotechnol 41:301–313

Chapter 16
Contribution of Proteomics Research to Understanding *Botrytis* Biology and Pathogenicity

Eva Liñeiro, Jesús Manuel Cantoral, and Francisco Javier Fernández-Acero

Abstract Recent work has clearly shown the capacity of proteomics-based methodologies to establish the roles played by specific proteins in different biological processes. Beyond the study of genes, it has been established that proteins are the relevant set to be analyzed in research aiming to solve specific biological questions. Proteomics approaches can be categorized according to three different methodologies; gel-based, mainly two-dimension gel electrophoresis (2-DE); gel free, based on liquid chromatography-mass spectrometry (LC-MS); and quantitative proteomics, by isobaric markers. Most of these methodologies have been applied to studies of the proteome of *Botrytis cinerea*. Since the publication of the first proteomics report on *Botrytis*, technological advances have accelerated the identification of global protein content. Clearly, the publication of the *B. cinerea* genome has been of tremendous value to the proteomics research community; this has supported the accurate identification, through MS, of this fungus' peptides. This landmark event has greatly facilitated the development of proteomics studies exploring the biology of the fungus; to date, mainly mycelium samples have been used. Only a few reports have aimed at the study of fractions of the total proteome, and all of these are focused on the secretome. The role of several particular proteins related to fungal pathogenicity, metabolism, biology, etc. has been elucidated, but the number of *Botrytis* proteins found, as a proportion of the total proteins predicted from the genome, remains below 10 %. There is much work to be done.

Keywords Subproteome • Secretome • 2-DE • LC-MS • Shotgun proteomics

E. Liñeiro • J.M. Cantoral • F.J. Fernández-Acero (✉)
Laboratory of Microbiology, Marine and Environmental Sciences Faculty,
Institute of Wine and Food Research, University of Cadiz,
Puerto Real, Cádiz 11510, Spain
e-mail: franciscojavier.fernandez@uca.es

S. Fillinger, Y. Elad (eds.), *Botrytis – the Fungus, the Pathogen and its Management in Agricultural Systems*, DOI 10.1007/978-3-319-23371-0_16

16.1 Introduction

For most of the last 30 years the study of phytopathogenic fungi has been undertaken from a classical viewpoint, through the use of molecular biology techniques based on the isolation and characterization of individual genes and proteins, one by one. In the late 1990s and early 2000s, the development of large-scale DNA sequencing technology produced a significant breakthrough, allowing the publication of many fungal genomic sequences. After the publication of the first fungal genome in 1996, (Dujon 1996), the development of functional genomic analyses and large-scale techniques such as transcriptomics, proteomics and, more recently, metabolomics, effectively transformed research into plant fungal pathology. This evolution, from the study of genomics to the application of these new research techniques, has led some authors to describe the current period as the "post-genomic era" (Fernandez-Acero et al. 2007a; James 1997; Kim et al. 2007). Currently, among the possibilities that "-omics" offer towards achieving a better understanding of *Botrytis cinerea* biology, high-throughput proteomics techniques, in conjunction with advances in bioinformatics, have been shown to be very useful and effective tools in the research of various aspects related to biological, metabolic, physiological and pathological processes, as we will describe in the course of this chapter.

Proteomics is the science concerned with the study of the structure, function and interaction of the whole set of proteins codified by an organism (Wilkins et al. 1995). It comprises a wide range of tools, techniques and technological approaches for the systematic, large-scale analysis of the proteome; this latter term has, in turn, been defined as the complete set of proteins expressed by an individual organism, a tissue, or a cell, etc., at a given time and under certain conditions. Proteomics studies may also be directed at studying a specific subset of proteins of an organism (referred to as a subproteome), such as specific organelles (Fernández-Acero et al. 2010; Garrido et al. 2011; Kim et al. 2007). Concerning fungal biology, the most relevant and studied subproteome is the secretome, which is defined as the combination of native secreted proteins and the cellular machinery involved in their secretion. This subproteome consists primarily of cell wall degrading enzymes (CWDEs), many of which are known virulence factors in these organisms (Fernández-Acero et al. 2010). The importance of this information for unravelling pathogenesis and the infection cycle is undisputable.

16.1.1 Technical Constraints: From Past to Present

Proteins are the final effectors for most of the processes occurring in the cell and, currently, proteomics methodologies based on Mass Spectrometry (MS) offer new tools for gaining insights into these processes. MS is an analytical technique used to determine the molecular mass of a compound by measuring the mass/charge ratio (m/z) of ions under vacuum conditions. When it was first discovered, MS could not be applied to the study of proteins, but in the late 1990s, the development of "Matrix-Assisted Laser Desorption/Ionization" (MALDI) (Hillenkamp et al. 1991) and

"Electrospray ionization" (ESI) (Fenn et al. 1989), enabled mass spectrometry to be applied to the study of proteins and peptides. MALDI and ESI are two soft ionization techniques that allow gas phase ions to be obtained from large biomolecules without excessive fragmentation, and these techniques were crucial for the initial development of proteomics (Dass 2000).

Unlike the genome, which remains constant under all circumstances, the proteome is a highly dynamic system that varies depending on the environmental, physiological and/or pathological conditions of the organism. Proteins are regulated in their activity and level in response to internal or external events, either as a result of a differential expression of genes or depending on post-transcriptional events such as differential RNA splicing/editing and post-translational modifications (PTMs) of the synthesized peptides. More than 200 different types of PTMs have been described (Minguez et al. 2012). Among the best known are phosphorylation, glycosylation and proteolysis, which can redirect the biological function of any synthesized peptide. It can be said that the proteome of a given organism is the expression of its phenotype. Proteomics studies are among the most appropriate approaches for analyzing the biological processes that take place in an organism, as well as for understanding the complex network of interactions involved in cellular functions.

The approaches used in proteomics studies are labor-intensive because they depend heavily on the specific requirements of each particular study. Due to the dynamic nature of the proteome, each change in the assayed conditions entails a dramatic change in protein synthesis, PTMs, etc., and, therefore, in the results and conclusions of the study, analysis or experiment. In addition the proteome of phytopathogenic fungi undergoes continuous changes over time and it is crucial to decide the right time or stage when biological samples should be taken (mycelia, conidia, sclerotia, etc.) and to repeat the sampling at the same point in all replicates carried out. Therefore a good experimental design is crucial for obtaining good final results (Fernandez-Acero et al. 2007a). That design will include the choice of the best methodology for sample preparation, harvesting, protein extraction and separation, MS analysis, protein identification and quantification, as well as data analysis and interpretation. Concerning filamentous fungi, a key limiting factor is the choice of an effective protein extraction method capable of breaking open their exceptionally robust cell wall and of solubilizing all mycelial proteins free from contaminants. Polysaccharides, lipids, pigments and other fungal metabolites are among the contaminants that can reduce the protein extraction ratio and/or disturb the subsequent separation of the proteins. This is particularly important because it is impossible to study proteins if the extraction method is unable to extract them from their natural location in the cell. Briefly, a fungal protein extraction protocol should comprise at least three stages: homogenization, protein precipitation and cleaning, and the protocol should be adapted to the particular requirement of the species being studied. Advances in fungal protein extraction procedures have been reviewed by several authors (Garrido et al. 2011; González Fernández et al. 2010).

Once the proteomics researcher has decided on a good protein extraction method, a wide range of possibilities for protein separation and MS identification is available. Basically these possibilities can be classified into two categories: gel-based and gel-free methods. The standard proteomics study of an organism has traditionally been based on the separation of the complex mix of proteins that make up its proteome by two-dimensional electrophoresis in polyacrylamide gel (2-DE), in combination with mass spectrometry identification. As we will see in detail in the next section, this is one of the most commonly used methods for proteomics research on *B. cinerea*. The 2-DE technique couples a first separation of proteins on the basis of their isoelectric point (first dimension) with a second separation according to their molecular mass (second dimension). The resulting gel is stained and digitalized. Once the resulting protein spot profile is analyzed and studied, proteins are picked directly from the gel and digested with a protease (typically trypsin), resulting in a set of peptides which are identified with a MALDI-TOF (Matrix-Assisted Laser Desorption/Ionization type – Time of Flight) mass spectrometry by Peptide Mass Fingerprinting (PMF) or with a MALDI- TOF/TOF by Peptide Fragmentation Fingerprinting (PFF), and data base searching. 2-DE techniques allow large-scale separation with enough resolution power to resolve complex mixtures of proteins. The visual perspective of the proteome thus obtained allows the easy discrimination of PTMs and protein isoforms; these techniques do, however, present limitations in terms of (i) co-migrating proteins, and spots produced by abundant proteins that may mask the more scarce proteins (due to the large dynamic range of proteins present in a cell) thus complicating the analysis; (ii) inadequate sensitivity, depending on the staining used, which can also lead to difficulties in detecting low-abundance proteins; and (iii) low reproducibility among replicates.

Some of these limitations have been overcome with the emergence of gel-free technologies based on liquid chromatography (LC) methods, coupled directly to electrospray ionization mass spectrometers (LC-ESI-MS or LC-ESI-MS/MS). LC methods allow different separation principles such ion affinity, molecular mass, hydrophobicity, etc. to be combined in an orthogonal way to achieve the highest resolution such as, for instance, a combination of strong cation exchange (SCX) followed by a high-resolution reverse phase separation of peptides. That technology is known as "multidimensional protein identification technology" (MudPIT). This methodology provides several advantages over the traditional 2-DE approach; it makes possible "shotgun proteomics" studies, in which large-scale identification of complex protein mixtures may be carried out by MS/MS simultaneously and immediately after their chromatographic separation in a fast and reliable way and with high sensitivity. MudPIT techniques are particularly important for identifying low abundance proteins and others that are often difficult to resolve in gels because of their distinctive physical properties such as extreme hydrophobicity, high or low molecular weight, or high or low pI values. The methodology gives a wide coverage of proteins and therefore allows a wide range of proteomes to be described. It has also facilitated the development of different experimental approaches like, for instance, the combination of one-dimensional electrophoresis in polyacrylamide gel (1-D) and LC-MS/MS (1D-LC-MS/MS), to describe complex proteomes. This latter gel-based technique, with a previously-described 2-DE analysis, are the most widespread approaches currently used in *B. cinerea* proteomics research and is yielding good results in term of sensitivity and number of proteins identified.

At this point in time when, all in all, thousands of fungal proteins can be separated and identified relatively quickly and easily, the challenge in *Botrytis* proteomics research is the application of new strategies aimed at improving the accurate quantitation of proteins. The methodology used to conduct a quantitative proteomics analysis is based on comparative analysis between the protein profiles obtained for the same organism, cell, etc., but under two different biological conditions, *e.g.* stimulated *vs.* unstimulated. Traditionally the mode of operation for carrying out a quantitative proteomics analysis is based on the analysis of the image obtained from 2-DE gels. In these studies proteins can be detected by staining them after 2-DE or by labeling them on lysine side chains before 2-DE with two different fluorescent cyanine dyes, each one having the same mass but different excitation and emission wavelengths. In the first approach, proteins from different samples are separated in different 2-DE gels, proteins are stained, the gels are digitalized and the obtained images are matched and compared with each other. In the second approach, however, proteins from different samples are fluorochrome-labeled, mixed and loaded onto the same 2-DE gel. The result is known as a "2-Dimensional fluorescence difference Gel Electrophoresis" (2-DIGE). After electrophoresis, the gel is digitalized using a fluorescence reader fitted with two different filters and two sets of images are obtained. Then images are overlaid and processed, revealing in a visual way the quantitative differences between the samples. Using this method, problems linked to reproducibility (local gel structure, running conditions, etc.) are avoided because the samples are run together. Since these approaches are gel-based they present the same limitations as mentioned above. The results rely on the size and intensity of the analyzed spots, and are highly dependent on the sensitivity and selectivity of the labeling-method used and the resolution of the processed image.

The progress made in recent years in LC-MS has provided a number of new, more reproducible and sensitive strategies for quantitative shotgun proteomics analyses based on the abundance or intensity of the analyzed peptides (MS) and/or fragments from these (MS/MS). These methods can be divided into two categories: labeled or label-free methods. In label-based approaches, proteins obtained from multiple biological samples under different conditions are tagged with stable-isotopic reagents whose physicochemical properties are identical except for their mass (one contains a light isotope and the other one contains a heavy isotope). Then samples are mixed and analyzed simultaneously (so reproducibility problems are eliminated) by LC-MS/MS using the isotopic-label as internal standard or references; this procedure gives accurate quantitative information based in peptide intensity. Label-based methods comprise different techniques which differ in their labeling chemistry and site of attachment, such as "Isotope-Coded Affinity Tag" (ICAT), "Isotope-Coded Protein Labeling" (ICPL), "Isobaric Tags for Relative and Absolute Quantification" (iTRAQ) and "Stable Isotope Labeling by Amino Acid in Cell Culture" (SILAC). Label-free approaches rely on the fact that the greater the abundance of a specific peptide the greater the number of spectral counts produced by that peptide; hence the number of spectral counts is linearly proportional to the concentration of the peptide measured. In a label-free approach peptides from different biological samples have to be compared across different LC-MS/MS runs, so results are less reproducible. To date (as we will see in next section) only label-free methods have been introduced, albeit gradually, into proteomics research on *B. cinerea*.

16.2 Proteomics Approaches to *B. cinerea*

16.2.1 Global Proteomics

Over the last 8 years an increasing number of original papers on *Botrytis cinerea* proteomics have been published; together these studies have reported a long list of proteins synthesized by this fungus in specific growth stages or assay conditions, and therefore provide valuable biological information, which will be discussed in the following sections. Most of these research reports deal with the study of the secretome, which has a particular relevance in phytopathogenic fungi, since it is considered a key element in the establishment and development of the infection cycle.

The initial proteomics study of *B. cinerea* was reported by Fernández-Acero et al. (2006). In this work, a protocol was optimized for protein extraction from *B. cinerea* mycelium using TCA-acetone protein precipitation, and the first proteome map of the fungus was described by 2-DE and MS analysis. Around 400 protein spots were resolved in a 2-DE Coomasie Blue Brillian (CBB) stained gel, covering the 5.4–7.7 pH and 14–85 kDa ranges. Twenty-two protein spots were identified by MALDI-TOF or ESI IT MS/MS, with a significant number of them related to virulence; these latter include Malate Dehydrogenase (MDH), Glyceraldehyde-3-Phosphate Dehydrogenase (GADPH) and cyclophilin. These findings, and others, showed the applicability of proteomics approaches as a starting point for the study of pathogenicity factors and for basic research on this plant pathogen in the post-genomic era. Following this initial paper, a second paper described an attempt to identify putative fungal virulence, and reported a comparative analysis carried out between two *B. cinerea* isolates (B.c 2100 and B05.10 strains) (Fernandez-Acero et al. 2007b) which differ in virulence and toxin production. It was carried out using the 2-DE MALDI TOF/TOF methodology. The resulting 2-DE protein profiles showed qualitative and quantitative differences between mycelial extracts from the two isolates, most of them associated with virulence factors. A total of 28 spots were identified with a significant number of isoforms of MDH, either over-expressed or exclusive to the more virulent strain (*B. cinerea* 2100). This protein plays a key role in the infection process, since it is involved in oxalic acid production (Lyon et al. 2004), pH decrease (Manteau et al. 2003), as well as in the biosynthesis and secretion of biotoxins (Durán-Patrón et al. 2004). In addition, four spots only present in *B. cinerea* 2100 were identified as a GADPH which, apart from its well-known role in the glycolytic cycle, has been reported to act as a virulence factor in several different organisms (Hernández et al. 2004; Alderete et al. 2001). These results lend support to the hypothesis that the different expression patterns revealed could be correlated with differences in virulence between strains, and again confirmed proteomics as an excellent tool for use as a first and direct step in the study of new fungal virulence factors in *B. cinerea*.

At that time, due to the absence of a protein and DNA database containing *B. cinerea* sequences, protein identification was achieved by MS *de novo* sequencing followed by MS BLAST alignment, yielding the identification against other species.

The emergence in 2007 of a data-base of *B. cinerea* proteins (http://www.broad.mit.edu) was a breakthrough, enabling researchers to apply a proteomics approach with a significant number of identified proteins. In the first study using this approach, the authors described the mycelial proteome of *B. cinerea* during cellulose degradation (Fernandez-Acero et al. 2009). Briefly, *B. cinerea* was grown in a minimal salt medium (MSM) supplemented with 1 % of carboxymethylcellulose (CMC) as a sole carbon source, in an attempt to simulate natural environmental conditions, based on the assumption that since cellulose is one of the major components of the plant cell wall, the use of CMC as a sole carbon source may reveal potential pathogenicity or virulence factors. Mycelial protein extracts were separated by 2-DE to obtain the proteome map. Two hundred and sixty-seven spots were selected for MALDI TOF/TOF MS analysis, resulting in 303 positive protein identifications, mostly representing non-annotated proteins. The authors classified these proteins into functional categories, showing the relevance of protein metabolism, modification processes, and oxidoreductase activity.

A breakthrough has recently taken place in the study of the *B. cinerea* mycelial proteome with the implementation of "shotgun" proteomics techniques. The adequacy of this methodology for mycelial *B. cinerea* protein identification and label-free quantification has been proven (Gonzalez-Fernandez et al. 2013). In this work the authors reported a comparative proteomics analysis of *B. cinerea* mycelium from two wild-type strains: B05.10 and T4 introducing label-free "shotgun" nUPLC–MSE methodology. In addition, they made the first comparative study between gel-based (2-DE) and gel-free/label-free (nUPLC-MS) approaches. A total of 225 (48 unique) and 170 (7 unique) protein species were identified by nUPLC–MSE in the B05.10 and T4 strains, respectively. Moreover, 129 protein were quantified in both strains. The authors concluded that the use of the label-free "shotgun" nUPLC–MS methodology to analyze the two *B. cinerea* wild-type strains isolated from different hosts allowed the quantification of differences in protein abundance. They obtained results complementary to the traditional gel-based approach (2-DE), which ultimately allowed the identification of strain-specific proteins. In short, the use of these two approaches provided comprehensive knowledge of the proteomes of *B. cinerea* strains.

Another interesting contribution concerning mycelial protein research has recently been made (Gonzalez-Fernandez et al. 2014). In this work, differences in the protein profiles of gel-based approaches among strains were used to carry out a comparative proteomics study of six wild-type strains. After image analysis, a total of 674 spots were considered, out of which 112 spots were strain-specific in this study. A total of 47 variable proteins were identified, some of them, such as malate dehydrogenase and peptidyl-prolyl cis–trans isomerase, had been reported as virulence factors. The main differences among strains were related to proteins involved in redox processes, such as mannitol dehydrogenase. These results again confirm the usefulness of proteomics approaches for gaining better understanding and insight into *Botrytis* phenotype variability.

16.2.2 B. cinerea *secretome*

The first proteomics-based study of secreted proteins in *B. cinerea* was reported in 2009 by Shah et al. (2009a). Since then, an increasing number of secretome studies have been carried out, resulting in the identification of secreted proteins under different assay conditions related either to growth conditions, to developmental stages or to fungal strains. To date, five different approaches to the study the secretome of *B. cinerea* under different growth conditions have been published. Three of them deal with changes in the proteome/secretome of *B. cinerea* as a result of changes in carbon source. Shah et al. (2009a), used a gel-free LC-MS/MS approach to investigate the secretome produced by *B. cinerea* B05.10 on different hosts, under *in vitro* growth conditions. These were performed by inoculating *B. cinerea* on a nutritive agar medium containing either filtered pulp of ripe strawberry, red tomato or whole *Arabidopsis* plants and covered with a sterile cellophane membrane. Significant changes in the relative abundances and in the composition of the secreted enzymes, in a substrate-dependent manner, were observed; overall 89 *B. cinerea* proteins were identified from all the growth conditions. Out of this total seven proteins were observed to be common to all the samples, thus implying the constitutive nature of their synthesis and secretion. In respect of functional differences, these 89 proteins comprised transport proteins, proteins well-characterized for carbohydrate metabolism, peptidases, oxido-/reductases, and pathogenicity factors. These data provide important insights into how *B. cinerea* may use secreted proteins for plant infection and colonization.

In a second work, (Shah et al. 2009b) a "shotgun" proteomics approach was used to study the *B. cinerea* secretome on three different carbon sources in liquid cultures: highly-esterified pectin, partially-esterified pectin, and sucrose, as sole carbon sources, in order to demonstrate the impact of the degree of pectin esterification in the plant cell wall on fungal secretion, mimicking ripe and unripe fruits. The two pectin sources simulate fungal interactions with the expected host nutrient source; sucrose was used to define those enzymes considered constitutively produced. A total of 126 proteins secreted by *B. cinerea* were identified, 67 of which were observed in all three growth conditions. Thirteen *B. cinerea* proteins with functions related to pectin degradation were identified in the two pectin growth conditions, while only four were identified on sucrose as carbon-source. These results indicate that the secretion of most of the pectinases depends on the carbon substrate used by the fungus for growth. However no major differences were found in protein secretion when *B. cinerea* was grown with highly- *vs* partially-esterified pectin. It is therefore likely that the activation of *B. cinerea* from the dormant state to active infection is not dependent solely on changes in the degree of esterification of the pectin component of the plant cell wall. A similar study using a 2-DE MALDI TOF/TOF approach was carried out by Fernández-Acero et al. (2010) to determine the differences in *B. cinerea* protein secretion induced by five carbon sources and plant-based elicitors: glucose, CMC, starch, pectin and tomato cell walls (TCW). This work showed a different degree of complexity of the fungal response moving from a state of constitutive fungal growth (by using glucose as a sole carbon source) towards more complex and possibly pathogenic secretory behaviour (induced by TCW). A total of 76 spots were identified yielding 95 positive hits that correspond to 56 unique proteins, including several known virulence factors (e.g. pectin methyl esterases, xylanases and proteases). These three proteomics studies are clear examples of how, by using modifications in the composition of culture media, the pathogenicity of *B. cinerea* can be induced or modified emulating *in vitro* conditions similar to those applicable *in planta*. The modification of culture media by plant compounds was used by Espino et al. (2010) to study the secretome proteins during the first 16 h, termed the "early secretome". By combining both approaches, 2-DE and LC-MS/MS, 116 proteins were identified, including several virulence factors. 2-DE was used to check the proteome profiles between the wild type and the mutant in an aspartic protease gene, and found that the specific spot disappeared from mutant gels.

Other interesting proteomics approaches to study the secretome from an environmental perspective have been reported. These works are aimed at studying the effects of metals stress and pH, respectively, on the secretome of *B. cinerea* using comparative proteomics based on 2-DE MALDI-TOF/TOF approaches. To define the effects of metals stress on the secretome of *B. cinerea*, four metals (copper, zinc, nickel and cadmium) were added to the culture media of the fungus (Cherrad et al. 2012) and a total of 116 protein spots were observed on 2-D gels; the findings indicate that the secretome signature seems to be metal-dependent. Fifty-five spots were associated with unique proteins, and functional classification revealed that the production of oxidoreductases and cell-wall degrading enzymes was modified in response to metals stress. These results clearly show that the fungus can adjust the production of secreted proteins in response to metals exposure. Finally the comparative analysis revealed that the accumulation of some secreted proteins increases specifically in response to one or to several metals. This proteomics study has thus identified a number of proteins that could be considered as potential biomarkers for monitoring pollution exposure.

Concerning the effect of ambient pH on the secretome of *B. cinerea*, Li et al. (2012) compared changes found in the secretome of *B. cinerea* grown in culture media buffered at pH 4 and 6. Forty-seven differential spots, corresponding to 21 unique proteins, were identified. Most of these proteins were cell wall degrading enzymes (CWDE) or proteases. At pH 4, more proteins related to proteolysis were induced, whereas most of the up-regulated proteins at pH 6 were CWDE. Analysis of gene expression using quantitative real-time PCR suggests that production of most of these proteins is regulated at the level of transcription. These findings indicate that *B. cinerea* can adjust its secretome profile in response to different ambient pH values.

All these observed changes in the profile of secreted proteins under different growth conditions (nutrient- or environment-dependent) are indicative of the versatility and adaptability of *B. cinerea* to different hosts and environmental conditions. These findings provide important information about the mechanisms employed by the fungus to access nutrients and to initiate its infectious cycle on a wide range of plant-hosts. Differences between the secretomes of six *B. cinerea* strains were analyzed by 1-DE coupled to LC-MS/MS (Gonzalez-Fernandez et al. 2014), and 51 variable proteins were identified, including endopolygalacturonases, aspartic proteases and the cerato-platanin protein. All of them have been reported before as virulence factors.

Even though plant components can usefully be incorporated in the cultures, the real situation can only be studied during actual plant-*Botrytis* interactions. However, in such studies, fungal proteins seem to disappear, masked by the huge amount of plant proteins (Mulema et al. 2011) or they are reduced to only a few spots (Cilindre et al. 2008). A deep analysis of tomato infected by *B. cinerea* was tackled by Shah et al. (2012). By using 1-DE plus LC-MS/MS, 79 *B. cinerea* proteins were found during the infection of three different tomato-ripening stages. This study allowed the identification of 15 common proteins from the fungus, including CWDE and other virulence factors.

16.3 Proteomics – A Long Future Ahead in the Biology of *Botrytis*

Proteomics-based approaches to the study of *B. cinerea* have shown, among others, the potent capacity of this science to identify accurately the many protein candidates involved in specific biochemical pathways or phytopathological events. In order to obtain a global perspective, all available *B. cinerea* protein identifications performed in the various different proteomics approaches were summarized and analyzed together. During this "experiment", we found several problems. Despite difficulties in comparing data published in different journals under different formats and those published in the pre- and post-genomic era (Fernandez-Acero et al. 2006; Amselem et al. 2011; Staats and Van Kan 2012), a file containing 682 non-redundant sequences was obtained from the 12 papers available. Taken together, proteomics research exploring the *Botrytis cinerea* proteome covered only a very small proportion (4.4–6.6 %) of the fungal genome, depending on the genome sequence and its annotation (gene prediction). This could mean that we are only at the beginning of a long distance race. There are several different explanations for this poor coverage. First there are, to date, only a few studies published, most of them based on 2-DE, in which, in the best cases, only hundreds of proteins are resolved. Nowadays, LC-based studies are identifying increasingly larger numbers of proteins. Another reason may be associated with the experimental approach used. Most of the published data has been obtained using general culture media; and some of them include plant components to induce specific fungal responses. Only a few studies have applied proteomics approaches directly to plant-*Botrytis* interactions, where fungal proteins only represent a small percentage compared to those of the plant components. Future proteomics approaches will increase this ratio by studying specific subproteomes, varying the components of the media to induce new sets of proteins or, finally, with wide application of gel-free proteomics technologies to increase the number of identified proteins. Recently, targeted proteomics was named the technology of the year by an important journal (Marx 2013). This approach will allow the search for specific proteins from a complex mixture, making it possible to find weakly represented proteins.

In order to measure the relevance of the identification within the overall biology of *Botrytis*, and check which are the principal groups of proteins that had been found, all identified sequences were categorized using the gene ontology classification (AgBase; http://www.agbase.msstate.edu/). **Cellular component** classification includes those proteins which are part of something larger, such as anatomic structures, organelles, etc. The first category found for the identified proteins corresponds to mycelial proteins such as intracellular, cell or cytosolic proteins, which made up around 36 % of all the identified proteins. This is quite a surprising result since most of the identifications were extracted from secretome publications. Proteins from extracellular regions appeared as the fifth category. However, when the whole content of the database was uploaded, extracellular regions accounted for less than 1 %. In previous studies (Fernández-Acero et al. 2010) it has been established that the protein sequences of the *B. cinerea* database that present secretion signals (Signal P) comprise around 15 % of the total, showing that these secreted proteins are included in other categories (e.g. endoplasmatic reticulum, protein complexes, etc.). This finding must be associated with the close relationship that secreted proteins have with the membranes during secretion processes.

Protein categorization by **biological processes** shows those protein involved in a series of events that take place in one or more distinct steps (Fig. 16.1). Major categories represented in identified proteins are concerned with metabolism (biosynthesis, catabolism, small molecule and carbohydrate metabolic processes) while the total genome presents major categories concerned with other biological processes (nitrogen, small molecules or transport metabolic processes). This is probably due to the use of synthetic culture media, which are normally artificially enriched. Other differences show that identified proteins present over-represented categories such as transport and cell cycle, cellular protein modification or DNA metabolic processes. The **molecular functions** of identified proteins (Fig. 16.2) describes the different catalytic or binding activities of the proteins that occur at a molecular level. A comparison of all the categories obtained between identified proteins and the whole genome makes it clear that most of the categories of identified proteins are still under-represented. Oxydoreductase activity percentages are similar but activities such as kinases or transmembrane transporters, crucial for the signal transduction pathways, are an average of 7.5 times less represented in the identified proteins than in the database.

The observed differences between the proteomics dataset and all predicted proteins may have several explanations. Mainly, most of the synthetic culture media used contain similar compounds, and these probably inhibit the synthesis of whole peptides that may be useful for the fungus during its necrotroph/hemi-biotroph/biotroph lifestyle (Tudzynski and Kokkelink 2009). Moreover, at present, there are only a few studies focused on the identification of fungal subproteomes, mainly the secretome. The *B. cinerea* proteomics community must develop new strategies for

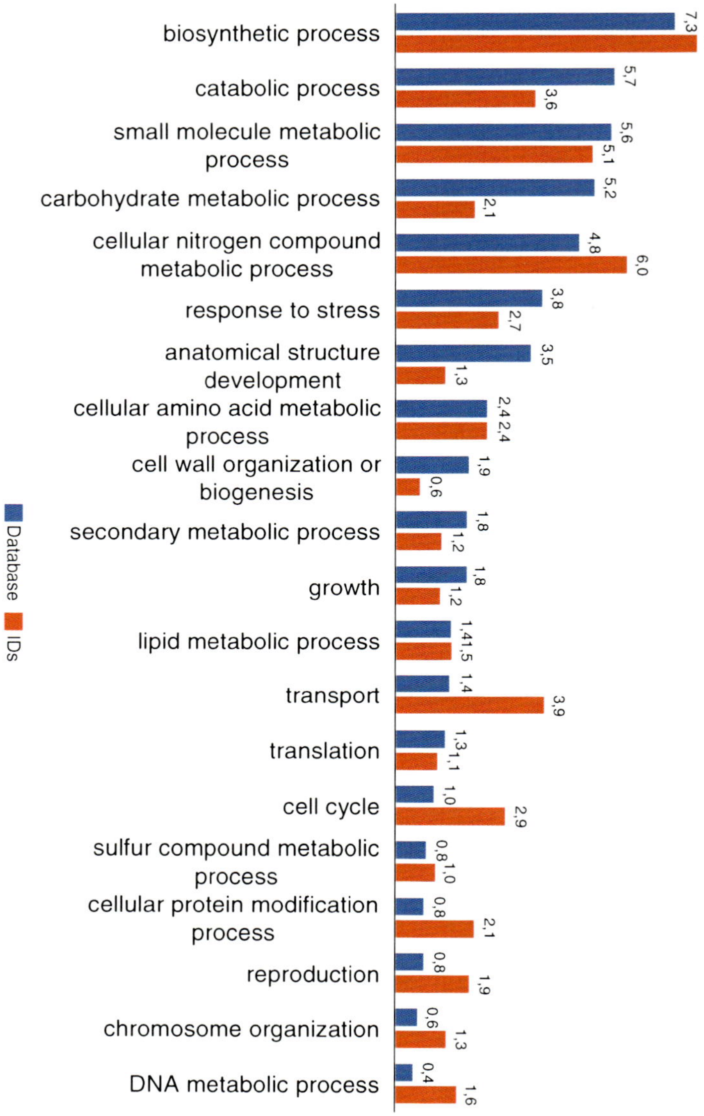

Fig. 16.1 Gene ontology classification by biological process. Percentages of *B. cinerea* proteins listed in the genome database vs. proteins from proteome datasets

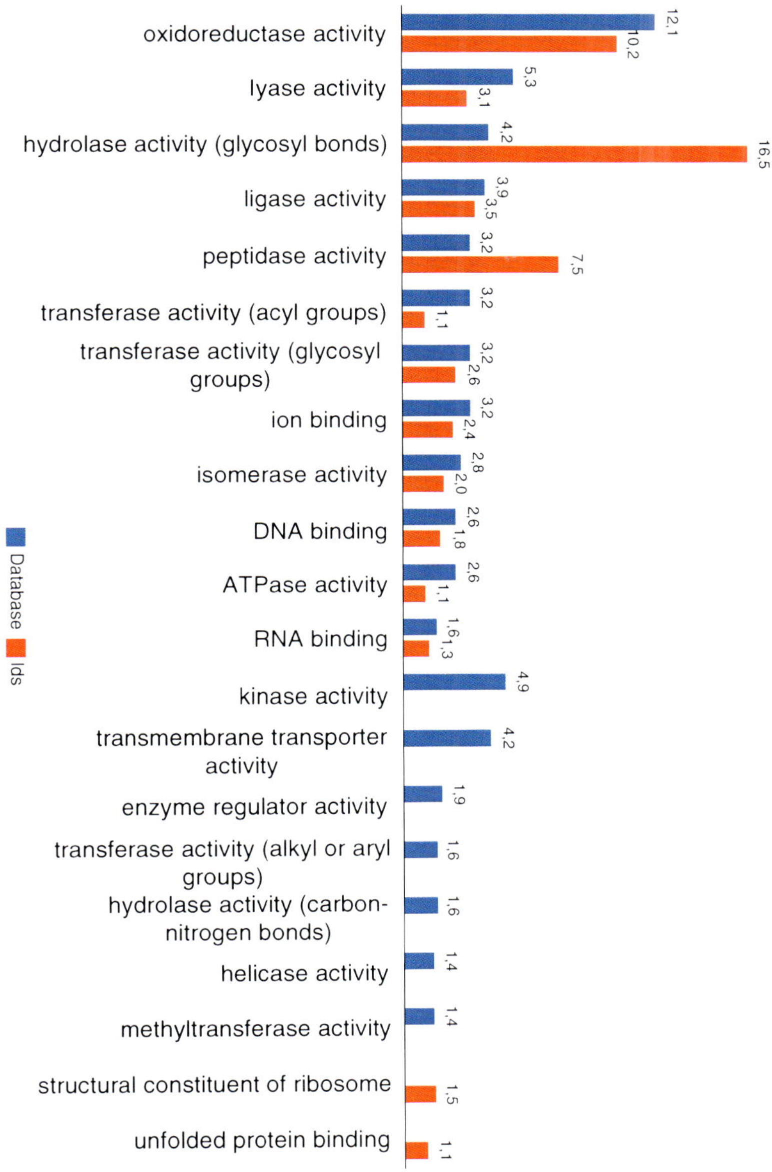

Fig. 16.2 Gene ontology classification by molecular function. Percentages of *B. cinerea* proteins listed in the genome database vs. proteins from proteome datasets

inducing the synthesis of new proteins by varying the media content, emulating natural conditions to collect under-represented proteins. New approaches to fungal subproteomes must be developed as well, allowing the study of those proteins specifically produced during discrete stages of the life cycle (conidia, appressoria, sclerotia, etc.). Such studies are now running in several laboratories. During the next few years, the proportion of genome coverage will probably increase accordingly.

16.4 Biological Relevance of Proteomics Approaches

As described above, proteomics approaches are making it possible to detect efficiently global protein production under specific conditions. They allow the detection of proteins involved in specific biological events. Moreover, in other biological systems where transcriptomics have been compared with proteomics analyses, it has been stated that the proteome is the "relevant" set for analysis (Rossignol et al. 2009). This conclusion is based on significant differences that had been described previously (Haider and Pal 2013; Hack 2004) between transcriptomics and proteomics data. This comparison showed a correlation between both datasets in the range from 0.46 to 0.76; thus sometimes, but not always, variations in protein abundances can be predicted from the transcriptome. Currently, there is no global analysis of this correlation in *B. cinerea*. However, one thing seems to be clear: if, after several controls, replicates and statistical analyses, one protein is found to be present in a specific context, it must be significant, independent of whether we are able to detect its mRNA, as has been widely discussed. This is one of the most important impacts that proteomics has had on our series of biological dogmas.

However, the main aim of the proteomics research activity in *B. cinerea* has been directed to the quest for those peptides/proteins involved in the different steps of the pathogenicity process. This aim has been present from the first papers (Fernandez-Acero et al. 2007b) showing that proteomics approaches were capable of detecting several candidates as virulence factors without previous exclusion criteria. It allows the recovery of dozens or hundreds of candidates from a single experiment. However, the downside to establishing a direct relationship between identified proteins and virulence or pathogenicity factors is the great amount of work that must subsequently be done in order to discriminate real virulence factors from other proteins. Normally, this discrimination is done by constructing knockout mutants of each candidate and evaluating the influence during plant infection. Due to the power of proteomics technologies to identify hundreds of potential candidates, the functional validation of the candidate genes to elucidate their specific roles in the pathogenicity process will rapidly become the bottleneck.

An example of this objective is the role of two secretome components. Aspartic proteases are the main components of the *B. cinerea* secretome (Fernández-Acero et al. 2010). A deep molecular analysis of these components revealed that deficient mutants show the same level of virulence (ten Have et al. 2010). However their usefulness in removing haze-forming proteins during wine fermentation has been highlighted (Van Sluyter et al. 2013). In another study, Espino et al. (2010) found a BcSpl1, a cerato-platanin, in the fungal secretome. Further molecular analysis has shown its phytotoxic activity and its role as a virulence factor (Frías et al. 2011). In spite of these advances, there are still hundreds of identified proteins waiting for molecular analysis, representing a source of potentially valuable biological information. Finally, the capacity of proteomics approaches to find virulence factors is indisputable, since new virulence factors usually appear during proteomics analyses.

16.4.1 Possible Applications of Proteomics in Botrytis

Another significant aspect for review is the capacity of proteomics approaches for revealing new therapeutic targets for fungicide design. At the moment, however, this potential is only a possibility. The basis of this idea is that most of the drug targets are proteins. The development of fungicides based on molecular information may open possibilities for the production of environmentally-friendly drugs, with low impact, and with high levels of species-specificity. The reasons why there is a lack of molecular-based fungicides are still unclear, even though the necessary scientific strategies and technologies seem now to exist. New proteomics-based technologies have shown their capacity to serve as effective tools for fungicide design (Fernández-Acero et al. 2011) given the discovery of new disease factors that may be candidates for new therapeutic targets. Tools such as peptide aptamers and RNA silencing may be future instruments for crop protection. In this context, the potential use of modified natural compounds to inhibit specific fungal activities has been revealed as a possibility. This approach has been termed "chemoproteomics", defined as the use of biological information to guide chemistry (Beroza et al. 2002), which is presented as a new methodology to accelerate drug discovery in the post-genomic era. In spite of this optimistic scenario, there are various serious problems in transferring these technologies to crop management; these include, for example, maintaining activity in plants, metabolite stability, and obviously the costs of development and production. None of these technologies has yet led to a recognizable fungicide development candidate (Tietjen and Schreier 2013).

16.5 Future Perspectives

Nowadays, the term "post-genomic era" is widely used to refer to the present stage of the molecular sciences. This term is mainly used to refer to advances in transcriptomics and high throughput mutagenesis, but "the name of the game is proteomics" (Brower 2001). However, for the community studying *B. cinerea*, it is still an open field where only a few milestones have been reached revealing the potential importance of proteomics; therefore many new problems, specifically those linked to fungal biology, remain to be solved.

To this end, the publication of the *B. cinerea* genome databases, after the publication of the genome sequences (Amselem et al. 2011; Staats and Van Kan 2013), has been crucial. This has enabled proteomics researchers to access directly the molecular information, making it possible for them to identify *B. cinerea* proteins directly, without the need to search for homologous proteins in other fungi or organisms. However, the slowness in developing specific bioinformatic tools is still a weakness, as these tools may help to understand the large set of proteins identified, avoiding the loss of relevant biological information. For example, specific fungal gene ontology queries and analyses of protein-protein interactions and differential pathways may improve the number of conclusions obtained from each approach. Moreover, the development of specific software to predict the role of a protein as a fungicide target may help to usher in a new era of directed fungicides. Most of these tools are a reality in other biological systems (Chen et al. 2007; Huan et al. 2007), suggesting that their use in *Botrytis* studies will soon become feasible.

Most of the proteomics approaches described so far with *Botrytis cinerea* only cover mycelia and/or the secretome. However the fungal life cycle includes many fungal stages, anatomical structures, organelles, etc. whose proteomes remain to be elucidated. The analysis of new subproteomes will help to find interesting new proteins that are needed to understand better the complexity of fungal biology. In this context, there are two other proteomics approaches that seem to be missing in *B. cinerea* – first, the analysis of post-translational modifications (PTMs) of fungal proteins and second, quantitative proteomics approaches. Around 200 different PTMs have been described. These modifications are the bridge between extracellular and intracellular environments, and are crucial for signal transduction cascades that direct fungal behavior and virulence, but only one paper has been published recently, on the phosphoproteome of *B. cinerea* (Davanture et al. 2014). As mentioned previously, the most recently developed proteomics technologies are also known as "quantitative proteomics"; this is related to the specific labeling of proteins with specific isobaric tags. As far as we know, this methodology has not yet been used for *B. cinerea*. It would allow the specific detection and quantification of differential proteins, thus revealing their role under assay conditions. We can assume that more subproteomes, PTMs and quantitative studies will be available soon. Another set of interesting proteins still unresolved are those involved directly in the plant pathogen interaction. These approaches represent a more realistic way of studying the fungal weapons deployed for invading plant cells and defeating the plant defense mechanisms. Some studies have been undertaken, but the fungal "hits" obtained are either few in number or non- existent. Greater success in the identification of fungal proteins may be obtained by selective extraction enrichment and/or labeling.

One of the most promising proteomics approaches to *B. cinerea* is "targeted proteomics". This set of strategies was selected as the method of the year 2013 by the journal Nature Methods (Marx 2013). In brief, it will enable the study of specific proteins by MS/MS from a whole set, allowing quantitative assays to specifically answer hypothesis-driven questions. The detection of specific peptide ions is made possible by focusing the MS instrument (typically, a triple quadrupole) for a preselected set of peptides. It will allow the specific detection of proteins in a wide set of samples, with a higher level of sensitivity. Sequential window acquisition of all theoretical fragment-ion spectra (SWATH), with improvement by independent data acquisition, will change the perspective of proteomics studies from one based on a large set of identified proteins to a smaller set of more relevant identifications. All these improvements should become publicly available within the next few years. They will transform proteomics technology into an indispensable and commonly-used tool for understanding the biology and virulence of *B. cinerea.*

Acknowledgements The authors gratefully acknowledge funding from the Spanish Government DGICYT – AGL2012-39798-C02-02 (www.micinn.es/portal/site/MICINN/). Eva Liñeiro was supported by a FPI grant from the University of Cadiz (2010-152). Special thanks are given to Celedonio Gonzalez (University of La Laguna) and Fiona McCarthy (PI of Agbase) for their guidance, support and patience in unravelling bioinformatics data.

References

Alderete JF, Millsap KW, Lehker MW et al (2001) Enzymes on microbial pathogens and *Trichomonas vaginalis*: molecular mimicry and functional diversity. Cell Microbiol 3(6):359–370

Amselem J, Cuomo CA, Van Kan JA, et al (2011) Genomic analysis of the necrotrophic fungal pathogens Sclerotinia sclerotiorum and Botrytis cinerea. PLoS Genet 7:e1002230

Beroza P, Villar HO, Wick MM et al (2002) Chemoproteomics as a basis for post-genomic drug discovery. Drug Discov Today 7(15):807–814

Brower V (2001) Proteomics: biology in the post-genomic era. EMBO Rep 2(7):558–560

Chen X, Fang Y, Yao L et al (2007) Does drug-target have a likeness? Methods Inf Med 46(3):360–363

Cherrad S, Girard V, Dieryckx C et al (2012) Proteomic analysis of proteins secreted by *Botrytis cinerea* in response to heavy metal toxicity. Metallomics 4(8):835–846

Cilindre C, Jegou S, Hovasse A et al (2008) Proteomics approach to identify champagne wine proteins as modified by *Botrytis cinerea* infection. J Proteome Res 7(3):1199–1208

Dass C (2000) Principles and practice of biological mass spectrometry, Wiley-Interscience series on mass spectrometry. Wiley, New York, 448 p

Davanture M, Dumur J, Bataillé-Simoneau N et al (2014) Phosphoproteome profiles of the phytopathogenic fungi *Alternaria brassicicola* and *Botrytis cinerea* during exponential growth in axenic cultures. Proteomics 14(13–14):1639–1645

Dujon B (1996) The yeast genome project: what did we learn? Trends Genet 12:263–270

Durán-Patrón R, Cantoral JM, Hernández-Galán R et al (2004) The biodegradation of the phytotoxic metabolite botrydial by its parent organism, *Botrytis cinerea*. J Chem Res 2004(6):441–443

Espino JJ, Gutiérrez-Sánchez G, Brito N et al (2010) The *Botrytis cinerea* early secretome. Proteomics 10(16):3020–3034

Fenn JB, Mann M, Meng CK et al (1989) Electrospray ionization for mass spectrometry of large biomolecules. Science 246:64–71

Fernandez-Acero FJ, Jorge I, Calvo E et al (2006) Two-dimensional electrophoresis protein profile of the phytopathogenic fungus *Botrytis cinerea*. Proteomics 6:S88–S96

Fernandez-Acero FJ, Carbu M, Garrido C et al (2007a) Proteomic advances in phytopathogenic fungi. Curr Proteome 4:79–88
Fernandez-Acero FJ, Jorge I, Calvo E et al (2007b) Proteomic analysis of phytopathogenic fungus *Botrytis cinerea* as a potentia tool for identifying pathogenicity factors, therapeutic targets and for basic research. Arch Microbiol 187:207–215
Fernandez-Acero FJ, Colby T, Harzen A et al (2009) Proteomic analysis of the phytopathogenic fungus *Botrytis cinerea* during cellulose degradation. Proteomics 9:2892–2902
Fernández-Acero FJ, Colby T, Harzen A et al (2010) 2-DE proteomic approach to the *Botrytis cinerea* secretome induced with different carbon sources and plant-based elicitors. Proteomics 10(12):2270–2280
Fernández-Acero FJ, Carbú M, El-Akhal MR et al (2011) Development of proteomics-based fungicides: new strategies for environmentally friendly control of fungal plant diseases. Int J Mol Sci 12(1):795–816. doi:10.3390/ijms12010795
Frías M, González C, Brito N (2011) BcSpl1, a cerato-platanin family protein, contributes to *Botrytis cinerea* virulence and elicits the hypersensitive response in the host. New Phytol 192(2):483–495
Garrido C, Cantoral JM, Carbú M et al (2011) New proteomic approaches to plant pathogenic fungi. Curr Proteomics 2011(4):306–315
González Fernández R, Prats E, Jorrin J (2010) Proteomics of plant pathogenic fungi. J Biomed Biotechnol 2010:1–36. doi:10.1155/2010/932527
Gonzalez-Fernandez R, Aloria K, Arizmendi JM et al (2013) Application of label-free shotgun nUPLC-MSE and 2-DE approaches in the study of *Botrytis cinerea* mycelium. J Proteome Res 12(6):3042–3056
Gonzalez-Fernandez R, Aloria K, Valero-Galvan J et al (2014) Proteomic analysis of mycelium and secretome of different *Botrytis cinerea* wild-type strains. J Proteome 97:195–221
Hack CJ (2004) Integrated transcriptome and proteome data: the challenges ahead. Brief Funct Genomic Proteomic 3(3):212–219. doi:10.1093/bfgp/3.3.212
Haider S, Pal R (2013) Integrated analysis of transcriptomic and proteomic data. Curr Genomics 14(2):91–110. doi:10.2174/1389202911314020003
Hernández R, Nombela C, Diez-Orejas R et al (2004) Two-dimensional reference map of *Candida albicans* hyphal forms. Proteomics 4:374–382
Hillenkamp F, Karas M, Beavis RC et al (1991) Matrix-associated laser desorption/ionization mass spectrometry of biopolymers. Anal Chem 63:A1193–A1202
Huan X, HangYang X, MingZhi L et al (2007) Learning the drug target-likeness of a protein. Proteomics 7(23):4255–4263
James P (1997) Protein identification in the post-genome era: the rapid rise of proteomics. Q Rev Biophys 30(4):279–331
Kim Y, Nandakumar MP, Marten MR (2007) Proteomics of filamentous fungi. Tren Biotechncl 25(9):395–400
Li B, Wang W, Zong Y et al (2012) Exploring pathogenic mechanisms of *Botrytis cinerea* secretome under different ambient pH based on comparative proteomic analysis. J Proteome Res 11(8):4249–4260
Lyon GD, Goodman BA, Williamson B (2004) Botrytis cinerea perturbs redox processes as an attack strategy in plants. In: Elad Y, Williamson B, Tudzynski P, Delen N (eds) Botrytis: biology, pathology and control. Kluwer Academic Publishers: Dordrecht, The Netherlands
Manteau S, Abouna S, Lambert B et al (2003) Differential regulation by ambient pH of putative virulence factor secretion by the phytopathogenic fungus *Botrytis cinerea*. FEMS Microbiol Ecol 43(3):359–366
Marx V (2013) Targeted proteomics. Nat Methods 10:19–22
Minguez P, Parca L, Diella F et al (2012) Deciphering a global network of functionally associated post-translational modifications. Mol Syst Biol 8:1–14
Mulema JMK, Okori P, Denby KJ (2011) Proteomic analysis of the *Arabidopsis thaliana-Botrytis cinerea* interaction using two-dimensional liquid chromatography. Afr J Biotechnol 10(76):17551–17563

Rossignol T, Kobi D, Jacquet-Gutfreund L et al (2009) The proteome of a wine yeast strain during fermentation, correlation with the transcriptome. J Appl Microbiol 107(1):47–55

Shah P, Atwood JA III, Orlando R et al (2009a) Comparative proteomic analysis of *Botrytis cinerea* secretome. J Proteome Res 8(3):1123–1130

Shah P, Gutierrez-Sanchez G, Orlando R, Bergmann C (2009b) A proteomic study of pectin-degrading enzymes secreted by *Botrytis cinerea* grown in liquid culture. Proteomics 9(11):3126–3135

Shah P, Powell ALT, Orlando R et al (2012) Proteomic analysis of ripening tomato fruit infected by *Botrytis cinerea*. J Proteome Res 11(4):2178–2192

Staats M, van Kan JA (2013) Genome update of Botrytis cinerea strains B05.10 and T4. Eukaryot Cell 11(11):1413–1414. doi:10.1128/EC.00164-12

Staats M, van Kan JA (2013) Genome update of Botrytis cinerea strains B05.10 and T4. Eukaryot Cell. 2012 Nov;11(11):1413–4. doi: 10.1128/EC.00164-12.

ten Have A, Espino JJ, Dekkers E et al (2010) The *Botrytis cinerea* aspartic proteinase family. Fungal Genet Biol 47(1):53–65

Tietjen K, Schreier PH (2013) New targets for fungicides. In: Modern methods in crop protection research. Wiley-VCH Verlag GmbH & Co. KGaA, Weinheim, pp 197–216

Tudzynski P, Kokkelink L (2009) *Botrytis cinerea*: molecular aspects of a necrotrophic life style. In: Deising HB (ed) The mycota. Springer, Berlin, pp 29–50

Van Sluyter SC, Warnock NI, Schmidt S et al (2013) Aspartic acid protease from *Botrytis cinerea* removes haze-forming proteins during white winemaking. J Agric Food Chem 61(40):9705–9711

Wilkins MR, Sanchez JC, Gooley AA et al (1995) Progress with proteome projects: why all proteins expressed by a genome should be identified and how to do it. Biotechnol Genet Eng Rev 13:19–50

Chapter 17
Overview of Plant Defence Systems: Lessons from Arabidopsis-*Botrytis cinerea* Systems Biology

Oliver Windram, Claire Stoker, and Katherine Denby

Abstract *Botrytis cinerea* is an important model system for studying the necrotrophic plant pathogen lifestyle, whilst also representing one of the most economically destructive agricultural pathogens. A key challenge to understanding the pathology of this virulent fungus involves unraveling host responses. These host responses involve complex regulatory mechanisms and multiple downstream defence processes. In addition, the pathogen is capable of manipulating cellular processes in the host to favour infection. In this chapter we will present recent advances in systems biology approaches, combining high-throughput 'omics technologies and computational/mathematical network inference techniques, which have been used to tease apart this complex host-pathogen interaction. We will also highlight novel systems approaches from other areas of plant pathology and plant science that can be applied to provide a more comprehensive understanding of plant defence against *B. cinerea*. We will conclude with the key challenges of understanding how both plant defence and pathogen attack are integrated, and translating knowledge from Arabidopsis to crop plants.

Keywords Regulatory network • Network modelling • Plant immunity • Plant-pathogen interaction • Systems biology

O. Windram
Grand Challenges in Ecosystems and the Environment, Silwood Park,
Department of Life Sciences, Imperial College London, London SL5 7PY, UK
e-mail: o.windram@imperial.ac.uk

C. Stoker
Midlands Integrative Bioscience Training Programme, School of Life Sciences,
University of Warwick, Conventry CV4 7AL, UK
e-mail: C.Stoker@warwick.ac.uk

K. Denby (✉)
School of Life Sciences and Warwick Systems Biology Centre,
University of Warwick, Conventry CV4 7AL, UK
e-mail: k.j.denby@warwick.ac.uk

S. Fillinger, Y. Elad (eds.), *Botrytis – the Fungus, the Pathogen and its Management in Agricultural Systems*, DOI 10.1007/978-3-319-23371-0_17

17.1 Overview of the Host Defence Response Against *Botrytis cinerea*

Initial pathogen recognition by a plant is facilitated by two distinct groups of molecules, pathogen-derived microbial associated molecular patterns (MAMPs) and host damage associated molecular patterns (DAMPs), via their recognition by host receptor-like kinases dubbed pattern recognition receptors (PRRs) (Boller and Felix 2009; Fig. 17.1 for an overview of pathogen perception and initial signaling responses). MAMPs are molecular components of pathogens whose highly conserved sequence is constrained due to their essentiality in microbial fitness, and which are absent from the host. Recognition of these signals initiates generalized basal defence, which curtails infection of non-host pathogens and helps to limit the extent of disease caused by virulent pathogens. Virulent pathogens are generally classified as those that cause plant disease. Some pathogens parasitize living host cells; loosely termed biotrophic pathogens these organisms, with their intimate relationship with host cells, have evolved complex gene-for-gene interactions with the host. Such microbes can suppress MAMP/DAMP triggered immune responses (MTI) using effector proteins that help promote disease. In certain cases the host plant has evolved specific resistance (*R*) genes whose products are capable of detecting the microbial effectors, or their activity, resulting in a heightened effector-triggered immunity (Jones and Dangl 2006). This heightened immune response commonly involves a hypersensitive response, cell death at infection sites, helping to contain and eliminate the pathogen.

Botrytis cinerea is an example of a necrotrophic pathogen meaning that it kills plant tissue and utilizes this dead tissue to support its growth. In order to engage in this lifestyle *B. cinerea* rapidly invades host tissues, actively suppressing expression of key defence components (Weiberg et al. 2013), and inducing necrosis within the first 24 h of infection on Arabidopsis (Windram et al. 2012). Once established, the fungus secretes necrosis-inducing factors effectively inducing cell death in advance of fungal growth (Govrin et al. 2006). The rapid onset of necrosis limits pathogen-host cell interaction and could indicate why gene-for-gene resistance has yet to be described for this pathogen (Chap. 12). However, variation in both host and pathogen genetic background alter susceptibility (Denby et al. 2004; Kliebenstein et al. 2005; Van Baarlen et al. 2007). With this in mind we will focus on the basal defence mechanisms impacting Arabidopsis susceptibility to *B. cinerea*.

Chitin functions as a MAMP during *B. cinerea* infection and is recognized via the receptor kinases CERK1 and LYM2 (Miya et al. 2007; Wan et al. 2008; Faulkner et al. 2013). The two chitin-perception pathways function independently with chitin-induced signaling via LYM2 leading to a reduction in inter-cellular molecular exchange through plasmodesmata, while CERK1 stimulates the more classical hallmarks of MTI: oxidative bursts and activation of mitogen activated protein kinase signaling cascade systems. Thus recognition of an essential fungal wall component initiates basal defence responses that help to limit the infection of this virulent pathogen.

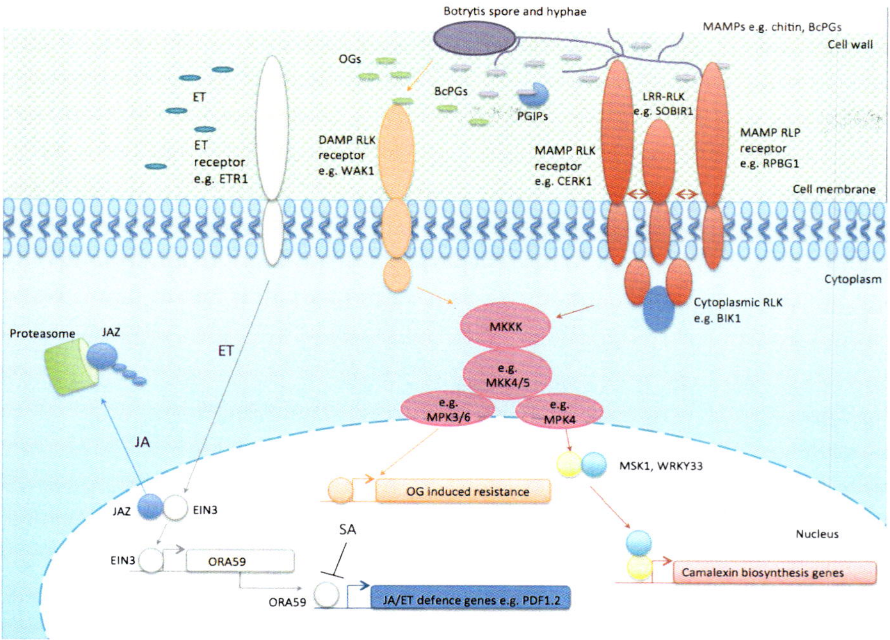

Fig. 17.1 Perception of the pathogen and initial signaling events during *B . cinerea* infection. Examples of microbe-associated molecular patterns (*MAMPs*) and damage-associated molecular patterns (*DAMPs*) present during *B. cinerea* infection are shown. Typical signaling pathways are shown indicating components that are known to play a role in the response to *B. cinerea* infection. MAMPs and DAMPs are perceived by receptor-like kinases (*RLKs*) and receptor-like proteins (*RLPs*) which typically associate with additional leucine-rich repeat (*LRR*) RLKs and/or cytoplasmic RLKs to transduce a signal to kinase cascades. Mitogen activated protein kinase (*MAPK*) kinase kinase (*MAPKKK*) phosphorylate MAPK kinases (*MAPKK*) which in turn phosphorylate MAPKs. MAPKs have been shown to be essential for DAMP-induced resistance and, in the case of MPK4, for activation of the transcription factor WRKY33 by phosphorylation of the MPK4-WRKY33-MPK substrate 1 (*MKS1*) complex. WRKY33 activates expression of camalexin biosynthetic genes, an important resistance determinant against many isolates of *B. cinerea*. Cross talk between the phytohormones ethylene (*ET*), salicylic acid (*SA*) and jasmonic acid (*JA*) is crucial in host defence responses and an example of this is outlined. After infection by *B. cinerea*, production of ET stabilizes the transcription factor (*TF*) EIN3. Production of JA triggers degradation of the repressive JAZ proteins by the proteasome, allowing the transcriptional cascade of JA and ET related defence genes downstream of EIN3 to proceed. SA represses this transcriptional cascade by inhibiting accumulation of ORA59 protein. In this diagram → indicates positive regulation and ⊣ indicates negative regulation

DAMPs are host-derived signals indicative of damage associated with the pathogen and also play a key role in recognition of *B. cinerea* infection. Oligogalaturonides (OGs) are released from the plant cell wall during infection due to the action of *B. cinerea* endopolygalacturonases (BcPGs) and host endopolygalacturonase inhibiting proteins (PGIPs). BcPGs degrade host cell wall pectin while PGIPs limit their activity resulting in the production of OGs which positively contribute to defence against *B. cinerea* (Ferrari et al. 2007). These DAMPs are recognized by a wall associated kinase, WAK1 (Brutus et al. 2010). The BcPGs also function as MAMPs themselves (Poinssot et al. 2003) and were recently shown to be recognized by the receptor-like protein RPBG1 (Zhang et al. 2014). Altered expression of *RPBG1* did not appear to alter susceptibility to *B. cinerea*, but priming plants expressing *RPBG1* with BcPGs reduced susceptibility to virulent *Hyaloperonospora arabidopsidis* suggesting that recognition led to enhanced defence responses. An additional uncharacterized protein (SCFE-1) from *Sclerotinia sclerotiorum* is recognized by Receptor-like Protein 30 (RLP30) and mutants lacking *RLP30* are more susceptible to both *B. cinerea* and *S. sclerotiorum* (Zhang et al. 2013b).

Following detection of these MAMPs and DAMPs by PRRs, additional receptor-like kinases are required to transduce the signal to downstream components and can integrate signaling through their interaction with multiple PRRs (Liebrand et al. 2014). The membrane bound receptor-like kinase BAK1 is required for signal transduction following detection of the elicitor SCFE-1 from *S. sclerotiorum* recognized by RLP30 (Zhang et al. 2013b). RLP30-mediated signaling also requires SOBIR1, another receptor-like kinase and transgenic plants with mutations in *BAK1* or *SOBIR1* show enhanced susceptibility to *S. sclerotiorum* and *B. cinerea*. RPBG1, which recognizes BcPG, interacts with SOBIR1 but not BAK1. In addition to the requirement for multiple membrane bound receptor-like kinases, cytoplasmic receptor-like kinases also play a role in transducing pathogen recognition signals within plant cells. Chitin perception requires the cytoplasmic receptor-like kinase BIK1 and the chitin receptor, CERK1, has been shown to interact with BIK1 (Zhang et al. 2010).

Typically these initial perception complexes at the cell membrane funnel into mitogen activated protein kinase (MAPK) signaling cascade systems, which are highly conserved within eukaryotes. At the top of the cascade MAPK kinase kinases are activated and phosphorylate MAPK kinases. These then phosphorylate and activate MAPK. While currently ill-defined, MAPK Kinase Kinases receive the initial input from PRR signaling complexes and MAPKs determine the output from the cascade through their activity on target proteins (Pitzschke et al. 2009). While upstream cascades can vary according to initial signals (for example, whether it is bacterial flagellin or fungal chitin that is being sensed) it appears clear that three terminal MAPKs: MPK3, MPK6 and MPK4 are key players in plant immunity (Rasmussen et al. 2012).

MPK3, MPK6 and MPK4 are all activated by SCFE-1 protein from *S. sclerotiorum* in a *RLP30*-, *BAK1*- and *SOBIR1*-dependent manner (Zhang et al. 2013b) demonstrating that this MAPK cascade is downstream of initial membrane perception. MPK3 and MPK6 are required for camalexin synthesis during *B. cinerea* infection. Accumulation of this metabolite is known to be a key factor influencing susceptibility of Arabidopsis to *B. cinerea* (Denby et al. 2004) and *mpk3* mutants show increased fungal growth (Ren et al. 2008). MPK6 activity, but not MPK3, is required for OG-induced resistance against *B. cinerea* (Galletti et al. 2011) revealing overlapping as well as specific MAPK signaling functions. The *Brassica napus* orthologue of MPK4 positively contributes to defence against *S. sclerotiorum* (Wang et al. 2009b), while in Arabidopsis MPK4 substrate MKS1 appears to negatively regulate defence against *B. cinerea* (Petersen et al. 2010). The positive role these MAP kinases play in *B. cinerea* defence is further supported by experiments showing that overexpression of the AP2C1 phosphatase, which inactivates MPK4 and MPK6, increases susceptibility (Schweighofer et al. 2007). In contrast plants lacking MAPK phosphatase 2 (MPK2) show increased susceptibility to *B. cinerea*. MPK2 can interact with both MPK3 and MPK6 and has been shown to dephosphorylate both kinases *in vitro* (Lee and Ellis 2007). However, it exhibits differential interaction with MPK3 and MPK6 in response to a fungal elicitor suggesting the balance of MAPK phosphorylation is important in defence (Lumbreras et al. 2010). Protein microarrays predict MAPK kinase kinases target multiple MAPK kinases, MAPK kinases target multiple MAPKs and, in turn, MAPK have many terminal phosphorylation targets (Popescu et al. 2009). A large number of potential signaling cascades are thus possible, and the variable impact of phosphatases on susceptibility to *B. cinerea* infection may result from integration of a number of upstream signaling events to determine the final components of the MAPK cascade.

Phosphorylation of MAPK signaling translates into a substantial transcriptional response to infection evidenced by the large numbers of differentially expressed genes following infection, for example (Tao et al. 2003; Windram et al. 2012). The final structure of the MAPK cascade may provide a mechanism to modulate activity of specific sets of transcription factors (TFs) and hence the transcriptional response. However, with the exception of WRKY33 described below, very little is known about how MAPK signaling induces transcriptional changes following *B. cinerea* infection. The TF WRKY33 is required for defence against *B. cinerea*. It is bound in a nuclear complex with MPK4 and MKS1. After infection, and activation of MPK4, phosphorylation of MKS1 releases MKS1/WRKY33 which goes on to activate expression of the camalexin biosynthetic genes and *WRKY33* itself in a feed-forward mechanism (Qiu et al. 2008). Overexpression of MKS1, however, increases susceptibility to *B. cinerea* in a WRKY33-independent manner which suggests that MKS1 is having a broader regulatory influence (Fiil et al. 2011).

Additional kinase signaling systems can also influence plant defence. Two component systems typically consist of a sensor domain coupled to a histidine kinase, which transmits a sensory detection signal to a response regulator domain via a phosphorelay. Response regulators typically target promoter regions altering target gene expression (Grefen and Harter 2004). The hybrid histidine kinase AHK5 contributes to *B. cinerea* defence highlighting that in addition to complex MAPK cascade systems more primitive two component systems are also involved in regulating basal defence against this pathogen, probably via altered transcription (Pham et al. 2012). How this histidine kinase is activated by *B. cinerea* infection, however, is unknown.

The scale of transcriptional reprogramming during defence is substantial; over 30 % of the host genome can undergo a dynamic shift. In the case of *B. cinerea* several hundred of these differentially regulated genes are themselves TFs highlighting the inherent complexity of this signal transduction system (Tao et al. 2003; Windram et al. 2012). A broad range of TFs from a number of families have been implicated in *B. cinerea* defence including eight WRKYs (Xu et al. 2006; AbuQamar et al. 2006; Lai et al. 2008; Chen et al. 2010; Birkenbihl et al. 2012), six AP2/ERFs (Berrocal-Lobo et al. 2002; Pré et al. 2008; Son et al. 2012; Zhao et al. 2012; Moffat et al. 2012; Maruyama et al. 2013), four TGAs (Zander et al. 2009; Windram et al. 2012), three NACs (Bu et al. 2008; Wang et al. 2009a) and two MYB TFs (Mengiste et al. 2003; Ramírez et al. 2011). However, very few direct regulatory targets of these TFs are known and how these regulators interact within a regulatory network (integrating positive and negative influences on defence) has not been elucidated.

Along with secondary metabolites such as camalexin that directly act as antimicrobials, the plant produces a number of signaling molecules that help to communicate the presence of infection, and shape the transcriptional response. These phytohormones play a central role in many aspects of plant defence signaling and we point to two excellent reviews on phytohormones and defence for in-depth information (Pieterse et al. 2012; Gimenez-Ibanez and Solano 2013). The phytohormones salicylic acid (SA), jasmonic acid (JA), ethylene, and abscisic acid (ABA) have long been known to play a role in defence against *B. cinerea* infection (Thomma et al. 1998, 1999; Audenaert et al. 2002). Generally, SA appears to play a positive role in early defence responses against *B. cinerea* (Ferrari et al. 2003) but this molecule also antagonizes JA signaling, while JA and ethylene positively regulate defence against *B. cinerea* (Thomma et al. 1998, 1999). ABA is better known for its potentiation of the abiotic stress response (Cutler et al. 2010), whilst its signaling effects tend to negatively impact defence against *B. cinerea* (Audenaert et al. 2002). Phytohormone perception triggers proteasomal degradation of repressor proteins leading to the activation of downstream TFs. These phytohormones have been shown in numerous studies to function both synergistically and antagonistically (Pieterse et al. 2012) with TFs functioning as key mediators of phytohormone-signaling interactions. For example, synergistic activity of JA and ET is mediated via the ET-stabilised TFs, EIN3 and EIL1 (Fig. 17.1). JA increases the transcriptional activity of EIN3 and EIL1 by alleviating binding of JA-sensitive JAZ repressors to these TFs (Zhu et al. 2011b). ERF1 and ORA59 are direct targets of EIN3 (Solano et al. 1998) and themselves induce expression of JA/ET dependent genes such as plant defensin 1.2 (Pré et al. 2008). SA antagonism of JA signaling is also mediated via TF activity. SA reduces the accumulation of ORA59, and the ORA59 binding site (GCC-box) was shown to be sufficient to mediate SA/JA antagonism (Van der Does et al. 2013).

The large-scale change within the cell and numerous points of interaction between signaling pathways during pathogen attack makes interpretation of data daunting. Furthermore despite abundant knowledge of components undergoing expression and activity shifts during infection the exact mechanisms and detailed topology of the regulatory networks manifesting these changes is poorly characterized. Of the several hundred TFs that are differentially regulated during infection only a handful of these have been experimentally validated to play a key role in immunity. Furthermore their interaction with each other and other cellular components, both physically and via transcriptional regulation, is for the most part uncharacterized. In this chapter we will review the progress in plant systems biology looking at technical advances facilitating the development of detailed information-rich models of the host defence response providing improved knowledge and understanding of this complex and fascinating interaction.

17.2 Dynamic Modelling of the Host Transcriptome During Infection

The importance of transcriptional regulation in defence against *B. cinerea* is clear. Regulators important in defence against the fungus have been identified and their substantial influence on the transcriptome is made evident in a number of transcriptome profiling studies (for example, AbuQamar et al. 2006; Ferrari et al. 2007; Rowe et al. 2010; Mulema and Denby 2011; Birkenbihl et al. 2012). Additionally these studies revealed that the number of responding genes varies significantly over the course of infection. Variable differential expression may arise from specific temporal regulation of many genes during infection. These studies also revealed evidence of transcriptional regulators providing cross-talk feedback between signaling hormone pathways (AbuQamar et al. 2006; Birkenbihl et al. 2012) and modulation of major cellular systems involving photosynthesis, protein synthesis and transport (Bilgin et al. 2010). However, the temporal order and regulatory dependence of these different responses during infection remained unclear. Furthermore, very little experimental evidence exists on the TF-DNA binding underlying these expression changes or information on how TF-DNA binding specificity is determined. The majority of TF characterization studies involve expression profiling of TF knockouts yielding both direct and indirect target gene expression changes complicating validation of specific TF-DNA interactions. This deficit in knowledge is further highlighted by the fact that despite there being over 2000 TFs in the Arabidopsis genome only a few hundred non-redundant TF-binding motifs have been described (Higo et al. 1999; Lescot et al. 2002; Matys et al. 2006; Jin et al. 2013). This lack of knowledge is one of the greatest challenges in elucidating gene regulatory networks in plants.

Our recent systems approach (Windram et al. 2012) attempts to answer many of the questions posed by previous work. The paper reports the generation of a high-resolution time series of the host transcriptome at 24 time points over 48 h after *B. cinerea* infection. The resolution of sampling enabled estimation of the time of first differential expression (TOFDE) of a gene in 30 min increments with differential expression observed for over a third of the Arabidopsis genome during this 48 h window. This data revealed a large number of genes that were diurnally regulated in the control leaves but interestingly the vast majority of these rhythmic profiles were overridden by a response to infection. Furthermore, the rhythmic expression of core clock genes was dampened following infection. This suggests influence by the pathogen or endogenous regulation of the central clock oscillator during the defence response. A similar observation was made following bacterial infection or treatment with flagellin (Zhang et al. 2013a) suggesting that clock-modulation may be part of the plant defence response. The relevance of this surprising transcriptional modulation of the core clock is yet unknown. It appears therefore that differential expression during infection is convoluted by diurnal rhythms but the high-resolution data of a time series allows us to distinguish rhythmic expression behavior from true infection-dependent differential expression.

The high-resolution time series provided an in-depth survey of the 48 h after infection giving a more comprehensive assessment of gene expression change than classical single snapshot or limited time point studies. These are prone to missing windows of transient expression change that may be critical to defence. Fungal growth stages consisted of an initial rapid growth phase followed by a lag phase during which lesion formation occurs, followed by a lesion expansion stage. The high-resolution time series revealed a major transcriptional shift occurred around 20–28 h after infection (HAI) coinciding with the lag phase in *B. cinerea* growth. A smaller peak in differential expression was also detected around ten HAI just prior to an earlier short lag phase, suggesting that the transcriptional response of the plant slows pathogen growth (albeit temporarily). This analysis highlights the utility of the time series approach in detecting narrow time windows of important gene expression changes due to the resolution of the data set and statistical power afforded by the scale of the data. Thus revealing the timing of these transcriptional changes allows us to highlight potential influence of early expression change on pathogen growth, as well as delineating the timing of transcriptional responses and hence chronological order of different defence processes.

The controlled vocabulary of gene ontologies (GOs) allows functional relationships between genes to be identified in an automated fashion (Ashburner et al. 2000; Maere et al. 2005) and GO enrichment of genes with the same TOFDE or within co-expressed clusters revealed a number of distinct processes which appeared to be activated or repressed during infection. During early infection stages phytohormone biosynthesis and signaling occurred in a distinct order suggesting a mechanism for cross-talk outcome based on the timing of gene expression. The enrichment analysis suggested that ethylene biosynthesis occurs early around 14 HAI followed by responses to ethylene and JA at 16 HAI. Activation of ABA catabolism occurs relatively late around 20 HAI along with negative regulation of ABA signaling and auxin biosynthesis at 22 HAI. This corresponds well with existing knowledge of the important role of ethylene and JA in promoting defence against *B. cinerea* (Thomma et al. 1998, 1999) and previous observations suggesting a negative role of ABA in defence against *B. cinerea* (Audenaert et al. 2002). It suggests that earlier repression of ABA signaling could enhance resistance.

The high-resolution gene expression profiles offer a unique opportunity to model the underlying transcriptional network by inferring regulatory relationships between genes based on temporal causality of gene expression profiles. This initially appears straight-forward but complex post translational control can lead to regulatory influence not directly captured within the data and the number of potential interactions is vast. We chose modelling methods that could accommodate regulatory uncertainty arising from post-translational control but such methods are limited in the number of expression profiles they can consider at one time. This is a consequence of noise within the data, a problem that can only be overcome by reducing experimental noise or increasing the number of measurements per gene, both intractable (Gershenfeld 1999). With this limitation, and size of the data set, only ~50–100 genes can be modeled simultaneously however there are nearly 10,000 differentially expressed genes. A logical solution to this is dimensionality reduction; this was accomplished through clustering of expression profiles. We therefore modeled clusters as individual nodes representative of gene modules within the network and inferred their interaction with each other from the time series data. In addition the causal structure identification method we chose to use allowed for the incorporation of pathogen growth time series data, which enabled the relationship between the host transcriptional response and fungal growth to be inferred. It is noteworthy that the fungus appeared to influence a large number of host clusters whilst only one host cluster was suggested to directly influence pathogen growth (Fig. 17.2). Also, while a few clusters appeared to act independently (right hand section) or upstream of pathogen growth, all but one terminal cluster was influenced by pathogen growth. This suggests that pathogen behavior significantly shapes the transcriptional defence response of the host. Part of this influence may be explained by the observation that

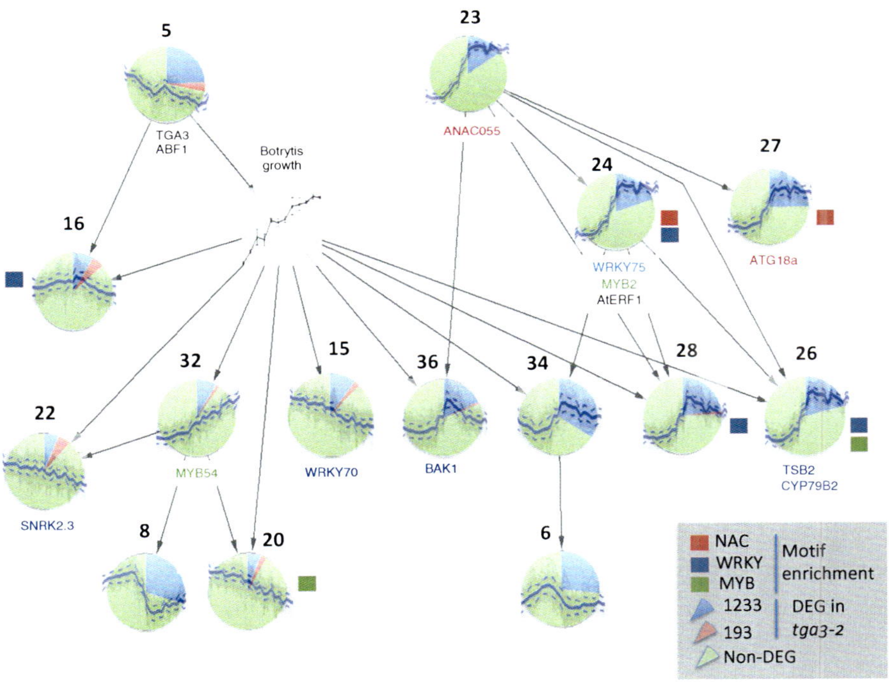

Fig. 17.2 Regulatory network inferred from time series expression data following *B . cinerea* infection of Arabidopsis leaves. *Numbered nodes* represent a cluster of genes co-expressed during *B. cinerea* infection. The expression profile of *B. cinerea* tubulin was used as a proxy for pathogen growth. Selected TFs present in clusters are indicated under nodes. *Coloured boxes* adjacent to nodes indicate TF binding motifs enriched in the promoter sequences of cluster genes, with TFs from the corresponding binding family highlighted in the same colour. The node *piecharts* indicate the proportion of cluster genes which are differentially expressed (DEG) in the *tga3-2* mutant compared to wildtype (193 are more likely to be direct targets) showing the downstream effect of knocking out *TGA3* on the network (Reprinted with permission from Windram et al. 2012)

B. cinerea produces small interfering RNAs that target and silence expression of a number of defence signaling components (Weiberg et al. 2013).

The clusters of co-expressed genes are represented by network nodes and within these are transcription factors that can potentially regulate target genes in other downstream clusters. The resulting network revealed some interesting features and hypotheses for experimental testing. Specific clusters were enriched for TF-binding motifs in the gene promoters. In several cases clusters upstream of such motif-enriched clusters contained a TF capable of binding the enriched motif providing evidence that the inferred relationship between clusters might be indicative of true regulation. One particular example involved a cluster containing the TF ANAC055. This cluster was predicted to regulate a downstream cluster enriched for genes containing NAC TF binding motifs in their promoters (Fig. 17.2). This suggests that ANAC055 may be responsible for regulation of these downstream genes. One of the downstream genes *ATG18a* is differentially regulated in *anac055* mutants indicating a regulatory influence on this target by ANAC055 (Hickman et al. 2013) and providing preliminary evidence for the accuracy of this network model. Only one cluster in the model appeared to regulate fungal growth and this cluster contained the SA signaling TF TGA3, known to be required for proper defence against the biotrophic pathogen *Pseudomonas syringae* (Kesarwani et al. 2007). T-DNA knockout mutants of *TGA3* have increased susceptibility to *B. cinerea* infection and expression profiling of *tga3* mutants identified a number of differentially expressed defence genes (Windram et al. 2012). Crucially, in the network model (Fig. 17.2) these genes were predominantly found in the TGA3 cluster itself and clusters downstream of TGA3, again lending support to the regulatory predictions. One potential downstream target of TGA3 (down regulated in the *tga3* knockout line compared to wildtype) was WRKY70, a TF integrating SA and JA signaling (Li et al. 2006). Overall network modeling provided a number of regulatory predictions which could not have been inferred without the high-resolution data, and highlighted TGA3 as an important early regulator of *B. cinerea* defence.

SA is an important defence hormone but its role in *B. cinerea* defence is not clear (Thomma et al. 1998; Ferrari et al. 2003, 2007). Early priming experiments suggested SA provides a defence advantage if priming occurs 24 h but not 48 h prior to infection (Zimmerli et al. 2001; Govrin and Levine 2002). Furthermore the SA-dependent hypersensitive response induced by avirulent bacterial pathogens led to enhanced susceptibility towards *B. cinerea* (Govrin and Levine 2000). This suggests that early SA signaling is important for *B. cinerea* defence but the role of SA signaling changes based on the timing and strength of the response. Network modeling of the time series data identified *TGA3* as a high level regulator influencing both pathogen growth and host gene expression. TGA3, like other members of its family, is dependent on the key SA signaling protein, non-expressor of PR1 (NPR1), for its activity (Johnson et al. 2003) with NPR1 being a transcriptional co-activator activated by SA. However, knockout analysis of *TGA3* revealed a positive influence on *B. cinerea* defence. A similar role has previously been shown for TGA2, TGA5 and TGA6 in a semi-redundant manner (Zander et al. 2009) but these three TFs also play a key role in JA and ET signaling. We hypothesise that TGA3 may be modulating early SA responses to repress initial pathogen growth but over time ET and JA signaling becomes more important and SA signalling is repressed. This may help to explain why TGA3 is actively down-regulated in respond to infection despite its positive influence on defence (Windram et al. 2012).

In summary the systems approach, using high-resolution data and network modeling, has facilitated the temporal analysis of host responses and pathogen growth, identified key regulators of the defence response against *B. cinerea* and provided regulatory hypotheses for experimental testing. Additional modelling and simulation algorithms have subsequently been used to indicate the most informative experiments to carry out and the TFs most likely to influence genes in biological pathways with a significant impact on resistance, such as camalexin biosynthesis (Denby et al. 2004).

17.3 Alternative Systems Approaches and What They Can Teach Us About Defence

To date the approach by Windram et al. (2012) outlined above is the only systems biology work on host defence against *B. cinerea* published. They used network inference from transcriptional time series, however, a number of groups have used other systems approaches to study different plant-pathogen interactions (see Windram et al. (2014) for more detailed descriptions of these studies). Some of these approaches are likely to be informative for understanding defence against *B. cinerea*; the pros and cons of each approach are summarized in Table 17.1. Co-expression networks are perhaps the most common network model in the literature and can be used to discover novel components of the defence response. The network is built typically from multiple genome-wide expression data sets and two nodes of the network (genes) are linked by an edge if their expression is correlated across these data sets. These networks do not capture regulatory relationships, but using the guilt-by-association principle (proximity in the network to known components of a process) novel genes involved in a process can be identified (Usadel et al. 2009). Many co-expression networks tend to be genome-wide, although the use of focused expression data is thought to enhance their power for gene discovery. One such network focused on defence responses in citrus (Zheng and Zhao 2013) and predicted groups of novel genes that were likely to influence resistance.

Such co-expression networks have been extended by approaches incorporating multiple different data types, not just expression data (for example, Lee et al. 2010), as well as data from other species. These functional association networks again use the guilt-by-association principle to predict gene functions. In rice, a genome-wide functional association network was built and probed for genes involved with defence (Lee et al. 2011). Genes closely linked to a set of known defence response genes were identified and several of these were shown experimentally to affect the immune response. An advantage of this functional association approach is that these net-

Table 17.1 Systems biology approaches for gene regulatory network elucidation. Different approaches that have been used to study plant-pathogen interactions, the advantages and disadvantages of each approach and key examples

Methodology	Advantages	Disadvantages	Key example
Co-expression networks	Genome-wide, can predict novel gene function	Generally need multiple data sets, hence more relevant for model organisms, predict gene function not regulation	Usadel et al. (2009)
Functional association networks	Genome-wide, cross species predictive ability	Extensive data sets still needed so more useful for model organisms, predict gene function not regulation	Lee et al. (2011)
Static regulatory network model – using mutant data	Can predict regulatory interactions, can generate a useful network from a limited data set, network nodes do not need to be transcriptionally regulated	Data sets available limit network size, interactions are undirected, network does not predict causal regulatory relationships	Sato et al. (2010)
Static regulatory network model – using extensive genome-wide data sets	Genome-wide, predicts regulatory relationships	Requires extensive data sets so only appropriate for model organisms	Carrera et al. (2009)
Dynamic regulatory network models – constrained by literature information	Predicts regulatory interactions, can simulate networks to predict effects of perturbations (e.g. gene knockouts)	Can only include known network components	Naseem et al. (2012)
Dynamic regulatory network models – inferred from data	Predicts regulatory interactions, can simulate networks to predict effects of perturbations (e.g. gene knockouts), can be inferred from a single data set, can include novel network components	Genome-wide networks computationally expensive, requires high-resolution time series expression data	Windram et al. (2012)

works appear to have cross-species predictive ability, for example, the rice network can predict gene functions in maize. However, extensive data is still required (although exploiting orthology between species) and these networks simply predict a role for a gene in a process, they do not shed light on how that gene functions or how it is regulated.

To engineer crops with durable enhanced defence against pathogens, it will not be sufficient to identify single genes whose presence and/or expression level alters resistance. Rather, it is necessary to understand how genes/proteins interact and function together in networks to regulate defence mechanisms, and be able to predict the effects of manipulating combinations of key genes from these networks. The ability to accurately predict the effect of genetic perturbations (i.e. phenotypic predictions from genotype) is crucial, and requires regulatory models. Sato et al. (2010) built a regulatory network using static (i.e. single time point) expression data from mutants lacking specific defence regulatory genes. They used 22 different Arabidopsis mutants and built a network based on the similarity of expression changes in those mutants (compared to wildtype) following infection with a bacterial pathogen, *P. syringae* pv *tomato* DC3000. Nodes (representing the genes mutated in each line) are connected in the network if they are predicted to regulate similar sets of genes.

In this aspect, although the network is not predicting a causal regulation between nodes, it is indicating a type of regulatory interaction not just co-expression. This type of network analysis limits the network to the nodes used in expression profiling, i.e. 22 genes in this case, but generates a regulatory framework from a relatively limited data set. Furthermore, an attraction of this approach is that the nodes themselves do not need to be transcriptionally regulated. The network topology relies on transcriptional changes downstream of these nodes, but the nodes themselves can be regulated in a post-transcriptional manner. Extending network models beyond transcriptional regulation will be crucial in accurately portraying plant defence responses. The network produced by Sato et al. (2010) was able to predict all the experimentally validated interactions between the 22 defence components and highlighted the prevalence of negative regulatory relationships between different sectors of the network.

Another network building approach from static gene expression data was taken by Carrera et al. (2009), although this was not specific to defence. They used a wide range of publicly available expression data from numerous treatments, tissue types and mutant genotypes to infer regulatory relationships between Arabidopsis transcription factors. Co-expression of the transcription factors (with the algorithm capturing both linear and non-linear relationships) was used to determine the overall network topology and subsequently the expression of a gene was then explained in terms of a weighted sum of expression of connecting genes (i.e. the upstream gene expression profile that could best explain the profile of the target gene would be given the most weight). The resulting network is genome-wide but required an extensive collection of public data. Interestingly, connectivity in this global network was increased in the subnetworks corresponding to biotic and abiotic stress, and genes in these subnetworks appeared to be regulated by a relatively small number of key regulators. One such regulator identified by Carrera et al. (2009) was ERF1, a transcription factor known to affect susceptibility to *B. cinerea*.

Ma et al. (2013) again used large public collections of expression data to infer a genome-wide co-expression network, but added to this using transcription factor binding motif information to build a regulatory network. This approach may be of more value as our knowledge of promoter motifs improves but currently we are only able to predict binding of a family of transcription factors to a motif, rather than an individual member of this family, and even that for only a handful of transcription factor families. Again this approach is only valid for model organisms where large collections of expression data and knowledge of transcription factor binding motifs exist.

A dynamic regulatory model focused on hormone signaling during plant defence was published by Naseem et al. (2012). The authors initially made a Boolean network from known literature interactions (Fig. 17.3). In Boolean networks the expression of a gene is usually classed as on or off and interactions are described using logic terms such as "AND" or "OR", i.e. where two genes were needed to activate a third, this would be an "AND" relationship. If either of the two genes could activate expression of the third, this would be an "OR" connection. The Boolean network included metabolic pathways of hormone synthesis and catabolism, as well as signaling pathways and the action of pathogen effectors, hence making an attempt to integrate components in both pathogen and host. Different levels of regulation are included such as levels of hormones, protein activity and gene expression. Converting this Boolean network into a dynamic network enabled the authors to simulate flux through the network after perturbing specific compo-

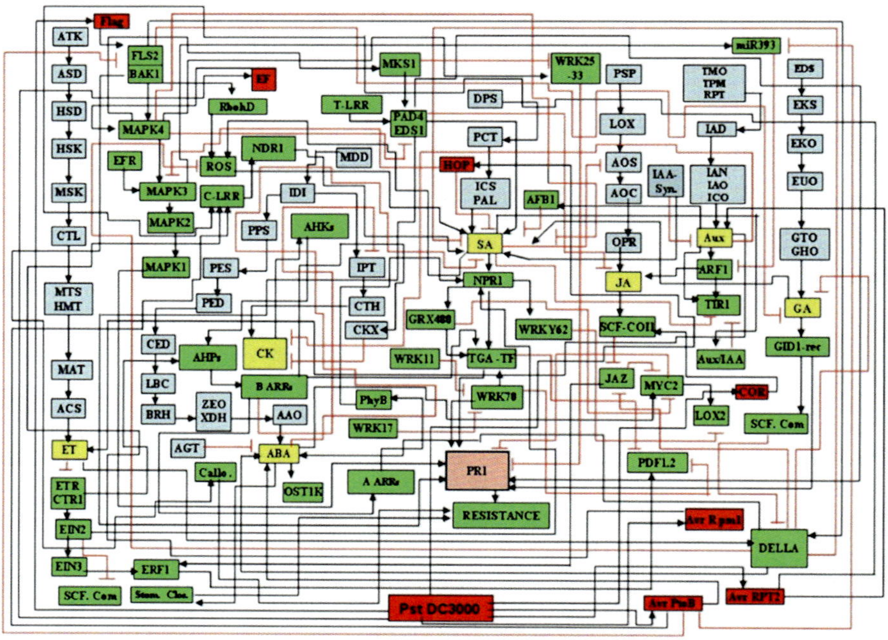

Fig. 17.3 Topology of the Boolean Network for hormone signaling during defence against *Pseudomonas syringae pv tomato* (Pst). Connectivity between nodes is based either on activation (→) or inhibition (—|). Node designation: *blue*, enzymes of hormone biosynthesis and degradation; *yellow*, active hormone molecules; *green*, host regulatory factors; *red*, Pst-originated pathogenicity factors responsible for triggering immunity in Arabidopsis; *pink*, PR-1, marker node for immunity against the infection of Pst in Arabidopsis. This network was derived from experimental evidence in the literature, converted into a dynamic network and used to simulate the effect on resistance of perturbations in specific network components (Reprinted with permission from Naseem et al. 2012)

nents (for example, presence or absence of a particular hormone and full or partial activation of a network component).

This *in silico* analysis made multiple predictions, one example being on the interaction of cytokinin with other defence hormones. Previous work had suggested both positive and negative impacts of cytokinin on SA signaling; this model predicted the interaction was positive and occurred downstream of SA synthesis. The advantages of this network modelling strategy are that specific predictions can be made for experimental testing, and the modeling includes different layers of regulation. However, it is only able to predict relationships and interactions between known components of plant immunity.

The approaches described above focus on transcriptional regulation of gene expression, however many post-transcriptional events are also involved. One way of investigating these events is via protein-protein interaction mapping and/or (post-translational modification) proteome analyses, but these are time-consuming processes that are not feasible in every organism. Hence a major challenge in systems biology is inference of protein-protein interactions and understanding the dynamic regulation of the proteome under different conditions. Prediction of protein-protein interactions often relies on the assumption that interacting protein domains are conserved across species. Using this assumption and experimental data from a range of model organisms, protein-protein interaction networks have been predicted for three plant species, Brassica, rice and Arabidopsis (Geisler-Lee et al. 2007; Zhu et al. 2011a; Yang et al. 2012). Although not focused on immunity, such genome-scale networks can be used to identify potential defence related interactions. A biotic stress network has been generated in rice by probing a Y2H cDNA library with known components of the defence response (Seo et al. 2011). Proteins with a large number of interactions were selected and several were shown to influence resistance to *Xanthomonas oryzae*.

Ultimately we will require models that can incorporate multiple levels of regulation (and events in both host and pathogen, see below). While we are still a long way from such networks, the modelling approaches discussed above will make valuable contributions to our understanding of the host defence response, how it is regulated and predictions on how to re-wire the network for enhanced resistance.

17.4 Integrating Pathogen and Host Responses and Future Prospects

Botytis cinerea like many other pathogens utilizes a number of virulence strategies to suppress host defence and enhance its pathogenicity (Chap. 12). The fungus produces phytotoxins and elicitors that act to promote cell death thereby enhancing disease (Van Kan 2006). Some isolates of the fungus can also produce ABA (Chap. 15) suggesting that the pathogen might employ antagonistic hormone signaling techniques to further its infection (Siewers et al. 2006). High-resolution transcriptome analysis of Arabidopsis during *B. cinerea* infection highlighted some unexpected transcriptional responses, for example genes such as TGA3 with a positive role in defence were downregulated during infection (Windram et al. 2012). One hypothesis was that the pathogen may influence host transcription. A more recent study provides additional evidence for this hypothesis. Weiberg et al. (2013) discovered a fascinating virulence capacity within *B. cinerea*. The fungus is capable of producing small interfering RNAs that can target and silence a number of host defence genes. siRNA targets included *MPK1*, *MPK2* and a *wall associated kinase*, with knockout analysis showing these genes played a positive role in defence against *B. cinerea* despite being downregulated during infection. This points to a complex intricate relationship between the pathogen and its host, and development of an accurate and comprehensive representation of plant defence will therefore require the incorporation of both pathogen and host data into a unified network model.

As mentioned above integration of pathogen growth data with host high resolution temporal transcriptome data can lead to informative network representations highlighting components modulating host defence (Windram et al. 2012). A natural development of this approach will be incorporation of pathogen transcriptome data. Modern transcriptome quantification methods such as RNA-seq can be used to profile both pathogen and host expression simultaneously. Furthermore with sufficient temporal resolution and utilizing conventional transcriptional modeling methods we can infer regulatory relationships between pathogen and host. Here again Bayesian methods are preferred for modeling of complex multi-dimensional regulation as these approaches can infer both direct and indirect interactions and take information not directly captured by transcriptome profiling into account (Penfold and Wild 2011).

Botrytis cinerea has an extremely broad host range (Chap. 20); transcriptome profiling of the fungus on different hosts will help to uncover the commonalities in the infection strategy that makes the pathogen so successful, but will also help highlight host responses limiting infection. We have used RNA-seq to profile *B. cinerea* gene expression during infection of Arabidopsis and tomato and seen that expression is highly correlated between the two different infections at early time points indicating a common infection strategy of the pathogen (unpublished data). This type of approach was used by Weiberg et al. (2013) to identify crucial small RNAs produced by *B. cinerea* during infection. Maintaining a focus on fungal siRNA expression will help elucidate some of the direct host pathogen interactions but it is likely that a large number of other pathogen factors like proteins and small molecules will also influence host defence.

A complementary strategy would be to use a similar approach to Sato and colleagues (2010) (outlined above) where regulatory interactions were inferred from single time point transcriptome data from knockout mutants of key defence regulators. The elegant simplicity of the method could easily incorporate pathogen data. The genetic tractability of *B. cinerea* combined with availability of known defence mutants in Arabidopsis offers up a unique opportunity to provide a more nuanced understanding of host-pathogen interaction using this approach. The regulatory mutant approach of Sato et al. (2010) provides an undirected network, however, methods have been developed in other areas of network biology which could be employed to establish directionality (Markowetz et al. 2007). Although extension of this method to include pathogen data is undoubtedly more challenging, combinations of methods are likely to give a more complete picture overall.

Apart from direct transcriptome interaction *B. cinerea* and its host also interact using other cellular components. In particular, certain MAMPs like BcPGs can trigger defence whilst a fungal xylanase can act to promote disease (Poinssot et al. 2003; Noda et al. 2010; Chap. 12). Recent developments in the field of plant defence modelling and the wider field of plant network biology, both in terms of techniques and reference data sets, present an opportunity to greatly enhance our understanding of host-pathogen protein-protein interaction. The AI-1 network is a binary protein-protein interaction network, elucidated using a yeast-2-hybrid approach consisting of ~6200 edges between ~2300 Arabidopsis proteins (Arabidopsis Interactome Mapping Consortium et al. 2011). An extension of this network was achieved using the same technology platform to elucidate interactions between the 8000 Arabidopsis proteins screened for AI-1 and an additional 552 proteins consisting of a mixture of host immune proteins as well as effectors from *P. syringae* and *H. arabidopsidis* (Mukhtar et al. 2011). Analysis of the resulting network highlighted a suite of novel proteins targeted by pathogen effectors and not previously identified as components of the defence response causing a major shift in how effectors were thought to function. However, effectors from the two distinct pathogens converged on a limited number of host targets and these targets tended to act as network hubs interacting with numerous other host proteins. This is in agreement with previous theoretical predictions that such hub proteins represent critical weak points in these networks (Albert et al. 2000).

It is likely that generating protein-protein interaction data for *B. cinerea* and host proteins would yield a wealth of knowledge that would vastly improve our understanding of the host-pathogen relationship. One caveat to this is selection of a protein subset for screening. Y2H approaches can yield useful information but exhaustive binary screening can be very resource- and time-intensive. Methods allowing establishment of a prioritization pipeline will vastly improve discovery of relevant interactions. In the AI-1 network interacting proteins were found to be much more highly co-expressed compared to random control samples (Arabidopsis Interactome Mapping Consortium et al. 2011) suggesting that existing transcriptome data could help prioritization. Additionally the *B. cinerea* genome is predicted to contain several hundred secreted proteins, several of which are differentially regulated within the first few hours of perception of a potential host (Espino et al. 2010; Leroch et al. 2013) suggesting a number of pathogen components for future host-pathogen protein-protein interaction studies.

Crucially computational methods have been developed to predict protein-protein interactions and one of the most exciting developments for plant immunity was the attempt to predict a cross-species interactome between a plant host and pathogen. A network was created containing over 3,000 interactions between ~1,400 Arabidopsis proteins and 119 proteins from *Ralstonia solanacearum* giving insight into potential points of interaction and hence manipulation (Li et al. 2011). Although these predictions were not experimentally validated the resulting network shared similar properties to the experimentally derived host-pathogen protein-protein interaction network of Mukhtar et al. (2011). In particular, pathogen effectors targeted host hub proteins suggesting that this in silico modeling approach might be a useful method to facilitate node selection for future experimental validation. Combining these types of networks with transcriptome models will provide a much more comprehensive understanding of the pathosystem helping highlight critical points within the network as foci for external intervention in the development of effective pest management strategies.

17.5 Translating Between Pathogens and from Model Organism to Crop Plants

As *B. cinerea* can infect a wide range of plants the true power of systems biology approaches will be realized by their application to crop plants. There are two basic strategies in translating systems biology to crops; (i) translating the networks identified in model organisms or (ii) translating the methodologies. The first approach requires network topologies to be conserved (or similar) between different plant species and the ability to determine true orthologues of genes in the networks. Predicting precise orthologues of genes between species is not always straightforward. Many crop plants are polyploid so there may be multiple genes with the same function, or one function may be split across several paralogues. Approaches such as synteny (the conservation of gene order along a chromosome), gene clustering (Field and Osbourn 2008) and conservation of regulatory elements within promoter sequences (Baxter et al. 2012) can help identify true functional orthologues; finding true orthologues with the same regulatory function is likely to still be a challenge. Clearly expression data could help identify true orthologues but depending on the species, and the availability of a high-quality reference genome, expression data may not be able to distinguish between alleles.

Currently it is difficult to predict how conserved regulatory networks are across species. Individual defence response components have been shown to be conserved and, for example, TFs from one plant can perform a similar function in another (Lyons et al. 2013), but whether the regulatory networks are conserved is not known. Cultivated brassicas represent the crop family most closely related to Arabidopsis yet we have little information on how complex regulatory relationships are conserved between Arabidopsis and other brassicas. A protein-protein interactome in *Brassica rapa* has been predicted from Arabidopsis data but it has yet to be experimentally verified (Yang et al. 2012). Functional association networks incorporating multiple data types from multiple species have been shown to be able to predict gene function and identify novel components of a biological process (Lee et al. 2010, 2011). The ability to share information across species improves the predictive ability of these networks, but it still requires fairly extensive data sets from the organism in question. Furthermore these networks are useful for gene discovery but do not help elucidate the regulation of a process.

If it is not feasible to translate network models across species, the alternative is to translate the methodology from Arabidopsis into crop plants. The feasibility of translating systems biology methodologies into crop plants depends on the technique in question. Approaches relying on large-scale publicly available data (see Table 17.1) are not feasible for most crops. However, an attraction of the network inference used by Windram et al. (2012) is that prediction of gene function and regulatory interaction can be made from a single time series data set. This can be supplemented with information from other related species using algorithms such as hierarchical Causal Structure Inference (Penfold et al. 2012). In this algorithm, transcriptome time series data from two different species can be used to jointly infer regulatory networks (Fig. 17.4). Information common to the two species will be shared via a hyperparent network to strengthen these interactions. Species-specific network predictions can override the hyperparent network or add to it. In this context, the hyperparent network resembles an ancestral network with additional connections or altered network connections arising in each species. This shared information approach can also be used across different pathogen infections within a single species. In this case, the hyperparent network represents the common defence response, which is decorated with interactions specific to each infection. The advan-

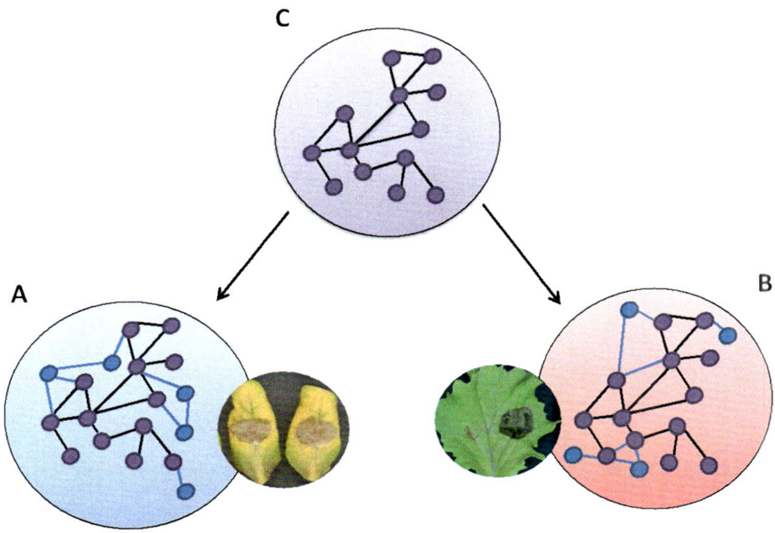

Fig. 17.4 Regulatory networks can be jointly learnt across species. In this example, the network mediating the transcriptional response to *B. cinerea* infection in Arabidopsis (**a**) and the network underlying the lettuce response to *B. cinerea* infection (**b**) are jointly inferred from transcriptome data in each species. This joint inference uses information common between the two species (**c**, the hyperparent network) to strengthen these interactions. Species-specific interactions can override the hyperparent network or supplement it. Species-specific interactions and nodes are shown in *blue*

tage of jointly learning networks in this manner, is that shared information is used to strengthen these network predictions and differences between networks (either between species or pathogens) can be easily identified. This approach has been used in modelling abiotic stress regulatory networks and predicted a transcriptional switch point in the regulation of responses to cold and osmotic shock (Penfold et al. 2012).

Perhaps the critical limiting factors in using systems biology approaches in crop plants are experimental factors. These include the experimental testing required to validate networks. For many crop plants, it is not simple or quick to generate transgenic lines in which specific genes are knocked out or overexpressed. This may be because of a lack of appropriate transformation methodology, the difficulties of specifically silencing a gene with closely related paralogues, or simply the time taken to complete a single generation. The combination of systems biology approaches with more traditional quantitative genetics may be one solution. For example, network modelling can predict key genes regulating the host defence response. Integrating this information with phenotypic (i.e. resistance) quantitative trait loci (QTL) identified from mapping populations may facilitate identification of the gene(s) underlying the QTL, and speed up the use of these traits in breeding programmes.

17.6 Conclusion

Network modelling and systems approaches can make a distinctive and valuable contribution to understanding plant-pathogen interactions and key control points determining the infection outcome. The control of plant defence responses is complex, involving multiple types and number of interacting components and multiple types of regulatory interaction between components. Modelling these interactions and their ultimate effect on plant immunity is not an easy task but is vital to our ability to fully understand and predict the outcome of a plant pathogen infection. Much of our work to date has been to perturb a single component of the defence network and monitor the effect on disease resistance phenotype and defence processes. When dealing with a regulatory network with multiple interactions between host components, let alone pathogen components, the information provided by these approaches is limited.

There are several challenges for creating comprehensive, accurate and useful network models. Firstly, there are many gaps in our knowledge that will require high-throughput screening approaches to fill. These include elucidation of DNA sequence motifs bound by different TFs and protein-protein interactions between host proteins and between host and pathogen proteins. A computational challenge is the development of methods that allow integration of the numerous modeling techniques and data types that capture regulatory information at multiple system levels such as the transcriptome, proteome and metabolome. Such a holistic model will undoubtedly provide superior predictive capability, as well as significant insight into the mechanisms of plant-pathogen interaction, whilst also revealing the optimal means to control plant disease. Accurate network models will enable large numbers of in silico experiments to be performed to identify perturbations of the network that enhanced disease resistance. Such perturbations could be used to manipulate the outcome of pathogen infection and engineer crops with increased disease resistance. Furthermore identifying the key interaction points between host and pathogen will drive additional strategies for enhancing resistance.

Another important hurdle to be overcome is the problem of beneficial perturbations of the defence network having unwanted deleterious effects on other plant processes. The most useful network models will therefore capture such broader interactions and enable these effects to be incorporated into any phenotypic predictions. The analysis of defence regulatory networks in species other than Arabidopsis, facilitated by the wealth of genome and transcriptome data being generated, will help understand the level of conservation between host defence responses and regulatory networks and to what extent information can be translated across crop plants. There are several challenges associated with generating network models in crop plants, both experimental and computational. It is vital that we work to overcome these to realize the potential impact systems biology techniques can have on breeding of crops with beneficial traits.

References

AbuQamar S, Chen X, Dhawan R et al (2006) Expression profiling and mutant analysis reveals complex regulatory networks involved in Arabidopsis response to *Botrytis* infection. Plant J 48:28–44

Albert R, Jeong H, Barabasi A (2000) Error and attack tolerance of complex networks. Nature 406:378–382

Arabidopsis Interactome Mapping Consortium, Dreze M, Carvunis AR et al (2011) Evidence for network evolution in an Arabidopsis interactome map. Science 333:601–607

Ashburner M, Ball CA, Blake JA et al (2000) Gene ontology: tool for the unification of biology. The gene ontology consortium. Nat Genet 25:25–29

Audenaert K, De Meyer GB, Höfte MM (2002) Abscisic acid determines basal susceptibility of tomato to *Botrytis cinerea* and suppresses salicylic acid-dependent signaling mechanisms. Plant Physiol 128:491–501

Baxter L, Jironkin A, Hickman R et al (2012) Conserved noncoding sequences highlight shared components of regulatory networks in dicotyledonous plants. Plant Cell 24:3949–3965

Berrocal-Lobo M, Molina A, Solano R (2002) Constitutive expression of ETHYLENE-RESPONSE-FACTOR1 in Arabidopsis confers resistance to several necrotrophic fungi. Plant J 29:23–32

Bilgin DD, Zavala JA, Zhu J et al (2010) Biotic stress globally downregulates photosynthesis genes. Plant Cell Environ 33:1597–1613

Birkenbihl RP, Diezel C, Somssich IE (2012) Arabidopsis WRKY33 is a key transcriptional regulator of hormonal and metabolic responses toward *Botrytis cinerea* infection. Plant Physiol 159:266–285

Boller T, Felix G (2009) A renaissance of elicitors: perception of microbe-associated molecular patterns and danger signals by pattern-recognition receptors. Annu Rev Plant Biol 60:379–406

Brutus A, Sicilia F, Macone A et al (2010) A domain swap approach reveals a role of the plant wall-associated kinase 1 (WAK1) as a receptor of oligogalacturonides. Proc Natl Acad Sci U S A 107:9452–9457

Bu Q, Jiang H, Li C-B et al (2008) Role of the *Arabidopsis thaliana* NAC transcription factors ANAC019 and ANAC055 in regulating jasmonic acid-signaled defense responses. Cell Res 18:756–767

Carrera J, Rodrigo G, Jaramillo A, Elena SF (2009) Reverse-engineering the *Arabidopsis thaliana* transcriptional network under changing environmental conditions. Genome Biol 10:R96

Chen L, Zhang L, Yu D (2010) Wounding-induced WRKY8 is involved in basal defense in Arabidopsis. Mol Plant-Microbe Interact 23:558–565

Cutler SR, Rodriguez PL, Finkelstein RR, Abrams SR (2010) Abscisic acid: emergence of a core signaling network. Annu Rev Plant Biol 61:651–679

Denby KJ, Kumar P, Kliebenstein DJ (2004) Identification of *Botrytis cinerea* susceptibility loci in *Arabidopsis thaliana*. Plant J 38:473–486

Espino JJ, Gutiérrez-Sánchez G, Brito N et al (2010) The *Botrytis cinerea* early secretome. Proteomics 10:3020–3034

Faulkner C, Petutschnig E, Benitez-Alfonso Y et al (2013) LYM2-dependent chitin perception limits molecular flux via plasmodesmata. Proc Natl Acad Sci U S A 110:9166–9170

Ferrari S, Plotnikova JM, De Lorenzo G, Ausubel FM (2003) Arabidopsis local resistance to *Botrytis cinerea* involves salicylic acid and camalexin and requires EDS4 and PAD2, but not SID2, EDS5 or PAD4. Plant J 35:193–205

Ferrari S, Galletti R, Denoux C et al (2007) Resistance to *Botrytis cinerea* induced in Arabidopsis by elicitors is independent of salicylic acid, ethylene, or jasmonate signaling but requires PHYTOALEXIN DEFICIENT3. Plant Physiol 144:367–379

Field B, Osbourn AE (2008) Metabolic diversification – independent assembly of operon-like gene clusters in different plants. Science 320:543–547

Fiil BK, Amselem J, Cuomo CA et al (2011) Constitutive expression of MKS1 confers susceptibility to *Botrytis cinerea* infection independent of PAD3 expression. Plant Signal Behav 6:1425–1427

Galletti R, Ferrari S, De Lorenzo G (2011) Arabidopsis MPK3 and MPK6 play different roles in asal and oligogalacturonide- or flagellin-induced resistance against *Botrytis cinerea*. Plant Physiol 157:804–814

Geisler-Lee J, O'Toole N, Ammar R et al (2007) A predicted interactome for Arabidopsis. Plant Physiol 145:317–329

Gershenfeld N (1999) The nature of mathematical modeling. Cambridge University Press, Cambridge

Gimenez-Ibanez S, Solano R (2013) Nuclear jasmonate and salicylate signaling and crosstalk in defense against pathogens. Front Plant Sci 4:72

Govrin EM, Levine A (2000) The hypersensitive response facilitates plant infection by the necrotrophic pathogen *Botrytis cinerea*. Curr Biol 10:751–757

Govrin EM, Levine A (2002) Infection of Arabidopsis with a necrotrophic pathogen, *Botrytis cinerea*, elicits various defense responses but does not induce systemic acquired resistance (SAR). Plant Mol Biol 48:267–276

Govrin EM, Rachmilevitch S, Tiwari BS et al (2006) An elicitor from *Botrytis cinerea* induces the hypersensitive response in *Arabidopsis thaliana* and other plants and promotes the gray mold disease. Phytopathology 96:299–307

Grefen C, Harter K (2004) Plant two-component systems: principles, functions, complexity and cross talk. Planta 219:733–742

Hickman R, Hill C, Penfold CA et al (2013) A local regulatory network around three NAC transcription factors in stress responses and senescence in Arabidopsis leaves. Plant J 75:26–39

Higo K, Ugawa Y, Iwamoto M, Korenaga T (1999) Plant cis-acting regulatory DNA elements (PLACE) database: 1999. Nucleic Acids Res 27:297–300

Jin J, Zhang H, Kong L et al (2013) PlantTFDB 3.0: a portal for the functional and evolutionary study of plant transcription factors. Nucleic Acids Res 42:D1182–D1187

Johnson C, Boden E, Arias J (2003) Salicylic acid and NPR1 induce the recruitment of trans-activating TGA factors to a defense gene promoter in Arabidopsis. Plant Cell 15:1846–1858

Jones JDG, Dangl JL (2006) The plant immune system. Nature 444:323–329

Kesarwani M, Yoo J, Dong X (2007) Genetic interactions of TGA transcription factors in the regulation of pathogenesis-related genes and disease resistance in Arabidopsis. Plant Physiol 144:336–346

Kliebenstein DJ, Rowe HC, Denby KJ (2005) Secondary metabolites influence Arabidopsis/Botrytis interactions: variation in host production and pathogen sensitivity. Plant J 44:25–36

Lai Z, Vinod K, Zheng Z et al (2008) Roles of Arabidopsis WRKY3 and WRKY4 transcription factors in plant responses to pathogens. BMC Plant Biol 8:68

Lee JS, Ellis BE (2007) Arabidopsis MAPK phosphatase 2 (MKP2) positively regulates oxidative stress tolerance and inactivates the MPK3 and MPK6 MAPKs. J Biol Chem 282:25020–25029

Lee I, Ambaru B, Thakkar P et al (2010) Rational association of genes with traits using a genome-scale gene network for *Arabidopsis thaliana*. Nat Biotechnol 28:149–156

Lee I, Seo Y-S, Coltrane D et al (2011) Genetic dissection of the biotic stress response using a genome-scale gene network for rice. Proc Natl Acad Sci U S A 108:18548–18553

Leroch M, Kleber A, Kleber A et al (2013) Transcriptome profiling of *Botrytis cinerea* conidial germination reveals upregulation of infection-related genes during the prepenetration stage. Eukaryot Cell 12:614–626

Lescot M, Déhais P, Thijs G et al (2002) PlantCARE, a database of plant cis-acting regulatory elements and a portal to tools for in silico analysis of promoter sequences. Nucleic Acids Res 30:325–327

Li J, Brader G, Kariola T, Palva ET (2006) WRKY70 modulates the selection of signaling pathways in plant defense. Plant J 46:477–491

Li Z-G, He F, Zhang Z, Peng Y-L (2011) Prediction of protein–protein interactions between *Ralstonia solanacearum* and *Arabidopsis thaliana*. Amino Acids 42:2363–2371

Liebrand TWH, Van den Burg HA, Joosten MHAJ (2014) Two for all: receptor-associated kinases SOBIR1 and BAK1. Trends Plant Sci 19:123–132

Lumbreras V, Vilela B, Irar S et al (2010) MAPK phosphatase MKP2 mediates disease responses in Arabidopsis and functionally interacts with MPK3 and MPK6. Plant J 63:1017–1030

Lyons R, Manners JM, Kazan K (2013) Jasmonate biosynthesis and signaling in monocots: a comparative overview. Plant Cell Rep 32:815–827

Ma S, Shah S, Bohnert HJ et al (2013) Incorporating motif analysis into gene co-expression networks reveals novel modular expression pattern and new signaling pathways. PLoS Genet 9:e1003840

Maere S, Heymans K, Kuiper M (2005) BiNGO: a Cytoscape plugin to assess overrepresentation of gene ontology categories in biological networks. Bioinformatics 21:3448–3449

Markowetz F, Kostka D, Troyanskaya OG, Spang R (2007) Nested effects models for high-dimensional phenotyping screens. Bioinformatics 23:i305–i312

Maruyama Y, Yamoto N, Suzuki Y et al (2013) The Arabidopsis transcriptional repressor ERF9 participates in resistance against necrotrophic fungi. Plant Sci 213:79–87

Matys V, Kel-Margoulis OV, Fricke E et al (2006) TRANSFAC and its module TRANSCompel: transcriptional gene regulation in eukaryotes. Nucleic Acids Res 34:D108–D110

Mengiste T, Chen X, Salmeron J, Dietrich R (2003) The BOTRYTIS SUSCEPTIBLE1 gene encodes an R2R3MYB transcription factor protein that is required for biotic and abiotic stress responses in Arabidopsis. Plant Cell 15:2551–2565

Miya A, Albert P, Shinya T et al (2007) CERK1, a LysM receptor kinase, is essential for chitin elicitor signaling in Arabidopsis. Proc Natl Acad Sci U S A 104:19613–19618

Moffat CS, Ingle RA, Wathugala DL et al (2012) ERF5 and ERF6 play redundant roles as positive regulators of JA/Et-mediated defense against *Botrytis cinerea* in Arabidopsis. PLoS One 7:e35995

Mukhtar MS, Carvunis AR, Dreze M et al (2011) Independently evolved virulence effectors converge onto hubs in a plant immune system network. Science 333:596–601

Mulema JMK, Denby KJ (2011) Spatial and temporal transcriptomic analysis of the *Arabidopsis thaliana-Botrytis cinerea* interaction. Mol Biol Rep 39:4039–4049

Naseem M, Philippi N, Hussain A et al (2012) Integrated systems view on networking by hormones in Arabidopsis immunity reveals multiple crosstalk for cytokinin. Plant Cell 24:1793–1814

Noda J, Brito N, Gonzalez C (2010) The *Botrytis cinerea* xylanase Xyn11A contributes to virulence with its necrotizing activity, not with its catalytic activity. BMC Plant Biol 10:38

Penfold CA, Wild DL (2011) How to infer gene networks from expression profiles, revisited. Interface Focus 1:857–870

Penfold CA, Buchanan-Wollaston V, Denby KJ, Wild DL (2012) Nonparametric Bayesian inference for perturbed and orthologous gene regulatory networks. Bioinformatics 28:i233–i241

Petersen K, Qiu J-L, Lütje J et al (2010) Arabidopsis MKS1 is involved in basal immunity and requires an intact N-terminal domain for proper function. PLoS One 5:e14364

Pham J, Liu J, Bennett MH et al (2012) Arabidopsis histidine kinase 5 regulates salt sensitivity and resistance against bacterial and fungal infection. New Phytol 194:168–180

Pieterse CMJ, Van der Does D, Zamioudis C et al (2012) Hormonal modulation of plant immunity. Annu Rev Cell Dev Biol 28:489–521

Pitzschke A, Schikora A, Hirt H (2009) MAPK cascade signalling networks in plant defence. Curr Opin Plant Biol 12:421–426

Poinssot B, Vandelle E, Bentéjac M et al (2003) The endopolygalacturonase 1 from *Botrytis cinerea* activates grapevine defense reactions unrelated to its enzymatic activity. Mol Plant-Microbe Interact 16:553–564

Popescu SC, Popescu GV, Bachan S et al (2009) MAPK target networks in *Arabidopsis thaliana* revealed using functional protein microarrays. Genes Dev 23:80–92

Pré M, Atallah M, Champion A et al (2008) The AP2/ERF domain transcription factor ORA59 integrates jasmonic acid and ethylene signals in plant defense. Plant Physiol 147:1347–1357

Qiu J-L, Fiil BK, Petersen K et al (2008) Arabidopsis MAP kinase 4 regulates gene expression through transcription factor release in the nucleus. EMBO J 27:2214–2221

Ramírez V, García-Andrade J, Vera P (2011) Enhanced disease resistance to *Botrytis cinerea* in myb46 Arabidopsis plants is associated to an early down-regulation of CesA genes. Plant Signal Behav 6:911–913

Rasmussen MW, Roux M, Petersen M, Mundy J (2012) MAP kinase cascades in Arabidopsis innate immunity. Front Plant Sci. doi:10.3389/fpls.2012.00169

Ren D, Liu Y, Yang K-Y et al (2008) A fungal-responsive MAPK cascade regulates phytoalexin biosynthesis in Arabidopsis. Proc Natl Acad Sci U S A 105:5638–5643

Rowe HC, Walley JW, Corwin J et al (2010) Deficiencies in jasmonate-mediated plant defense reveal quantitative variation in *Botrytis cinerea* pathogenesis. PLoS Pathog 6:e1000861

Sato M, Tsuda K, Wang L et al (2010) Network modeling reveals prevalent negative regulatory relationships between signaling sectors in Arabidopsis immune signaling. PLoS Pathog 6:e1001011

Schweighofer A, Kazanaviciute V, Scheikl E et al (2007) The PP2C-type phosphatase AP2C1, which negatively regulates MPK4 and MPK6, modulates innate immunity, jasmonic acid, and ethylene levels in Arabidopsis. Plant Cell 19:2213–2224

Seo Y-S, Chern M, Bartley LE et al (2011) Towards establishment of a rice stress response interactome. PLoS Genet 7:e1002020

Siewers V, Kokkelink L, Smedsgaard J, Tudzynski P (2006) Identification of an abscisic acid gene cluster in the grey mold *Botrytis cinerea*. Appl Environ Microbiol 72:4619–4626

Solano R, Stepanova A, Chao Q, Ecker JR (1998) Nuclear events in ethylene signaling: a transcriptional cascade mediated by ETHYLENE-INSENSITIVE3 and ETHYLENE-RESPONSE-FACTOR1. Genes Dev 12:3703–3714

Son GH, Wan J, Kim HJ et al (2012) Ethylene-responsive element-binding factor 5, ERF5, is involved in chitin-induced innate immunity response. Mol Plant-Microbe Interact 25:48–60

Tao Y, Xie Z, Chen W et al (2003) Quantitative nature of Arabidopsis responses during compatible and incompatible interactions with the bacterial pathogen *Pseudomonas syringae*. Plant Cell 15:317–330

Thomma BP, Eggermont K, Penninckx IA et al (1998) Separate jasmonate-dependent and salicylate-dependent defense-response pathways in Arabidopsis are essential for resistance to distinct microbial pathogens. Proc Natl Acad Sci U S A 95:15107–15111

Thomma BP, Eggermont K, Tierens KF, Broekaert WF (1999) Requirement of functional ethylene-insensitive 2 gene for efficient resistance of Arabidopsis to infection by *Botrytis cinerea*. Plant Physiol 121:1093–1102

Usadel B, Obayashi T, Mutwil M et al (2009) Co-expression tools for plant biology: opportunities for hypothesis generation and caveats. Plant Cell Environ 32(12):1633–1651

Van Baarlen P, Woltering EJ, Staats M, Van Kan JAL (2007) Histochemical and genetic analysis of host and non-host interactions of Arabidopsis with three *Botrytis* species: an important role for cell death control. Mol Plant Pathol 8:41–54

Van der Does D, Leon-Reyes A, Koornneef A et al (2013) Salicylic acid suppresses jasmonic acid signaling downstream of SCFCOI1-JAZ by targeting GCC promoter motifs via transcription factor ORA59. Plant Cell 25:744–761

Van Kan J (2006) Licensed to kill: the lifestyle of a necrotrophic plant pathogen. Trends Plant Sci 11:247–253

Wan J, Zhang X-C, Neece D et al (2008) A LysM receptor-like kinase plays a critical role in chitin signaling and fungal resistance in Arabidopsis. Plant Cell 20:471–481

Wang X, Basnayake BMVS, Zhang H et al (2009a) The Arabidopsis ATAF1, a NAC transcription factor, is a negative regulator of defense responses against necrotrophic fungal and bacterial pathogens. Mol Plant-Microbe Interact 22:1227–1238

Wang Z, Mao H, Dong C et al (2009b) Overexpression of *Brassica napus* MPK4 enhances resistance to *Sclerotinia sclerotiorum* in oilseed rape. Mol Plant-Microbe Interact 22:235–244

Weiberg A, Wang M, Lin FM et al (2013) Fungal small RNAs suppress plant immunity by hijacking host RNA interference pathways. Science 342:118–123

Windram O, Madhou P, McHattie S et al (2012) Arabidopsis defense against *Botrytis cinerea*: chronology and regulation deciphered by high-resolution temporal transcriptomic analysis. Plant Cell 24:3530–3557

Windram O, Penfold CA, Denby KJ (2014) Network modeling to understand plant immunity. Annu Rev Phytopathol 52:93–111

Xu X, Chen C, Fan B, Chen Z (2006) Physical and functional interactions between pathogen-induced Arabidopsis WRKY18, WRKY40, and WRKY60 transcription factors. Plant Cell 18:1310–1326

Yang J, Osman K, Iqbal M et al (2012) Inferring the *Brassica rapa* interactome using protein-protein interaction data from *Arabidopsis thaliana*. Front Plant Sci 3:297

Zander M, La Camera S, Lamotte O et al (2009) *Arabidopsis thaliana* class-II TGA transcription factors are essential activators of jasmonic acid/ethylene-induced defense responses. Plant J 61:200–210

Zhang J, Li W, Xiang T et al (2010) Receptor-like cytoplasmic kinases integrate signaling from multiple plant immune receptors and are targeted by a *Pseudomonas syringae* effector. Cell Host Microbe 7:290–301

Zhang C, Xie Q, Anderson RG et al (2013a) Crosstalk between the circadian clock and innate immunity in Arabidopsis. PLoS Pathog 9:e1003370

Zhang W, Zhang W, Fraiture M et al (2013b) Arabidopsis receptor-like protein30 and receptor-like kinase suppressor of BIR1-1/EVERSHED mediate innate immunity to necrotrophic fungi. Plant Cell 25:4227–4241

Zhang L, Kars I, Essenstam B et al (2014) Fungal endopolygalacturonases are recognized as microbe-associated molecular patterns by the arabidopsis receptor-like protein RESPONSIVENESS TO BOTRYTIS POLYGALACTURONASES1. Plant Physiol 164:352–364

Zhao Y, Wei T, Yin K-Q et al (2012) Arabidopsis RAP2.2 plays an important role in plant resistance to *Botrytis cinerea* and ethylene responses. New Phytol 195:450–460

Zheng Z-L, Zhao Y (2013) Transcriptome comparison and gene coexpression network analysis provide a systems view of citrus response to "Candidatus Liberibacter asiaticus" infection. BMC Genomics 14:27

Zhu P, Gu H, Jiao Y et al (2011a) Computational identification of protein-protein interactions in rice based on the predicted rice interactome network. Genomics Proteomics Bioinforma 9:128–137

Zhu Z, An F, Feng Y et al (2011b) Derepression of ethylene-stabilized transcription factors (EIN3/EIL1) mediates jasmonate and ethylene signaling synergy in Arabidopsis. Proc Natl Acad Sci U S A 108:12539–12544

Zimmerli L, Métraux JP, Mauch-Mani B (2001) beta-Aminobutyric acid-induced protection of Arabidopsis against the necrotrophic fungus *Botrytis cinerea*. Plant Physiol 126:517–523

Chapter 18
Hitting the Wall: Plant Cell Walls During *Botrytis cinerea* Infections

Barbara Blanco-Ulate, John M. Labavitch, Estefania Vincenti, Ann L.T. Powell, and Dario Cantu

Abstract The cell wall is among the first structures that *Botrytis cinerea* encounters when colonizing plant tissues. From the perspective of *B. cinerea* as an infecting pathogen, host cell walls are potential sources of nutrients, but intact walls are also barriers that limit advancement of growing fungal hyphae beyond the initial sites of penetration. Plant cell walls are polysaccharide-rich extracellular matrices that surround individual cells. The architecture and composition of cell walls vary among plant species and organs. The shapes and attributes of organs are determined by the arrangements of the macromolecules that compose cell walls. Walls are synthesized, remodeled and disassembled as cells divide, differentiate, expand, and expire. Metabolic, developmental and external events, including infections by pathogens, alter the properties and components of plant cell walls. This chapter focuses on the cell walls of host plant tissues during infections by *B. cinerea*. The expression and the polysaccharide targets of *B. cinerea* and plant genes predicted to encode proteins that could modify plant cell walls as a consequence of infection are described. The impacts of these proteins on the properties of walls are discussed, noting potential alterations to extracellular anti-pathogen and pathogen-related defence proteins associated with the wall matrix.

Keywords Carbohydrate-Active enZymes (CAZymes) • Pectins • Hemicellulose-cellulose network • Pathogen perception • Defence responses

---

B. Blanco-Ulate • D. Cantu (✉)
Department of Viticulture and Enology, University of California, Davis, CA 95616, USA
e-mail: dacantu@ucdavis.edu

J.M. Labavitch • E. Vincenti • A.L.T. Powell
Department of Plant Sciences, University of California, Davis, CA 95616, USA

S. Fillinger, Y. Elad (eds.), *Botrytis – the Fungus, the Pathogen and its Management in Agricultural Systems*, DOI 10.1007/978-3-319-23371-0_18

18.1 Introduction

Botrytis cinerea can enter host plant tissues deliberately by breaching the cuticle using enzymes and physical means, but stomata or incidental surface breaks provide options for the fungus to enter its hosts opportunistically (reviewed in Van Kan 2006; Chap. 12). Regardless of the penetration strategy that *B. cinerea* uses, the cell wall matrix is among the first plant structures encountered by the pathogen.

Plant cell walls are potentially rich sources of nutrients for the pathogen but walls are also barriers that can limit expansion of the fungus into and through host tissues. *B. cinerea*'s access to the more diverse and readily metabolized nutrient compounds in the cellular cytoplasmic compartments can be thwarted by the host cell wall. Host cell walls that are recalcitrant to disassembly or walls that support appropriate anti-pathogen responses, limit growth of *B. cinerea* beyond the initial sites of infection (Cantu et al. 2008a). Therefore, it is not surprising that the breakdown of polysaccharide linkages within the plant cell wall is a prominent feature of the initial contact between the growing fungal hyphae and host cells. Digestion of host cell walls continues throughout infection, resulting in cell death and the extensive maceration of the host tissues, which are hallmarks of diseases caused by *B. cinerea*.

While it is clear that plant cell walls can contribute to resistance as well as susceptibility, the complexity of the architecture and composition of walls explains why confrontations between *B. cinerea* and plant hosts are diverse, dynamic and interactive. Determining which cell wall polysaccharides are important targets for disassembly by *B. cinerea* and the mechanisms it uses to advance within plant tissues, might lead to the development of resistant fruit and ornamental crops, and guide strategies for efficient disease management.

18.2 The Plant Cell Wall

The structures of plant cell walls have been extensively reviewed (Harris and Stone 2009; Voragen et al. 2009; Burton et al. 2010; Scheller and Ulvskov 2010; Keegstra 2010; Cosgrove and Jarvis 2012). Primary cell walls are composed of polysaccharides, structural proteins and ions. Secondary cell walls have additional simple and polymerized phenylpropanoids that contribute to their rigidity, and the proportions of the major polysaccharides in secondary walls differ from those in primary cell walls. The heterogeneity of the walls of different cell types from diverse plant species makes it challenging to describe a general structure for all plant cell walls. Nevertheless, models based on comprehensive analyses of structural and compositional properties of leaf and fruit cell walls provide an excellent starting point to study the characteristics of the host tissues that *B. cinerea* encounters (Keegstra et al. 1973; Carpita and Gibeaut 1993; McQueen-Mason and Cosgrove 1995; Scheller and Ulvskov 2010; Hayashi and Kaida 2011; Park and Cosgrove 2012).

The most accepted model of the primary cell wall structure corresponds to the tethered-network model, which includes two "co-extensive" polysaccharide networks. These networks are (1) cellulose microfibrils cross-linked to one another via hemicelluloses (also called cross-linking glycans), embedded within (2) simple and branched pectin polysaccharides (Carpita and Gibeaut 1993; Cosgrove 2001; Hayashi and Kaida 2011). The overall strength and rigidity of the plant cell wall depends on the integrity of the hemicellulose-cellulose microfibril network (Harris and Stone 2009). The pectin network influences the wall's porosity and provides structural coherence (Ishii et al. 2001). The proportions of cellulose, hemicelluloses and pectins change during development and vary depending on the plant tissue and species. Figure 18.1 illustrates the most common features of primary plant cell walls.

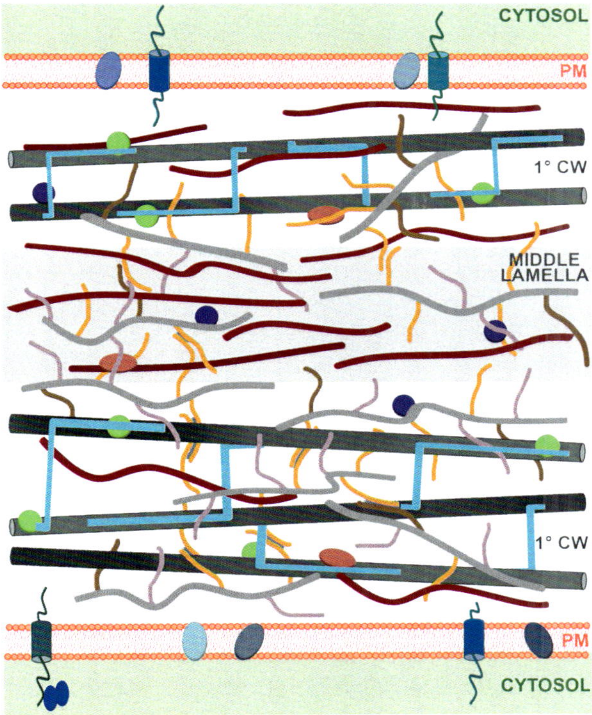

Fig. 18.1 Schematic model of the primary cell wall structure of a dicot plant based on the tethered-network model. The primary cell wall (1° CW) is composed mainly of cellulose, hemicelluloses, pectins and structural proteins. Cellulose microfibrils are represented as slate-gray rods, while hemicelluloses (i.e., xyloglucans and xylans) are the cyan-colored connectors that join the cellulose microfibrils together. The middle lamella is a pectin-rich matrix between two adjoining cells. Two major classes of pectin backbones are illustrated: homogalaturonan backbones are *dark-red lines*, and rhamnogalacturonan-I (RG-I) backbones are *gray lines*. Three types of RG-I side-branches are shown: branched arabinan (*pink lines*), linear galactan (*orange lines*) and branched, "type-I"- arabinogalactan (*brown lines*). Cell wall structural proteins (e.g., glycoproteins and extensins) are depicted as *purple* and *green circles*. Other proteins associated with the wall (e.g., PR-proteins) are shown as *orange ovals*. Trans-membrane wall-associated and receptor-like kinase proteins (WAKs and RLKs, respectively), which have parts that are in the plasma membrane (PM) and extensions into the cell wall, are depicted as blue ovals and cylinders

Cellulose is composed of identical, long, unbranched glucans that are tightly bound together in microfibrils by multiple hydrogen (H)-bonds. Cellulose microfibrils are rigid and are the major strength-conferring elements of walls (Table 18.1; Harris and Stone 2009). Hemicellulose polysaccharides are H-bonded to the surface of the cellulose microfibrils; they maintain the cellulose fibrils in fixed positions relative to one another. Two major types of hemicellulosic polymers are found in the primary cell walls of angiosperm plants: xyloglucans (XyGs) are found in dicots, while xylans are mostly detected in monocots and in the secondary cell walls of dicots (Table 18.1). The synthesis and disassembly of hemicelluloses is critical for plant cell growth and expansion (McNeil et al. 1984; Scheller and Ulvskov 2010). Other hemicelluloses, such as mannans (i.e., linear mannans, galactomannans, and galactoglucomannans) are important energy stores for seeds, but can be found in the cell walls of other plant tissues (Table 18.1; Scheller and Ulvskov 2010; Albersheim et al. 2010).

The pectin polymers in plant cell walls are complex and diverse macromolecules that may contain as many as 17 different monosaccharide constituents joined by more than 20 different linkages (Voragen et al. 2009). Besides their roles in determining the porosity of the wall, pectins influence a variety of physiological and cellular processes including growth and expansion of cells and cell-to-cell adhesion. Pectins influence or are integral components of signaling, pH and ion balances in the apoplast, seed imbibition, leaf abscission, fruit softening and responses to pathogens (Ridley et al. 2001; Mohnen 2008). The major pectins are homogalacturonan (HG), rhamnogalacturonan (RG-I and RG-II; Table 18.1) and are present in the walls of dicots and non-graminaceous monocots. Pectins are particularly abundant in the middle lamella and at the corners of cells (Mohnen 2008). The cell walls of graminaceous monocots, such as cereals, have fewer pectins than dicot cell walls, and the complexity of these pectins is reduced (Jarvis et al. 1988). In general, *B. cinerea* is not an important pathogen of cereals, and it remains unclear whether the reduced pectin in cereal walls is one of the reasons why *B. cinerea* growth is not supported. Other monocots, such as onions, are infected by *B. allii*, *B. aclada*, *B. byssoidea*, and *B. squamosa* and tulips are infected by *B. tulipae* and *B. gesneriana*, but the role of the host tissue cell wall composition is not known. Lilies, another monocot, can be infected similarly by *B. cinerea* and *B. elliptica* (Hsieh et al. 2001).

Structural proteins, such as glycoproteins, contribute to the establishment and maintenance of the structural features of plant cell walls (Rose and Lee 2010, Fig. 18.1), although the details of their associations with the components of the wall and the ways they accomplish their structural roles are not well understood. Glycoproteins have been associated with wall strength, assembly of cell wall polysaccharides, regulation of cell growth, and responses to biotic and abiotic stresses (Showalter 1993; Nguema-Ona et al. 2013a). The major groups of these proteins are identified based on their amino acid contents; for example, glycine- and proline-rich proteins, hydroxyproline-rich glycoproteins (HRGPs), and arabinogalactan proteins (AGPs; Cassab 1998; Jamet et al. 2006). Characterizations of the chemical compositions and associations between constituents of cell walls are the primary approaches used to study how *B. cinerea* sequentially digests the wall polysaccharides of its hosts to successfully invade tissues.

Table 18.1 Chemical composition of the main polysaccharides in plant cell walls

Cell wall polysaccharides	Sugar components	Backbone	Side-groups and other features
Pectins			
Homogalacturonan (HG)	Galacturonic acid (GalA), Methyl ester (Me) to GalA carboxyl groups (variable frequency), Acetyl ester (Ac) to GalA 2- or 3-OH groups (variable), and Xylose (Xyl), rarely	Linear array of GalA residues α-1,4-linked to one another	Xyl α-1,3-linked to GalA residues of the backbone (rare side-chain)
Rhamnogalacturonan-I (RG-I)	GalA and Rhamnose (Rha) in backbone, GalA residues can be Me- and Ac-esterified (as for HG): Galactose (Gal) and Arabinose (Ara) in side chains associated with backbone Rha	Repeating disaccharide of Rha α-linked to the 4-OH of GalA, GalA, in turn, α-linked to the 2-OH group of Rha	Galactan (Gal) residues β-1,4-linked to one another; Arabinan (Ara) residues α-1,5-linked to each other with occasional branches α-1,3-linked; and Arabinogalactans, consisting of β-1,4-linked Gal residues occasionally substituted at the 3-OH with α-linked Ara residues
Rhamnogalacturonan-II (RG-II)	GalA with variable Me content (backbone) and GalA, Glucuronic acid (GlcA), Aceric acid, Rha, Ara, Fucose (Fuc) (some Fuc with a Me ether at the 2-OH), Gal, Apiose (Api), 2-keto-3-deoxy-D-manno-octulosonic acid (Kdo) and 2-keto-3-deoxy-D-lyxo-heptulosaric acid (Dha) in the four side branches	Linear array of GalA residues α-1,4-linked to one another	Numerous forms. The RG-II backbone is decorated with four different side branches, two of which are branched (O'Neill et al. 2004). Dimers of RG-II monomers associate when the Api residues of two RG-II monomers form a covalent diester cross-link with borate

(continued)

Table 18.1 (continued)

Cell wall polysaccharides	Sugar components	Backbone	Side-groups and other features
Hemicelluloses			
Xyloglucans (XyGs)	Glucosyl residues (Glc) form the backbone. Single residues of Xyl, Gal, Fuc and Ara are side-groups. Acetyl esters have been identified at the 6-OH group of the Glc residues in the XyG backbone and at the 5-OH of Ara residues and the 6-OH of Gal residues in XyG side-branches	The fundamental XyG backbone consists of Glc that are β-linked to the 4-OH of the next backbone Glc residue	The main side-group is single Xyl residues that are α-linked to the 6-OH groups of >50 % of Glc residues in the backbone. The Xyl side-groups can be substituted with Gal, β-linked to the Xyl 2-OH group. In some XyGs the Gal residue then is substituted with a Fuc residue, α-linked to its 2-OH group. In other XyGs, Ara residues may be α-linked to the 2-OH position of a Xyl side-group or to an unsubstituted Glc residue
Xylans	The backbones are composed of Xyl. Single Ara or GlcA residues are found as side-groups	Xyl residues b-linked to the 4-OH of the next backbone residue	Most side-branches are Ara residues that are α-linked to the 3-OH groups of Xyl residues in the backbone, as in monocots, or to the 2-OH groups, as in dicots. An additional xylan side-group is a single GlcA residue α-linked to a few of the 2-OH groups of backbone Xyl residues
Mannans, galactomannans and galactoglucomannans	Mannose (Man) and Glc may compose the backbones. Gal is present in the side-branches	Man residues b-linked to the 4-OH of the next sugar residue in the backbone. Galactoglucoman-nans with backbones composed of alternating stretches of b-linked Man and Glc residues	Galactomannans can have single Gal residues that are α-linked to the 6-OH of backbone Man residues. Side-groups in Galactoglucomannans are single Gal residue side units that are α-linked to the 6-OH groups of Man residues
Cellulose	Glc residues	Backbone is Glc residues b-1,4-linked to one another	Multiple backbones are H-bonded and assembled into microfibrils

18.3 Decomposition of Plant Cell Walls by *B. cinerea*

Botrytis cinerea produces and secretes proteins and enzymes, which modify host cell walls, in order to overcome the physical constraints imposed by the wall and to release sugars to sustain its own energy needs (Chaps. 12 and 16). The enzymatic cleavage of relatively few linkages between cell wall polysaccharides, which are important for maintaining the integrity of the polysaccharide networks, might facilitate the intercellular expansion of hyphae and, thus, enhance the pathogen's ability to access cellular resources. On the other hand, use of the wall's sugar constituents for energy might require the complete breakdown of a variety of polysaccharides into monosaccharides. Here we provide details about the proteins and enzymes that *B. cinerea* expresses when infecting various host tissues.

18.3.1 **Botrytis cinerea** *CAZy Proteins*

The diverse group of enzymes and proteins that affect cell walls are usually referred to as cell wall modifying proteins. The CAZy (Carbohydrate-Active enZymes; http://www.cazy.org) database "describes the families of structurally-related catalytic and carbohydrate-binding modules (or functional domains) of enzymes that degrade, modify, or create glycosidic bonds" (Cantarel et al. 2009). Therefore, CAZy proteins are associated with modifications and breakdown of cell wall polysaccharides.

Among the approximately 16,000 genes in the genome of *B. cinerea* (strain B05.10; according to Amselem et al. 2011), 1,155 are predicted to encode CAZy proteins, and 275 of these have secretion signal peptide sequences that suggest that they function in the apoplast (Blanco-Ulate et al. 2014). Most of these putative secreted *B. cinerea* CAZy proteins (49 %) are glycoside hydrolases (GHs) and the most abundant GH subfamily is GH28, which includes polygalacturonases (PGs) that target pectins in plant cell walls (14 % of all GHs). The next most abundant CAZy family (23 %) is the carbohydrate-binding proteins (CBMs). Carbohydrate esterases (CEs, 17 %), glycolsyltransferases (GTs, 8 %) and polysaccharide lyases (PLs, 3 %) are less abundant (Blanco-Ulate et al. 2014). The potentially secreted proteins encoded by 88 of these CAZy genes have been detected in *B. cinerea*-infected tomato fruit and/or *B. cinerea* grown in culture (Shah et al. 2009a, b, 2012; Espino et al. 2010; Fernández-Acero et al. 2010; Li et al. 2012). The activities and potential roles of most of these enzymes during *B. cinerea*-plant interactions have not been confirmed.

A core set of 229 CAZy genes encoding proteins with secretion signals was detected in *B. cinerea*-infected lettuce leaves, ripe tomato fruit and ripe grape berries (Blanco-Ulate et al. 2014). Nine PGs and eight pectin/pectate lyases (PLs/PELs) are in this core set and are the most abundantly expressed genes in all hosts evaluated. Because *B. cinerea* is a generalist pathogen with a broad host-range, these proteins are likely to be the primary enzymatic machinery utilized by the fungus as it penetrates and invades diverse host tissues. Other cell wall modifying proteins might be produced only in specific conditions. These proteins might provide adaptations for diverse host tissues and enable different stages of infection, such as, during penetration or during active fungal growth (Blanco-Ulate et al. 2014).

Although CAZy proteins secreted by *B. cinerea* are expected to target plant cell wall substrates, they could also remodel the fungal cell wall as the pathogen grows and develops, or they may degrade host cellular contents, including starch and glycosylated compounds (e.g., glycosylated proteins or secondary metabolites) (Faure 2002; Klis et al. 2010; Sha et al. 2009b). CAZy proteins could act on more than one polysaccharide substrate. The conditions in infected tissues and/or the availability of substrates may alter the kinetics and substrate preferences of their activities (Eklöf and Brumer 2010). Identifying common *B. cinerea* CAZy proteins or families expressed during infections of different host tissues predicts plant cell wall polysaccharides that are inevitable targets in multiple host tissues. Determining which CAZy genes are uniquely expressed on particular hosts reveals that *B. cinerea* makes adjustments to its virulence strategies in order to adapt to conditions in each host.

18.3.2 Cell Wall Polysaccharide Targets of B. cinerea *CAZy Proteins*

18.3.2.1 Pectins

Pectin breakdown during *B. cinerea* infections may increase the cell wall porosity and may facilitate the degradation of other classes of wall polysaccharides by enhancing access of other fungal enzymes to their substrates. Growth of *B. cinerea* may be enabled by its metabolism of sugars released from hydrolyzed pectins (Zhang et al. 2013). Pectins appear to be the main cell wall targets during *B. cinerea* infections, regardless of the host tissue or species (Blanco-Ulate et al. 2014). Enzymes that target pectin backbones include, PGs and rhamnogalacturonases (RGaes; GH28) and pectin/pectate lyases (PL/PELs; PL1 and PL3). Pectin methylesterases (PMEs; CE8) and rhamnogalacturonan esterases (CE12) might cooperate in the effective degradation of pectin backbones (Zhang and Van Kan 2013).

Polygalacturonases hydrolyze the backbone of homogalacturonan (HG; Table 18.1). Exo-PGs remove one D-galacturonic acid monomer at a time from the non-reducing ends of HG; endo-PGs can hydrolyze the polymer at internal sites and release oligogalacturonides or pectin-derived oligomers (PDOs) (Jayani et al. 2005). Of the 11 secreted PGs predicted in the *B. cinerea* genome, at least 5 are likely exo-PGs and 6 are endo-PGs (Blanco-Ulate et al. 2014). Most of the endo-PGs (i.e., PG1-6) have been extensively characterized (Wubben et al. 2000; Reignault et al. 2000; ten Have et al. 2001; Kars et al. 2005a). *B. cinerea* produces particular endo-PGs depending on the conditions in the host and the stage of infection (Wubben et al. 2000; Reignault et al. 2000; ten Have et al. 2001; Kars et al. 2005a;

Blanco-Ulate et al. 2014). PG1 is constitutively expressed, PG3 is preferentially produced in acidic conditions, PG4 and PG6 are induced by monomers of galacturonic acid, and PG4 is inhibited by glucose (Kars et al. 2005a; Zhang and Van Kan 2013). Kars et al. (2005a) determined that these PGs have specific substrate preferences and produce distinct pectin-degradation product profiles. For example, PG1, PG2 and PG4 are more effective in the depolymerization of demethylesterified HGs.

The endo-PGs expressed during *B. cinerea* infections have been investigated by characterizing the virulence of single-gene knockout mutants (i.e., *Δpg1-6*) and a double knockout mutant, *Δpg1Δpg2*. The PGs appear to have redundant functions because none of the mutants in single genes result in the total loss of *B. cinerea* virulence. The *Δpg1* mutant is less virulent on leaves of tomato, broad bean, tobacco and Arabidopsis, as well as on apple fruit and unripe tomato fruit (ten Have et al. 1998; Zhang and Van Kan 2013; Blanco-Ulate et al. pers. obs.), supporting the conclusion that it has a major role during plant infections (ten Have et al. 1998, 2001; Kars et al. 2005a). The virulence of the *Δpg2* mutant is reduced in infections of tomato and broad bean leaves and unripe tomato fruit (Kars et al. 2005a; Blanco-Ulate et al. pers. obs.), but not on tobacco or Arabidopsis leaves (Zhang and Van Kan 2013). Both PG1 and PG2 are required for *B. cinerea* infections in certain plant tissues, as the double mutant *Δpg1Δpg2* is avirulent in unripe tomato fruit, and the other PGs do not seem to complement the missing endo-PG activity (Blanco-Ulate and Dario Cantu, pers. obs.). The single mutants *Δpg3*, *Δpg4*, *Δpg5* and *Δpg6* are as virulent as the wild-type strain in leaves of different plant hosts (Joubert et al. 2007; Zhang and Van Kan 2013), but their ability to cause disease in other plant organs (e.g., stems or fruit) has not been investigated.

Analyses of knockout mutants in putative exo-PGs might provide further information on the cooperative roles of different PGs. In infected bean leaves, *B. cinerea* exo-PG activity increases early in infection, but endo-PGs increase only after host penetration has occurred (Kapat et al. 1998). The temporal deployment and synergistic functions of exo- and endo-PGs has not been broadly analyzed on different host tissues.

PLs and PELs utilize a β-elimination mechanism rather than hydrolysis to degrade HGs. PLs act on HG backbones with high degrees of methylesterification, while PELs are more efficient on HGs with low levels of methylesterification and are calcium-dependent (Jayani et al. 2005). Four genes encoding PLs and four encoding PELs are annotated in the *B. cinerea* genome. The expression of genes encoding putative PLs/PELs is not as high as that of PG genes (PL1/PL3 versus GH28), suggesting that PLs/PELs assist PGs rather than being the primary enzymes for the decomposition of HGs or that PME action (discussed below) reduces the need for enzymes that can digest a methylesterified HG backbone. Mutants in *B. cinerea* PLs or PELs have not been developed, thus the impact of these enzymes on virulence is not known.

Methylation and acetylation of the HG backbones can impact the activity of *B. cinerea* endo-PGs and PELs (Kars et al. 2005a). PMEs catalyze the specific demethylesterification of HGs, releasing methanol and acidic HG. PMEs tend to act randomly on methylesterified HG backbones, releasing protons that may favor the activity of particular endo-PGs (Micheli 2001). The *B. cinerea* genome encodes three putative secreted PMEs. PME1 and PME2 have been described; they are constitutively expressed and are stable in a wide range of pHs and temperatures (Reignault et al. 1994, 2000; Valette-Collet et al. 2003; Kars et al. 2005b). No pectin acetylesterases have been identified in the *B. cinerea* genome, although an RG-I acetylesterase has been noted (discussed below).

Botrytis cinerea PME activity may not always be essential for virulence. Mutations in *PME1* and *PME2* did not affect *B. cinerea* virulence on tomato and grapevine leaves or pear and tomato fruit (Kars et al. 2005b; Blanco-Ulate et al. pers. obs.), although PME1 seems to be necessary for successful infections of apple fruit (Valette-Collet et al. 2003). Kars et al. (2005b) proposed that PME3 or other putative PMEs could compensate for the lack of PME activity in the *Δpme1Δpme2* mutant or that PMEs are not necessary for endo-PG action *in planta*. However, it is possible that *B. cinerea* relies on plant PMEs to demethylesterify the HG backbones to make them amenable for subsequent degradation (Raiola et al. 2011; Blanco-Ulate et al. pers. obs.).

The *B. cinerea* genome encodes six possible secreted rhamnogalacturonan hydrolases (RGases), which could cleave rhamnogalacturonan-I (RG-I; Table 18.1) pectin backbones (Blanco-Ulate et al. 2014). The galacturonic acids that make up the RG-I backbone can be methylated or acetylated. Deacetylation by rhamnogalacturonan acetylesterases (RGAEs) is essential for the subsequent action of RGases (Mølgaard et al. 2000). Only one RGAE is predicted in the *B. cinerea* genome and it appears to be relevant for plant cell wall degradation, as it is expressed in multiple hosts (Blanco-Ulate et al. 2014). Trimming of the side branches of the RG-I pectin backbone is an important pre-requisite for hydrolysis (Mutter et al. 1998). Although RG-I is a major part of the "hairy" (i.e., highly branched) regions of pectins in plant cell walls, it is not as abundant as pectins with HG-backbones, which could explain why expression of RGases and RGAEs is low compared to expression of genes encoding PGs and PLs/PELs.

The rhamnogalacturonan-II (RG-II, Table 18.1) pectins are complex and recalcitrant to hydrolysis. RG-II is a major component of the lees in red wine (Vidal et al. 2000); presumably, it accumulates there because the diversity of sugars and glycosidic linkages in its side-groups make its digestion a relatively expensive challenge for microbes.

The *B. cinerea* genome encodes other enzymes that could degrade the diverse side-branches of pectins. These enzymes could cut entire side-branches from the backbone, internally cleave side-branches or remove terminal residues (Zhang and Van Kan 2013). Among these putative genes are four α-arabinofuranosidases (GH GH51, GH54|CBM42 and GH62|CBM13), three β-galactosidases (GH2 and GH35) and two α-L-1,5-arabinanases (GH43 and GH93).

Ara1, an endo-α-L-1,5-arabinanase, cleaves linear arabinans present in RG-I, but is not able to cut branched arabinans (Nafisi et al. 2014). The *Δara1* knockout *B. cinerea* mutant is not capable of degrading 1,5-arabinan *in vitro* and displays a delay in secondary lesion formation when infecting Arabidopsis leaves. However, no differences in virulence in comparison to the wild-type strain were observed when this mutant was introduced on tobacco or tomato leaves. Nafisi et al. (2014) concluded that the role of *ara1* depends on the plant host.

18.3.2.2 Hemicelluloses

Multiple hemicellulose-modifying enzymes are encoded in *B. cinerea*'s genome (Blanco-Ulate et al. 2014). Cleavage of hemicelluloses may loosen the cell wall by disrupting the hemicellulose-cellulose microfibril network (Fig. 18.1). Relaxation of the wall network may facilitate access to targets by other enzymes of the pathogen. XyG backbones (Table 18.1) are hydrolyzed by endo-acting β-1,4-glucanases or β-glucosidases, which also cleave cellulose (Gilbert 2010). The *B. cinerea* genome encodes one candidate XyG-specific β-glucanase (GH12) and six β-glucosidases (GH3; Blanco-Ulate et al. 2014).

XyG endo-transglycosylases/hydrolases (XTHs) cleave the XyG backbones of hemicelluloses using two mechanisms: (1) XyG endo-transglycosylase (XET) non-hydrolytically cleaves and re-ligates shortened XyG polymers, and (2) XyG endo-hydrolase (XEH) irreversibly hydrolyses the XyG backbone (Eklöf and Brumer 2010). Plant XETs participate in loosening of cell walls (Rose et al. 2002). The *B. cinerea* genome has six candidate XTHs (GH16), and two are expressed in diverse plant tissues (Blanco-Ulate et al. 2014).

Xylans and mannans are present in the primary and secondary walls of many *B. cinerea's* hosts, but they tend to be less abundant than XyGs. Five β-xylanases (GH10, GH10|CBM1, GH11 and GH11|CBM1) and three β-xylosidases (GH43) are predicted in the *B. cinerea* genome (Blanco-Ulate et al. 2014). Expression of the *xyn11A* gene (GH11), encoding a putatively secreted endo-β-1,4-xylanase, is detected when *B. cinerea* infects lettuce leaves and ripe fruit (tomato and grape berries). The *Δxyn11a* mutant is less virulent in tomato leaves and table grape berries (Brito et al. 2006). However, the contribution of Xyn11A to virulence does not depend on its xylan-cleaving activity; it is related to the necrosis in the host caused by the xylanase protein itself (Noda et al. 2010). The involvement of other xylanases in plant cell wall degradation or necrotizing activities has not been studied.

Some of the side-branches along the XyG and xylan backbones (Table 18.1) contribute to the overall strength of the hemicellulose-cellulose microfibril network (Pauly et al. 2013). Removal of these groups might affect the hemicellulose cross-linking properties and favor the breakdown of the hemicellulose backbones. Enzymes that remove the side-branches in these polysaccharides, include β-xylosidases (GH31), α-L-fucosidases (GH95) and other enzymes that are equivalent or similar to those acting on the side groups of pectins as previously described. Furthermore, because the strength of XyG binding to cellulose depends on the integrity of the XyG's side chains modifications of these during pathogenesis can have an impact on wall porosity, hence enzyme access to other wall substrates.

18.3.2.3 Cellulose

Botrytis cinerea can degrade cellulose *in vitro* and *in vivo* (Verhoeff et al. 1983). Enzymes that might be involved in the degradation of cellulose are expressed by *B. cinerea* during infections (Blanco-Ulate et al. 2014); these include ten endo-β-1,4-glucanases (GH5, GH5|CBM1 and GH45), three cellobiohydrolases (GH6 and GH7), and the six β-glucosidases (GH3). Espino et al. (2005), demonstrated that a mutant with a deletion in *cel5A*, another endo-β-1,4-glucanase, is able to infect tomato leaves and gerbera petals. This gene is among the group of CAZy genes commonly expressed in lettuce leaves and fruit tissues. Evaluating the virulence of this mutant in other hosts may provide some information about the importance of this enzyme in *B. cinerea*-lettuce or *B. cinerea*-fruit interactions.

18.3.2.4 Glycans Attached to Proteins

Four putative α-mannosidases, which could release mannose from complex high-mannose N-glycans attached to structural cell wall proteins, have been predicted in the *B. cinerea* genome (Blanco-Ulate et al. 2014). The activity of fungal α-mannosidases in the disassembly of plant cell walls requires further investigation.

18.3.3 Vesicle Transport of Cell Wall Modifying Enzymes

Botrytis cinerea's secretion of cell wall modifying enzymes is required for virulence. Without the essential exocytosis machinery, *B. cinerea* is not able to deliver cell wall modifying enzymes to targets in plant hosts and, therefore, would not be expected to advance with infections. The *sas1* gene encodes a Rab GTPase, which might be required for vesicle docking and fusion, and, thus, may play a central role in the secretory pathway. A knockout mutant in *sas1* displays suppressed hyphal growth, decreased sporulation and reduced virulence on tomato and apple fruit (Zhang et al. 2014).

The deletion of the *sas1* gene causes the accumulation of transport vesicles at the hyphal tip, a significant reduction of extracellular proteins (e.g., glycoside hydrolases and proteases) and a decrease in PG and xylanase activity when *B. cinerea* is grown in culture. Because expression of genes encoding the main endo-PGs and a xylanase is unaffected in the *Δsas1* mutant, it is likely, that the secretion, not the synthesis, of these hydrolases and proteases is hampered in this mutant (Zhang et al. 2014).

18.4 *Botrytis cinerea* Infections Influence Cell Wall Modifications by Endogenous Plant Enzymes and Proteins

Botrytis cinerea may time its infections to a particular developmental stage of the host, when the plant tissues or organs are more susceptible (e.g., as tissues senesce or fruit ripen), or, it may actively induce susceptibility in hosts. For example,

B. cinerea's infections accelerate the ripening of unripe fruit and activate disassembly of the host cell wall polysaccharide matrix, as described in Chap. 19. *B. cinerea-derived* molecules that induce or suppress the expression of cell wall modifying proteins by the plant host have not been definitively identified but may include plant hormone analogues, small RNAs, protein effectors and/or pathogen- and damage-associated molecular patterns (PAMPs and DAMPs, respectively).

18.4.1 Plant Cell Wall Degrading Enzymes and Proteins

The cellulose microfibrils of plant cell walls are targeted for degradation during development (e.g., fruit ripening, organ abscission) by the activity of endo-β-1,4-glucanases (EGs) (Minic and Jouanin 2006). Suppressed expression of two tomato EG genes, *SlCel1* and *SlCel2*, reduces the susceptibility of leaves and tomato fruit to *B. cinerea*. The absence of EG activity may limit *B. cinerea* growth and promote the activation of defence responses. Enhanced callose deposition and expression of defence genes, such as *SlPR1* and *SlLoxD*, were observed when *B. cinerea* infects fruit from the double EG-suppressed line (Flors et al. 2007; Finiti et al. 2013).

Mutations in *AtKOR1*, a membrane-bound EG in Arabidopsis, correlate with improved resistance to *B. cinerea* infections. *Atkor1-1* mutant plants deposit high levels of callose in response to pathogens. However, the accumulation of callose in the *Atkor-1* mutants may impact signaling networks and hormone homeostasis. High levels of abscisic acid (ABA) and jasmonic acid (JA) were detected in infected *Atkor1-1* plants (Finiti et al. 2013). ABA and JA have roles in plant susceptibility or resistance to *B. cinerea* (reviewed in Chap. 19).

Expansins are extracellular plant proteins involved in the loosening of the hemicellulose-cellulose microfibril network during cell expansion (i.e., the acid-growth response) and during ripening (i.e., softening). Host plant expansins might facilitate access of *B. cinerea's* cell wall modifying enzymes (e.g., PGs, PLs, EGs et al.) to their substrates (Cantu et al. 2008b). Induction of the ABA-dependent expansin-like gene *AtEXLA2* correlates with increased susceptibility of Arabidopsis leaves to *B. cinerea* (Abuqamar et al. 2013). Leaves of the knockout *exla2* mutant are more resistant to *B. cinerea* and *Alternaria brassicicola*. The absence or suppression of *AtEXLA2* expression activates immune responses via a cyclopentenone oxylipin-signaling pathway. Whether reducing *AtEXLA2* expression leads to reduced cell wall porosity, has not been determined, but may be another reason why *exla2* mutants are more resistant to necrotrophic infections (Abuqamar et al. 2013).

Once healthy tomato fruit begin to ripen, an expansin gene, *SlExp1*, is up-regulated (Rose and Bennett 1999). Concurrently, the expression of a gene encoding a plant PG, *SlPG2A*, is also strongly induced (Bennett and Labavitch 2008). *B. cinerea* precociously activates the expression of *SlExp1* and *SlPG2A* once it establishes an interaction with unripe fruit (Cantu et al. 2009). This observation leads to the conclusion that *B. cinerea* infections of unripe tomato fruit can over-ride endogenous fruit host ripening programs, especially the expression of the cell wall modifying proteins (Cantu et al. 2009; Blanco-Ulate et al. pers. obs.). When either *SlExp1* or *SlPG2A* expression is suppressed in tomato fruit, no change in susceptibility to *B. cinerea* is observed, although pectin depolymerization is reduced (Cooper et al. 1998; Cantu et al. 2008b). However, when the expression of both *SlExp1* and *SlPG2A* is compromised, the typical ripening-associated increase in *B. cinerea* susceptibility of fruit decreases (Cantu et al. 2008b) and these fruit are firmer. The composition and architecture of fruit cell walls may directly impact the ability of *B. cinerea* to grow because less fungal growth is seen in cultures containing cell walls extracted from uninfected *SlPG2A*- and *SlExp1*-suppressed fruit than in cultures with cell walls from control fruit (Cantu et al. 2008b). The activities of *SlPG2A* and *SlExp1* may influence access by other fruit and fungal proteins to their polysaccharide substrates in the cell wall.

Expansin-like microbial proteins, such as swollenin, have been identified in the cellulolytic fungus *Trichoderma reesei*. Swollenin has an N-terminal fungal type cellulose-binding domain connected by a linker region to its expansin-like domain. This protein is able to disrupt the hemicellulose-cellulose microfibril network without producing detectable amounts of reducing sugars. At present, *B. cinerea* proteins with expansin activities have not been identified. However, *B. cinerea* abundantly produces a cerato-platanin protein (BC1G_02163), which contains an expansin-like domain (Frias et al. 2011, 2013; Martellini et al. 2013).

In Arabidopsis leaves, two cellulose synthase genes, *AtCeSA1* and *AtCeSA3*, have roles in responses to *B. cinerea* (Windram et al. 2012). Plants with mutations of *CeSA3* gene display decreased susceptibility to *B. cinerea*, possibly due to the induction of JA and ethylene synthesis and signaling (Ellis et al. 2002). A mutant in *AtCeSA1* also exhibits increased expression of a JA-inducible gene (i.e., *VEGETATIVE STORAGE PROTEIN*, *VSP1*), suggesting overproduction of JA in this mutant as well. In wild-type Arabidopsis plants, *AtCeSA1* and *AtCeSA3* are down-regulated by *B. cinerea* infections, possibly as part of the plant's effort to limit pathogen growth (Windram et al. 2012). Mutation of a secondary cell wall regulator, *MYB46*, enhances Arabidopsis resistance to *B. cinerea* (Ramirez et al. 2011a). The *myb46* knockout mutants show a rapid reduction in the expression of cellulose synthase genes, including *AtCeSA1* and *AtCeSA3*, after *B. cinerea* infections. This observation indicates that the timing of *AtCeSA* repression may be important to control *B. cinerea* spread in Arabidopsis leaves (Ramirez et al. 2011b).

18.4.2 Modifications in Plant Cell Wall Polysaccharides

Botrytis cinerea infections of Arabidopsis leaves alter the expression of plant host PME genes (Abuqamar et al. 2006). The enhanced expression and activity of AtPME3 has been linked to increased susceptibility to *B. cinerea* and the necrotrophic bacterium *Pectobacterium carotovorum* (Raiola et al. 2011). In response to *B. cinerea*, plant PME expression increases in fruit. The tomato *SlPMEU1* is up-regulated during infections of unripe and ripe tomato fruit, and *SlPME2* increases slightly at the early stages of infections of ripe fruit (Blanco-Ulate and Dario Cantu, pers. obs.).

Plant PMEs may act as susceptibility factors by cooperating with *B. cinerea* PMEs in the extensive demethylesterification of HG backbones that facilitates their further breakdown by fungal or host enzymes (Lionetti et al. 2012). The importance of plant and fungal PMEs during *B. cinerea* infections may depend on the conditions in the host.

Elevated acetylation of pectins and XyGs in the cell walls of Arabidopsis is associated with susceptibility to *B. cinerea*. A knockout mutation in the *REDUCED WALL ACETYLATION2* (*AtRWA2*) gene, which encodes an *O*-acetyltransferase, results in decreased levels of acetylated cell wall polymers and increased tolerance to *B. cinerea* (Manabe et al. 2011). Because *O*-acetylation could interfere with the hydrolysis of polysaccharides by microbial enzymes (Selig et al. 2009), it is unclear why the *Atrwa2* Arabidopsis mutant is more resistant to *B. cinerea* (Manabe et al. 2011).

18.5 Pathogen Perception, Defences and Host Cell Walls

When *B. cinerea* confronts the cell wall matrix of its host, the ensuing alterations and disassembly of the wall polysaccharides have secondary consequences for the host-pathogen interaction. Changes in the host cell wall may also evoke a switch in *B. cinerea*'s infection strategy. DAMPs, such as fragments generated by digestion of the wall polysaccharides, or *B. cinerea* cell wall modifying proteins that are PAMPs, may trigger defence responses, including fortifications of the host wall. Wall-associated kinases and receptors, which protrude from the host plasma membrane into the extracellular cell wall space, may "sense" that the integrity of the wall has been compromised and relay that information to the cellular cytoplasmic compartment. Host defence proteins, such as pathogenesis-related (PR) proteins that are exported to the apoplast may reside in the wall matrix and they may become dislodged by infections and, thus, become ineffectual.

18.5.1 Host Wall-Associated Plant Defence Proteins

Plant receptors that sense the integrity of the plant cell wall or perceive PAMPs and other effectors protrude from their anchor points in the plasma membrane into the apoplast (Fig. 18.1). How these receptors and signal transmitting proteins participate in host resistance and susceptibility is not entirely known, although several have been shown to be important for resistance or susceptibility to *B. cinerea* in vegetative organs, especially in Arabidopsis. Since the composition of cell walls differs depending on the plant tissues and species (Keegstra 2010), exposure to PAMPs or microbial effectors may be influenced significantly by the wall's composition and architecture. Access to *B. cinerea*-derived signals may be promoted especially as walls are destroyed by *B. cinerea* or disassembled during plant developmental processes, such as ripening and senescence.

Knowing that *B. cinerea* can activate immune responses, suggests that plasma membrane-anchored extracellular receptors may participate in responses to *B. cinerea* (Lai and Mengiste 2013). The PG1 protein itself is recognized as a PAMP by an Arabidopsis RBPG1 receptor, which complexes with a Leucine-rich Repeat Receptor-like Kinase (LRR-RLK) that is anchored in the plasma membrane (Zhang et al. 2014).

The extracellular domains of Wall-Associated Kinases (WAKs) interact via carbohydrate binding domains with the pectins in primary cell walls (Fig. 18.1; Decreux and Messiaen 2005; Kohorn and Kohorn 2012) and bind PDOs (Brutus et al. 2010; De Lorenzo et al. 2011). In Arabidopsis leaves, *B. cinerea* small RNAs target a WAK-coding gene (*At5g50290*; Weiberg et al. 2013). Since WAKs bind pectins and pectin degradation is key for *B. cinerea* infections, WAKs might signal the host cell wall that its integrity has been compromised by a fungal attack.

Another plasma membrane anchored protein, SlCOBRA, that has an extracellular cellulose binding domain is expressed abundantly early in tomato fruit development and then declines, in contrast to most WAKs. When SlCOBRA is suppressed, fruit tend to crack extensively and have altered cellulose synthesis and cell wall architecture, thus over-expression of SlCOBRA improves the postharvest performance of fruit. Expression of tomato cell wall modifying genes, WAKs and other RLKs, such as *Theseus 1* and *Lectin Receptor-like Kinase* is up-regulated by the suppression of SlCOBRA expression; however, it is not clear how SlCOBRA impacts other dynamic changes in plant cell walls (Cao et al. 2012). Also, it is not known whether expression of SlCOBRA or other RLKs like *Theseus 1* or LRR-RLK homologues is altered in response to *B. cinerea*.

In leaves, the cytoplasmic protein, *B. cinerea*-Induced Kinase (BIK), is involved in PAMP triggered immunity to *B. cinerea* through phosphorylation and signaling that includes ethylene (Laluk et al. 2011). BIK interacts with BRASSINOSTEROID INSENSITIVE 1- ASSOCIATED KINASE 1 (BAK1) to signal for PAMP triggered immunity (Mengiste 2012). BAK1 has an extracellular domain that protrudes into the cell wall space. BAK1 receptors are involved with FLAGELLIN SENSITIVE2 (FLS2) by their common association with BIK1. FLS2 is a receptor of flagellin, an extracellular structural protein produced by bacteria (Chinchilla et al. 2006). Exposure to flagellin results in responses similar to those to fungal elicitors (e.g., chitin, xylanases, ergosterol, mannose-rich glycopeptides) and the wound hormone peptide, systemin (Scheer and Ryan 1999; Meindl et al. 2000). While in Arabidopsis many of the responses to *B. cinerea* may be signaled through the WAK1 receptor, PAMP perception by the BAK1/BIK1 system may also transmit information about infections.

Plant PR proteins induced in response to pathogens have diverse potential antipathogen functions and they accumulate in response to *B. cinerea* (Díaz et al.

2002; reviewed in Chap. 19). Since PR proteins are located in the apoplast, changes in the plant cell walls due to *B. cinerea* infections may impact the efficacy of these anti-pathogen proteins.

18.5.2 Oligosaccharides

PDOs accumulate in the cell wall as a consequence of the pathogen-induced breakdown of pectin polysaccharides. PDOs can be recognized by WAK receptors that, then, signal plant responses, including defences (Hahn et al. 1981; Ridley et al. 2001; Galletti et al. 2011). Only de-esterified PDOs induce defence responses that enhance resistance to *B. cinerea* (Spadoni et al. 2006; Osorio et al. 2008), suggesting that WAKs may discriminate classes of PDOs. For example, over-expression of the PME, *FaPE1*, in strawberry fruit leads to the accumulation of defence-eliciting PDOs (Osorio et al. 2008, 2011). During *B. cinerea*-plant interactions, the perception of PDOs could cause increases in ethylene and reactive oxygen species (ROS) (Campbell and Labavitch 1991; Bellincampi et al. 2000; Galletti et al. 2008), changes in membrane polarization and ion fluxes (Mathieu et al. 1991; Thain et al. 1995), elevated expression of defence-related genes and accumulation of phytoalexins (Davis et al. 1986).

Signaling for plant responses to pathogens through PDOs involves mitogen-activated protein kinases (MAPKs; Chap. 17), at least in Arabidopsis leaves. MPK3 and MPK6 are activated by PDOs, flg22 and other PAMPs. Mutations in *MPK3* increase susceptibility of leaves and while a knockout of *MPK6* does not affect basal resistance, it does suppress PDO- and flg22-induced resistance to *B. cinerea*. This observation suggests that MPK6 may be more important for responses to *B. cinerea* (Galletti et al. 2011). Weinberg et al. (2013) demonstrated in Arabidopsis that *B. cinerea* releases small RNAs that reduce expression of MAPKs involved in defence.

18.5.3 Proteins That Interfere with B. cinerea–Induced Cell Wall Disassembly

One class of plant proteins associated with the cell wall that is known to affect the ability of *B. cinerea* to infect fruit and leaves is the PG inhibiting proteins, PGIPs. Defence-activating PDOs are thought to accumulate to higher levels when pathogen PGs are inhibited by plant PGIPs *in vitro* (De Lorenzo et al. 1994; Reymond et al. 1995; De Lorenzo and Ferrari 2002; Casasoli et al. 2009). PDOs, including those that result from PG and PL activities, have been isolated from developing *B. cinerea* lesions on infected tomato fruit (An et al. 2005). PGIPs can bind pectins and when pectin disassembly is reduced by suppression of the ripening PG in tomato fruit, the PGIP protein is retained in the cell wall matrix (Powell pers. obs.). Expression of

PGIP-encoding genes is high in unripe fruit and declines as fruit ripen; however, although the protein is stable, it is probably less tightly associated with the looser wall matrix of ripe fruit (Powell et al. 2000). PGIPs inhibit microbial PGs, including some but not all of *B. cinerea*'s PGs (Sharrock and Labavitch 1994; Joubert et al. 2007). Particular amino acids in the PGIP LRRs determine which PGs are inhibited (Stotz et al. 2000). PGIP expression is induced by *B. cinerea* and other pathogens. Over-expression of PGIPs reduces *B. cinerea* growth on ripe tomato fruit (Powell et al. 2000). In some situations, expression of PGIPs can cause alterations in the polysaccharides within the wall matrix (Nguema-Ona et al. 2013b), further altering the extracellular environment that *B. cinerea* encounters.

Proteins that are inhibitors of plant PMEs (called PMEIs), limit *B. cinerea* growth in vegetative tissues (Lionetti et al. 2007, 2014). Since tomato fruit PMEs are abundantly induced by *B. cinerea*, induced PMEIs may participate in plant responses to the infections.

Plants express other proteins that inhibit fungal host cell wall modifying enzymes. Tomato fruits produce an endo-β-glucanase inhibitor, XEGIP, which suppresses the activities of xyloglucan β-1,4 endoglucanases (XEGs) of the fungus *Aspergillus aculeatus* (Qin et al. 2003; York et al. 2004). Expression of XEGIP decreases during fruit ripening. Other glucanase inhibitors are serine proteases; loss of the Arabidopsis Unusual serine Protease Inhibitor (UPI) results in increased resistance to *B. cinerea* (Laluk et al. 2011).

18.5.4 Plant Cell Wall Fortifications

Fortifications of host plant cell walls are common responses that can limit pathogen progress. Localized structural reinforcements of the cell wall are usually accomplished by secretion of cross-linking phenolic compounds and subsequent accumulation of apoplastic polymers that are recalcitrant to degradation (Underwood 2012; Finiti et al. 2013). During *B. cinerea* infections, the localized fortifications of host cell walls can restrain fungal growth and impede degradation of wall components that serve as nutrient sources for the pathogen (Van Baarlen et al. 2007).

Asselbergh et al. (2007) proposed that deposition of phenolic compounds and cross-linkage of cell wall structural proteins may limit *B. cinerea* infections of leaves in the *sitiens* tomato mutant, an ABA-deficient tomato line. The *sitiens* leaves also have increased methylesterification of the pectins in their walls, which may limit the activity of *B. cinerea* endo-PGs and PLs. Thus, the higher levels of methylesterified pectins and the wall reinforcements may result in the improved resistance of *sitiens* mutant leaves to *B. cinerea* (Curvers et al. 2010). In another study, Arabidopsis mutants with low levels of phenolic monomers (i.e., monolignols) failed to mount efficient defences against *B. cinerea* infections (Lloyd et al. 2011).

In young tomato fruit, accumulation of the polymers callose and suberin has been linked to the formation of "ghost spots" (i.e., small necrotic lesions, usually surrounded by a white halo), which appear to restrain *B. cinerea* growth (de Leeuw 1985).

In addition, Cantu et al. (2009) demonstrated that suberin and lignin are deposited and H_2O_2 accumulates when unripe tomato fruit display resistance to *B. cinerea*.

When *B. allii* attempts to penetrate the epidermal cells of onions, granular deposits of reaction material (RM) form on the cell wall. Feruloyl-3′-methoxytyramine (FMT) and feruloyltyramine (FT) are the main components of the RM; while other phenolics, such as coumaroyl glucose, coumaroyltyramine (CT) and 2-hydroxy-2-(4-hydroxyphenyl) ethylferulate, are minor constituents. The formation of RM is associated with early increases in peroxidase activity. FMT, FT or CT are not antifungal compounds *per se*, thus these phenolics probably contribute to resistance because they prevent or retard cell wall degradation (McLusky et al. 1999).

18.6 Outlook for Improving Resistance by Altering Host Cell Walls

The plant cell wall is a complex and dynamic structure that provides important functions for the integrity of plant tissues. *B. cinerea* has evolved a wide array of virulence mechanisms that target multiple components of the cell walls of its plant hosts. The confrontation between *B. cinerea* and its hosts alters structural and biochemical aspects of the plant cell walls. *B. cinerea*'s virulence activities have substantial impacts on host cell wall integrity and ultimately on whether the outcome is resistance or susceptibility (Fig. 18.2).

Multiple approaches using plant and *B. cinerea* mutants as well as transgenic plants have identified processes targeting the plant cell wall by enzymes expressed by the host and by *B. cinerea*. However, comprehensive studies of the progressive disassembly and remodeling of plant cell wall polysaccharides during plant development or *B. cinerea* infections, but also on the interaction between plant and fungal

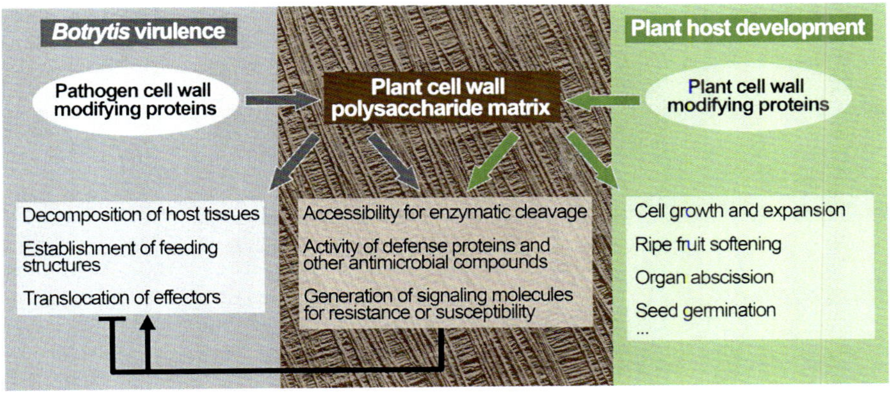

Fig. 18.2 Summary of the cell wall implications during plant-*B. cinerea* interactions (Reproduced and adapted, with permission, from Cantu et al. 2008a)

enzymes and cell wall components, are still needed to expand our knowledge of the molecular basis of plant susceptibility and *B. cinerea* virulence. Forward screens of genetic variants in natural, mapping, or mutagenized populations in plants and in *B. cinerea* may identify master regulators of processes that impact the cell walls of plant hosts. Reverse genetic approaches may be beneficial once the sequence of degradation of key cell wall targets and recognition of the direct and indirect consequences of wall modification in particular hosts is characterized. Particular attention should be paid to the developmental stage of the potentially susceptible host tissues in order to develop strategies that will effectively limit damage caused by *B. cinerea*.

References

AbuQamar S, Chen X, Dhawan R et al (2006) Expression profiling and mutant analysis reveals complex regulatory networks involved in Arabidopsis response to *B. cinerea* infection. Plant J 48:28–44

AbuQamar S, Ajeb S, Sham A et al (2013) A mutation in the *expansin-likeA2* gene enhances resistance to necrotrophic fungi and hypersensitivity to abiotic stress in *Arabidopsis thaliana*. Mol Plant Pathol 14:813–827

Albersheim P, Darvill A, Roberts K et al (2010) Biochemistry of cell wall molecules. In: Plant cell walls: from chemistry to biology. Garland Science, Taylor and Francis Group, LLC, New York.

Amselem J, Cuomo CA, Van Kan JA et al (2011) Genomic analysis of the necrotrophic fungal pathogens *Sclerotinia sclerotiorum* and *B. cinerea cinerea*. PLoS Genet 7:e1002230. doi:10.1371/journal.pgen.1002230

An HJ, Lurie S, Greve LC, Rosenquist D et al (2005) Determination of pathogen-related enzyme action by mass spectrometry analysis of pectin breakdown products of plant cell walls. Anal Biochem 338:71–82

Asselbergh B, Curvers K, Franca SC et al (2007) Resistance to *B. cinerea* in *sitiens*, an abscisic acid-deficient tomato mutant, involves timely production of hydrogen peroxide and cell wall modifications in the epidermis. Plant Physiol 144:1863–1877

Bellincampi D, Dipierro N, Salvi G et al (2000) Extracellular H_2O_2 induced by oligogalacturonides is not involved in the inhibition of the auxin-regulated *rolB* gene expression in tobacco leaf explants. Plant Physiol 122:1379–1386

Bennett AB, Labavitch JM (2008) Ethylene and ripening-regulated expression and function of fruit cell wall modifying proteins. Plant Sci 175:130–136

Blanco-Ulate B, Morales-Cruz A, Amrine KC et al (2014) Genome-wide transcriptional profiling of *B. cinerea* genes targeting plant cell walls during infections of different hosts. Front Plant Sci 5:435. doi:10.3389/fpls.2014.00435

Brito N, Espino J, González C (2006) The endo-β-1,4-xylanase Xyn11A is required for virulence in *B. cinerea*. Mol Plant Microbe Interact 19:25–32

Brutus A, Sicilia F, Macone A et al (2010) A domain swap approach reveals a role of the plant wall-associated kinase 1 (WAK1) as a receptor of oligogalacturonides. Proc Natl Acad Sci U S A 107:9452–9457

Burton RA, Gidley MJ, Fincher GB (2010) Heterogeneity in the chemistry, structure and function of plant cell walls. Nat Chem Biol 6:724–732

Campbell AD, Labavitch JM (1991) Induction and regulation of ethylene biosynthesis by pectic oligomers in cultured pear cells. Plant Physiol 97:699–705

Cantarel BL, Coutinho PM, Rancurel C et al (2009) The carbohydrate-active enzymes database (CAZy): an expert resource for glycogenomics. Nucleic Acids Res 37:233–238

Cantu D, Vicente AR, Labavitch JM et al (2008a) Strangers in the matrix: plant cell walls and pathogen susceptibility. Trends Plant Sci 13:610–617

Cantu D, Vicente A, Greve L et al (2008b) The intersection between cell wall disassembly, ripening and fruit susceptibility to *B. cinerea*. Proc Natl Acad Sci U S A 105:859–864

Cantu D, Blanco-Ulate B, Yang L et al (2009) Ripening-regulated susceptibility of tomato fruit to *B. cinerea* requires NOR but not RIN or ethylene. Plant Physiol 150:1434–1449

Cao Y, Tang X, Giovannoni J et al (2012) Functional characterization of a tomato COBRA-like gene functioning in fruit development and ripening. BMC Plant Biol 12:211

Carpita NC, Gibeaut DM (1993) Structural models of primary cell walls in flowering plants: consistency of molecular structure with the physical properties of the walls during growth. Plant J 3:1–30

Casasoli M, Federici L, Spinelli F et al (2009) Integration of evolutionary and desolvation energy analysis identifies functional sites in a plant immunity protein. Proc Natl Acad Sci U S A 106:7666–7671

Cassab GI (1998) Plant cell wall proteins. Annu Rev Plant Biol 49:281–309

Chinchilla D, Bauer Z, Regenass M et al (2006) The Arabidopsis receptor kinase FLS2 binds flg22 and determines the specificity of flagellin perception. Plant Cell 18:465–476

Cooper W, Bouzayen M, Hamilton A et al (1998) Use of transgenic plants to study the role of ethylene and polygalacturonase during infection of tomato fruit by *Colletotrichum gloeosporioides*. Plant Pathol 47:308–316

Cosgrove DJ (2001) Wall structure and wall loosening. A look backwards and forwards. Plant Physiol 125:131–134

Cosgrove DJ, Jarvis MC (2012) Comparative structure and biomechanics of plant primary and secondary cell walls. Front Plant Sci 3:204. doi:10.3389/fpls.2012.00204

Curvers K, Seifi H, Mouille G et al (2010) ABA-deficiency causes changes in cuticle permeability and pectin composition that influence tomato resistance to *B. cinerea cinerea*. Plant Physiol 154:847–860

Davis KR, Darvill AG, Albersheim P et al (1986) Host-pathogen interactions XXIX. Oligogalacturonides released from sodium polypectate by endopolygalacturonic acid lyase are elicitors of phytoalexins in soybean. Plant Physiol 80:568–877

De Leeuw G (1985) Deposition of lignin, suberin and callose in relation to the restriction of infection by *B. cinerea* in ghost spots of tomato fruits. J Phytopathol 112:143–152

De Lorenzo G, Ferrari S (2002) Polygalacturonase-inhibiting proteins in defence against phytopathogenic fungi. Curr Opin Plant Biol 5:295–299

De Lorenzo G, Cervone F, Bellincampi D et al (1994) Polygalacturonase, PGIP and oligogalacturonides in cell-cell communication. Biochem Soc Trans 22:394–397

De Lorenzo G, Brutus A, Savatin DV et al (2011) Engineering plant resistance by constructing chimeric receptors that recognize damage-associated molecular patterns (DAMPs). FEBS Lett 585:1521–1528

Decreux A, Messiaen J (2005) Wall-associated kinase WAK1 interacts with cell wall pectins in a calcium-induced conformation. Plant Cell Physiol 46:268–278

Diaz J, ten Have A, Van Kan AL (2002) The role of ethylene and wound signaling in resistance of tomato to *B. cinerea*. Plant Physiol 129:1341–1345

Eklöf JM, Brumer H (2010) The XTH gene family: an update on enzyme structure, function, and phylogeny in xyloglucan remodeling. Plant Physiol 153:456–466

Ellis C, Karafyllidis I, Wasternack C, Turner JG (2002) The Arabidopsis mutant cev1 links cell wall signaling to jasmonate and ethylene responses. Plant Cell 14(7):1557–1566

Espino J, Brito N, Noda J et al (2005) *B. cinerea* endo-ß-1, 4-glucanase Cel5A is expressed during infection but is not required for pathogenesis. Physiol Mol Plant Pathol 66:213–221

Espino J, Gutiérrez-Sánchez G, Brito N et al (2010) The *B. cinerea* early secretome. Proteomics 10:3020–3034

Faure D (2002) The family-3 glycoside hydrolases: from housekeeping functions to host-microbe interactions. Appl Environ Microbiol 68:1485–1490

Fernández-Acero FJ, Colby T, Harzen A et al (2010) 2-DE proteomic approach to the *B. cinerea cinerea* secretome induced with different carbon sources and plant-based elicitors. Proteomics 10:2270–2280

Finiti I, Leyva M, López-Cruz J et al (2013) Functional analysis of endo-1, 4-β-glucanases in response to *B. cinerea* and *Pseudomonas syringae* reveals their involvement in plant–pathogen interactions. Plant Biol 15:819–831

Flors V, Leyva M, Vicedo B et al (2007) Absence of the endo-β-1, 4-glucanses Cel1 and Cel2 reduces susceptibility to *B. cinerea* in tomato. Plant J 52:1027–1040

Frías M, González C, Brito N (2011) Spl1, a cerato-platanin family protein, contributes to *B. cinerea cinerea* virulence and elicits the hypersensitive response in the host. New Phytol 192:483–495

Frías M, Brito N, González C (2013) The *B. cinerea* cerato-platanin Spl1 is a potent inducer of systemic acquired resistance (SAR) in tobacco and generates a wave of salicylic acid expanding from the site of application. Mol Plant Pathol 14:191–196

Galletti R, Denoux C, Gambetta S et al (2008) The AtrbohD-mediated oxidative burst elicited by oligogalacturonides in Arabidopsis is dispensable for the activation of defence responses effective against *B. cinerea*. Plant Physiol 148:1695–1706

Galletti R, Ferrari S, De Lorenzo G (2011) Arabidopsis MPK3 and MPK6 play different roles in basal and oligogalacturonide-or flagellin-induced resistance against *B. cinerea*. Plant Physiol 157:804–814

Gilbert HJ (2010) The biochemistry and structural biology of plant cell wall deconstruction. Plant Physiol 153:444–455

Hahn MG, Darvill AG, Albersheim P (1981) Host-pathogen interactions XIX. The endogenous elicitor, a fragment of a plant cell wall polysaccharide that elicits phytoalexin accumulation in soybeans. Plant Physiol 68:1161–1169

Harris PJ, Stone BA (2009) Chemistry and molecular organization of plant cell walls. In: Biomass recalcitrance. Blackwell Publishing Ltd,, Oxford, pp 61–93

Hayashi T, Kaida R (2011) Functions of xyloglucan in plant cells. Mol Plant 4:17–24

Hsieh TF, Huang JW, Hsiang T (2001) Light and scanning electron microscopy studies on the infection of oriental lily leaves by *Botrytis elliptica*. Eur J Plant Pathol 107:571–581

Ishii T, Matsunaga T, Hayashi N (2001) Formation of rhamnogalacturonan II-borate dimer in pectin determines cell wall thickness of pumpkin tissue. Plant Physiol 126:1698–1705

Jamet E, Canut H, Boudart G et al (2006) Cell wall proteins: a new insight through proteomics. Trends Plant Sci 11:33–39

Jarvis MC, Forsyth W, Duncan HJ (1988) A survey of the pectic content of nonlignified monocot cell walls. Plant Physiol 88:309–314

Jayani RS, Saxena S, Gupta R (2005) Microbial pectinolytic enzymes: a review. Process Biochem 40:2931–2944

Joubert DA, Kars I, Wagemakers L et al (2007) A polygalacturonase-inhibiting protein from grapevine reduces the symptoms of the endopolygalacturonase PG2 from *B. cinerea* in *Nicotiana benthamiana* leaves without any evidence for *in vitro* interaction. Mol Plant Microbe Interact 20:392–402

Kapat A, Zimand G, Elad Y (1998) Effect of two isolates of *Trichoderma harzianum* on the activity of hydrolytic enzymes produced by *B. cinerea*. Physiol Mol Plant Pathol 52:127–137

Kars I, Krooshof GH, Wagemakers L et al (2005a) Necrotizing activity of five *B. cinerea* endopolygalacturonases produced in *Pichia pastoris*. Plant J 43:213–225

Kars I, McCalman M, Wagemakers L et al (2005b) Functional analysis of *B. cinerea* pectin methylesterase genes by PCR-based targeted mutagenesis: *pme1* and *pme2* are dispensable for virulence of strain B05.10. Mol Plant Pathol 6:641–652

Keegstra K (2010) Plant cell walls. Plant Physiol 154:483–486

Keegstra K, Talmadge KW, Bauer WD et al (1973) The structure of plant cell walls: III. A model of the walls of suspension-cultured sycamore cells based on the interconnections of the macromolecular components. Plant Physiol 51:188–197

Klis FM, Brul S, De Groot PWJ (2010) Covalently linked wall proteins in ascomycetous fungi. Yeast 27:489–493

Kohorn BD, Kohorn SL (2012) The cell wall-associated kinases, WAKs, as pectin receptors. Front Plant Sci 3:88. doi:10.3389/fpls.2012.00088

Lai Z, Mengiste T (2013) Genetic and cellular mechanisms regulating plant responses to necrotrophic pathogens. Curr Opin Plant Biol 16:505–512

Laluk K, Luo H, Chai M et al (2011) Biochemical and genetic requirements for function of the immune response regulator B. CINEREA-INDUCED KINASE1 in plant growth, ethylene signaling, and PAMP-triggered immunity in Arabidopsis. Plant Cell 23:2831–2849

Li B, Wang W, Zong Y, Qin G et al (2012) Exploring pathogenic mechanisms of *B. cinerea* secretome under different ambient pH based on comparative proteomic analysis. J Proteome Res 11:4249–4260

Lionetti V, Raiola A, Camardella L et al (2007) Overexpression of pectin methylesterase inhibitors in Arabidopsis restricts fungal infection by *B. cinerea*. Plant Physiol 143:1871–1880

Lionetti V, Cervone F, Bellincampi D (2012) Methyl esterification of pectin plays a role during plant–pathogen interactions and affects plant resistance to diseases. J Plant Physiol 169:1623–1630

Lionetti V, Raiola A, Cervone F et al (2014) Transgenic expression of pectin methylesterase inhibitors limits tobamovirus spread in tobacco and Arabidopsis. Mol Plant Pathol 15:265–274

Lloyd AJ, Allwood WJ, Winder CL et al (2011) Metabolomic approaches reveal that cell wall modifications play a major role in ethylene-mediated resistance against *B. cinerea*. Plant J 67:852–868

Manabe Y, Nafisi M, Verhertbruggen Y et al (2011) Loss-of-function mutation of REDUCED WALL ACETYLATION2 in Arabidopsis leads to reduced cell wall acetylation and increased resistance to *B. cinerea cinerea*. Plant Physiol 155:1068–1078

Martellini F, Faoro F, Carresi L et al (2013) Cerato-populin and cerato-platanin, two non-catalytic proteins from phytopathogenic fungi, interact with hydrophobic inanimate surfaces and leaves. Mol Biotechnol 55:27–42

Mathieu Y, Armen K, Xia H et al (1991) Membrane responses induced by oligogalacturonides in suspension-cultured tobacco cells. Plant J 1:333–343

McLusky SR, Bennett MH, Beale MH et al (1999) Cell wall alterations and localized accumulation of feruloyl-3′-methoxytyramine in onion epidermis at sites of attempted penetration by *B. allii* are associated with actin polarisation, peroxidase activity and suppression of flavonoid biosynthesis. Plant J 17:523–534

McNeil M, Darvill AG, Fry SC et al (1984) Structure and function of the primary cell walls of plants. Annu Rev Biochem 53:625–663

McQueen-Mason SJ, Cosgrove DJ (1995) Expansin mode of action on cell walls. Analysis of wall hydrolysis, stress relaxation, and binding. Plant Physiol 107:87–100

Meindl T, Boller T, Felix G (2000) The bacterial elicitor flagellin activates its receptor in tomato cells according to the address–message concept. Plant Cell 12:1783–1794

Mengiste T (2012) Plant immunity to necrotrophs. Annu Rev Phytopathol 50:267–294

Micheli F (2001) Pectin methylesterases: cell wall enzymes with important roles in plant physiology. Trends Plant Sci 6(9):414–419

Minic Z, Jouanin L (2006) Plant glycoside hydrolases involved in cell wall polysaccharide degradation. Plant Physiol Biochem 44:435–449

Mohnen D (2008) Pectin structure and biosynthesis. Curr Opin Plant Biol 11:266–277

Mølgaard A, Kauppinen S, Larsen S (2000) Rhamnogalacturonan acetylesterase elucidates the structure and function of a new family of hydrolases. Structure 8:373–383

Mutter M, Renard CM, Beldman G et al (1998) Mode of action of RG-hydrolase and RG-lyase toward rhamnogalacturonan oligomers. Characterization of degradation products using RG-rhamnohydrolase and RG-galacturonohydrolase. Carbohydr Res 311:155–164

Nafisi M, Stranne M, Zhang L et al (2014) The endo-arabinanase Ara1 is a novel host-specific virulence factor of the necrotic fungal phytopathogen *B. cinerea*. Mol Plant Microbe Interact. doi:10.1094/MPMI-02-14-0036-R

Nguema-Ona E, Vicré-Gibouin M, Cannesan MA et al (2013a) Arabinogalactan proteins in root–microbe interactions. Trends Plant Sci 18:440–449

Nguema-Ona E, Moore JP, Fagerström AD et al (2013b) Overexpression of the grapevine *PGIP1* in tobacco results in compositional changes in the leaf arabinoxyloglucan network in the absence of fungal infection. BMC Plant Biol 13:1–15

Noda J, Brito N, Gonzalez C (2010) The *B. cinerea* xylanase Xyn11A contributes to virulence with its necrotizing activity, not with its catalytic activity. BMC Plant Biol 10:38. doi:10.1186/1471-2229-10-38

O'Neill MA, Ishii T, Albersheim P, Darvill AG (2004) Rhamnogalacturonan II: structure and function of a borate cross-linked cell wall pectic polysaccharide. Annu Rev Plant Biol 55:109–139

Osorio S, Castillejo C, Quesada MA et al (2008) Partial demethylation of oligogalacturonides by pectin methyl esterase 1 is required for eliciting defence responses in wild strawberry (*Fragaria vesca*). Plant J 9:43–55

Osorio S, Alba R, Damasceno CM et al (2011) Systems biology of tomato fruit development: combined transcript, protein, and metabolite analysis of tomato transcription factor (*nor*, *rin*) and ethylene receptor (*Nr*) mutants reveals novel regulatory interactions. Plant Physiol 157:405–425

Park YB, Cosgrove DJ (2012) A revised architecture of primary cell walls based on biomechanical changes induced by substrate-specific endoglucanases. Plant Physiol 158:1933–1943

Pauly M, Gille S, Liu L et al (2013) Hemicellulose biosynthesis. Planta 238:627–642

Powell ALT, Van Kan J, ten Have A et al (2000) Transgenic expression of pear *PGIP* in tomato limits fungal colonization. Mol Plant-Microbe Interact 13:942–950

Qin Q, Bergmann CW, Rose JK et al (2003) Characterization of a tomato protein that inhibits a xyloglucan-specific endoglucanase. Plant J 34:327–338

Raiola A, Lionetti V, Elmaghraby I et al (2011) Pectin methylesterase is induced in Arabidopsis upon infection and is necessary for a successful colonization by necrotrophic pathogens. Mol Plant Microbe Interact 24:432–440

Ramirez V, García-Andrade J, Vera P (2011a) Enhanced disease resistance to *B. cinerea* in myb46 Arabidopsis plants is associated to an early down-regulation of CesA genes. Plant Signal Behav 6:91191–91193

Ramirez V, Agorio A, Coego A et al (2011b) MYB46 modulates disease susceptibility to *B. cinerea* in Arabidopsis. Plant Physiol 155:1920–1935

Reignault P, Mercier M, Bompeix G, Boccara M (1994) Pectin methylesterase from Botrytis cinerea: physiological, biochemical and immunochemical studies. Microbiology 140(12):3249–3255

Reignault P, Kunz C, Delage N et al (2000) Host-and symptom-specific pectinase isozymes produced by *B. cinerea*. Mycol Res 104:421–428

Reymond P, Grünberger S, Paul et al (1995) Oligogalacturonide defence signals in plants: large fragments interact with the plasma membrane *in vitro*. Proc Natl Acad Sci U S A 92:4145–4149

Ridley BL, O'Neill MA, Mohnen D (2001) Pectins: structure, biosynthesis, and oligogalacturonide-related signaling. Phytochemistry 57:929–967

Rose JKC, Bennett AB (1999) Cooperative disassembly of the cellulose-xyloglucan network of plant cell walls: parallels between cell expansion and fruit ripening. Trends Plant Sci 4:176–183

Rose JKC, Lee SJ (2010) Straying off the highway: trafficking of secreted plant proteins and complexity in the plant cell wall proteome. Plant Physiol 153:433–436

Rose JK, Braam J, Fry SC et al (2002) The XTH family of enzymes involved in xyloglucan endotransglucosylation and endohydrolysis: current perspectives and a new unifying nomenclature. Plant Cell Physiol 43:1421–1435

Scheer JM, Ryan CA (1999) A 160-kD systemin receptor on the surface of *Lycopersicon peruvianum* suspension-cultured cells. Plant Cell 11:1525–1535

Scheller HV, Ulvskov P (2010) Hemicelluloses. Annu Rev Plant Biol 61:263–289

Selig MJ, Adney WS, Himmel ME et al (2009) The impact of cell wall acetylation on corn stover hydrolysis by cellulolytic and xylanolytic enzymes. Cellulose 16:711–722

Shah P, Gutierrez-Sanchez G, Orlando R et al (2009a) A proteomic study of pectin-degrading enzymes secreted by *B. cinerea* grown in liquid culture. Proteomics 9:3126–3135

Shah P, Atwood JA, Orlando R et al (2009b) Comparative proteomic analysis of *B. cinerea* secretome. J Proteome Res 8:1123–1130

Shah P, Powell AL, Orlando R et al (2012) Proteomic analysis of ripening tomato fruit infected by *B. cinerea*. J Proteome Res 11:2178–2192

Sharrock KR, Labavitch JM (1994) Polygalacturonase inhibitors of Bartlett pear fruits: differential effects on *B. cinerea* polygalacturonase isozymes, and influence on products of fungal hydrolysis of pear cell walls and on ethylene induction in cell culture. Physiol Mol Plant Pathol 45:305–319

Showalter AM (1993) Structure and function of plant cell wall proteins. Plant Cell 5:9–23

Spadoni S, Zabotina O, Di Matteo A et al (2006) Polygalacturonase-inhibiting protein interacts with pectin through a binding site formed by four clustered residues of arginine and lysine. Plant Physiol 141:557–564

Stotz HU, Bishop JG, Bergmann C et al (2000) Identification of target amino acids that affect interactions of fungal polygalacturonases and their plant inhibitors. Physiol Mol Plant Pathol 56:117–130

ten Have A, Mulder W, Visser J et al (1998) The endopolygalacturonase gene *pg1* is required for full virulence of *B. cinerea*. Mol Plant Microbe Interact 11:1009–1016

ten Have A, Breuil WO, Wubben JP et al (2001) *Botrytiscinerea* endopolygalacturonase genes are differentially expressed in various plant tissues. Fungal Genet Biol 33:97–105

Thain J, Gubb I, Wildon D (1995) Depolarization of tomato leaf cells by oligogalacturonide elicitors. Plant Cell Environ 18:211–214

Underwood W (2012) The plant cell wall: a dynamic barrier against pathogen invasion. Front Plant Sci 3:85. doi:10.3389/fpls.2012.00085

Valette-Collet O, Cimerman A, Reignault P et al (2003) Disruption of *B. cinerea* pectin methylesterase gene *pme1* reduces virulence on several host plants. Mol Plant-Microbe Interact 16:360–267

Van Baarlen P, Legendre L, Van Kan JL (2007) Plant defence compounds against *B. cinerea* infection. In: Elad Y, Williamson B, Tudzynski P, Delen N (eds) *Botrytis*: biology, pathology and control. Springer, Dordrecht, pp 143–161

Van Kan JAL (2006) Licensed to kill: the lifestyle of a necrotrophic plant pathogen. Trends Plant Sci 11:247–253

Verhoeff K, Lihm J, Scheffer R et al (1983) Cellulolytic activity of *B. cinerea in vitro* and *in vivo*. J Phytopathol 106:97–103

Vidal S, Doco T, Williams P et al (2000) Structural characterization of the pectic polysaccharide rhamnogalacturonan II: evidence for the backbone location of the aceric acid-containing oligoglycosyl side chain. Carbohydr Res 326:277–294

Voragen AJ, Coenen G-J, Verhoef R et al (2009) Pectin, a versatile polysaccharide present in plant cell walls. Struct Chem 20:263–275

Weiberg A, Wang M, Lin FM, Zhao H, Zhang Z, Kaloshian I, Huang H-D, Jin H (2013) Fungal small RNAs suppress plant immunity by hijacking host RNA interference pathways. Science, 342(6154):118–123

Windram O, Madhou P, McHattie S et al (2012) Arabidopsis defence against *B. cinerea*: chronology and regulation deciphered by high-resolution temporal transcriptomic analysis. Plant Cell 24:3530–3557

Wubben JP, Ten Have A, Van Kan J et al (2000) Regulation of endopolygalacturonase gene expression in *B. cinerea* by galacutonic acid, ambient pH and carbon catabolite, repression. Curr Genet 37:152–157

York WS, Qin Q, Rose JK (2004) Proteinaceous inhibitors of endo-β-glucanases. BBA Protein Proteomics 1696:223–233

Zhang L, Van Kan JAL (2013) *B. cinerea* mutants deficient in d-galacturonic acid catabolism have a perturbed virulence on *Nicotiana benthamiana* and Arabidopsis, but not on tomato. Mol Plant Pathol 14:19–29

Zhang L, Hua C, Stassen JH, Chatterjee S, Cornelissen M, van Kan JAL (2013) Genome-wide analysis of pectate-induced gene expression in Botrytis cinerea: Identification and functional analysis of putative d-galacturonate transporters. Fungal Genet Biol. doi:10.1016/j.fgb.2013.10.002

Zhang L, Kars I, Essenstam B et al (2014) Fungal endopolygalacturonases are recognized as microbe-associated molecular patterns by the arabidopsis receptor-like protein RESPONSIVENESS TO B. CINEREA POLYGALACTURONASES1. Plant Physiol 164:352–364

Chapter 19
Ripening of Tomato Fruit and Susceptibility to *Botrytis cinerea*

Barbara Blanco-Ulate, Estefania Vincenti, Dario Cantu, and Ann L.T. Powell

Abstract *Botrytis cinerea* causes gray mold decay in the ripe fruit of many plant species leading to significant economic losses for the producers, distributers and consumers of fresh and stored produce. This chapter summarizes current knowledge about the biology of a model fleshy fruit, tomato, during *B. cinerea* infections. The information presented emphasizes how ripening regulation and events in the host influence resistance when fruit are unripe and susceptibility when fruit are ripe. Fruit ripening regulators (e.g., transcription factors, epigenetic modifiers and hormones) and events unique to ripening that can impact the susceptibility of tomato fruit to *B. cinerea* are discussed. Understanding the processes in the fruit that underlie the shift from resistance to susceptibility during ripening and resolving how *B. cinerea* modifies its strategies of infection in response to the developmental changes of the host may guide efforts to improve the resistance of fruit to *B. cinerea* and other fungal pathogens.

Keywords Developmentally-regulated susceptibility • Fruit ripening • Plant hormones • Fruit metabolism • Plant defences

19.1 Introduction

The interplay between tomato fruit and *Botrytis cinerea* is a prototype for climacteric fruit-necrotroph interactions (Powell et al. 2000; Flors et al. 2007; Cantu et al. 2008, 2009; Weiberg et al. 2013; Blanco-Ulate et al. 2013). Tomato fruit have been used extensively as a model system to understand the development and ripening of fleshy fruit organs. The acquisition of susceptibility to *B. cinerea* is a hallmark of tomato

---

B. Blanco-Ulate • D. Cantu
Department of Viticulture and Enology, University of California, Davis, CA 95616, USA

E. Vincenti • A.L.T. Powell (✉)
Department of Plant Sciences, University of California, Davis, CA 95616, USA
e-mail: alpowell@ucdavis.edu

S. Fillinger, Y. Elad (eds.), *Botrytis – the Fungus, the Pathogen and its Management in Agricultural Systems*, DOI 10.1007/978-3-319-23371-0_19

fruit ripening: unripe fruit are resistant to gray mold and become increasingly susceptible as they ripen. The abundance of molecular and genetic resources that have described the biology of fruit development and ripening provide powerful experimental tools to understand the intersection between fruit ripening and susceptibility to *B. cinerea*. Complete genome sequences are available as well as large collections of spontaneous and induced mutations and high-quality "omics" datasets are frequently published (Carrari et al. 2006; Arie et al. 2007; Cantu et al. 2009; Klee and Giovannoni 2011; Tomato Genome Consortium 2012; Osorio et al. 2012; Seymour et al. 2013; Blanco-Ulate et al. 2014). Fresh and prepared tomato fruit products are economically important and fungal pathogens, such as *B. cinerea*, cause losses of billions of dollars worldwide (Narayanasamy 2006). Understanding the mechanisms that make ripe fruit susceptible will help breeders and postharvest technologists develop more effective approaches to reduce losses caused by *B. cinerea*.

19.1.1 Fruit Ripening

Ripening is a complex and coordinated sequence of inter related biochemical and physiological processes that transform the appearance, composition and characteristics of unripe fruit into a ripe fruit suitable for consumption by animals and humans (Brady 1987; Osorio et al. 2013; Seymour et al. 2013). Ripening is regulated by multiple interacting mechanisms and occurs in an organized and developmentally determined manner (Seymour et al. 2013). In tomato fruit, ripening begins at the distal end of the fruit (Nguyen et al. 2014). Complex positive and negative determinative processes occur at the onset of ripening and impact how ripening progresses. Networks of transcription factors constitute the core regulators of ripening, while hormones and other molecules act downstream to activate particular ripening pathways (Klee and Giovannoni 2011; Seymour et al. 2013). In addition, epigenetic and post-transcriptional modifications influence the expression of master regulators and specific ripening-related genes (Mohorianu et al. 2011; Karlova et al. 2013; Seymour et al. 2013; Zhong et al. 2013; Liu et al. 2015).

19.1.2 Unripe Fruit Become Susceptible During Ripening

Since tomato fruit become more susceptible as they ripen (Fig. 19.1), it had been thought that susceptibility is an inherent outcome of ripening (Prusky 1996; Klee and Giovannoni 2011). However, studies of the regulation of ripening and susceptibility have led to the conclusions that only selected ripening events and pathways are required to facilitate *B. cinerea* infections, and that *B. cinerea* modifies its infection strategy as fruit ripen (Cantu et al. 2008, 2009; Blanco-Ulate et al. 2013, 2014).

In the context of fruit-*B. cinerea* interactions, it is critical to distinguish between the processes of normal ripening and those that are needed to render ripe fruit susceptible. It is also necessary to determine how *B. cinerea* impacts ripening by

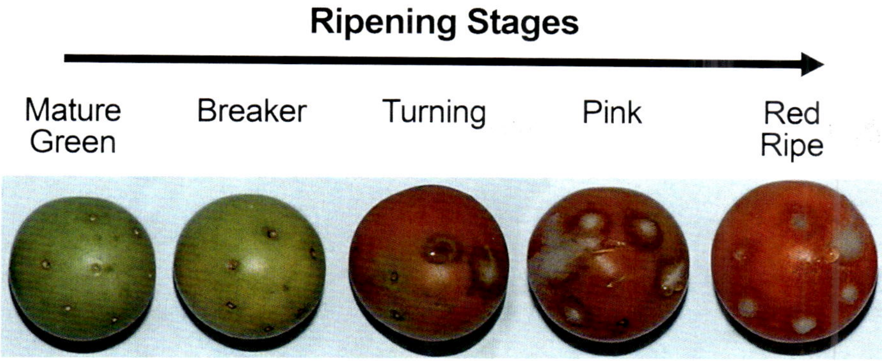

Fig. 19.1 Susceptibility of tomato fruit to *B. cinerea* as a function of ripening stage. Tomato fruit (cv. Ailsa Craig) at Mature Green, Breaker, Turning, Pink and Red Ripe stages at 3 days after the introduction of *B. cinerea* spores into wound sites

exploiting and diverting regulators and processes during fruit development. On unripe tomato fruit, *B. cinerea* accelerates selected aspects of ripening, thereby enhancing host susceptibility (Cristescu et al. 2002; Cantu et al. 2009; Blanco-Ulate et al. 2013). How *B. cinerea* promotes susceptibility in fruit has not yet been determined, but toxins, plant hormone analogues, small RNAs and/or pathogen- and damage-associated molecular patterns (PAMPs and DAMPs, respectively) generated by *B. cinerea* during infections may be involved.

19.2 Control of Fruit Ripening and Susceptibility to *B. cinerea*

19.2.1 Transcriptional Regulators

Transcription factors are the central regulators of fruit ripening (Seymour et al. 2013). The intricate and dynamic relationships between the ripening-regulating transcription factors indicate that multiple coordinated inputs are addressed as fruit develop. These modulators could be perturbed by *B. cinerea* infections. Among these, three tomato transcription factors, RIPENING-INHIBITOR (RIN), COLORLESS NON-RIPENING (CNR) and NON-RIPENING (NOR), control ripening up-steam of hormones and other regulatory molecules. Mutations in *CNR*, *RIN* and *NOR* result in fruit that do not ripen (Vrebalov et al. 2002; Manning et al. 2006; Seymour et al. 2013). Because susceptibility to *B. cinerea* differs among the fruit of the three non-ripening mutants, it has been concluded that *B. cinerea* infections require only specific aspects of ripening (Cantu et al. 2009).

The *rin* mutation results in full-sized firm tomato fruit that remain green (i.e. carotenoid compounds do not accumulate and chlorophylls are only slowly degraded). *rin* fruit do not produce or respond to ethylene, a plant hormone involved

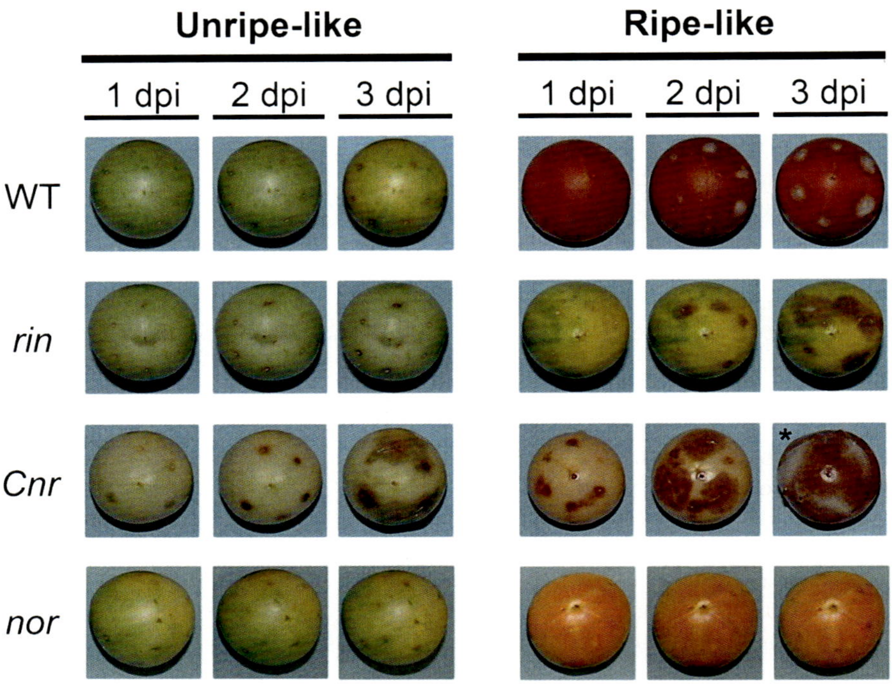

Fig. 19.2 Susceptibility of non-ripening tomato fruit mutants to *B. cinerea*. Tomato fruit of wild-type (WT, cv. Ailsa Craig) and three isogenic non-ripening mutants (*rin*, *Cnr* and *nor*) at equivalent ripening stages (Unripe-like: 31 days post-anthesis [dpa] and Ripe-like: 42 dpa) after inoculation with *B. cinerea* (1–3 days post-inoculation [dpi]). * *Cnr* red ripe-like fruit are dark at 3 dpi due to excessive *B. cinerea* growth

in ripening (Vrebalov et al. 2002). *rin* fruit show the same susceptibility to *B. cinerea* as wild-type fruit (Fig. 19.2); they are resistant when unripe but become susceptible as they complete maturation. Thus, RIN is not needed for the resistance of unripe fruit or the ripening-associated development of susceptibility (Cantu et al. 2009). RIN activates many aspects of ripening including: (1) ethylene synthesis and perception (2) cell wall disassembly; (3) carotenoid formation; (4) volatile aroma biosynthesis; (5) generation of ATP; and (6) suppression of auxin responses. RIN binds to the promoters of other ripening regulators, including *CNR* and *NOR* (Martel et al. 2011; Qin et al. 2012; Fujisawa et al. 2013; Zhong et al. 2013), and RIN regulation of ripening genes depends on CNR (Martel et al. 2011; Zhuo et al. 2013).

In contrast to *rin* and wild-type fruit, *Cnr* mutant fruit are hyper-susceptible to *B. cinerea*, even at the unripe stage (Fig. 19.2). The *Cnr* mutation leads to smaller colorless non-ripening fruit (Fraser et al. 2001; Eriksson et al. 2004; Manning et al. 2006). The pericarp of the *Cnr* mutant fruit is mealy due to biochemical changes in the cell walls that cause substantial loss of cell-to-cell adhesion (Orfila et al. 2002; Eriksson et al. 2004; Ordaz-Ortiz et al. 2009). Susceptibility-related genes targeted directly by CNR are unclear (Eriksson et al. 2004; Seymour et al. 2008), although it is known that the *Cnr* mutation results in the up-regulation of pathogenesis-related (PR) gene expression (Eriksson et al. 2004). Intriguingly, the disruption of normal ripening and the constitutive activation of PR genes in the *Cnr* mutant fruit is not sufficient to inhibit *B. cinerea* infections and in fact, infections are precociously robust, perhaps due to disturbances in adhesion between cells.

In contrast, fruit with the *nor* mutation, which also do not undergo most of the changes associated with ripening, are entirely resistant to *B. cinerea* infection at both the unripe-like and ripe-like stages (Fig. 19.2, Cantu et al. 2009). Thus, NOR promotes susceptibility and/or inhibits resistance during ripening. Osorio et al. (2011) identified some of the ripening processes that are reduced by the mutation of *NOR*, such as: (1) ethylene synthesis and perception, (2) expression of the ripening-related polygalacturonase, (3) levels of hexoses, carotenoid compounds and organic acids.

Transcription factors that influence fruit ripening downstream of RIN, CNR, and NOR, or interact with them to repress or activate ripening, may influence *B. cinerea* infections of fruit. The susceptibility of fruit with altered expression of the negative ripening regulators: SlMADS1, a tomato homolog of APETALLA SlAP2a and the ethylene response factor SlERF6 (Chung et al. 2010; Karlova et al. 2011, 2013; Lee et al. 2012; Dong et al. 2013), or the positive regulators, SlNAC4, TOMATO AGAMOUS-LIKE 1 (SlTAGL1), the tomato FRUITFULL homologs (SlFUL1 and SlFUL2) and the HD-Zip homeobox transcription factor SlHB1 (Lin et al. 2008; Shima et al. 2013; Zhu et al. 2014; Fujisawa et al. 2014) has not been evaluated. Other transcription regulators related to fruit development, but not directly involved in ripening, could also impact the outcome of fruit-*B. cinerea* interactions. A Golden2-like (SlGLK2) and KNOX (SlTKN4 and SlTKN2) transcription factors and a response-regulator factor (SlAPRR2-like) regulate chloroplast development and chlorophyll accumulation prior to the onset of ripening (Powell et al. 2012; Pan et al. 2013; Nadakudti et al. 2014; Nguyen et al. 2014). *SlGLK2* over-expression apparently favors *B. cinerea* infections of unripe fruit, hypothetically because of the increased sugar contents of the fruit (A.L.T. Powell pers. comm.).

19.2.2 Epigenetic Changes and Small RNAs

When plants and pathogens interact, epigenetic modifications occur. These epigenetic changes may include DNA methylation and modifications of histone tails that lead to reprogramming of the transcription of genes involved in defense responses (Dowen et al. 2012; Yang et al. 2013). Epigenetic changes could be a means of inducing susceptibility in ripening fruit.

Throughout fruit development, epigenetic reprogramming occurs, but overall DNA methylation declines once fruit begin to ripen (Zhong et al. 2013; Liu et al. 2015). Notably, the promoters of tomato genes under the control of RIN are demethylated (Zhong et al. 2013). Expression of *CNR* is epigenetically controlled (Manning et al. 2006; Zhong et al. 2013). When seeds enclosed in the fruit flesh have developed full viability and the fruit have reached their maximum size, the *CNR* promoter is demethylated and subsequently a cascade of ripening events occurs. Although, CNR is not the only catalyst of tomato fruit ripening, its activity coincides with the demethylation of the promoters of genes regulated by RIN (Zhong et al. 2013; Chen et al. 2015). Whether *B. cinerea* perturbs this form of ripening regulation has not been explored, although *B. cinerea* does influence host gene expression for susceptibility (Cantu et al. 2009).

Plant small RNAs (sRNAs), such as microRNAs (miRNAs) and small interfering RNAs (siRNAs), are important means of regulating development, defences and epigenetic modifications (Chen 2009; Ruiz-Ferrer and Voinnet 2009). In fruit, sRNAs influence the expression of transcription factors or proteins that contribute to key aspects of ripening (Moxon et al. 2008; Itaya et al. 2008; Mohorianu et al. 2011; Zuo et al. 2012, 2013; Karlova et al. 2013).

miRNAs coordinate the onset and progression of ripening by integrating both transcriptional and hormonal controls (Moxon et al. 2008; Zuo et al. 2012; Karlova et al. 2013). For example, in tomato, miR156 and miR172 suppress the expression of *CNR* and *SlAP2a*, respectively (Karlova et al. 2013). Other miRNAs influence the expression of ripening genes downstream of these regulators, such as genes involved in ethylene synthesis and perception (Moxon et al. 2008; Mohorianu et al. 2011; Zuo et al. 2012, 2013; Karlova et al. 2013). siRNAs are highly abundant in tomato fruit and particular clusters of siRNAs might regulate early development and ripening. Most siRNA targets have not been validated yet, but may include genes involved in lipid and polysaccharide metabolism, flavor and aroma compound biosynthesis and hormone signaling (Itaya et al. 2008; Mohorianu et al. 2011; Zuo et al. 2013).

sRNAs may be involved in controlling fruit resistance and susceptibility. Studies on the roles of plant sRNA in the regulation of defence responses to *B. cinerea* have been done in vegetative tissues but not in fruit (Jin et al. 2012). Nevertheless, some plant miRNAs with putative roles in disease resistance (e.g., miR6022, miR6027 and miRZ8) have been detected in fruit (Karlova et al. 2013; Zuo et al. 2013). *B. cinerea* expresses sRNAs to regulate its own development and to suppress host immune responses (Weiberg et al. 2013). During infections of Arabidopsis and tomato leaves, *B. cinerea* introduces its sRNAs into the host where they act as effectors, which hijack plant sRNA-mediated mechanisms to silence genes involved in biotic stress signaling (Weiberg et al. 2013). *B. cinerea* sRNAs have been detected during fruit infections (Weiberg et al. 2013), but no connections have been drawn yet between them and the susceptibility of ripe fruit.

19.2.3 Hormones

Regulation involving the synthesis and perception of hormones impacts basal resistance and mediates plant defence responses. Interactions among hormone pathways provide the plant with adaptable defence strategies against different types of attackers, but these might also be vulnerable to exploitation by pathogens to favor infections (Van der Ent and Pieterse 2012). Plant hormones coordinate a variety of ripening events downstream of the main transcriptional controllers of ripening. In tomato fruit, *B. cinerea* infections cause transcriptional reprogramming of multiple plant hormone networks and depending on the developmental stage of the fruit, either resistance or susceptibility results (Blanco-Ulate et al. 2013).

19.2.3.1 Ethylene

Since the gaseous hormone, ethylene, is essential for the normal ripening of climacteric fruit (e.g., tomato, banana, apple, and peach), failure to synthesize or perceive ethylene prevents ripening (Klee and Giovannoni 2011; Pech et al. 2012). Ethylene influences the resistance and the susceptibility of plant tissues to pathogens. In cooperation with other hormones, ethylene activates defences against diverse pathogens, but ethylene also induces senescence or ripening, which facilitate susceptibility and infections by necrotrophs (Diaz et al. 2002; Van Loon et al. 2006; Cantu et al. 2009; Nambeesan et al. 2012; Rasul et al. 2012; Van der Ent and Pieterse 2012).

Two systems, 1 and 2, are used to synthesize ethylene via the conversion of S-adenosyl methionine (SAM) to 1-aminocyclopropane-1-carboxylic acid (ACC) (Baur et al. 1971). Which system is used, depends on whether fruit ripen with climacteric or non-climacteric mechanisms and the developmental stage of the fruit. System 1 produces low levels of ethylene due to auto-inhibition of its biosynthetic genes and is active throughout early fruit development. Only System 1 is active when non-climacteric fruit ripen but System 2 is required and predominates when climacteric fruit ripen. System 2 generates high levels of ethylene by positive feedback regulation of its ethylene biosynthetic genes (Yokotani et al. 2009; Klee and Giovannoni 2011; Pech et al. 2012). In tomato fruit, the switch from System 1 to System 2 requires the transcription factor RIN and ethylene (Martel et al. 2011; Pech et al. 2012).

Infections of unripe fruit by *B. cinerea* could induce ethylene production under System 1 and consequently trigger plant defense responses, but could also prematurely activate System 2 and the consequent climacteric ripening or, in non-climacteric fruit, initiate senescence or ripening pathways, which result in enhanced susceptibility that overthrows ethylene-induced defences (Marcos et al. 2005; Swartzberg et al. 2008; Cantu et al. 2009; Blanco-Ulate et al. 2013). When *B. cinerea* infects unripe tomato fruit, expression of two *ACS* genes from System 2, *SlACS1a* and *SlACS2*, is precociously activated, suggesting that the pathogen activates ethylene synthesis by System 2 (Blanco-Ulate et al. 2013). *B. cinerea* itself is capable of synthesizing ethylene from methionine via the α-keto-γ-methylthiobutyric acid (KMBA) pathway (Chagué et al. 2002). Ethylene production by *B. cinerea* during infections (Cristescu et al. 2002) may stimulate the production of host-derived ethylene and, thereby, promote premature ripening or senescence.

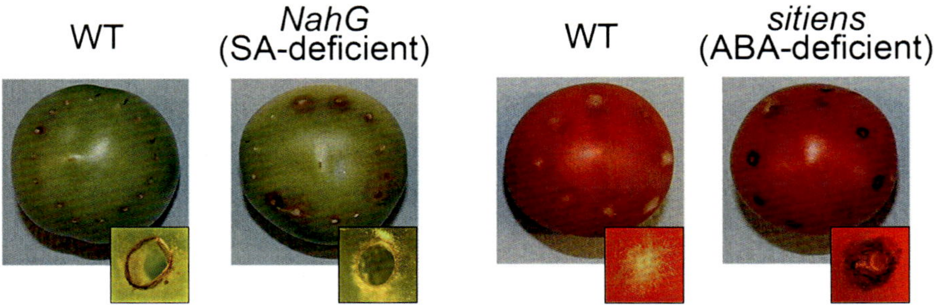

Fig. 19.3 Susceptibility of salicylic acid (SA) and abscisic acid (ABA) deficient tomato fruit to *B. cinerea*. Inoculated tomato fruit (3 dpi) of wild-type (WT, cv. Moneymaker), *NahG* (SA-deficient) and *sitiens* (ABA-deficient). Insets show a magnification of one of the inoculation sites. Unripe wild-type and ripe *sitiens* fruit have dark necrotic rings that limit infection symptoms; whereas unripe *NahG* fruit or ripe WT fruit have no necrotic zone (Reproduced, with permission, from Blanco-Ulate et al. 2013)

19.2.3.2 Salicylic Acid

When examined in leaves, salicylic acid (SA) is primarily associated with basal host defenses, effector-triggered immunity and local and systemic acquired resistance against biotrophic and hemi-biotrophic pathogens (Durrant and Dong 2004; Vlot et al. 2009). Because of mutual regulation, SA might enhance susceptibility to necrotrophs by antagonizing jasmonic acid (JA) and auxin signaling (Glazebrook 2005; Beckers and Spoel 2006; Spoel and Dong 2008). SA synthesis and signaling occur in fruit, but no roles for this hormone in regulating fruit development have been defined.

In unripe tomato fruit, SA responses involving an NPR1-independent pathway appear to be associated with resistance to *B. cinerea*. SA induction of Mitogen-Activated Protein Kinase (MAPK) signaling via MAPK3 and MAPK6 may establish efficient defences against *B. cinerea* before ripening, while NPR-1 related mechanisms or the absence of MAPK may be counterproductive for defence strategies later in ripening (Blanco-Ulate et al. 2013). Evidence that SA is important for unripe fruit defenses comes from the increased susceptibility of unripe fruit from the *NahG* tomato line, which does not accumulate SA (Fig. 19.3, Brading et al. 2000; Cantu et al. 2009; Blanco-Ulate et al. 2013). Ripe *NahG* and wild-type fruit are equally susceptible, indicating that SA-related defences are unimportant in ripe fruit (Blanco-Ulate et al. 2013).

19.2.3.3 Jasmonates

JA and its derivatives are well-known participants in the induction of plant responses to insects and necrotrophic pathogens (Glazebrook 2005; Browse 2009). As in infections of leaves, *B. cinerea* infections of tomato fruit increase JA biosynthetic gene expression (Wasternack 2007; Browse 2009; Blanco-Ulate et al. 2013).

JAs are associated with the control of normal fruit ripening, possibly because JAs act synergistically with ethylene (Peña-Cortés et al. 2004).

The efficiency of JA-mediated defence responses depends on interactions with other hormone pathways and signals derived from the pathogen itself (Rowe et al. 2010). For example, JA responses are inhibited by the NON-EXPRESSOR OF *PR* GENES 1 (NPR1), a key component of the SA signaling network, which prioritizes responses against biotrophs (Pieterse et al. 2009; Van der Ent and Pieterse 2012). Infections by *B. cinerea* activate the NPR1-mediated pathway, which antagonizes JA-derived defences and promote susceptibility (El Oirdi et al. 2011). Ethylene and JA-mediated defences can act synergistically to impede necrotrophic infections in leaves (Glazebrook 2005). The simultaneous activation of ethylene and JA pathways impairs the NPR1-mediated suppression of JA-dependent defences, which results in the co-existence of resistance mechanisms against biotrophs and necrotrophs (Spoel and Dong 2008; Leon-Reyes et al. 2010).

In unripe tomato fruit, the opportune activation of ethylene, SA and JA-mediated pathways could establish efficient defences to *B. cinerea*. However, as unripe fruit begin to ripen, ethylene-mediated induction of ripening and/or senescence probably results in a shift towards increased susceptibility by over-riding the defence pathways regulated by these hormones (Blanco-Ulate et al. 2013).

19.2.3.4 Abscisic Acid

Abscisic acid (ABA)-mediated responses are essential for tolerance to abiotic stress (Fujita et al. 2006). ABA induces susceptibility to pathogens by antagonizing the SA- and JA/ethylene-dependent defences (Spoel and Dong 2008; Yasuda et al. 2008; Sánchez-Vallet et al. 2012). The effect of ABA for the outcome of particular plant-pathogen interactions depends on the organisms, the developmental stage of the host and/or the environmental conditions in which the interaction occurs (Mauch-Mani and Mauch 2005; Ton et al. 2009; Robert-Seilaniantz et al. 2011). ABA regulates aspects of ripening in climacteric as well as non-climacteric fruit (Zhang et al. 2009; Koyama et al. 2010; Jia et al. 2011; Soto et al. 2013). In tomato fruit, ABA activates ripening because it enhances the expression of ethylene biosynthetic genes from System 2 (Zhang et al. 2009). ABA also triggers senescence in vegetative tissues and possibly in fruit (Lee et al. 2011).

ABA contributes to the ripening-associated susceptibility of tomato fruit to *B. cinerea*. Ripe tomato fruit from the *sitiens* mutant, which does not synthesize ABA (Harrison et al. 2011), are less susceptible to *B. cinerea* than wild-type fruit (Fig. 19.3). A localized necrotic response that is typical of resistant unripe fruit develops in infected ripe fruit (Blanco-Ulate et al. 2013). Leaves from the *sitiens* tomato mutant also are more resistant to *B. cinerea* infections (Asselbergh et al. 2007; Curvers et al. 2010; Seifi et al. 2013a), although it is not clear whether the resistance mechanisms in *sitiens* leaves and fruit are similar. Characterization of infected leaves suggest that increased cell viability, induction of defence-related genes and increased oxidative burst limit *B. cinerea* growth (Curvers et al. 2010; Seifi et al. 2014).

19.2.3.5 Auxins and Brassinosteroids

Normal fruit development requires auxins for growth and sugar metabolism (Sagar et al. 2013) and once ripening begins, auxins are inactivated (Seymour et al. 2013). Indole-3-acetic acid (IAA), the most abundant auxin in plants, decreases just before ripening starts, and IAA-aspartate, the inactive conjugate, increases (Böttcher et al. 2010, 2013; Kumar et al. 2012). Auxin has been shown to be both a susceptibility or resistance factor depending on the host tissue (Sharon et al. 2007; Fu and Wang 2011). IAA induces susceptibility to *B. cinerea* in Arabidopsis leaves by antagonizing oligosaccharide-induced resistance (Savatin et al. 2011), but in tomato leaves and eggplant fruit, pre-treatments with IAA or naphthaleneacetic acid (i.e. NAA, a synthetic auxin) reduce *B. cinerea* infections (Sharon et al. 2007). *B. cinerea* can produce auxins (Sharon et al. 2007), but whether this fungus produces these hormones *in planta* and how this production impacts the susceptibility of fruit is currently unknown. Auxins affect brassinosteroid levels (Halliday 2004) and along with ethylene, IAA and ABA affect fruit ripening (Vardhini and Rao 2002; Symons et al. 2012; Zaharah et al. 2012), but whether they also are involved in determing outcomes of infections in fruit as they are in leaves is not known (Laluk et al. 2011; Lin et al. 2013; De Bruyne et al. 2014).

19.3 Ripening Fruit Properties and Susceptibility to *B. cinerea*

19.3.1 Tissue Integrity and Physical Barriers

Tissue softening is an integral aspect of ripening in many fruit. Fruit softening occurs primarily as a result of the activities of plant cell wall modifying enzymes that disassemble the networks of polysaccharide in the wall. The composition and thickness of the fruit cuticle and the decline in cellular turgor pressure during ripening also influence the texture of the fruit flesh (Shakel et al. 1991; Brummell and Harpster 2001; Shi et al. 2013; Lara et al. 2014). Softening compromises the integrity of the physical barriers that fruit have against pathogens, and thereby, increases susceptibility to pathogen infections. Furthermore, softened fruit are more vulnerable to physical damage, such as bruising or cracking, which may provide entry points for pathogens (Brummell and Harpster 2001).

Cuticles are composite of cutin and waxes that cover the epidermis of aerial plant organs, including fruit (Yeats and Rose 2013; Yeats et al. 2014). When reaching the plant surface, *B. cinerea* can secrete an assortment of cutinases and lipases to breach the cuticle and penetrate the host (Van Kan et al. 1997; Reis et al. 2005; Van Kan 2006; Leroch et al. 2013). The permeability of Arabidopsis and tomato leaf cuticles has been correlated with enhanced resistance to *B. cinerea*. Enhanced permeability of plant cuticles may favor the generation, movement and perception of pathogen-induced signals (e.g., reactive oxygen species, oligosaccharides, cutin monomers), which activate defence responses (Chassot et al. 2007; Bessire et al. 2007; Curvers et al. 2010; L'Haridon et al. 2011). However, defects in cuticle development, apart from permeability, could compromise defences against bacteria and *B. cinerea*, as reported in leaves of the Arabidopsis mutant *glabra1* (Xia et al. 2010).

When infecting tomato fruit, *B. cinerea* expresses a putative cutinase (*cut-like1*) (Blanco-Ulate et al. 2014). Since mutations in the previously characterized *B. cinerea cutA* gene do not reduce the pathogen's virulence in tomato fruit (Van Kan et al. 1997), the requirement of other *B. cinerea* cutinases, such as *cut-like1*, for successful infections should be studied.

The properties of the cuticles of ripe tomato fruit influence *B. cinerea* infections. The fruit cuticles of the *delayed fruit deterioration* (*dfd*) tomato mutant are stiffer and have higher wax contents than wild-type cuticles. *dfd* mutant fruit have minimal loss of water due to transpiration, substantially elevated cellular turgor and reduced susceptibility to *B. cinerea* (Saladie et al. 2007). The lipid composition of fruit cuticles is regulated by ripening (Kosma et al. 2010). Fruit from *cutin deficient* mutants (*cd1-3*) are more susceptible to *B. cinerea* (Isaacson et al. 2009).

The implications of the disassembly or degradation of plant cell walls during *B. cinerea* infections of vegetative and fruit tissues are examined in detail in Chap. 18. Here we highlight a few relevant points to relate fruit ripening and cell wall integrity with the susceptibility of ripened fruit to *B. cinerea*. As fruit ripen, the structure and composition of their cell walls change. Hemicellulose and cellulose networks loosen and are broken down, structural proteins associated with the cell wall are released or are no longer synthesized and pectins are degraded. These changes allow the walls around cells in the pericarp flesh and epidermis to expand and become hydrated, which increases the porosity of the extracellular matrix (Redgwell et al. 1997; Brummell and Harpster 2001; Vicente et al. 2007). The changes in fruit cell wall architecture during ripening may disrupt the deposition and localization of defence-related proteins and antimicrobial compounds, and is likely to favor the pathogen-induced degradation of wall polysaccharides. In unripe fruit, cells are enclosed within compacted walls. When infecting unripe tomato fruit, *B. cinerea* expresses a wide-range of cell wall modifying enzymes but *B. cinerea* also accelerates disassembly of the host cell wall (Cantu et al. 2009; Blanco-Ulate et al. 2014).

19.3.2 Metabolism and Metabolites

Transcription factors and other regulators of ripening influence basic metabolites, such as sugars, amino acids and organic acids, but these metabolites in turn also exert control over fruit development itself. For example, the accumulation of sucrose is needed for ABA synthesis and normal ripening in strawberries (Jia et al. 2013).

19.3.2.1 Sugars

The main apoplastic sugars in tomato fruit are glucose and fructose; sucrose is found in lower concentrations (Ruan et al. 1996; 2012). Soluble galacturonic acid, a monosaccharide derived from pectin breakdown, is detected in the ripe tomato fruit apoplast (Chun and Huber 1998). One might expect that soluble sugars in fruit cells can be carbon nutrients for pathogens. However, in tomato fruit, sugar accumulation is not a major factor in susceptibility, since *Cnr* mutant fruit, which are very susceptible to *B. cinerea*, have low levels of soluble solids, largely sugars (B. Blanco-Ulate pers. obs.). The susceptibility of *Cnr* mutant fruit may be mostly promoted by changes in the cell wall architecture. However, increased sugars in fruit with enhanced chloroplast development may be a factor that enhances susceptibility of ripe fruit (A.L.T. Powell, pers. obs.).

In response to microbial infections, plant cells increase extracellular invertases to enhance cleavage of sucrose to hexoses in the apoplast, ensuring that sufficient sugars are available as energy sources for the synthesis of defence responses (Doidy et al. 2012). Hexose sugars released by disassembly of the cell wall can also act as signals to induce the expression of defence-related genes and to repress photosynthesis (Roitsch et al. 2003) and sugars may function as primers for plant innate immunity (Bolouri et al. 2010; Bolouri and EndeVan den 2012). SlLIN5, an extracellular tomato fruit invertase, has been shown to alter hormone synthesis and signaling networks (e.g., ABA, JA, ethylene) and responses to pathogens (Zanor et al. 2009). Infections of unripe tomato fruit by *B. cinerea* induce the expression of *SlLIN5* and *SlLIN6*, which presumably alter hexose accumulation and impact JA-mediated responses (Cantu et al. 2009; Hyun et al. 2011).

B. cinerea has hexose transporters involved in the uptake of fructose (i.e. Ftr1/Hxt13) and galacturonic acid (Hxt15 and Hxt19). Mutants in these genes, including the double mutant *Δhxt15/Δhxt19*, show normal development and virulence on leaves when compared to the wild-type strain, suggesting that additional hexose transporters or other mechanisms may transport sugars (Doehlemann et al. 2005; Zhang et al. 2014). The role of *B. cinerea*'s hexose transporters during infections of fruit has yet to be described. *B. cinerea* has two functional hexokinases, Hxk1, and the glucokinase, Glk1. These enzymes phosphorylate hexoses and may participate in carbon source sensing. Hxk1 is needed during *B. cinerea* infections of tomato and apple fruit, plant organs, which have higher levels of fructose compared to vegetative tissues (Rui and Hahn 2007).

Galacturonic acid released by the disassembly of pectin polysaccharides (e.g., as fruit soften or are decomposed by the fungus) is metabolized by *B. cinerea*. Mutants in each step of the galacturonic acid catabolism pathway, *Δgar1/Δgar2*, *Δlgd1* and *Δlga1*, have reduced virulence on tobacco and Arabidopsis leaves (Zhang and Van Kan 2013), but no differences in virulence has been observed between the mutants and the wild-type strain on tomato leaves or apple and pepper fruit (Zhang et al. 2011). Preliminary results indicate that only the *Δlgd* mutant, affected in galactonate dehydratase activity, is less virulent in tomato fruit (A. L. T. Powell and B. Blanco-Ulate, pers. obs.).

19.3.2.2 pH

Most ripe fruit (e.g., tomato, strawberry, grape, apple, cherry and apricot) are acidic (Manteau et al. 2003). As tomato fruit ripen, the pH of the apoplastic fluid decreases from 6.7 in unripe to 4.4 in ripe fruit (Almeida and Huber 1999). The acidic environment of the fruit favors the production of proteases, polygalacturonases, laccases, extracellular polysaccharides and oxalic acid by *B. cinerea* (Manteau et al. 2003). However, acidification can also activate plant defences as reactive oxygen species are generated (Schaller and Oecking 1999; Nurnberger and Scheel 2001), so changes in intra- and extra-cellular pH alter host responses as well as infection functions of *B. cinerea* (Prusky et al. 2013).

19.3.2.3 Organic Acids

As tomato fruit develop before ripening, carboxylic acids from the TCA cycle raise the osmotic potential that drives cell growth and they are major sources of energy during ripening (Liu et al. 2007; Morgan et al. 2013). Increased citric acid in ripe fruit contributes to the synthesis of important biochemical intermediates and energy-rich molecules (Morgan et al. 2013). A role of citric acid during plant-*B. cinerea* interactions has not been described.

Malic acid/malate regulates starch biosynthesis and the accumulation of total soluble solids during tomato fruit ripening (Centeno et al. 2011). Ripe tomato fruit with high levels of malic acid (i.e. by suppression of malate dehydrogenase) have elevated water loss by transpiration, significantly reduced soluble sugars (glucose, fructose and sucrose) and enhanced susceptibility to *B. cinerea*. The altered osmotic potential in these lines seems to be the cause of both the water loss phenotype and the consequent pathogen susceptibility of the fruit (Centeno et al. 2011), although the direct influence of malate on *B. cinerea* virulence and metabolism has not been determined.

Oxalic acid (OA) accumulates as tomato fruit ripen. Changes in the concentrations of OA can cause changes in the levels of citric, malic and ascorbic acids. OA is a strong acid and a powerful Ca^{2+} chelator (Chakraborty et al. 2013). *B. cinerea* secretes significant amounts of OA when infecting leaves (Van Kan 2006). OA modifies the host environment and contributes to pathogenesis by: (1) acidification that favors *B. cinerea* virulence factors production; (2) strong Ca^{2+} chelation that weakens pectin interactions within the cell wall and inhibits some defence signaling pathways; and (3) inhibition of callose deposition. OA also may alter the production of ROS in a pH-dependent manner, inactivating host responses or, alternatively, inducing programmed cell death and susceptibility to *B. cinerea* (Prusky et al. 2013).

19.3.2.4 Amino Acids

Efficient regulation of host carbon/nitrogen metabolism is required for the accurate activation of defence-related pathways and to avoid unnecessary consumption of energy. In tomato leaves, increased cell viability and effective resistance against *B. cinerea* infections is achieved by replenishment of the TCA cycle by γ-aminobutyric acid (GABA) and glutamate metabolism (Seifi et al. 2013a, b). *B. cinerea* can alter primary amino acid metabolism in leaf hosts to induce susceptibility. It has been demonstrated that induction of asparagine synthetase after *B. cinerea* infections of tomato leaves promotes susceptibility by accumulation of asparagine, a rich source of nitrogen for the pathogen and by facilitating pathogen-induced host senescence (Seifi et al. 2014).

The levels of free amino acids increase as fruit ripen and GABA and glutamate are the most abundant amino acids in tomato fruit (Boggio et al. 2000; Rolin et al. 2000; Akihiro et al. 2008; Sorrequieta et al. 2010). The highest levels of GABA are in unripe fruit and they decrease rapidly later in development and during ripening (Carrari et al. 2006; Mounet et al. 2007). GABA is a non-protein amino acid that regulates the cytosolic pH, protects against oxidative stress and contributes to resistance against pathogens and insects (Akihiro et al. 2008).

Glutamate is the main free amino acid in ripe tomato fruit and provides the characteristic 'umami taste' of tomato (Bellisle 1999). The metabolism of glutamate is important for: (1) the transport of nitrogen via the glutamine synthetase and glutamine-oxoglutarate aminotransferase cycle, (2) cellular redox and (3) the reprograming of TCA-dependent energy (Seifi et al. 2013b). The changes in amino acid metabolism in tomato fruit during *B. cinerea* infections have not yet been investigated, but will be informative for determining whether similar mechanisms to those in leaves are present in resistant unripe fruit.

19.3.2.5 Reactive Oxygen Species, Antioxidants and Nitric Oxide

B. cinerea causes an oxidative burst as it penetrates the cuticle layer and colonizes hosts (Van Kan 2006). Hydrogen peroxide (H_2O_2) accumulates in the plasma membrane of invading hyphae and in the fungal cell wall (Schouten et al. 2002). *B. cinerea* not only produces reactive oxygen species (ROS) itself but also induces free radical production by the host tissue (Chap. 14). Plants commonly respond to infections with a rapid oxidative burst that induces localized cell death (e.g., the hypersensitive response), which can help to constrain infections by biotrophic pathogens, but also susceptibility to *B. cinerea* in some settings such as unripe fruit (Cantu et al. 2008, 2009). On susceptible hosts, this oxidative burst potentiates infections especially by necrotrophic pathogens, including *B. cinerea* (Lyon et al. 2007).

During tomato fruit ripening, the levels of ROS, lipid peroxidation and protein oxidation increase, while the expression and activity of the principal scavenging enzymes, such as superoxide dismutase, catalase and those associated with the ascorbate-glutathione cycle decrease (Lacan and Baccou 1998; Jimenez et al. 2002; Qin et al. 2009). In ripe tomato fruit, the accelerated accumulation of antioxidants, such as ascorbic acid and carotenoid pigments (e.g., lycopene, γ-carotene and β-carotene), may partially neutralize the increase in oxidative processes during ripening (Jimenez et al. 2002; Mondal et al. 2004).

Even though *B. cinerea* may take advantage of ROS production by the fruit to generate widespread oxidative stress and promote susceptibility (Heller and Tudzynski 2011; Chap. 14), precisely timed and localized generation of ROS can limit *B. cinerea* infections (Asselbergh et al. 2007). On unripe tomato fruit, localized accumulation of H_2O_2 is detected as a consequence of *B. cinerea* inoculation, but this response is not observed as result of mechanical wounding or in infected ripe tomato fruit, unless ABA is lacking (see above). The accumulation of H_2O_2 however correlated with resistance in unripe fruit (Cantu et al. 2008, 2009).

The accumulation of free radicals during *B. cinerea* infections depletes the antioxidant pool in the host tissues (Lyon et al. 2007). Thus, it can be anticipated that increasing antioxidants in fruit may limit the oxidative stress generated by *B. cinerea* infections and promote resistance (Van Baarlen et al. 2007). Interestingly, tomato fruit that are purple due to the engineered accumulation of anthocyanins, which have antioxidant activities (Butelli et al. 2008), are resistant to *B. cinerea* at all ripening stages and the fruit senesce extremely slowly. This delay in senescence and softening-related events in the purple tomato fruit probably contributes to their improved resistance (Zhang et al. 2013).

Nitric oxide (NO) is a diffusible free radical that delays fruit ripening and senescence in both climacteric and non-climacteric fruit by inhibiting ethylene pathways (Singh et al. 2013), but NO also influences H_2O_2 levels (Małolepszá and Różalska 2005). In general, endogenous levels of NO are significantly higher in unripe fruit than in ripe fruit (Leshem et al. 1998; Leshem 2000). During plant-pathogen interactions, NO cooperates with other signaling agents (e.g., Ca^{2+}, SA, JA and ethylene-mediated pathways) to induce programmed cell death and systemic acquired resistance (Mur et al. 2006; Bellin et al. 2013). In tomato fruit, MAPK kinases are involved in NO signaling during infections. In unripe tomato fruit, up-regulation of the transcripts encoding SlMAPK1, SlMAPK2 and SlMAPK3 is observed in response to *B. cinerea* infections and might be involved in establishing NO-induced responses (Zheng et al. 2014). However, it is not resolved whether NO is involved in the resistance of unripe tomato fruit to *B. cinerea* infections.

19.3.3 Defences in Ripening Fruit

Most plant tissues utilize some form of preformed and inducible defences to protect themselves against pathogen attack. The resistance of unripe fruit is associated with preformed antimicrobial compounds, inducible phytoalexins and PR proteins, which mostly accumulate in the cuticle or cell walls. During fruit ripening, most of these compounds either decline or become ineffective, perhaps also by losing their association with the wall (Prusky et al. 2013).

19.3.3.1 Secondary Metabolites

The leaves and peel of unripe tomato fruit are rich in the toxic steroidal alkaloid α–tomatine, a saponin that disrupts plasma membranes, causing leakage of electrolytes and depolarization of membrane potential (Itkin et al. 2011, 2013; Mintz-Oron et al. 2008). α-tomatine levels decline as tomato fruit ripen (Friedman 2002) and might inhibit growth of *B. cinerea* mycelia in unripe tomato fruit (Verhoeff and Liem 1975), however most *B. cinerea* strains can detoxify α–tomatine by hydrolysis of its sugar adduct (Quidde et al. 1998). Thus, α-tomatine may partially restrict *B. cinerea* infections of unripe tomato fruit, but it is not the only mechanism fruit use for resistance. Early studies reported that the sesquiterpene, rishitin, is produced in unripe tomato fruit as a consequence of *B. cinerea* infection and that it may boost the deleterious effects of α-tomatine on the fungus (Glazener and Wouters 1981).

Inoculations of tomato fruit with *B. cinerea*, up-regulate three hydroxycinnamoyl-CoA:tyramine N-(hydroxycinnamoyl) transferases (THTs) (Cantu et al. 2009), which are required for the production of *p*-coumaroyloctopamine and *p*-coumaroylnoradrenaline, two secondary metabolites with antimicrobial qualities (von Roepenack-Lahaye et al. 2003). The levels and effects of these metabolites during tomato fruit-*B. cinerea* interactions are not known.

In vitro and *in planta* studies have shown that volatile compounds produced by ripe fruit can inhibit *B. cinerea* growth (Vaughn and Gardner 1993; Vaughn et al. 1993). The major flavor volatiles, hexanal, 1-hexanol, (*E*)-2-hexen-1-ol, (*Z*)-6-nonenal, (*E*)-3-nonen-2-one, methyl salicylate and methyl benzoate are used for antifungal postharvest treatments of strawberry, blackberry and grape berries (Chap. 11; Archbold et al. 1997).

19.3.3.2 Pathogenesis-Related Proteins

Infections by *B. cinerea* induce the expression and accumulation of host PR proteins in plant cell walls and vacuoles (Van Loon and Van Strien 1999). Some PR proteins hydrolyze fungal cell walls (e.g., chitinases, β-1,3-glucanases, mannanases), inhibit fungal enzymes (e.g., polygalacturonase-inhibitor proteins) or have contact toxicity (e.g., thaumatin-like proteins and thionins) and others signal for defences (e.g., PR-1 and the defensin PDF1.2) (Van Loon and Van Strien 1999; Van Loon et al. 2006; Van Baarlen et al. 2007).

In contrast to other plant organs, uninfected fruit accumulate high concentrations of certain PR-like proteins (e.g., chitinases) as a consequence of ripening (Derckel et al. 1998; Cantu et al. 2009). During tomato fruit-*B. cinerea* interactions, PR proteins are induced both by ripening and infections. A β-1,3-glucanase gene (e.g., *TomQ'a*) is induced 1 day post-inoculation when unripe fruit is infected but no induction occurs in infections of ripe fruit, or as consequence of ripening (Cantu et al. 2009). Later in infections (3 days post-inoculation), enhanced expression and accumulation of PR-proteins, such as the putative thaumatin-like protein *SlPRP-23*, the possible chitinase, *SlChi-like1*, and the β-1,3-glucanases, *TomQ'b* and *TomB13GLUB*, was detected in both unripe and ripe fruit (Cantu et al. 2009; Shah et al. 2012). PR-proteins in fruit may not be effective in controlling *B. cinerea* infections since the *Cnr* mutation induces PR gene expression in fruit but the fruit are hyper-susceptible to *B. cinerea*. It has not been resolved whether the retention of PR proteins is impacted by disassembly of the host cell walls, by for example, ripening or *B. cinerea* infections.

19.3.4 Botrytis cinerea *Toxins in Fruit*

Plant cells, including those in fruit, are killed by *B. cinerea* not only through its extensive disassembly of the host cell wall matrix and its active release of ROS, but also through its production of toxins that lead to death of host cells (Chap. 15). *B. cinerea* can secrete toxic bicyclic sesquiterpenes (Collado et al. 1996; Rebordinos et al. 1996). Botrydial is a well-studied *B. cinerea* sesquiterpene, but other toxins with a botryane skeleton have also been described (Duran-Patron et al. 1999). From *in vitro* and *in planta* phytotoxicity assays, botrydial has the highest activity and effectively induces chlorosis and cell collapse in hosts (Colmenares et al. 2002). Botrytidial was detected when *B. cinerea* infected the ripe fruit of sweet pepper (Deighton et al. 2001). Whether fruit have the means to cope with the toxicity of botrydial is not known and more research is needed to understand whether these compounds have a role in susceptibility of fruit.

19.4 Perspectives

The aim of understanding the increase in susceptibility that accompanies ripening is to improve the quality of ripened fruit. However, it is clear that the ripening fruit-*B. cinerea* interaction is dynamic, involving multiple inputs and responses by the host as well as the pathogen (Fig. 19.4). By understanding what distinguishes ripening

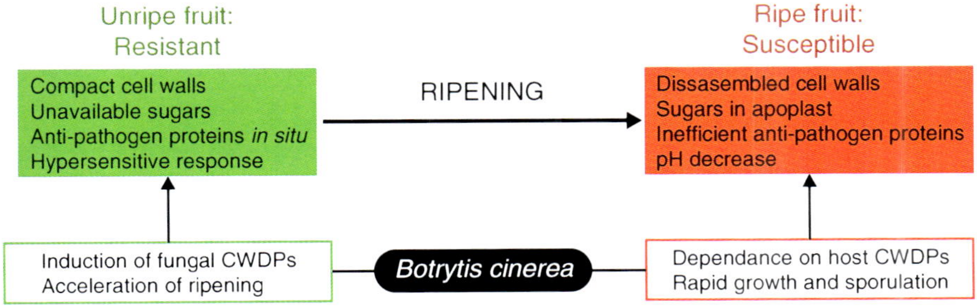

Fig. 19.4 Model depicting host and pathogen-derived regulators during ripening tomato fruit infections by *B. cinerea*. Events and conditions in unripe and ripe fruit that impact susceptibility are enumerated. Differences in *B. cinerea* infection strategies on unripe and ripe fruit are also shown

from susceptibility, new fruit varieties can be developed which ripen acceptably but are not predisposed to *B. cinerea* infections. Targeted reverse genetic approaches and evaluations of large mapping populations are needed to genetically dissect these complex traits in the fruit hosts. It will be especially important to identify how *B. cinerea* interferes with the regulation of these traits that occurs when healthy fruit ripen and how *B. cinerea* adapts to changing conditions in the host. Examining susceptibility in ripening fruit provides novel information about events that occur uniquely in fruit, but not in vegetative tissues, and has expanded knowledge about the versatility of *B. cinerea*-plant interactions.

References

Akihiro T, Koike S, Tani R et al (2008) Biochemical mechanism on GABA accumulation during fruit development in tomato. Plant Cell Physiol 49:1378–1389

Almeida DPF, Huber DJ (1999) Apoplastic pH and inorganic ion levels in tomato fruit: a potential means for regulation of cell wall metabolism during ripening. Physiol Plant 105:506–512

Archbold DD, Hamilton-Kemp TR, Barth MM et al (1997) Identifying natural volatile compounds that control gray mold (*Botrytis cinerea*) during postharvest storage of strawberry, blackberry and grape. J Agric Food Chem 45:4032–4037

Arie T, Takahashi H, Kodama M et al (2007) Tomato as a model plant for plant-pathogen interactions. Plant Biotechnol 24:135–147

Asselbergh B, Curvers K, Franca SC et al (2007) Resistance to *Botrytis cinerea* in *sitiens*, an abscisic acid-deficient tomato mutant, involves timely production of hydrogen peroxide and cell wall modifications in the epidermis. Plant Physiol 144:1863–1877

Baur AH, Yang SF, Pratt HK et al (1971) Ethylene biosynthesis in fruit tissues. Plant Physiol 47:696–699

Beckers GJM, Spoel SH (2006) Fine-tuning plant defence signalling: salicylate versus jasmonate. Plant Biol 8:1–10

Bellin D, Asai S, Delledonne M et al (2013) Nitric oxide as a mediator for defence responses. Mol Plant Microbe Interact 26:271–277

Bellisle F (1999) Glutamate and the UMAMI taste: sensory, metabolic, nutritional and behavioural considerations. A review of the literature published in the last 10 years. Neurosci Biobehav Rev 23:423–438

Bessire M, Chassot C, Jacquat AC et al (2007) A permeable cuticle in Arabidopsis leads to a strong resistance to *Botrytis cinerea*. EMBO J 26:2158–2168

Blanco-Ulate B (2014) Developmentally Regulated Susceptibility of Tomato Fruit to Botrytis cinerea.PhD dissertation, University of California, Davis

Blanco-Ulate B, Vincenti E, Powell AL et al (2013) Tomato transcriptome and mutant analyses suggest a role for plant stress hormones in the interaction between fruit and *Botrytis cinerea*. Front Plant Sci 4:142

Blanco-Ulate B, Morales-Cruz A, Amrine KCH et al (2014) Genome-wide transcriptional profiling of *Botrytis cinerea* genes targeting plant cell walls during infections of different hosts. Front Plant Sci 5:435

Boggio SB, Palatnik JF, Heldt HW et al (2000) Changes in amino acid composition and nitrogen metabolizing enzymes in ripening fruits of *Lycopersicon esculentum* mill. Plant Sci 159:125–133

Bolouri Moghaddam MR, Van den Ende W (2012) Sugars and plant innate immunity. J Exp Bot 63:3989–3998

Bolouri-Moghaddam MR, Le Roy K, Xiang L et al (2010) Sugar signalling and antioxidant network connections in plant cells. FEBS J 277:2022–2037

Böttcher C, Keyzers RA, Boss PK et al (2010) Sequestration of auxin by the indole-3-acetic acid-amido synthetase GH3-1 in grape berry (*Vitis vinifera* L.) and the proposed role of auxin conjugation during ripening. J Exp Bot 61:3615–3625

Böttcher C, Burbidge CA, Boss PK et al (2013) Interactions between ethylene and auxin are crucial to the control of grape (*Vitis vinifera* L.) berry ripening. BMC Plant Biol 13:222

Brading PA, Hammond-Kosack KE, Parr A, Jones JD (2000) Salicylic acid is not required for Cf-2- and Cf-9-dependent resistance of tomato to Cladosporium fulvum. Plant J 23:305–318

Brady CJ (1987) Fruit ripening. Annu Rev Plant Physiol 38:155–178

Browse J (2009) Jasmonate passes muster: a receptor and targets for the defence hormone. Annu Rev Plant Biol 60:183–205

Brummell DA, Harpster MH (2001) Cell wall metabolism in fruit softening and quality and its manipulation in transgenic plants. Plant Mol Biol 47:311–339

Butelli E, Titta L, Giorgio M et al (2008) Enrichment of tomato fruit with health-promoting anthocyanins by expression of select transcription factors. Nat Biotechnol 26:1301–1308

Cantu D, Vicente A, Greve L et al (2008) The intersection between cell wall disassembly, ripening and fruit susceptibility to *Botrytis cinerea*. Proc Natl Acad Sci U S A 105:859–864

Cantu D, Blanco-Ulate B, Yang L et al (2009) Ripening-regulated susceptibility of tomato fruit to *Botrytis cinerea* requires NOR but not RIN or ethylene. Plant Physiol 150:1434–1449

Carrari F, Baxter C, Usadel B et al (2006) Integrated analysis of metabolite and transcript levels reveals the metabolic shifts that underlie tomato fruit development and highlight regulatory aspects of metabolic network behavior. Plant Physiol 142:1380–1396

Centeno DC, Osorio S, Nunes-Nesi A et al (2011) Malate plays a crucial role in starch metabolism, ripening and soluble solid content of tomato fruit and affects postharvest softening. Plant Cell 23:162–184

Chague V, Elad Y, Barakat R et al (2002) Ethylene biosynthesis in *Botrytis cinerea*. FEMS Microbiol Ecol 40:143–149

Chakraborty N, Ghosh R, Ghosh S et al (2013) Reduction of oxalate levels in tomato fruit and consequent metabolic remodeling following overexpression of a fungal oxalate decarboxylase. Plant Physiol 162:364–378

Chassot C, Nawrath C, Métraux J-P (2007) Cuticular defects lead to full immunity to a major plant pathogen. Plant J 49:972–980

Chen X (2009) Small RNAs and their roles in plant development. Annu Rev Cell Dev Biol 25:21–44

Chen W, Kong J, Qin C et al (2015) Requirement of CHROMOMETHYLASE3 for somatic inheritance of the spontaneous tomato epimutation *Colourless non-ripening*. Sci Rep 5:9192

Chun J-P, Huber DJ (1998) Polygalacturonase-mediated solubilization and depolymerization of pectic polymers in tomato fruit cell walls: regulation by pH and ionic conditions. Plant Physiol 117:1293–1299

Chung MY, Vrebalov J, Alba R, Lee J et al (2010) A tomato (*Solanum lycopersicum*) *APETALA2/ERF* gene, *SlAP2a*, is a negative regulator of fruit ripening. Plant J 64:936–947

Collado IG, Hernandez Galan R, Prieto V et al (1996) Biologically active sesquiterpenoid metabolites from the fungus *Botrytis cinerea*. Phytochemistry 41:513–517

Colmenares AJ, Duran-Patron RM, Hernandez-Galan R et al (2002) Four new lactones from *Botrytis cinerea*. J Nat Prod 65:1724–1726

Cristescu SM, De Martinis D, Te Lintel Hekkert S et al (2002) Ethylene production by *Botrytis cinerea* in vitro and in tomatoes. Appl Environ Microbiol 68:5342–5350

Curvers K, Seifi H, Mouille G et al (2010) ABA-deficiency causes changes in cuticle permeability and pectin composition that influence tomato resistance to *Botrytis cinerea*. Plant Physiol 154:847–860

De Bruyne L, Hofte M, De Vleesschauwer (2014) Connecting growth and defense: the emergingroles of brassinosteroids and gibberellins in plant innate immunity. Mol Plant 7:943–959

Deighton N, Muckenschnabel I, Colmenares AJ et al (2001) Botrydial is produced in plant tissues infected by *Botrytis cinerea*. Phytochemistry 57:689–692

Derckel J-P, Audran J-C, Haye B et al (1998) Characterization, induction by wounding and salicylic acid and activity against *Botrytis cinerea* of chitinases and β-1,3-glucanases of ripening grape berries. Physiol Plant 104:56–64

Diaz J, ten Have A, Van Kan AL (2002) The role of ethylene and wound signaling in resistance of tomato to *Botrytis cinerea*. Plant Physiol 129:1341–1351

Doehlemann G, Molitor F, Hahn M (2005) Molecular and functional characterization of a fructose specific transporter from the gray mold fungus *Botrytis cinerea*. Fungal Genet Biol 42:601–610

Doidy J, Grace E, Kuhn C et al (2012) Sugar transporters in plants and in their interactions with fungi. Trends Plant Sci 17:413–422

Dong T, Hu Z, Deng L et al (2013) A tomato MADS-box transcription factor, SlMADS1, acts as a negative regulator of fruit ripening. Plant Physiol 163:1026–1036

Dowen RH, Pelizzola M, Schmitz RJ et al (2012) Widespread dynamic DNA methylation in response to biotic stress. Proc Natl Acad Sci U S A 109:E2183–E2191

Duran-Patron R, Hernandez-Galan R, Rebordinos LG et al (1999) Structure-activity relationships of new phytotoxic metabolites with the botryane skeleton from *Botrytis cinerea*. Tetrahedron 55:2389–2400

Durrant WE, Dong X (2004) Systemic acquired resistance. Annu Rev Phytopathol 42:185–209

El Oirdi M, El Rahman TA, Rigano L et al (2011) *Botrytis cinerea* manipulates the antagonistic effects between immune pathways to promote disease development in tomato. Plant Cell 23:2405–2421

Eriksson EM, Bovy A, Manning K et al (2004) Effect of the colorless non-ripening mutation on cell wall biochemistry and gene expression during tomato fruit development and ripening. Plant Physiol 136:4184–4197

Flors V, Leyva M, Vicedo B et al (2007) Absence of the endo-β-1,4-glucanses Cel1 and Cel2 reduces susceptibility to *Botrytis cinerea* in tomato. Plant J 52:1027–1040

Fraser PD, Bramley P, Seymour GB (2001) Effect of the *Cnr* mutation on carotenoid formation during tomato fruit ripening. Phytochemistry 58:75–79

Friedman M (2002) Tomato glycoalkaloids: role in the plant and in the diet. J Agric Food Chem 50:5751–5780

Fu J, Wang S (2011) Insights into auxin signaling in plant-pathogen interactions. Front Plant Sci 2:74

Fujisawa M, Nakano T, Shima Y et al (2013) A large-scale identification of direct targets of the tomato MADS Box transcription factor RIPENING INHIBITOR reveals the regulation of fruit ripening. Plant Cell 25:371–386

Fujisawa M, Shima Y, Nakagawa H et al (2014) Transcriptional regulation of fruit ripening by tomato FRUITFULL homologs and associated MADS box proteins. Plant Cell 26:89–101

Fujita M, Fujita Y, Noutoshi Y et al (2006) Crosstalk between abiotic and biotic stress responses: a current view from the points of convergence in the stress signaling networks. Curr Opin Plant Biol 9:436–442

Glazebrook J (2005) contrasting mechanisms of defence against biotrophic and necrotrophic pathogens. Annu Rev Phytopathol 43:205–227

Glazener JA, Wouters CH (1981) Detection of rishitin in tomato fruits after infection with *Botrytis cinerea*. Physiol Plant Pathol 19:243–248

Halliday K (2004) Plant hormones: the interplay of brassinosteroids and auxin. Current Biol 14:R1008–R1010

Harrison E, Burbidge A, Okyere JP et al (2011) Identification of the tomato ABA-deficient mutant *sitiens* as a member of the ABA-aldehyde oxidase gene family using genetic and genomic analysis. Plant Growth Regul 64:301–309

Heller J, Tudzynski P (2011) Reactive oxygen species in phytopathogenic fungi: signaling, development and disease. Annu Rev Phytopathol 49:369–390

Hyun T, Eom S, Rim Y et al (2011) Alteration of the expression and activation of tomato invertases during *Botrytis cinerea* infection. Plant Omics 4:413–417

Isaacson T, Kosma DK, Matas AJ et al (2009) Cutin deficiency in the tomato fruit cuticle consistently affects resistance to microbial infection and biomechanical properties, but not transpirational water loss. Plant J 60:363–377

Itaya A, Bundschuh R, Archual AJ et al (2008) Small RNAs in tomato fruit and leaf development. BBA Genet Regul Mech 1779:99–107

Itkin M, Rogachev I, Alkan N et al (2011) GLYCOALKALOID METABOLISM1 is required for steroidal alkaloid glycosylation and prevention of phytotoxicity in tomato. Plant Cell 23:4507–4525

Itkin M, Heinig U, Tzfadia O et al (2013) Biosynthesis of antinutritional alkaloids in solanaceous crops is mediated by clustered genes. Science 341:175–179

Jia H-F, Chai Y-M, Li C-L et al (2011) Abscisic acid plays an important role in the regulation of strawberry fruit ripening. Plant Physiol 157:188–199

Jia H-F, Wang Y, Sun M et al (2013) Sucrose functions as a signal involved in the regulation of strawberry fruit development and ripening. New Phytol 198:453–465

Jimenez A, Creissen G, Kular BP et al (2002) Changes in oxidative processes and components of the antioxidant system during tomato fruit ripening. Planta 214:751–758

Jin W, Wu F, Xiao L et al (2012) Microarray-based analysis of tomato miRNA regulated by *Botrytis cinerea*. J Plant Growth Regul 31:38–46

Karlova R, Rosin F, Busscher-Lange J et al (2011) Transcriptome and metabolite profiling show that APETALA2a is a major regulator of tomato fruit ripening. Plant Cell 23:923–941

Karlova R, Van Haarst JC, Maliepaard C et al (2013) Identification of microRNA targets in tomato fruit development using high-throughput sequencing and degradome analysis. J Exp Bot 64:1863–1878

Klee HJ, Giovannoni JJ (2011) Genetics and control of tomato fruit ripening and quality attributes. Annu Rev Genet 45:41–59

Kosma DK, Parsons EP, Isaacson T et al (2010) Fruit cuticle lipid composition during development in tomato ripening mutants. Physiol Plant 139:107–117

Koyama K, Sadamatsu K, Goto-Yamamoto N (2010) Abscisic acid stimulated ripening and gene expression in berry skins of the Cabernet Sauvignon grape. Funct Integr Genomic 10:367–381

Kumar R, Agarwal P, Tyagi A et al (2012) Genome-wide investigation and expression analysis suggest diverse roles of auxin-responsive GH3 genes during development and response to different stimuli in tomato (*Solanum lycopersicum*). Mol Genet Genomics 287:221–235

L'Haridon F, Besson-Bard A, Binda M et al (2011) A permeable cuticle is associated with the release of reactive oxygen species and induction of innate immunity. PLoS Pathog 7:e1002148

Lacan D, Baccou J-C (1998) High levels of antioxidant enzymes correlate with delayed senescence in nonnetted muskmelon fruits. Planta 204:377–382

Lara I, Belge B, Goulao LF (2014) The fruit cuticle as a modulator of postharvest quality. Postharvest Biol Technol 87:103–112

Laluk K, Luo H, Chai M et al (2011) Biochemical and genetic requirements for the function of the immune response regulator BOTRYTIS-INDUCED KINASE1 in plant growth, ethylene signaling and PAMP-triggered immunity in Arabidopsis. Plant Cell 23:2831–2849

Lee IC, Hong SW, Whang SS et al (2011) Age-dependent action of an ABA-inducible receptor kinase, RPK1, as a positive regulator of senescence in Arabidopsis leaves. Plant Cell Physiol 52:651–662

Lee JM, Joung J-G, McQuinn R et al (2012) Combined transcriptome, genetic diversity and metabolite profiling in tomato fruit reveals that the ethylene response factor SlERF6 plays an important role in ripening and carotenoid accumulation. Plant J 70:191–204

Leon-Reyes A, Du Y, Koornneef A et al (2010) Ethylene signaling renders the jasmonate response of Arabidopsis insensitive to future suppression by salicylic acid. Mol Plant Microbe Interact 23:187–197

Leroch M, Kleber A, Silva E et al (2013) Transcriptome profiling of *Botrytis cinerea* conidial germination reveals upregulation of infection-related genes during the prepenetration stage. Eukaryot Cell 12:614–626
Leshem YY (2000) Nitric oxide in plants. Occurrence, function and use. Kluwer Academic Publishers, Dordrecht, pp 1–154
Leshem YY, Wills RBH, Ku VV-V (1998) Evidence for the function of the free radical gas -nitric oxide (NO•)- as an endogenous maturation and senescence regulating factor in higher plants. Plant Physiol Biochem 36:825–833
Lin W, Lu D, Gao X et al (2013) Inverse modulation of plant immune and brassinosteroid signaling pathways by the receptor-like cytoplasmic kinase BIK1. Proc Natl Acad Sci U S A 110:21114–12119
Lin Z, Hong Y, Yin M et al (2008) A tomato HD-Zip homeobox protein, LeHB-1, plays an important role in floral organogenesis and ripening. Plant J 55:301–310
Liu Y, Liu Q, Xiong J et al (2007) Difference of a citrus late-ripening mutant (*Citrus sinensis*) from its parental line in sugar and acid metabolism at the fruit ripening stage. Sci China C Life Sci 50:511–517
Liu R, How-Kit A, Stammitti L et al (2015) A DEMETER-like DNA demethylase governs tomato fruit ripening. Proc Natl Acad Sci USA 112:10804-10809
Lyon G, Goodman B, Williamson B (2007) *Botrytis cinerea* perturbs redox processes as an attack strategy in plants. In: Elad Y, Williamson B, Tudzynski P, Delen N (eds) *Botrytis*: biology, pathology and control. Springer, Dordrecht, pp 119–141
Małolepszá U, Różalska S (2005) Nitric oxide and hydrogen peroxide in tomato resistance. Nitric oxide modulates hydrogen peroxide level in *o*-hydroxyethylorutin-induced resistance to *Botrytis cinerea* in tomato. Plant Physiol Biochem 43:623–635
Manteau S, Abouna S, Lambert B, Legendre L (2003) Differential regulation by ambient pH of putative virulence factor secretion by the phytopathogenic fungus Botrytis cinerea. FEMS Microbiol Ecol 43:359–366
Manning K, Tor M, Poole M et al (2006) A naturally occurring epigenetic mutation in a gene encoding an SBP-box transcription factor inhibits tomato fruit ripening. Nat Genet 38:948–952
Marcos JF, González-Candelas L, Zacarías L (2005) Involvement of ethylene biosynthesis and perception in the susceptibility of citrus fruits to *Penicillium digitatum* infection and the accumulation of defence-related mRNAs. J Exp Bot 56:2183–2193
Martel C, Vrebalov J, Giovannoni J (2011) The tomato (*Solanum lycopersicum*) MADS-box transcription factor RIN interacts with promoters involved in numerous ripening processes in a CNR dependent manner. Plant Physiol 157:1568–1579
Mauch-Mani B, Mauch F (2005) The role of abscisic acid in plant–pathogen interactions. Curr Opin Plant Biol 8:409–414
Mintz-Oron S, Mandel T, Rogachev I et al (2008) Gene expression and metabolism in tomato fruit surface tissues. Plant Physiol 147:823–851
Mohorianu I, Schwach F, Jing R et al (2011) Profiling of short RNAs during fleshy fruit development reveals stage-specific sRNAome expression patterns. Plant J 67:232–246
Mondal K, Sharma NS, Malhotra SP et al (2004) Antioxidant systems in ripening tomato fruits. Biol Plant 48:49–53
Morgan MJ, Osorio S, Gehl B et al (2013) Metabolic engineering of tomato fruit organic acid content guided by biochemical analysis of an introgression line. Plant Physiol 161:397–407
Mounet F, Lemaire-Chamley M, Maucourt M et al (2007) Quantitative metabolic profiles of tomato flesh and seeds during fruit development: complementary analysis with ANN and PCA. Metabolomics 3:273–288
Moxon S, Jing R, Szittya G et al (2008) Deep sequencing of tomato short RNAs identifies microRNAs targeting genes involved in fruit ripening. Genome Res 18:1602–1609
Mur LA, Kenton P, Atzorn R et al (2006) The outcomes of concentration-specific interactions between salicylate and jasmonate signaling include synergy, antagonism and oxidative stress leading to cell death. Plant Physiol 140:249–262

Nadakudti SS, Holdsworth WL, Klein CL et al (2014) *KNOX* genes influence a gradient of fruit chloroplast development through regulation of *GOLDEN2-LIKE* expression in tomato. Plant J 78:1022–1033

Narayanasamy P (2006) Integrated systems for the management of postharvest diseases. In: Postharvest pathogens and disease management. Wiley, Hoboken, pp 537–554

Nambeesan S, Abuqamar S, Laluk K et al (2012) Polyamines attenuate ethylene-mediated defence responses to abrogate resistance to *Botrytis cinerea* in tomato. Plant Physiol 158:1034–1045

Nguyen CV, Vrebalov JT, Gapper NE et al (2014) Tomato GOLDEN2-LIKE transcription factors reveal molecular gradients that function during fruit development and ripening. Plant Cell 26:585–601

Nürnberger T, Scheel D (2001) Signal transmission in the plant immune response. Trends Plant Sci 6:372–379

Ordaz-Ortiz JJ, Marcus SE, Knox JP (2009) Cell wall microstructure analysis implicates hemicellulose polysaccharides in cell adhesion in tomato fruit pericarp parenchyma. Mol Plant 2:910–921

Orfila C, Huisman M, Willats W et al (2002) Altered cell wall disassembly during ripening of *Cnr* tomato fruit: implications for cell adhesion and fruit softening. Planta 215:440–447

Osorio S, Alba R, Damasceno CM et al (2011) Systems biology of tomato fruit development: combined transcript, protein and metabolite analysis of tomato transcription factor (*nor*, *rin*) and ethylene receptor (*Nr*) mutants reveals novel regulatory interactions. Plant Physiol 157:405–425

Osorio S, Alba R, Nikoloski Z et al (2012) Integrative comparative analyses of transcript and metabolite profiles from pepper and tomato ripening and development stages uncovers species-specific patterns of network regulatory behavior. Plant Physiol 159:1713–1729

Osorio S, Scossa F, Fernie AR (2013) Molecular regulation of fruit ripening. Front Plant Sci 4:198

Pan H, Bradley G, Pyke K et al (2013) Network inference analysis identifies and *APRR2-Like* gene linked to pigment accumulation in tomato and pepper fruits. Plant Physiol 161:1476–1485

Pech J-C, Purgatto E, Bouzayen M et al (2012) Ethylene and fruit ripening. In: McManus MT (ed) Annual plant reviews, vol 44: The plant hormone ethylene. Wiley-Blackwell, Oxford, pp 275–304

Peña-Cortés H, Barrios P, Dorta F et al (2004) Involvement of jasmonic acid and derivatives in plant response to pathogen and insects and in fruit ripening. Plant Growth Regul 23:246–260

Pieterse CM, Leon-Reyes A, Van der Ent S, Van Wees SC (2009) Networking by small-molecule hormones in plant immunity. Nat Chem Biol 5:308–316

Powell ALT, Van Kan JAL, ten Have A et al (2000) Transgenic expression of pear PGIP in tomato limits fungal colonization. Mol Plant Microbe Interact 13:942–950

Powell AL, Nguyen CV, Hill T et al (2012) Uniform ripening encodes a Golden 2-like transcription factor regulating tomato fruit chloroplast development. Science 336:1711–1715

Prusky D (1996) Pathogen quiescence in postharvest diseases. Annu Rev Phytopathol 34:413–434

Prusky D, Alkan N, Mengiste T et al (2013) Quiescent and necrotrophic lifestyle choice during postharvest disease development. Annu Rev Phytopathol 51:155–176

Qin G, Meng X, Wang Q et al (2009) Oxidative damage of mitochondrial proteins contributes to fruit senescence: a redox proteomics analysis. J Proteome Res 8:2449–2462

Qin G, Wang Y, Cao B et al (2012) Unraveling the regulatory network of the MADS box transcription factor RIN in fruit ripening. Plant J 70:243–255

Quidde T, Osbourn AE, Tudzynski P (1998) Detoxification of α-tomatine by *Botrytis cinerea*. Physiol Mol Plant Pathol 52:151–165

Rasul S, Dubreuil-Maurizi C, Lamotte O et al (2012) Nitric oxide production mediates oligogalacturonide-triggered immunity and resistance to *Botrytis cinerea* in *Arabidopsis thaliana*. Plant Cell Environ 35:1483–1499

Rebordinos L, Cantoral JM, Prieto MV et al (1996) The phytotoxic activity of some metabolites of *Botrytis cinerea*. Phytochemistry 42:383–387

Redgwell RJ, MacRae E, Hallett I et al (1997) *In vitro* and *in vivo* swelling of cell walls during fruit ripening. Planta 203:162–172

Reis H, Pfiffi S, Hahn M (2005) Molecular and functional characterization of a secreted lipase from *Botrytis cinerea*. Mol Plant Pathol 6:257–267

Robert-Seilaniantz A, Grant M, Jones JDG (2011) Hormone crosstalk in plant disease and defence: more than just jasmonate-salycilate antagonism. Annu Rev Phytopathol 49:317–343

Roitsch T, Balibrea M, Hofmann M et al (2003) Extracellular invertase: key metabolic enzyme and PR protein. J Exp Bot 54:513–524

Rolin D, Baldet P, Just D et al (2000) NMR study of low subcellular pH during the development of cherry tomato fruit. Funct Plant Biol 27:61–69

Rowe HC, Walley JW, Corwin J et al (2010) Deficiencies in jasmonate-mediated plant defence reveal quantitative variation in *Botrytis cinerea* pathogenesis. PLoS Pathog 6:e1000861

Ruan Y, Patrick JW, Bouzayen M, Osorio S, Fernie A (2012) Molecular regulation of seed and fruit set. Trends Plant Sci 17:656–666

Ruan Y-L, Patrick J, Brady C (1996) The composition of apoplast fluid recovered from intact developing tomato fruit. Aust J Plant Phys 23:9–13

Rui O, Hahn M (2007) The *Botrytis cinerea* hexokinase, Hxk1, but not the glucokinase, Glk1, is required for normal growth and sugar metabolism and for pathogenicity on fruits. Microbiology 153:2791–2802

Ruiz-Ferrer V, Voinnet O (2009) Roles of plant small RNAs in biotic stress responses. Annu Rev Plant Biol 60:485–510

Sagar M, Chervin C, Mila I et al (2013) SlARF4, an auxin response factor involved in the control of sugar metabolism during tomato fruit development. Plant Physiol 161:1362–1374

Saladie M, Matas AJ, Isaacson T et al (2007) A reevaluation of the key factors that influence tomato fruit softening and integrity. Plant Physiol 144:1012–1028

Sánchez-Vallet A, López G, Ramos B et al (2012) Disruption of abscisic acid signalling constitutively activates Arabidopsis resistance to the necrotrophic fungus *Plectosphaerella cucumerina*. Plant Physiol 160:2109–2124

Savatin DV, Ferrari S, Sicilia F et al (2011) Oligogalacturonide-auxin antagonism does not require posttranscriptional gene silencing or stabilization of auxin response repressors in Arabidopsis. Plant Physiol 157:1163–1174

Schaller A, Oecking C (1999) Modulation of plasma membrane H^+-ATPase activity differentially activates wound and pathogen defence responses in tomato plants. Plant Cell 11:263–272

Schouten A, Tenberge KB, Vermeer J et al (2002) Functional analysis of an extracellular catalase of *Botrytis cinerea*. Mol Plant Pathol 3:227–238

Seifi HS, Curvers K, De Vleesschauwer D et al (2013a) Concurrent overactivation of the cytosolic glutamine synthetase and the GABA shunt in the ABA-deficient *sitiens* mutant of tomato leads to resistance against *Botrytis cinerea*. New Phytol 199:490–504

Seifi HS, Van Bockhaven J, Angenon G et al (2013b) Glutamate metabolism in plant disease and defence: friend or foe? Mol Plant Microbe Interact 26:475–485

Seifi HS, De Vleesschauwer D, Aziz A et al (2014) Modulating plant primary amino acid metabolism as a necrotrophic virulence strategy: the immune-regulatory role of asparagine synthetase in *Botrytis cinerea*-tomato interaction. Plant Signal Behav 9:e27995

Seymour GB, Poole M, Manning K et al (2008) Genetics and epigenetics of fruit development and ripening. Curr Opin Plant Biol 11:58–63

Seymour GB, Østergaard L, Chapman NH et al (2013) Fruit development and ripening. Annu Rev Plant Biol 64:219–241

Shackel KA, Greve C, Labavitch JM et al (1991) Cell turgor changes associated with ripening in tomato pericarp tissue. Plant Physiol 97:814–816

Shah P, Powell AL, Orlando R, Bergmann C, Gutierrez-Sanchez G (2012) Proteomic analysis of ripening tomato fruit infected by Botrytis cinerea. J Proteome Res 11:2178–2192

Sharon A, Elad Y, Barakat R et al (2007) Phytohormones in *Botrytis*-plant interactions. In: Elad Y, Williamson B, Tudzynski P, Delen N (eds) *Botrytis*: Biology, Cathology and Control. Springer, Dordrecht, pp 163–179

Shi JX, Adato A, Alkan N et al (2013) The tomato SlSHINE3 transcription factor regulates fruit cuticle formation and epidermal patterning. New Phytol 197:468–480
Shima Y, Kitagawa M, Fujisawa M et al (2013) Tomato FRUITFULL homologues act in fruit ripening via forming MADS-box transcription factor complexes with RIN. Plant Mol Biol 82:427–438
Singh Z, Khan AS, Zhu S et al (2013) Nitric oxide in the regulation of fruit ripening: challenges and thrusts. Stewart Postharvest Rev 9:1–11
Sorrequieta A, Ferraro G, Boggio SB et al (2010) Free amino acid production during tomato fruit ripening: a focus on L-glutamate. Amino Acids 38:1523–1532
Soto A, Ruiz KB, Ravaglia D et al (2013) ABA may promote or delay peach fruit ripening through modulation of ripening- and hormone-related gene expression depending on the developmental stage. Plant Physiol Biochem 64:11–24
Spoel SH, Dong X (2008) Making sense of hormone crosstalk during plant immune responses. Cell Host Microbe 3:348–351
Swartzberg D, Kirshner B, Rav-David D et al (2008) *Botrytis cinerea* induces senescence and is inhibited by autoregulated expression of the *IPT* gene. Eur J Plant Pathol 120:289–297
Symons GM, Davies C, Shavrukov Y et al (2012) Grapes on steroids. Brassinosteroids are involved in grape berry ripening. Plant Physiol 140:150–158
Tomato Genome Consortium (2012) The tomato genome sequence provides insights into fleshy fruit evolution. Nature 485:635–641
Ton J, Flors V, Mauch-Mani B (2009) The multifaceted role of ABA in disease resistance. Trends Plant Sci 14:310–317
Van Baarlen P, Legendre L, Kan JL (2007) Plant defence compounds against *Botrytis* infection. In: Elad Y, Williamson B, Tudzynski P, Delen N (eds) *Botrytis*: Biology, Pathology and Control. Springer, Dordrecht, pp 143–161
Van der Ent S, Pieterse CMJ (2012) Ethylene: multi-tasker in plant–attacker interactions. Annu Plant Rev 44:343–377
Van Kan JAL (2006) Licensed to kill: the lifestyle of a necrotrophic plant pathogen. Trends Plant Sci 11:247–253
Van Kan JAL, Van't Klooster JW, Wagemakers CAM et al (1997) Cutinase A of *Botrytis cinerea* is expressed, but not essential, during penetration of gerbera and tomato. Mol Plant Microbe Interact 10:30–38
Van Loon LC, Van Strien EA (1999) The families of pathogenesis-related proteins, their activities and comparative analysis of PR-1 type proteins. Physiol Mol Plant Pathol 55:85–97
Van Loon LC, Rep M, Pieterse CMJ (2006) Significance of inducible defence-related proteins in infected plants. Annu Rev Phytopathol 44:135–162
Vardhini VB, Rao SSR (2002) Acceleration of ripening of tomato pericarp discs by brassinosteroids. Phytochemistry 61:843–847
Vaughn SF, Gardner HW (1993) Lipoxygenase-derived aldehydes inhibit fungi pathogenic on soybean. J Chem Ecol 19:2337–2345
Vaughn SF, Spencer GF, Shasha BS (1993) Volatile compounds from raspberry and strawberry fruit inhibit postharvest decay fungi. J Food Sci 58:793–796
Verhoeff K, Liem JI (1975) Toxicity of tomatine to *Botrytis cinerea*, in relation to latency. J Phytopathol 82:333–338
Vicente AR, Saladié M, Rose JKC et al (2007) The linkage between cell wall metabolism and fruit softening: looking to the future. J Sci Food Agric 87:1435–1448
Vlot AC, Dempsey DMA, Klessig DF (2009) Salicylic acid, a multifaceted hormone to combat disease. Annu Rev Phytopathol 47:177–206
von Roepenack-Lahaye E, Newman M-A, Schornack S et al (2003) *p*-Coumaroylnoradrenaline, a novel plant metabolite implicated in tomato defence against pathogens. J Biol Chem 278:43373–43383
Vrebalov J, Ruezinsky D, Padmanabhan V et al (2002) A MADS-Box gene necessary for fruit ripening at the tomato ripening-inhibitor (*RIN*) locus. Science 296:343–346

Wasternack C (2007) Jasmonates: an update on biosynthesis, signal transduction and action in plant stress response, growth and development. Ann Bot 100:681–697

Weiberg A, Wang M, Lin F-M et al (2013) Fungal small RNAs suppress plant immunity by hijacking host RNA interference pathways. Science 342:118–123

Xia Y, Yu K, Navarre D et al (2010) The *glabra1* mutation affects cuticle formation and plant responses to microbes. Plant Physiol 154:833–846

Yang L-P, Fang Y-Y, An C-P et al (2013) C2-mediated decrease in DNA methylation, accumulation of siRNAs and increase in expression for genes involved in defence pathways in plants infected with beet severe curly top virus. Plant J 73:910–917

Yasuda M, Ishikawa A, Jikumaru Y et al (2008) Antagonistic interaction between systemic acquired resistance and the abscisic acid–mediated abiotic stress response in Arabidopsis. Plant Cell 20:1678–1692

Yeats TH, Rose JKC (2013) The formation and function of plant cuticles. Plant Physiol 163:5–20

Yeats TH, Huang W, Chatterjee S et al (2014) Tomato Cutin Deficient 1 (CD1) and putative orthologs comprise an ancient family of cutin synthase-like (CUS) proteins that are conserved among land plants. Plant J 77:667–675

Yokotani N, Nakano R, Imanishi S et al (2009) Ripening-associated ethylene biosynthesis in tomato fruit is autocatalytically and developmentally regulated. J Exp Bot 60:3433–3442

Zaharah S, Singh Z, Symons G et al (2012) Role of brassinosteroids, ethylene, abscisic acid and indole-3-acetic acid in mango fruit ripening. J Plant Growth Regul 31:363–372

Zanor MI, Osorio S, Nunes-Nesi A et al (2009) RNA interference of LIN5 in tomato confirms its role in controlling Brix content, uncovers the influence of sugars on the levels of fruit hormones and demonstrates the importance of sucrose cleavage for normal fruit development and fertility. Plant Physiol 150:1204–1218

Zhang L, Van Kan JAL (2013) *Botrytis cinerea* mutants deficient in d-galacturonic acid catabolism have a perturbed virulence on *Nicotiana benthamiana* and Arabidopsis, but not on tomato. Mol Plant Pathol 14:19–29

Zhang M, Yuan B, Leng P (2009) Cloning of *9-cis-epoxycarotenoid dioxygenase* (*NCED*) gene and the role of aba on fruit ripening. Plant Signal Behav 4:460–463

Zhang L, Thiewes H, Van Kan JAL (2011) The d-galacturonic acid catabolic pathway in *Botrytis cinerea*. Fungal Genet Biol 48:990–997

Zhang Y, Butelli E, De Stefano R et al (2013) Anthocyanins double the shelf life of tomatoes by delaying overripening and reducing susceptibility to gray mold. Curr Biol 23:1094–1100

Zhang L, Hua C, Stassen JH et al (2014) Genome-wide analysis of pectate-induced gene expression in *Botrytis cinerea*: identification and functional analysis of putative d-galacturonate transporters. Fungal Genet Biol 72:182–191. http://dx.doi.org/10.1016/j.fgb.2013.10.002

Zheng Y, Hong H, Chen L et al (2014) LeMAPK1, LeMAPK2 and LeMAPK3 are associated with nitric oxide-induced defence response against *Botrytis cinerea* in the *Lycopersicon esculentum* fruit. J Agric Food Chem 62:1390–1396

Zhong S, Fei Z, Chen Y-R et al (2013) Single-base resolution methylomes of tomato fruit development reveal epigenome modifications associated with ripening. Nat Biotechnol 31:154–159

Zhu M, Chen G, Zhou S et al (2014) A new tomato NAC (NAM/ATAF1/2/CUC2) transcription factor, SlNAC4, functions as a positive regulator of fruit ripening and carotenoid accumulation. Plant Cell Physiol 55:119–135

Zuo J, Zhu B, Fu D et al (2012) Sculpting the maturation, softening and ethylene pathway: the influences of microRNAs on tomato fruits. BMC Genomics 13:7

Zuo J, Fu D, Zhu Y et al (2013) SRNAome parsing yields insights into tomato fruit ripening control. Physiol Plant. 149:540–553

Chapter 20
Plant Hosts of *Botrytis* spp.

Yigal Elad, Ilaria Pertot, Alba Marina Cotes Prado, and Alison Stewart

Abstract A list of known plant species that serve as hosts of *Botrytis* spp. is presented. *Botrytis* species are reported to attack a total of 596 genera of vascular plants (tracheophytes), representing over 1400 plant species, although the host range is most probably much wider since there are limited reports of diseases on wild plants. *B. cinerea* was found on 586 genera, while the other species have narrower host ranges. Of the 596 genera, the majority (580) belongs to the Spermatophyta, 15 to the Pteridophyta and only 1 to the Lycopodiophyta. In the Spermatophyta, one genus belongs to Cycadophyta (cycads), 18 genera to the Pinophyta (conifers), one genus to the Ginkgophyta (ginkgo), and 562 genera to the Magnoliophyta (also known as flowering plants or Angiospermae). Geographically, *Botrytis* spp. occur wherever their host plants are grown, ranging from tropical and subtropical areas to cold temperate zones of Northern and Southern globe regions including regions where plants are grown in extreme cold weather or where agriculture is practiced in the desert.

Keywords Infection • Agricultural crops • Host species • Host range • Grey mould • Rot • Blight • Symptoms • Botanical family

---

Y. Elad (✉)
Department of Plant Pathology and Weed Research, Agricultural Research Organization, The Volcani Center, Bet Dagan 50250, Israel
e-mail: elady@volcani.agri.gov.il

I. Pertot
Fondazione Edmund Mach (FEM), Research and Innovation Centre, via Mach 1, S., Michele all'Adige 38010, TN, Italy

A.M. Cotes Prado
Colombian Corporation for Agricultural Research (CORPOICA), AA 240142, Las Palmas, parque central Bavaria, Bogotá, Colombia

A. Stewart
Marrone Bio Innovations, 1540 Drew Avenue, Davis, CA 95618, USA

S. Fillinger, Y. Elad (eds.), *Botrytis – the Fungus, the Pathogen and its Management in Agricultural Systems*, DOI 10.1007/978-3-319-23371-0_20

20.1 Introduction

Estimating the number of plant hosts of *Botrytis* spp. is a difficult task. Jarvis (1977) reported more than 200 plant hosts infected by *B. cinerea*. He based his estimation on reports such as MacFarlane (1968), that listed 235 hosts of *B. cinerea* from records in the Review of Applied Mycology, Anderson (1924) who reported more than 100 hosts from Alaska including three genera of Pteridophytes and a species of the Bryophyte Lunularia, Baker (1946) who listed 36 hosts, mostly ornamental crops, from California, Dingley (1969) who listed 69 hosts from New Zealand and Conners (1967) who listed 163 hosts in Alaska, Canada, and Greenland from the Canadian Plant Disease Survey. Grigaliunaite (2001) listed eight *Botrytis* spp. (*B. allii*, *B. anthophila*, *B. cinerea*, *B. elliptica*, *B. fabae*, *B. gladiolorum*, *B. paeoniae*, *and B. tulipae*) recorded on 687 host-plant taxa (677 plant species and 10 cultivars) that are present in Lithuania. The first attempt to compile a world wide list of *Botrytis* spp. hosts is presented in the current chapter. The reports, found in various sources and lists (for example, Elad et al. 2014; Farr et al. 1989; Moor 1959; Anonimous; Palmucci et al. 1997; Smith et al. 2009; ATCC 2015; Chase 1997) and indicate infection of seeds, seedlings and all parts of mature plants, represent important diseases that inflict major damage to agricultural crops, sporadic observations of diseases caused by *Botrytis* spp. or diseases of no economical importance.

Geographically, *Botrytis* spp. occur wherever their host plants are grown, ranging from tropical and subtropical areas to cold temperate zones of Northern and Southern globe regions including regions where plants are grown in extreme cold weather or where agriculture is practiced in the desert. The majority of records originate from economic hosts grown in cool-temperate and warm-temperate zones. Jarvis (1977) reported that *B. cinerea* even occurs as a snow mold. Various plant hosts and all *Botrytis* spp. are detailed in previous *Botrytis* books (Beever and Weeds 2004; Coley-Smith et al. 1980; Elad et al. 2004; Verhoeff et al. 1992; Chap. 2 this book).

20.2 Spread of *Botrytis* spp. Hosts amongst Families of the Plantae Kingdom

Botrytis spp. are reported to attack a total of 596 genera of vascular plants (tracheophytes) (Table 20.1), although the host range is most probably much wider, since there are limited reports of diseases on wild plants. *B. cinerea* was found on 586 genera, while the other species have narrower host ranges, limited to one or very few genera for *B. aclada* (syn. *B. allii*), *B. anemone*, *B. byssoidea*, *B. calthae*, *B. caroliniana*, *B. convoluta*, *B. croci*, *B. douglasii*, *B. elliptica*, *B. fabae*, *B. fabiopsis*, *B. ficariarum*, *B. galanthina*, *B. gladiolorum*, *B. globosa*, *B. hyacinthi*, *B. narcissicola*, *B. paeoniae*, *B. pelargonii*, *B. polyblastis*, *B. porri*, *B. ranunculi*, *B. sinoallii*, *B. squamosa*, *B. sphaerosperma*, *B. tulipae* (Table 20.1, Chap. 6). Of these 596 genera, the majority (580) belongs to the division of seed plants, Spermatophyta, 15 to the division of flowerless plants, the Pteridophyta and only 1 to the Lycopodiophyta (division of spore-bearing vascular plants). No members of the Bryophyta (nonvascular plants = mosses and liverworts) have been reported to be hosts of *B. cinerea* in nature, although the fungus can produce symptoms on *Physcomitrella patens* following artificial inoculation under laboratory conditions (Ponce de León et al. 2007).

In the Spermatophyta, one genus belongs to the subdivision Cycadophyta (cycads), 18 genera to the Pinophyta (conifers), one genus to the Ginkgophyta (ginkgo), and 562 genera to the Magnoliophyta (also known as flowering plants or Angiospermae). In the Magnoliophyta, the majority of the genera belongs to the eudicots (419), while the rest belong to the monocots (114) or the unranked clades Magnoliales, Piperales, Piperales, Austrobaileyales (4). *Botrytis* spp. are reported on 42 orders of Magnoliophyta (Austrobaileyales, Piperales, Laurales, Magnoliales, Acorales, Alismatales, Liliales, Asparagales, Arecales, Commelinales, Poales, Zingiberales, Ranunculales, Proteales, Buxales, Saxifragales, Caryophyllales, Vitales, Celastrales, Oxalidales, Malpighiales, Cucurbitales, Fabales, Fagales, Rosales, Geraniales, Myrtales, Crossosomatales, Brassicales, Malvales, Sapindales, Cornales, Ericales, Garryales, Gentianales, Lamiales, Solanales, Aquifoliales, Asterales and Dipsacales) corresponding to 33 % of the total number of Families, while the pathogen is not recorded on 16 Orders (Amborellales, Nymphaeales, Chloranthales, Canellales, Petrosaviales, Ceratophyllales, Trochodendrales, Gunnerales, Berberidopsidales, Santalales, Zygophyllales, Picramniales, Huerteales, Escalloniales, Bruniales and Paracryphiales), which represents 68 % of the Families. Twelve Families have more than 10 genera where *Botrytis* spp. are reported as pathogens (Asteraceae, Fabaceae, Rosaceae, Lamiaceae, Asparagaceae, Apiaceae, Araceae, Brassicaceae, Orchidaceae, Poaceae, Ranunculaceae, and Malvaceae, with 59, 31, 23, 22, 18, 18, 17, 15, 14, 13, 12, and 10 genera, respectively). The genera with the highest number of species where *Botrytis* spp. are reported include crops widely cultivated for production of food or feed and ornamental plants (i.e. *Actinidia*, *Brassica*, *Cattleya*, *Cereus*, *Chrysanthemum*, *Citrus*, *Eucalyptus*, *Geranium*, *Prunus*, *Vaccinium*, etc., see table for more details). Generally Table 20.1 lists more than 1,400 plant species that are associated with *Botrytis* spp. but there are missing plant spp. in the listed genera because of missing reports so it can be assumed that the number of plant taxa is significantly higher.

No specific correlation between the occurrence of *Botrytis* spp. with the phylogeny of Plantae can be observed. *Botrytis* spp. seem more related to the presence of soft and tender tissues rather than a specific clade of plants. The most common types of disease caused by *Botrytis* species are grey mould and rots reported in 78 % of the cases listed in the table below followed by blight (25 %) and spots or lesions (2.2 %). However, there may be several symptom types typical to each plant species and these are not necessarily reported in the table. Most of the reports on *B. cinerea* and *Botrytis* spp. were not ascertained by molecular phylogeny, thus it is possible that in the future we might face multiple new species after a thorough phylogenic analysis as suggested in Chap. 6.

Table 20.1 Plant hosts of *Botrytis* spp.

Genus	Species	Common name[a]	Family[b]	*Botrytis* species	Disease[c]
Abelmoschus	*A. esculentus*	Okra, ladies' fingers, bhindi, bamia, ochro	*Malvaceae*	*cinerea*	GM
Abies	*A. alba*	European silver fir	*Pinaceae*	*cinerea*	GM
	A. concolor	White-, Colorado white-fir		*cinerea*	GM
	A. grandis	Grand fir, lowland fir		*cinerea*	GM
	A. koreana	Korean fir		*cinerea*	GM
	A. nordmanniana	Nordmann-, Caucasian- fir		*cinerea*	GM
	A. procera	Noble-, Christmas tr.- fir		*cinerea*	GM
Abutilon	*A.* spp.	Indian mallow, velvet leaf	*Malvaceae*	*cinerea*	GM
	A. megapotamicum	Trailing abutilon		*cinerea*	GM
	A. theophrasti	Velvetleaf		*cinerea*	GM
Acacia	*A.* spp.	Acacia, thorn tr., whistling thorn, wattle	*Fabaceae*	*cinerea*	GM
Acalypha	*A. hispida*	Chenille, cola de zorro roja(S)	*Euphorbia*	*cinerea*	GM
	A. wilkesiana	Flamengueira, Fijian fire-, fire dragon-, Beefsteak- pl.		*cinerea*	GM
Acer	*A.* spp.	Maple	*Sapinda*	*cinerea*	GM
	A. negundo	Boxelder maple, ash-leaved maple, maple ash		*cinerea*	GM
Achillea	*A.* spp.	Yarrow	*Asteraceae*	*cinerea*	GM
	A. filipendulina	Fernleaf yarrow		*cinerea*	GM
	A. millefolium	Milfoil, yarrow		*cinerea*	GM
	A. ptarmica	Sneezewort yarrow		*cinerea*	GM
Aconitum	*A.* spp.	Aconite, monkshood, wolf's-, leopard's-bane	*Ranuncula*	*cinerea*	GM
	A.×cammarum	Aconitum hybrid		*cinerea*	GM
Acorus	*A. calamus*	Sweet flag, calamus	*Acoraceae*	*cinerea*	GM
	A. gramineus	Golden dwarf sweet flag		*cinerea*	GM

Actinidia	*A. arguta*	Baby kiwifruit, hardy kiwi	*Actinidia*	*cinerea*	Fr r, GM
	A. chinensis	Golden kiwifruit		*cinerea*	Fr r, GM
	A. coriacea	Chinese egg gooseberry		*cinerea*	Fr r, GM
	A. deliciosa	Fuzzy kiwifruit		*cinerea*	Fr r, GM
	A. kolomikta	Kolomikta vine, arctic kiwifruit, arctic beauty		*cinerea*	Fr r, GM
	A. melanandra	Red kiwifruit		*cinerea*	Fr r, GM
	A. polygama	Silver vine		*cinerea*	Fr r, GM
	A. purpurea	Purple kiwifruit		*cinerea*	Fr r, GM
Adiantum	*A.* sp.	Adiantum fern	*Pteridaceae*	*cinerea*	Bl
	A. adiantum capillusveneris	Capelvenere		*cinerea*	Bl
	A. raddianum var. *raddianum*	Maidenhair fern		*cinerea*	Bl
Aechmea	*A. fasciata*	Silver vase, urn pl., lengua de suegra(S), bromeliad fasciada(S)	*Bromelia*	*cinerea*	GM
Aeschynanthus	*A. pulcher*	Lipstick vine	*Gesneria*	*cinerea*	GM
Aesculus	*A. hippocastanum*	Horse chestnut, conker tr.	*Sapinda*	*cinerea*	GM
	A. pavia	Red buckeye, firecracker pl.		*cinerea*	GM
Agave	*A.* spp.	Agave, century pl.	*Agavaceae*	*cinerea*	GM
Ageratum	*A.* spp.	Floss fl.	*Asteraceae*	*cinerea*	GM
	A. houstonianum = A. mexicanum	Floss fl., bluemink, pussy foot, blueweed		*cinerea*	GM
Aglaonema	*A.* spp.	Chinese evergreen, Aglaonema(S)	*Araceae*	*cinerea*	GM
	A. commutatum	Chinese evergreen		*cinerea*	GM
	A. stenophyllum	Aglaonema(S)		*cinerea*	GM
Ajuga	*A.* spp.	Bugleweed	*Lamiaceae*	*cinerea*	GM
	A. reptans	Bugle, blue bugle, bugleherb, carpetweed		*cinerea*	GM

(continued)

Table 20.1 (continued)

Genus	Species	Common name[a]	Family[b]	*Botrytis* species	Disease[c]
Alcea	*A.* spp.	Hollyhock	*Malvaceae*	*cinerea*	GM
	A. rosea = *Althaea chinensis*	Common hollyhock		*cinerea*	GM
Alchemilla	*A. mollis*	Lady's mantle	*Rosaceae*	*cinerea*	GM
Aldrovanda	*A. vesiculosa*	Waterwheel pl., rueda de agua(S)	*Droseraceae*	*cinerea*	GM
Ailanthus	*A. altissima*	Tr. of heaven	*Simarouba*	*cinerea*	GM
Allamanda	*A.* spp.	Allamanda	*Apocyna*	*cinerea*	GM
	A. cathartica	Golden trumpet, yellow allamanda		*cinerea*	GM
	A. blanchetii	Purple allamanda		*cinerea*	GM
Allium	*A. ampeloprasum* var. *porrum*	Leek	*Amaryllida*	*cinerea*	Brown stain, lf & umbel bl
				porri	Neck r
				squamosa	Lf & umble bl, neck r
	A. angulosum	Mouse garlic		*aclada =allii*	Neck & bulb r
	A. caeruleum= A. azureum	Blue globe-, blue ornmental onion		*aclada =allii*	Neck & bulb r
	A. cepa	Onion, cebolla(S)		*aclada= allii types AI AII*	Neck & bulb r, scape & umbel bl
				porri	Neck r
				globosa	Neck r
				sinoallii	Bl, GM
				squamosa	As above
	A. cepa var. *aggregatum*	Shallot		*aclada= allii*	As above
				byssoidea	Mycelial neck r
				cinerea	As above
	A. fistulosum	Japanese bunching-, Welsh- onion		*Aclada =allii*	Neck & bulb r, scape & umbel bl

				byssoidea	As above
	A. flavum	Yellow-flowered garlic		*Aclada =allii*	Neck & bulb r
	A. odorum	Chinese chive		*Aclada =allii*	Neck & bulb r
	A. sativum	Garlic, ajo (S)		*Aclada =allii*	Neck & bulb r
				byssoidea	As above
				cinerea	As above
				porri	Neck r
				sinoallii	Bl, GM
				squamosa	As above
	A. schoenoprasum	Chives		*byssoidea*	Lf bl
	A. triquetrum	Three-cornered leek		*Aclada =allii*	As above
				byssoidea	As above
				cinerea	As above
				porri	Neck r
				sphaerosperma	Bl, LSs
				squamosa	As above
	A. tuberosum	Garlic-, Chinese- chive		*byssoidea*	Lf bl
				sinoallii	Bl, GM
	A. ursinum	Ramsons, buckrams, wild-, broad-leaved-, wood-, bear's- garlic		*globosa*	Neck r
Alnus	*A.* spp.	Alder	*Betulaceae*	*cinerea*	GM
	A. glutinosa	Common-, black-, European- alder		*cinerea*	GM
	A. rubra	Red alder, Oregon alder		*cinerea*	GM
Alocasia	*A.* spp.	Alocasia, African mask	*Begoniacea*	*cinerea*	Bl
	A. robusta	Alocasia		*cinerea*	Bl

(continued)

Table 20.1 (continued)

Genus	Species	Common name[a]	Family[b]	*Botrytis* species	Disease[c]
Aloe	*A. versa*	Penca sábila, Chinese-, Indian-aloe	*Xanthorrhoea*	*cinerea*	Bl
Alpinia	*A. purpurata*	Red ginger	*Zingibera*	*cinerea*	GM
Alstromeria	*A.* spp.	Alstroemeria	*Alstroemeria*	*cinerea*	GM
	A. aurea	Lily of the Incas		*cinerea*	GM
	A. aurantiaca	Peruvian-, princess- lily		*cinerea*	GM
	A. caryophyllacea	Brazilian lily		*cinerea*	GM
	A. haemantha	Purplespot parrot lily		*cinerea*	GM
	A. ligtu	Lily of the Nile		*cinerea*	GM
	A. pelegrina	Inca lily, Pelegrina, Mari-posa de Los Molles(S)		*cinerea*	GM
	A. psittacina	White-edged Peruvian lily, white alstroemeria		*cinerea*	GM
	A. pulchella	Parrot lily, parrot Fl., NZ Christmas bell		*cinerea*	GM
	A.×hybrida	Peruvian Lily		*cinerea*	GM
Althaea	*A.* sp.	Common or rose mallow	*Malvaceae*	*cinerea*	GM
	A. officinalis	Marsh mallow		*cinerea*	GM
Amaranthus	*A. hypochondriacus*	Amaranth	*Amarantha*	*cinerea*	GM
	A. hybridus	Amaranth		*cinerea*	GM
Amaryllis	*A. belladonna*	Amaryllis (hippeastrum)	*Amaryllida*	*cinerea*	GM
Abelmoschus	*A. moschatus*	Muskdana abelmosk, ambrette seeds, bamia moschata, ornamental okra	*Malvaceae*	*cinerea*	GM
Amelanchier	*A. alnifolia*	Saskatoon serviceberry, alder-leaved shadbush	*Rosaceae*	*cinerea*	GM
	A. lamarckii	Serviceberry, snowy mespilus, June berry		*cinerea*	GM
	A. alnifolia×A. stolonifera	Serviceberry		*cinerea*	GM
Ammi	*A. majus*	Bishop's weed	*Apiaceae*	*cinerea*	GM

Amorphophallus	*A.* sp.	Zaminkand	*Araceae*	*cinerea*	GM
	A. konjac=A. rivieri	Devil's tongue, corpse fl.		*cinerea*	GM
	A. bulbifer			*cinerea*	GM
	A. paeoniifolius			*cinerea*	GM
Amygdalus	*A. communis= Prunus dulcis*	Almond	*Rosaceae*	*cinerea*	Green fr r
Anacyclus	*A.* sp.	Anacyclus	*Asteraceae*	*cinerea*	GM
Anagallis	*A. arvensis*	Scarlet pimpernel, red pimpernel, red chickweed	*Primulaceae*	*cinerea*	GM
Anemone	*A. coronaria*	Anemone	*Ranuncula*	*anemone*	Bl
				cinerea	Collar r
	A. hupehensis	Japanese anemone		-"-	-"-
Anethum	*A. graveolens*	Dill, eneldo(S)	*Apiaceae*	*cinerea*	GM
Angelica	*A. archangelica*	Angelica	*Apiaceae*	*cinerea*	GM
	A. keiskei	Ashitaba (Japanese), (="Tomorrow's leaf")		*cinerea*	GM
Anigozanthos	*A. manglesii*	Kangaroo paws	*Haemodora*	*cinerea*	GM
Annona	*A.* spp.	Anona, anón(S)	*Annonaceae*	*cinerea*	GM
	A. cherimola	Chirimuya, custard apple-tree, anona, chirimolla(S)		*cinerea*	GM
	A. muricata	Soursop, guanábana(S)		*cinerea*	GM
	A. squamosa	Sugar apple, sweetsop		*cinerea*	GM
Anthurium	*A. spp.*	Anthurium	*Araceae*	*cinerea*	GM
Antirrhinum	*A. majus*	Dragon fl., snapdragon	*Scrophularia*	*cinerea*	Bl
Anthriscus	*A. cerefolium*	Larixchervil	*Apiaceae*	*cinerea*	GM
Aphelandra	*A. justicia*	Zebra Pl.	*Acantha*	*cinerea*	GM
	A. squarrosa	Zebra pl., Afelandra(S)			
Apium	*A. graveolens* var. *dulce*	Celery	*Apiaceae*	*cinerea*	GM
	A. graveolens var. *rapaceum*	Celeriac		*cinerea*	GM

(continued)

Table 20.1 (continued)

Genus	Species	Common name[a]	Family[b]	*Botrytis* species	Disease[c]
Aquilegia	*A.* spp.	Columbine, Granny's bonnet	*Ranuncula*	*cinerea*	GM
	A. bertolonii	Bertoloni columbine		*cinerea*	GM
	A. canadensis	Canada-, eastern red-columbine		*cinerea*	GM
	A. discolor	Columbine, Granny's bonnets		*cinerea*	GM
	A. flabellata	Dwarf fan-shaped colum-bine		*cinerea*	GM
	A. vulgaris	European-, common- columbine, Granny's nightcap		*cinerea*	LSs, GM
Arabidopsis	*A. thaliana*	Thale cress, mouse-ear cress, arabidopsis	*Brassica*	*cinerea*	R
Arabis	*A. alpina*	Rock cress, Arabis	*Brassica*	*cinerea*	GM
	A. blepharophylla	Arabis		*cinerea*	GM
Arachis	*A. hypogaea*	Peanut, maní(S)	*Fabaceae*	*cinerea*	Bl
Araucaria	*A.* spp.	Araucaria	*Araucaria*	*cinerea*	GM
	A. araucana	Monkey-puzzle, pehuén		*cinerea*	GM
Arisaema	*A.* sp.	Jack-in-the-pulpit, Indian turnip, wake robin	*Araceae*	*cinerea*	GM
Aristolochia	*A. elegans*	Calico fl.	*Aristolochia*	*cinerea*	GM
	A. macrophylla	Pipevine Dutchman's pipe		*cinerea*	GM
Armeniaca	*A. vulgaris = Prunus armeniaca*	Apricot	*Rosaceae*	*cinerea*	GM
Armeria	*A.* sp.	Lady's cushion, thrift, sea pink	*Plumbagina*	*cinerea*	GM
Aronia	*A.* spp.	Chokeberries	*Rosaceae*	*cinerea*	GM
Artemisia	*A. dracunculus*	Tarragon	*Asteraceae*	*cinerea*	GM
Artocarpus	*A. altilis*	Artocarpus, breadfruit	*Moraceae*	*cinerea*	GM
	A. heterophyllus	Jack Fr.		*cinerea*	GM
	A. odoratissimus	Madang, Johey oak		*cinerea*	GM
	A. schmidtiana	Artemisia		*cinerea*	Melting out

Asclepias	*A. curassavica*	Asclepias, Mexican butter-fly weed	*Apocyna*	*cinerea*	GM
	A. tuberosa	Butterfly weed, milkweed		*cinerea*	GM
Ascocenda	*A.* sp.	Orchid	*Orchida*	*cinerea*	GM
Aspalathus	*A. linearis*	Rooibos	*Fabaceae*	*cinerea*	GM
Asparagus	*A. densiflorus*	Asparagus	*Asparaga*	*cinerea*	GM, shoot Bl
	A. myriocladius	Asparagus		*cinerea*	-"-
	A. officinalis	Asparagus		*cinerea*	-"-
Aspidistra	*A. elatior*	Cast-iron pl.		*cinerea*	GM
Asplenium	*A.* spp.	Asplenium fern	*Asplenia*	*cinerea*	Bl
Aster	*A.* spp.	Aster	*Asteraceae*	*cinerea*	GM
	A. callistephus	Aster		*cinerea*	GM
	A. amellus	Aster, European Michael-mas daisy		*cinerea*	GM
Astilbe	*A.* spp.	False goat's beard, false spirea	*Saxifraga*	*cinerea*	GM
	A. chinensis	Chinese astilbe		*cinerea*	GM
	A. glaberrima	Glaucous rock-, smooth rock- astilbe		*cinerea*	GM
	A. simplicifolia	False spirea, Plumeflower meadowsweet		*cinerea*	GM
	A.×arendsii	Fanal astilbe, astilbe arendsii		*cinerea*	GM
	A.×crispa	Astilbe crispa		*cinerea*	GM
	A.×japonica	Japanese astilbe		*cinerea*	GM
	A.×simplicifolia	–		*cinerea*	GM
Athyrium	*A. niponicum*	Athyrium, Japanese painted fern	*Asplenia*	*cinerea*	GM
Aubrieta	*A.* spp.	Aubretia, rock cress	*Brassica*	*cinerea*	GM
	A. deltoidea	Lilacbush, purple rock-, rainbow rock-cress		*cinerea*	GM
Aucuba	*A. japonica*	Spotted-, Japanese- laurel, gold dust pl.	*Garryaceae*	*cinerea*	GM

(continued)

Table 20.1 (continued)

Genus	Species	Common name[a]	Family[b]	*Botrytis* species	Disease[c]
Avena	*A. saliva var. nuda*	Naked oat	*Poaceae*	*cinerea*	GM
	A. saliva	Dwarf oat		*cinerea*	GM
Basella	*B. rubra*	Malabar nightshade	*Basellaceae*	*cinerea*	GM
Begonia	*B.* spp.	Begonia	*Begoniaceae*	*cinerea*	GM
	B. coccinea	Wing begonia		*cinerea*	GM
	B. grandis	Hardy begonia		*cinerea*	GM
	B. obliqua	Begonia obliqua		*cinerea*	GM
	B. semperflorens	Wax begonia		*cinerea*	GM
	B.×lucerna hybrid	Begonia lucerna		*cinerea*	GM
	B.×tuberhybrida	Tuberous begonia		*cinerea*	GM
Belamcanda	*B.* sp.	Blackberry lily	*Iridaceae*	*cinerea*	GM
	B. chinensis	Blackberry-, leopard- lily		*cinerea*	GM
Bellis	*B. perennis*	English daisy	*Asteraceae*	*cinerea*	GM
Berberis	*B.* spp.	Barberry	*Berberida*	*cinerea*	GM
	B. darwinii	Darwin's barberry		*cinerea*	GM
	B. dictyophylla	Netleaf barberry		*cinerea*	GM
	B. julianae	Wintergreen-, Chinese- barberry		*cinerea*	GM
	B. thunbergii	Japanese-, red- Thun-berg's- barberry		*cinerea*	GM
	B. verruculosa	Warty barberry		*cinerea*	GM
Bergenia	*B.* spp.	Pigsqueak elephant-eared saxifrage, elephant's ears	*Saxifraga*	*cinerea*	Bl
	B. cordifolia	Heart-leaf bergenia		*cinerea*	Bl
	B. crassifolia	Leather bergenia, winter-blooming bergenia		*cinerea*	Bl

Beta	*B. vulgaris* ssp. *vulgaris* convar. *cicla*	Leaf beet	*Amarantha*	*cinerea*	SR
	B. v. ssp. *v. convar. cicla.* var. *cicla*	Spinach beet		*cinerea*	SR
	B. v. ssp. *v. convar. cicla.* var. *flavescens*	Chard		*cinerea*	SR
	B. v. ssp. *vulgaris* convar. *vulgaris*	Tuberous beet		*cinerea*	SR
	B. vulgaris ssp. *vulgaris* var. *vulgaris*	Garden beet		*cinerea*	SR
	B. v. ssp. *v. convar. vulgaris* var. *altissima*	Sugar beet		*cinerea*	SR
	B. v. ssp. *v. convar. vulgaris* var. *crassa*	Mangelwurzel		*cinerea*	SR
	B. v. var. *esculenta*	Chard, red chard		*cinerea*	GM
Betula	*B. pubescens* = *Betula alba*	European-, downy-birch	*Betulaceae*	*cinerea*	Bl
Beaucarnea	*B.* sp.	Beaucarnea	*Asparaga*	*cinerea*	GM
	B. gracilis	Slender beaucarnea		*cinerea*	GM
Billbergia	*B. nutans*	Queen's tears	*Bromelia*	*cinerea*	GM
Blechnum	*B.* sp.	Hard fern	*Polypodia*	*cinerea*	Bl
Borago	*B. officinalis*	Starfl.	*Boragina*	*cinerea*	GM
Bouvardia	*B.×hybrid*	Bouvardia	*Rubiaceae*	*cinerea*	GM

(continued)

Table 20.1 (continued)

Genus	Species	Common name[a]	Family[b]	*Botrytis* species	Disease[c]
Brassica	*B.* spp.		*Brassica*	*cinerea*	GM
				pseudo-cinerea	GM
	B. campestris	Komatsuna		*cinerea*	GM, Lf bl
	B. caulorapa = *B. oleracea* var. *caulorapa*	Kohlrabi		*cinerea*	GM
	B. juncea	Chinese mustard		*cinerea*	GM
	B. napus var. *napobrassica* = *B. napobrassica*, *B. napus* subsp. *rapifera*	Rutabaga, swede (Swedish turnip), turnip, yellow turnip, neep		*cinerea*	GM
	B. napus var. *napus*	Rapeseed, rape, oilseed rape, rapa, canola(S)		*cinerea*	GM
	B. napus var. *oleifera*	Spring rape		*cinerea*	GM
	B. oleracea var. *acephala*	Kale, collard greens		*cinerea*	GM
	B. oleracea var. *capitata*	White-, red-cabbage, Repollo(S)		*cinerea*	GM
	B. oleracea var. *Botrytis*	Caulifl., Coliflor(S)		*cinerea*	GM
	B. oleracea var. *gemmifera*	Brussels sprouts		*cinerea*	GM
	B. oleracea var. *italica*	Broccoli		*cinerea*	GM
	B. oleracea var. *sabauda*	Savoy cabbage		*cinerea*	GM
	Brassica rapa, subsp. *chinensis*	Chinese cabbage, bok choi		*cinerea*	GM
	B. rapa subsp. *rapini* = *B. campestris esculenta*	Turnip		*cinerea*	GM
	B. rapa subsp. *nipposinica*	Mizuna Japanese-, potherb-mustard		*cinerea*	GM
	B. rapa subsp. *pekinensis*	Chinese-, Nappa-, celery- cabbage		*cinerea*	GM
	Brassica rapa subsp. *rapa*	Rapini, broccoli raab, broccoli rabe		*cinerea*	GM
Brassolaeliocattleya	*B. hybrid*	Blc. Orchid	*Orchida*	*cinerea*	GM

Bromus	*B. tectorum*	Downy brome	*Poaceae*	*cinerea*	GM
Browallia	*B. speciosa*	Sapphire Fl.	*Solanaceae*	*cinerea*	GM
Brunfelsia	*B. americana*	Lady-of-the-night	*Solanaceae*	*cinerea*	GM
	B. australis	Yesterday-today-tomorrow		*cinerea*	GM
	B. pauciflora	Kiss me quick, Brazil rain tr.		*cinerea*	GM
Buddleia	*B.* spp.	Butterfly bush	*Scrophularia.*	*cinerea*	GM
	B. alternifolia	Alternate-leaved butterfly bush		*cinerea*	GM
	B. davidii	Summer lilac, butterfly-bush, orange eye		*cinerea*	GM
	B. globosa	Orange ball buddleja		*cinerea*	GM
Bou-gainvillea	*B.* spp.	Buganvilla, bugambilia (S), Napoleón(S), veranera(S)	*Nyctagina*	*cinerea*	GM
	B. spectabilis	Great bougainvillea		*cinerea*	GM
Bupleurum	*B.* sp.	Bupleurum	*Apiaceae*	*cinerea*	GM
	B. rotundifolium	Hares ear		*cinerea*	GM
Cajanus	*C. cajan*	Pigeonpea	*Fabaceae*	*cinerea*	GM
Caladium	*C. bicolor*	Caladium	*Araceae*	*cinerea*	GM
				ricini	Bl, SR
Calamagrostis	*C.* spp.	Feather reed grass, small reed, reed grass	*Poaceae*	*cinerea*	GM
	C. epigeios	Chee reedgrass		*cinerea*	GM
	C.×acutiflora	Calamagrostis hybrid, feather reedgrass		*cinerea*	GM
Calathea	*C.* spp.	Calathea, peacock pl.	*Maranta*	*cinerea*	GM
	C. makoyana	peacock pl., cathedral windows		*cinerea*	GM
Calceolaria	*C.* sp.	Slipper fl., lady's purse, pocketbook fl.	*Calceolaria*	*cinerea*	GM
	C. crenatiflora	Pocketbook pl.		*cinerea*	Bl

(continued)

Table 20.1 (continued)

Genus	Species	Common name[a]	Family[b]	*Botrytis* species	Disease[c]
Caladium	*C.* spp.	Elephant ear, heart of Jesus, angel wings	*Araceae*	*cinerea*	GM
	C. bicolor	Fancy leaf caladium		*cinerea*	GM
	C. humboldtii	Angel wings, fancy leaf-, miniature- caladium		*cinerea*	GM
	C. lindenii	Angel wings caladium, Indian-kale; spoonflower		*cinerea*	GM
	C. marmoratum	Elephant ear, heart of Jesus angel wings		*cinerea*	GM
	C. paradoxum	Caladium		*cinerea*	GM
	C. schomburgkii	Elephant's ear, angel wings, lance-leaf caladium, heart of Jesus		*cinerea*	GM
	C. ternatum	–		*cinerea*	GM
	C.×hortulanem	Caladium hybrid		*cinerea*	GM
Calendula	*C. officinalis*	Calendula, pot marigold, Caléndula(S)	*Asteraceae*	*cinerea*	Bl
Callistephus	*C. chinensis*	Lavender Matsumoto, China aster, annual aster	*Asteraceae*	*cinerea*	GM
Calluna	*C. vulgaris*	Common heather, ling, heather	*Ericaceae*	*cinerea*	Bl
Calocedrus	*C. decurrens*	Incense cedar	*Cupressa*	*cinerea*	GM
Caltha	*C. palustris*	Kingcup, marsh marigold	*Ranuncula*	*cinerea*	GM
				calthae	Lesions
Camassia	*C. cusickii*	Cusick's camass	*Asparaga.*	*cinerea*	GM
	C. leichtlinnii	Leichtlin's camass		*cinerea*	GM
Camelina	*C. sativa*	False flax	*Brassica.*	*cinerea*	GM

Camellia	*C. sinensis*	Tea	*Theaceae*	*cinerea*	GM
	C. japonica	Japanese camellia, rose of winter		*cinerea*	GM
	C. oleifera	Oil-seed camellia, tea oil camellia		*cinerea*	GM
	C. reticulata	Reticulata camellia		*cinerea*	GM
	C. sasanqua	Christmas camellia, yuletide camellia		*cinerea*	GM
	C.×hiemalis	Camelia hybrid		*cinerea*	GM
	C.×japonica	Camelia hybrid		*cinerea*	GM
	C.×sasanqua	Camelia hybrid		*cinerea*	GM
	C.×williamsii	Camelia hybrid		*cinerea*	GM
Campanula	*C. carpatica*	Bell fl.	*Campanula*	*cinerea*	Bl
	C. medium	Campanilla de Canterbury (S), Canterbury bells		*cinerea*	Bl
	C. persicifolia	Mell fl.		*cinerea*	Bl
	C. rapunculus	Rampion bell fl., rampion, rover bell fl.		*cinerea*	GM
Cannabis	*C. sativa*	Cannabis, marijuana, hemp, marihuana(S)	*Cannaba*	*cinerea*	GM
Canna	*C.* spp.	Canna lily	*Cannaceae*	*cinerea*	GM
	C. discolor	Edible canna		*cinerea*	GM
	C. indica	Saka siri, Indian shot, canna		*cinerea*	GM
	C. tropicana	Tropicanna canna		*cinerea*	GM
Capsicum	*C. annuum*	Pepper, Chile dulce(S)	*Solanaceae*	*cinerea*	GM, ghost spots
	C. baccatum	Bishop's-hat, Christmas bell, Brown's-, Peruvian- pepper		-"-	-"-
	C. chinense	Yellow lantern chili		-"-	-"-
	C. frutescens	African chillies		-"-	-"-
	C. pubescens	Tree pepper, rocoto		-"-	-"-

(continued)

Table 20.1 (continued)

Genus	Species	Common name[a]	Family[b]	*Botrytis* species	Disease[c]
Carica	*C. papaya*	Papaya	*Caricaceae*	*cinerea*	GM
	C. pubescens = Vasconcellea pubescens	Mountain papaya		*cinerea*	GM
	C.×heilbornii var. *pentagona*	Babaco, Mountain Papaya, Champagne Fr.		*cinerea*	GM
Carpinus	*C. betulus*	European-, common- hornbeam	*Betulaceae*	*cinerea*	GM
Carum	*C. carvi* var. *annuum*	Annual-, common caraway, meridian fennel, Persian cumin	*Apiaceae*	*cinerea*	GM
Caryopteris	*C.* sp.	Blue mist spirea, bluebeard	*Lamiaceae*	*cinerea*	GM
	C.×clandonensis	Caryopteris hybrid		*cinerea*	GM
Carthamus	*C. tinctorius*	Safflower	*Asteraceae*	*cinerea*	GM, head r
Castanea	*C. crenata*	Japanese chestnut	*Fagaceae*	*cinerea*	GM
	C. dentata	American chestnut		*cinerea*	GM
	C. mollissima	Chinese chestnut		*cinerea*	GM
	C. sativa	European-, sweet- chestnut		*cinerea*	GM
Catalpa	*C. bignonioides*	Indian bean tr., southern catalpa	*Bignonia*	*cinerea*	GM
Catharanthus	*C. roseus= Vinca rosea*	Vinca, Madagascar periwinkle	*Apocyna*	*cinerea*	Bl, GM
Castilleja	*C. miniata*	Giant red Indian paintbrush	*Scrophularia*	*cinerea*	GM
Cattleya	*C.* spp.	Cattleya, orquidea(S)	*Orchida*	*cinerea*	Pe Bl
	C. acklandiae	Lady Ackland's cattleya		*cinerea*	Pe Bl
	C. amethystoglossa	Amethest lipped cattleya		*cinerea*	Pe Bl
	C. araguaiensis	Araguaia cattleya		*cinerea*	Pe Bl
	C. aurantiaca	Orange cattleya		*cinerea*	Pe Bl
	C. aurea	Golden yellow cattleya		*cinerea*	Pe Bl
	C. bicolor	Bicolored cattleya		*cinerea*	Pe Bl
	C. bowringiana	Bowring's cattleya		*cinerea*	Pe Bl

	C. deckeri	Decker's cattleya		*cinerea*	Pe Bl
	C. dolosa	Dolose cattleya		*cinerea*	Pe Bl
	C. dormaniana	Dorman's cattleya		*cinerea*	Pe Bl
	C. dowiana	Dow's cattleya		*cinerea*	Pe Bl
	C. eldorado	Cattleya of the golden land		*cinerea*	Pe Bl
	C. elongata	Elongated cattleya		*cinerea*	Pe Bl
	C. forbesii	Forbes' cattleya		*cinerea*	Pe Bl
	C. gaskelliana	Gaskell's cattleya		*cinerea*	Pe Bl
	C. granulosa	Granulose cattleya		*cinerea*	Pe Bl
	C. guatemalensis	Guatemalan cattleya		*cinerea*	Pe Bl
	C. guttata	Spotted cattleya		*cinerea*	Pe Bl
	C. hardyana	Hardy's cattleya		*cinerea*	Pe Bl
	C. harrisoniana	Harrison's cattleya		*cinerea*	Pe Bl
	C. intermedia	Intermediate cattleya		*cinerea*	Pe Bl
	C. iricolor	Rainbow-colored cattleya		*cinerea*	Pe Bl
	C. jenmanii	Jenman's cattleya		*cinerea*	Pe Bl
	C. kerrii	Kerr's cattleya		*cinerea*	Pe Bl
	C. labiata	Crimson, ruby lipped cattleya		*cinerea*	Pe Bl
	C. lawrenceana	Sir Trevor Lawrence's cattleya		*cinerea*	Pe Bl
	C. leopoldii	Tiger striped cattleya		*cinerea*	Pe Bl
	C. loddigesii	Loddiges' cattleya		*cinerea*	Pe Bl
	C. lueddemanniana	Lueddemann's cattleya		*cinerea*	Pe Bl
	C. luteola	Pale yellow cattleya		*cinerea*	Pe Bl
	C. maxima	Greatest cattleya		*cinerea*	Pe Bl
	C. mendelii	Mendel's cattleya		*cinerea*	Pe Bl

(continued)

Table 20.1 (continued)

Genus	Species	Common name[a]	Family[b]	*Botrytis* species	Disease[c]
	C. mooreana	Moore's cattleya		*cinerea*	Pe Bl
	C. mossiae	Mrs. Moss' cattleya		*cinerea*	Pe Bl
	C. nobilior	Noble cattleya		*cinerea*	Pe Bl
	C. patinii	Patin's cattleya		*cinerea*	Pe Bl
	C. percivaliana	Percival's-, Christmas cattleya		*cinerea*	Pe Bl
	C. porphyroglossa	Purple lipped cattleya		*cinerea*	Pe Bl
	C. quadricolor	Four-colored cattleya		*cinerea*	Pe Bl
	C. rex	King of the cattleyas		*cinerea*	Pe Bl
	C. schilleriana	Consul Schiller's cattleya		*cinerea*	Pe Bl
	C. schofieldiana	Schofeld's cattleya		*cinerea*	Pe Bl
	C. schroderae	Baroness Schroeder's cattleya		*cinerea*	Pe Bl
	C. skinneri	Skinner's cattleya		*cinerea*	Pe Bl
	C. tenuis	Slender stemmed cattleya		*cinerea*	Pe Bl
	C. trianae	Dr. Triana's cattleya, Orquidea de navidad(S)		*cinerea*	Pe Bl
	C. velutina	Velvety cattleya		*cinerea*	Pe Bl
	C. violacea	Superba of the Orinoco		*cinerea*	Pe Bl
	C. walkeriana	Walker's cattleya		*cinerea*	Pe Bl
	C. warneri	Warner's cattleya		*cinerea*	Pe Bl
	C. warscewiczii	Warscewicz's cattleya		*cinerea*	Pe Bl
	C. patrocinii	Patrocinio's cattleya		*cinerea*	Pe Bl
Caulophyllum	*C. thalictroid es*	Blue cohosh, squaw-, papoose- root	*Berberida*	*cinerea*	GM

Celosia	*C.* spp.	Cockscomb	*Amarantha*	*cinerea*	GM
	C. argentea	Plumed cockscomb, silver cock's comb		*cinerea*	GM
	C. cristata	Cockscomb		*cinerea*	GM
	C. isetii	Cockscomb		*cinerea*	GM
	C. leptostachya	Yellow toreador		*cinerea*	GM
	C. nitida	Slender celosia		*cinerea*	GM
	C. odorata	Canterbury bells		*cinerea*	GM
	C. plumosa	Cockscomb, brain celosia		*cinerea*	GM
	C. palmeri	Palmer's cock's-comb		*cinerea*	GM
	C. nitida	West Indian cock's-comb		*cinerea*	GM
	C. spicata	Flamingo feather		*cinerea*	GM
	C. trigyna	Wool fl.		*cinerea*	GM
	C. virgata	Albahaca(S)		*cinerea*	GM
Celtis	*C. occidentalis*	Common hackberry	*Cannaba*	*cinerea*	GM
Centaurea	*C. cyanus*	Corn fl., bachelor's button, bluebottle	*Asteraceae*	*cinerea*	GM
	C. montana	Mountain bluet		*cinerea*	Bl
	C. solstitialis	Yellow starthistle		*cinerea*	GM
Cephalotus	*C. folicularis*	Fly-catcher pl., Jarrito enano(S)	*Cephalota*	*cinerea*	GM
Cerasus	*C. avium* = *Prunus avium*	Sweet cherry	*Rosacea*	*cinerea*	Green fr r
	C. vulgaris = *P. cerasus*	Sour cherry		*cinerea*	-"-

(continued)

Table 20.1 (continued)

Genus	Species	Common name[a]	Family[b]	*Botrytis* species	Disease[c]
Cereus	*C.* spp.	Queen of the night, Peruvian apple cactus	*Cactaceae*	*cinerea*	Bl
	C. adelmarii	–		*cinerea*	Bl
	C. aethiops	–		*cinerea*	Bl
	C. albicaulis	–		*cinerea*	Bl
	C. argentinensis	–		*cinerea*	Bl
	C. bicolor	–		*cinerea*	Bl
	C. braunii	–		*cinerea*	Bl
	C. cochabambensis	–		*cinerea*	Bl
	C. comarapanus	–		*cinerea*	Bl
	C. fernambucensis	–		*cinerea*	Bl
	C. fricii	–		*cinerea*	Bl
	C. haageanus	–		*cinerea*	Bl
	C. hankeanus	–		*cinerea*	Bl
	C. hexagonus	–		*cinerea*	Bl
	C. hildmannianus	–		*cinerea*	Bl
	C. horrispinus	–		*cinerea*	Bl
	C. huilunchu	–		*cinerea*	Bl
	C. insularis	–		*cinerea*	Bl
	C. jamacaru	–		*cinerea*	Bl
	C. kroenleinii	–		*cinerea*	Bl
	C. lamprospermus	–		*cinerea*	Bl
	C. lanosus	–		*cinerea*	Bl
	C. mirabella	–		*cinerea*	Bl
	C. mortensenii	–		*cinerea*	Bl
	C. pachyrhizus	–		*cinerea*	Bl

	C. phatnospermus	–		*cinerea*	Bl
	C. repandus	–		*cinerea*	Bl
	C. roseiflorus	–		*cinerea*	Bl
	C. saddianus	–		*cinerea*	Bl
	C. spegazzinii	–		*cinerea*	Bl
	C. stenogonus	–		*cinerea*	Bl
	C. tacuaralensis	–		*cinerea*	Bl
	C. trigonodendron	–		*cinerea*	Bl
	C. uruguayanus	–		*cinerea*	Bl
	C. validus	–		*cinerea*	Bl
	C. vargasianus	–		*cinerea*	Bl
Cercis	*C. canadensis*	Redbud	*Fabaceae*	*cinerea*	GM
	C. siliquastrum	Redbud, Judas tr., arbol del amor(S)		*cinerea*	GM
Chamaedorea	*C. elegans*	Parlor palm	*Arecaceae*	*cinerea*	GM
Chamaecyparis	*C.* sp.	False cypress	*Cupressa*	*cinerea*	GM
Chamaemelum	*C. nobile = Anthemis nobilis*	Roman-, English-, garden- chamomile, babunag	*Asteraceae*	*cinerea*	GM
Chamelaucium	*C. uncinatum*	Geraldton wax, wax fl.	*Myrtaceae*	*cinerea*	Bl, GM
Chenopodium	*C. quinoa*	Quinoa	*Amarantha*	*cinerea*	GM
Chlorophytum	*C.* sp.	Spider pl.	*Asparaga*	*cinerea*	GM
	C. comosum	Spider pl., airplane pl., hen-and-chickens		*cinerea*	GM
Chrysanthemoides	*C. monilifera*	Boneseed, bitou bush, tick berry	*Asteraceae*	*cinerea*	GM

(continued)

Table 20.1 (continued)

Genus	Species	Common name[a]	Family[b]	*Botrytis* species	Disease[c]
Chrysanthemum	*C.* spp.	Chrysanthemum	*Asteraceae*	*cinerea*	GM, Bl
	C. carinatum	Painted daisy		*cinerea*	GM, Bl
	C. coccineum	Pyrethrum, painted daisy		*cinerea*	GM, Bl
	C. coronarium	Edible chrysanthemum, chrysanthemum greens		*cinerea*	GM, Bl
	C. frutescens	Marguerite daisy		*cinerea*	GM, Bl
	C. majus	Costmary, alecost, balsam herb, bible leaf		*cinerea*	GM, Bl
	C. maximum	Shasta daisy		*cinerea*	GM, Bl
	C. morifolium	Florist's daisy, Hardy garden mum		*cinerea*	GM, Bl
	C. multicaule= Coleostephus myconis	Yellow daisy		*cinerea*	GM, Bl
	C. parthenium	Feverfew, bachelor's buttons, featherfew		*cinerea*	GM, Bl
	C. segetum	Corn marigold, corn daisy		*cinerea*	GM, Bl
	C. sizzle	Sizzle yellow		*cinerea*	GM, Bl
	C.×superbum	Shasta daisy		*cinerea*	GM, Bl
Cicer	*C. arietinum*	Chickpea	*Fabaceae*	*cinerea*	GM
Cichorium	*C. endivia*	Endive, curled endive	*Asteraceae*	*cinerea*	GM
	C. intybus var. *sativum*	Red-, common-, root- chicory		*cinerea*	GM
Cissus	*C. antartica*	Kangaroo vine	*Vitaceae*	*cinerea*	GM
	C. rhombifolia	Grape ivy		*cinerea*	GM
	C. striata	Grape ivy		*cinerea*	GM
Citrofortunella	*C.* spp.	Calamondin	*Rutaceae*	*cinerea*	GM
Citrullus	*C. lanatus = C. vulgaris*	Watermelon, citron	*Cucurbita*	*cinerea*	GM

Citrus	*C*. spp.	Citrus	*Rutaceae*	*cinerea*	Blo and twig Bl, gummosis, GM, r
	C. australasica	Australian finger lime		*cinerea*	GM
	C. glaberrima	Kinukawa		*cinerea*	GM
	C. japonica 'Margarita'	Oval kumquat			
	C. junos	Yuzu		*cinerea*	GM
	C. limetta	Sweet lime, sweet lemon, sweet limetta		*cinerea*	GM
	C. limon = *C. medica* var. *limon*, *C. limonum*	Lemon		*cinerea*	GM
	C. maxima	Grapefruit		*cinerea*	GM
	C. medica	Citron		*cinerea*	GM
	C.×meyeri	Meyers-, Chinese dwarf- lemon		*cinerea*	GM
	C. natsudaidai	Kawano Natsudaidai		*cinerea*	GM
	C. paradisi	Triumph		*cinerea*	GM
	C. reticulata	Mandarin, ponkan		*cinerea*	GM
	C. sinensis var. *brasiliensis*	Morita navel		*cinerea*	GM
	C. tangelo	Seminole		*cinerea*	GM
	C. tankan	Tankan		*cinerea*	GM
	C. trifoliate = *Poncirus trifoliata*	Trifoliate orange		*cinerea*	GM
	C. unshiu	Cold hardy-, Satsuma-mandarin		*cinerea*	GM
	C. unshiu	Imamura-unshu		*cinerea*	GM
	C.×aurantium	Bitter-, Seville-, sour -, bigarade -, marmalade- orange		*cinerea*	GM
	C.×limon	Eureka lemon		*cinerea*	GM
	C. reticulata×C. sinensis	Ellendale, murcott		*cinerea*	GM

(continued)

Table 20.1 (continued)

Genus	Species	Common name[a]	Family[b]	*Botrytis* species	Disease[c]
	C.×sinensis	Orange		*cinerea*	GM
	C.×tangelo = C. paradisi×C. reticulata	Tangelo		*cinerea*	GM
	C. grandis hybrid	Pomelo		*cinerea*	GM
	Citrus×paradisi	Grapefruit		*cinerea*	GM
	Citrus×paradisi var. Ruby red	Pink grapefruit		*cinerea*	GM
	Citrange×kumquat	Citrangequat		*cinerea*	GM
Clarkia	*C. rubicunda*	Godetia, farewell-to-spring, satin fl.	*Onagraceae*	*cinerea*	Bl
Clematis	*C.* spp.	Clematis	*Ranuncula*	*cinerea*	GM
				elliptica	Bl
	C. integrifolia	Clematis Mongolian bells		-"-	-"-
	C. montana	Anemone clematis		-"-	-"-
	C. terniflora	Sweet autumn clematis		-"-	-"-
	C. texensis	Crimson-, scarlet- clematis		-"-	-"-
	C. viticella	Virgin's bower		*cinerea*	GM
Clerodendrum	*C. splendens*	Flaming glorybower	*Verbena*	*cinerea*	GM
	C. thomsoniae	Bleeding heart		*cinerea*	Bl
Clivia	*C. miniata*	Clivia, Natal-, bush- lily	*Amaryllida*	*cinerea*	GM
Codiaeum	*C. variegatum = Croton variegatum*	Croton, garden-, varie-gated- croton	*Euphorbia*	*cinerea*	Bl
Coffea	*C. arabica*	Arabica coffee	*Rubiaceae*	*cinerea*	Warty berry
	C. canephora	Robusta coffee		*cinerea*	-"-

Colchicum	*C.* spp.	Autumn crocus	*Colchica*	*cinerea*	GM
				elliptica	LSs, tip bl
				tulipae	-"-
				narcissicola	-"-
	C. agrippinum	Autumn crocus		-"-	-"-
	C. autumnale	Autumn snowflake		-"-	-"-
	C. byzantinum	Naked lady, autumn crocus		-"-	-"-
	C. speciosum	Autumn crocus		-"-	-"-
Colocasia	*C. esculenta*	Taro, eddoe	*Araceae*	*cinerea*	Cormel r
Conioselinum	*C.* sp.	Hemlock-parsley	*Apiaceae*	*cinerea*	GM
Consolida	*C.* spp.	Larkspur	*Ranuncula*	*cinerea*	GM
	C. ajacis= C. ambigua, Delphinium ajacis	Doubtful knight's spur, rocket larkspur		*cinerea*	GM
Convallaria	*C. majalis*	Lily of the valley	*Asparaga*	*cinerea*	GM
				paeoniae	GM, Bl
Corchorus	*C. olitorius*	Jew's mallow	*Malvaceae*	*cinerea*	GM
Cordyline	*C.* spp.	Cordyline	*Asparaga*	*cinerea*	GM
Coreopsis	*C. grandiflora*	Tickweed, calliopsis	*Asteraceae*	*cinerea*	Bl
	C. lanceolata	Lanceleaf, coreopsis, pot of gold, longstalk		*cinerea*	Bl
	C. verticillata	Whorled-, thread leaf- coreopsis, thread-leaved tickseed		*cinerea*	Bl

(continued)

Table 20.1 (continued)

Genus	Species	Common name[a]	Family[b]	*Botrytis* species	Disease[c]
Coriandrum	*C. sativum*	Coriander, cilantro(S), Chinese parsley, dhania	*Apiaceae*	*cinerea*	GM
Cornus	*C.* spp.	Dogwood	*Cornaceae*	*cinerea*	GM
	C. canadensis	Canadian dwarf cornels, bunchberries		*cinerea*	GM
	C. florida	Flowering dogwood		*cinerea*	Bl, GM
	C. kousa	Kousa dogwood		*cinerea*	GM
	C. nuttallii	Pacific dogwood		*cinerea*	GM
	C. sanguinea	Common dogwood		*cinerea*	GM
	C. suecica	Eurasian dwarf cornels		*cinerea*	GM
Coronopus	*C. sativus*	Saffron crocus	*Iridaceae*	*cinerea*	GM
Corydalis	*C.* spp.	Greek korydalís, crested lark	*Papavera*	*cinerea*	GM
Corylus	*C.* spp.	Hazel	*Betulaceae*	*cinerea*	GM
	C. avellana	Common hazel		*cinerea*	GM
	C. colurna	Turkish hazel		*cinerea*	GM
	C. maxima	Filbert		*cinerea*	GM
	C. americana	American hazelnut		*cinerea*	GM
Cosmos	*C.* spp.	Cosmos, Mexican aster	*Asteraceae*	*cinerea*	GM
	C. bipinnatus	Garden cosmos		*cinerea*	GM
	C. sulphureus = Bidens sulphurea	Sulfur-, yellow- cosmos		*cinerea*	GM
Costus	*C. guanaiensis*	Cana de India, costus(S)	*Costaceae*	*cinerea*	GM
Cotinus	*C.* spp.	Smoketree, smoke bush	*Anacardia*	*cinerea*	GM
	C. coggygria = Rhus cotinus	Eurasian smoke tr., smoke tr., smoke bush		*cinerea*	GM
	C. obovatus	American smoketree		*cinerea*	GM
Cotoneaster	*C.* spp.	Cotoneaster	*Rosaceae*	*cinerea*	GM
Crambe	*C. abyssinica*	Crambe	*Brassica*	*cinerea*	GM

Craspedia	*C. globosa*	Billy buttons	*Asteraceae*	*cinerea*	Bl
	C. uniflora	Yellow Billy-buttons		*cinerea*	Bl
Crassula	*C.* sp.	Crassula	*Crassula*	*cinerea*	GM
	C. ovata	Popular jade pl.		*cinerea*	GM
Crataegus	*C.* spp.	Hawthorn, thornapple, May-tr., whitethorn,	*Rosaceae*	*cinerea*	GM
	C. mexicana	Tejocotes		*cinerea*	GM
	C. monogyna	Common hawthorn, single-seeded hawthorn		*cinerea*	GM
	C. pinnatifida	Chinese hawthorn		*cinerea*	GM
Crocosmia	*C. aurea*	Falling stars, Valentine Fl., Montbretia	*Iridaceae*	*cinerea*	GM
	C. pottsii	Pott's Montbretia		*cinerea*	GM
	C.×crocosmiiflora	Crocosmia, Montbretia		*cinerea*	GM
Crocus	*C. sativus*	Pott's Montbretia	*Iridaceae*	*cinerea*	GM
				elliptica	LSs
				croci	Bl
Crotalaria	*C. medicaginea*	Trefoil rattlepod	*Fabaceae*	*cinerea*	GM
Cryptanthus	*C. bromeliads*	Earth stars	*Apiaceae*	*cinerea*	GM
Cucumis	*C. melo*	Melon, muskmelon, cantaloupe, casaba(S), melón tuna(S)	*Cucurbita*	*cinerea*	GM
	C. sativus	Cucumber, pepino(S)		*cinerea*	GM
Cucurbita	*C. pepo*	Summer squash	*Cucurbita*	*cinerea*	GM
	C. pepo	Pumpkin, vegetable marrow, courgettes		*cinerea*	GM
	C. maxima	Ornamental gourd		*cinerea*	GM
	C. moschata	Winter squash		*cinerea*	GM
Cuminum	*C. cyminum*	Cumin	*Apiaceae*	*cinerea*	GM
Cunninghamia	*C. lanceolata*	China fir	*Cupressa*	*cinerea*	GM

(continued)

Table 20.1 (continued)

Genus	Species	Common name[a]	Family[b]	*Botrytis* species	Disease[c]
Cuphea	*C. hyssopifolia*	False heather	*Lythraceae*	*cinerea*	GM
Cupressus	*C. macrocarpa*	Monterey cypress	*Cupressa*	*cinerea*	GM
	C. nootkatensis = *Xanthocyparis nootkatensis*	Nootka-, yellow-, Alaska- cypress		*cinerea*	GM
	C. semper-virens	Mediterranean-, Italian-, Tuscan-, graveyard-cypress, pencil pine		*cinerea*	GM
	C.×leylandii	Leyland cypress		*cinerea*	GM
Cyathea = *Alsophila*	*C. pungens* = *A. procera*	Spiny treefern, alsophila	*Cyathea*	*cinerea*	GM
Cycas	*C.* spp.	Cycas	*Cycadaceae*	*cinerea*	GM
	C. revoluta	Sago palm		*cinerea*	GM
Cyclamen	*C. persicum*	Persian cyclamen, violeta de los Alpes(S)	*Primulaceae*	*cinerea*	GM
Cydonia	*C. oblonga*	Quince	*Rosaceae*	*cinerea*	GM
Cymbalaria	*C. muralis*	Ivy-leaved toadflax, Kenilworth ivy	*Scrophularia*	*cinerea*	GM
Cymbidium	*C.* sp.	Cymbidium	*Orchida*	*cinerea*	GM
Cynara	*C. cardunculus* subsp. *scolymus* = *C. scolymus*	Artichoke, globe artichoke, artichoke thistle, cardone, cardoni, carduni, cardi	*Asteraceae*	*cinerea*	GM
Cynodon	*C. dactylon*	Bermuda-, Bahama-, dog's tooth-, Indian doab	*Poaceae*	*cinerea*	GM
Cynoglossum	*C. officinale*	Hound's tongue	*Boragina*	*cinerea*	GM
Cyphomandra	*C. betacea* = *Solanum betaceum*	Tr. tomato, tamarillo(S), Tomate de árbol(S)	*Solanaceae*	*cinerea*	GM
Cyperus	*C. papyrus*	Papyrus sedge, paper reed, Nile grass, papiro(S)	*Cyperaceae*	*cinerea*	GM
Cyrtanthus	*C. mackenii*	Ifafa lily	*Amaryllida*	*cinerea*	GM
Cytisus	*C. scoparius*	Scotch broom	*Fabaceae*	*cinerea*	GM
Dahlia	*D.* sp.	Garden dahlia	*Asteraceae*	*cinerea*	Fl. Bl

Daphne	*D. odora*	Winter daphne	*Thymelaea*	*cinerea*	GM
Dasylirion	*D. serratifolium*	Sandpaper sotol	*Asparaga*	*cinerea*	LSs & Bl
Datura	*D.* sp.	Thorn apple	*Solanaceae*	*cinerea*	GM
Daucus	*D. carota subsp. sativus = D. maximus*	Carrot	*Apiaceae*	*cinerea*	GM
Davallia	*D.* spp.	Rabbit foot fern	*Polypodia*	*cinerea*	GM
Delphinium	*D. ajacis = Consolida ambigua*	Delphinium	*Ranuncula*	*cinerea*	GM
	D. ambiguum	Delphinium		*cinerea*	GM
Dendranthema	*D.×grandiflorum*	Garden mum, garden chrysanthemum	*Asteraceae*	*cinerea*	Bl
Dendrobium	*D.* sp.	Dendrobium orchid	*Orchida*	*cinerea*	Blo Bl
Dianthus	*D. caryophyllus*	Carnation	*Caryophylla*	*cinerea*	GM
	D. zeyheri	Wild pink dianthus, African carnation		*cinerea*	GM
Didiscus	*D. caeruleus*	Blue lace flower	*Apiaceae*	*cinerea*	GM
Dieffenbachia	*D.* sp.	Dumb cane	*Araceae*	*cinerea*	GM
Digitalis	*D.* spp.	Foxglove	*Scrophularia*	*cinerea*	GM
	D. purpurea	Common-, purple-foxglove, lady's glove		*cinerea*	GM
Dimorphotheca	*D.* spp.	Cape marigold	*Calenduleae*	*cinerea*	GM
	D. aurantiaca	African daisy		*cinerea*	GM
	D. ecklonis	Cape marguerite, Van Staden's river daisy		*cinerea*	GM
	D. sinuata	Salmon African daisy		*cinerea*	GM
	D. pluvialis	Cape marigold, African daisy, star-of-the-Veldt		*cinerea*	GM
Dionaea	*D. muscipula*	Venus flytrap, Venus atrapamoscas(S)	*Droseraceae*	*cinerea*	GM
Diospyros	*D. kaki = D. chinensis*	Persimmon	*Ebenaceae*	*cinerea*	GM
Dioscorea	*D. zingiberensis*	Wild-, Atlantic- yam, China-, colic- root	*Dioscorea*	*cinerea*	GM
Disporum	*D. pullum*	Fairy bells	*Colchica*	*cinerea*	GM

(continued)

Table 20.1 (continued)

Genus	Species	Common name[a]	Family[b]	*Botrytis* species	Disease[c]
Dracaena	*D.* spp.	Dracaena, dragon tr.	*Asparaga*	*cinerea*	GM
	D. draco	Dragon's-blood		*cinerea*	GM
	D. fragans	Cornstalk dracaena, palo de agua(S)		*cinerea*	GM, Bl
	D. ombet	Gabal Elba dragon tr.		*cinerea*	GM
	D. sanderana	Belgian evergreen		*cinerea*	GM
Drosera	*D.* sp.	Sundews	*Droseraceae*	*cinerea*	GM
Dryopteris	*D.* spp.	Wood-, male-fern	*Asplenia*	*cinerea*	Bl
Echeveria	*E.* sp.	Echeveria	*Crassula*	*cinerea*	GM
	E. elegans	Mexican snow ball, white Mexican rose		*cinerea*	GM
Echinacea	*E.* sp.	Cone fl.	*Asteraceae*	*cinerea*	Bl
	E. angustifolia	Cone fl.		*cinerea*	Bl
	E. pallida var. *angustifolia*	Pale purple coneflower		*cinerea*	Bl
	E. purpurea	Eastern purple cone fl.		*cinerea*	Bl
Echinops	*E. ritro*	Globe thistle	*Asteraceae*	*cinerea*	Bl
Echinopsis	*E.* sp.	Hedgehog-, sea-urchin-cactus, easter lily	*Cactaceae*	*cinerea*	GM
	E. chamaecereus = *Chamaecereus silvestrii*	Peanut cactus		*cinerea*	GM
Echium	*E. plantagineum*	Purple viper's bugloss, Paterson's curse	*Boragina*	*cinerea*	GM
Epidendrum	*E.* spp.	Epidendrum	*Orchida*	*cinerea*	Blo Bl
	E. cinnabarinum	Vermilion epidendrum		*cinerea*	-"-
	E. ibaguense	Ground-rooting, fire-star orchid		*cinerea*	-"-
	E. nocturnum	Night scented-, nocturnal-epidendrum		*cinerea*	-"-
	E. radicans	Fire-star orchid, ground-rooting epidendrum		*cinerea*	-"-
	E. secundum	Lopsided star orchid		*cinerea*	-"-

Epilobium	*E.* spp.	Willowherb, fireweed	*Onagraceae*	*cinerea*	Bl
	E. angustifolium	Willowherb, fireweed		*cinerea*	Bl
	E. fleischeri	Alppihorsma, Alpine willowherb		*cinerea*	Bl
Epiphyllum	*E.* spp.	Orchid-, leaf-cacti	*Cactaceae*	*cinerea*	Bl
	E. anguliger	Orchid-, leaf-cacti		*cinerea*	Bl
	E. oxypetalum	Night queen		*cinerea*	Bl
Epipremnum	*E. aureus= E. pinnatum*	Pothos, devils ivy, scindapsus	*Araceae*	*cinerea*	GM
Episcia	*E.* spp.	Flame	*Gesneria*	*cinerea*	GM
	E. cupreata	Flame violet		*cinerea*	GM
	E. reptans	Flame violet		*cinerea*	GM
Eremurus	*E. robustus*	Giant desert candle	*Liliaceae*	*cinerea*	Bl
Erica	*E.* sp.	Heath, heather	*Ericaceae*	*cinerea*	GM
Erigeron	*E.* spp.	Fleabane	*Asteraceae*	*cinerea*	GM
Eriobotrya	*E. japonica= Mespilus japonica*	Loquat	*Rosaceae*	*cinerea*	Blo Bl
Eryngium	*E.* spp.	Eryngium, eryngo, sea holly	*Apiaceae*	*cinerea*	GM
Erysimum	*E. cheiri =. Cheiranthus cheiri*	Wallflower	*Brassica*	*cinerea*	GM
Erythronium	*E.* spp.	Fawn-, trout-lily	*Liliaceae*	*cinerea*	GM
				elliptica	Bl
	E. albidum	White trout lily		-"-	-"-
	E. denscanis	Dog's tooth violet		-"-	-"-
	E. japonicum	Katakuri		-"-	-"-
Eschscholzia	*E. californica*	California poppy	*Papavera*	*cinerea*	GM
	E. caespitosa	Collarless California-, tufted-, foothill- poppy		*cinerea*	GM
	E. lobbii	Frying pans		*cinerea*	GM

(continued)

Table 20.1 (continued)

Genus	Species	Common name[a]	Family[b]	*Botrytis* species	Disease[c]
Eucalyptus	*E.* spp.	Eucalyptus, gum	*Myrtaceae*	*cinerea*	GM
				fusca	Bl
	E. alba	White-, poplar-gum		*cinerea*	GM
	E. benthami	Camden white gum		*cinerea*	GM
	E. botryoides	Bangalay, southern mahogany		*cinerea*	GM
	E. camaldulensis	River-, Murray- red gum		*cinerea*	GM
	E. cinerea	Argyle apple, silver dollar		*cinerea*	GM
	E. citriodora	Lemon scented gum		*cinerea*	GM
	E. cladocalyx	Sugar gum		*cinerea*	GM
	E. cordata	Silver dollar eucalyptus, heart-leaved silver gum		*cinerea*	GM
	E. deglupta	Mindanao gum, rainbow eucalyptus		*cinerea*	GM
	E. delegatensis	Alpine ash, gum-topped stringybark, white-top		*cinerea*	GM
	E. fastigata	Brown barrell, cut-tail		*cinerea*	GM
	E. fraxinoides	White mountain ash		*cinerea*	GM
	E. globulus ssp. *Bicostata* = *E. globulus* var. *bicostata*	Tasmanian blue-, southern blue-gum, eurabbie		*cinerea*	GM
	E. gunnii	Cider gum		*cinerea*	GM
	E. leucoxylon	Yellow-, blue-gum		*cinerea*	GM
	E. nitens = *E. goniocalyx* var. *nitens*	Shining gum		*cinerea*	GM
	E. ovata	Swamp-, black-gum		*cinerea*	GM
	E. regnans	Giant-, mountain-, Victorian-ash, swamp gum		*cinerea*	GM
	E. pilularis	Blackbutt		*cinerea*	GM
				fusca	Bl
	E. saligna	Sydney blue gum		*cinerea*	GM

	E. sideroxylon	Mugga, red-, Mugga- Ironbark		*cinerea*	GM
	E. sieberiana = *E. sieberi*	Coast ash, black ash		*cinerea*	GM
	E. smithii	Gully gum		*cinerea*	GM
	E. urophylla	Rose gum, flooded gum		*cinerea*	GM
Eucharis	*E.×grandiflora* = *E. mastersii*	Amazon lily	*Amaryllida*	*cinerea*	GM
	E. grandiflora	Amazon lily		*cinerea*	GM
	E. amazonica	Amazon-, eucharist-lily		*cinerea*	GM
Eucomis	*E. vandermerwei*	Pineapple lily	*Asparaga*	*hyacinthi*	GM
Euonymus	*E.* spp.	Spindle, spindle tr.	*Celastra*	*cinerea*	GM
	E. europaeus	Spindle, European-, common-spindl		*cinerea*	GM
	E. japonicus	Japanese spindle		*cinerea*	GM
Eupatorium	*E. cannabinum*	Hemp-agrimony	*Asteraceae*	*cinerea*	GM
Euphorbia	*E. pulcherrima*	Poinsettia, flor de pascua(S), nochebuena(S), stella di natale	*Euphorbia*	*cinerea*	GM, Bl
				ricini	GM, Bl
Eustoma	*E. grandiflorum* = *Lisianthus russellianus*	Lisianthus, lisianto(S)	*Gentiana*	*cinerea*	GM
Evolvulus	*E. nuttalianus*	Shaggy dwarf morning glory	*Convolvula*	*cinerea*	GM
Exacum	*E. affine*	Arabian-, German-Persian-violet	*Gentiana*	*cinerea*	Bl
Fagus	*F. sylvatica*	Copper-, purple-, European-, common-beech	*Fagaceae*	*cinerea*	GM
Feijoa	*F. sellowiana* = *Acca sellowiana*	Feijoa, pineapple guava, guavasteen	*Myrtaceae*	*cinerea*	GM
Fatsia	*F. japonica*	Fatsi, Japanese aralia, glossy-leaved paper pl.	*Araliaceae*	*cinerea*	GM
Felicia	*F. amelloides*	Felicia pinwheel	*Asteraceae*	*cinerea*	GM

(continued)

Table 20.1 (continued)

Genus	Species	Common name[a]	Family[b]	*Botrytis* species	Disease[c]
Ficaria	*F. verna*	Fig buttercup, lesser celandine	*Ranuncula*	*ficariarum*	–
Ficus	*F. auriculata*	Timla fig	*Moraceae*	*cinerea*	GM
	F. benjamina	Weeping fig		*cinerea*	Bl
	F. carica	Common fig, breva(S)		*cinerea*	Limb Bl, dieback
	F. elastica	rubber-tr., -fig, -bush, rubber pl.		*cinerea*	GM
	F. lyrata	Fiddle-leaf fig		*cinerea*	GM
Fittonia	*F.* sp.	Nerve pl., mosaic pl.	*Acantha*	*cinerea*	GM
Foeniculum	*F. vulgare*	Fennel	*Apiaceae*	*cinerea*	GM
Forsythia	*F.* sp.	Forsythia	*Oleaceae*	*cinerea*	GM
	F.×intermedia	Border forsythia		*cinerea*	GM
Fortunella	*F.* spp.	Kumquat	*Rutaceae*	*cinerea*	GM
Fragaria	*F.×ananassa*	Strawberry, fresa(S)	*Rosaceae*	*cinerea*	GM, lf Bl, dry crown r
				caroliniana	GM
				pseudocinerea	GM
	F. vesca	Wild strawberry		*cinerea*	GM
	F. ananassa ×F. moschata hybrids = F.×ananachata	Garden strawberry		*cinerea*	GM
Fraxinus	*F. excelsior*	Ash	*Oleaceae*	*cinerea*	GM
Freesia	*F. refracta×F. leichtlinii*	Freesia hybrid	*Iridaceae*	*cinerea*	GM
Fuchsia	*F.×hybrida*	Fuchsia	*Onagraceae*	*cinerea*	Bl
	F. excorticata	Tree-, New Zealand- fuchsia, kotukutuku (M_ori)		*cinerea*	Bl
	F. magellanica	Hummingbird-, hardy- fuchsia		*cinerea*	Bl
Galanthus	*G.* spp.	Snowdrops	*Amaryllida*	*galanthina*	GM
	G. nivalis	Snowdrops		*galanthina*	GM

Galega	*G. orientalis*	Eastern-, fodder- galega, oriental goat's rue	*Fabaceae*	*cinerea*	GM
Gardenia	*G.* spp.	Gardenia	*Rubiaceae*	*cinerea*	GM
	G. jasminoides	Common gardenia, cape jasmine		*cinerea*	GM
Gaura	*G. lindheimeri*	Wand flower		*cinerea*	Bl
Gaylussacia	*G.* spp.	Huckleberry	*Ericaceae*	*cinerea*	GM
	G. baccata	Black huckleberry		*cinerea*	GM
Gazania	*G. rigens*	Clumping gazania	*Asteraceae*	*cinerea*	GM
	G. rigens var. *leucolaena*	Trailing gazania		*cinerea*	GM
Gentiana	*G.* spp.	Gentian	*Gentiana*	*cinerea*	GM
Geranium	*G.* spp.	Crane's-bill	*Geraniaceae*	*cinerea*	Bl
	G. clarkei	Clark's geranium		*cinerea*	Bl
	G. endressii	Endres's cranesbill		*cinerea*	Bl
	G. fremontii	Fremont's geranium		*cinerea*	Bl
	G. ibericum	Caucasus geranium		*cinerea*	Bl
	G. macrorrhizum	Bigroot cranesbill, bigroot geranium		*cinerea*	Bl
	G. maculatum	Wild geranium		*cinerea*	Bl
	G. maderense	Giant herb robert		*cinerea*	Bl
	G. platypetalum	Broad-petaled geranium		*cinerea*	Bl
	G. pratense	Meadow cranesbill		*cinerea*	Bl
	G. psilostemon	Armenian cranesbill		*cinerea*	Bl
	G. renardii	Renard geranium		*cinerea*	Bl
	G. sanguineum	Bloody cranesbill		*cinerea*	Bl
	G. subcaulescens	Grey cranesbill		*cinerea*	Bl
	G. sylvaticum	Wood cranesbill		*cinerea*	Bl
	G.×magnificum	Showy geranium		*cinerea*	Bl

(continued)

Table 20.1 (continued)

Genus	Species	Common name[a]	Family[b]	*Botrytis* species	Disease[c]
Gerbera	*G. jamesonii*	Gerbera, gerbera-, Transvaal-, African- daisy	*Asteraceae*	*cinerea*	Bl, GM
Gilia	*G.* spp.	Gilia	*Polemonia*	*cinerea*	GM
Ginkgo	*G. biloba*	Maidenhair tr.	*Ginkgoa*	*cinerea*	GM
Gladiolus	*G.* spp.	Gladioli, sword lily	*Iridaceae*	*cinerea*	GM
				gladiolorum	Lf Bl
				elliptica	Fire
	G. grandiflorus	Grandiflora gladiolus		-"-	-"-
	G. cardinalis×G. tristis	Colville's gladiolus		-"-	-"-
	G. dalenii×G. oppositiflorus	Gladiolus		-"-	-"-
	G. nanus	Dwarf gladiolus		-"-	-"-
Glycosmis	*G. parviflora*	Flower axistree	*Rutaceae*	*cinerea*	GM
	G. pentaphylla	Orangeberry, gin berry		*cinerea*	GM
Gloriosa	*G.* sp.	Flame-, fire-, glory-, superb-, climbing- lily	*Colchica*	*cinerea*	GM
	G. superba =*G. rothschildiana*	Flame-, climbing-, glory-, creeping-, gloriosa-, fire- lily		*cinerea*	GM
Glycine	*G. max*	Soybean, soya bean	*Fabaceae*	*cinerea*	GM
Gomphrena	*G. serrata*	Amaranth, prostrate globe-, coastal globe- amaranth	*Amarantha*	*cinerea*	GM
Gossypium	*G. hirsutum*	Cotton	*Malvaceae*	*cinerea*	GM
Grevilea	*G.* spp.	Grevilea, spider fl., silky oak, toothbrush pl.	*Proteaceae*	*cinerea*	GM
	G. mucronulata	Green spider fl., green grevillea		*cinerea*	GM
	G. pteridifolia	Silky grevillea, Darwin silky oak		*cinerea*	GM
	G. rosmarinifolia	Rosemary grevillea		*cinerea*	GM
	G. robusta	Silky-, Australian silky- oak		*cinerea*	GM

Guzmania	*G.* spp.	Guzmania	*Bromelia*	*cinerea*	GM
	G. lingulata	Scarlet star		*cinerea*	GM
Gymnocarpium	*G.* spp.	Western-, common-, northern- oak fern	*Asplenia*	*cinerea*	Bl
Gynura	*G. aurantiaca*	Purple passion vine	*Asteraceae*	*cinerea*	Bl
Gypsophila	*G. paniculata*	Gypsophila, baby's-breath	*Caryophylla*	*cinerea*	GM
Hatiora	*H. gaertneri* = *Rhipsalidopsis gaertneri*	Easter-, Whitsun- cactus	*Cactaceae*	*cinerea*	GM
	H. rosea = *Rhipsalidopsis rosea*	Easter cactus		*cinerea*	GM
Hebe	*H. parviflora* = *Veronica parviflora*	Shrubby veronica	*Scrophularia*	*cinerea*	GM
Hedera	*H. canariensis*	Canarian ivy	*Araliaceae*	*cinerea*	GM
	H. helix	English ivy		*cinerea*	GM
Helenium	*H. flexuosum* = *H. nudiflorum*	Purple-headed sneezeweed	*Asteraceae*	*cinerea*	GM
Helianthemum	*H.* spp.	Rock rose, sunrose, rushrose, frostweed	*Cistaceae*	*cinerea*	Bl
Heliconia	*H. bihai*	Red palulu, heliconia(S)	*Heliconia*	*cinerea*	GM
Helleborus	*H. orientalis*	Hellebores	*Ranuncula*	*cinerea*	Bl
Helianthus	*H. annuus*	Sunflower	*Asteraceae*	*cinerea*	Head r, GM
	H. tuberosus	Sunchoke, sunroot, Je-rusalem artichoke		*cinerea*	GM
	H. tuberosus×*H. macrophyllus*	Topin		*cinerea*	GM
Heliotropium	*H.* spp.	Heliotrope	*Boragina*	*cinerea*	GM
	H. arborescens	Garden heliotrope		*cinerea*	GM

(continued)

Table 20.1 (continued)

Genus	Species	Common name[a]	Family[b]	*Botrytis* species	Disease[c]
Hemerocallis	*H.* spp.	Daylily	*Xanthorrhoea*	*cinerea*	GM
				deweyae	Spring sickness
				elliptica	Fire
	H. fulva	Tawny-, railroad-, roadside- daylily		*cinerea*	GM
	H. lilioasphodelus= H. flava	Lemon-lily		*cinerea*	GM
Heuchera	*H.* spp.	Alumroot, coral bells	*Saxifraga*	*cinerea*	GM
	H. americana	American alumroot		*cinerea*	GM
	H. elegans	Urnflower alumroot		*cinerea*	GM
Hibiscus	*H.* spp.	Hibiscus	*Malvaceae*	*cinerea*	Bl, GM
	H. canabinus	Kenaf		*cinerea*	Bl, GM
	H. sabdariffa	Roselle, sour tea		*cinerea*	Bl, GM
	H. syriacus	Rose-of-Sharon		*cinerea*	GM
	H. rosasinensis	Chinese-, Hawaiian- hibiscus, China rose		*cinerea*	Bl, GM
Hippophae	*H. rhamnoides*	Common sea-buckthorn	*Elaeagna*	*cinerea*	GM
Hordeum	*H. vulgare*	Barley	*Poaceae*	*cinerea*	GM
Hosta	*H. fortunei* var. *hyacinthina*	Plantain lily, hosta	*Asparaga*	*cinerea*	Bl
Howea	*H.* spp.	Howea palm	*Arecaceae*	*cinerea*	GM
	H. forsteriana	Kentia-, thatch- palm		*cinerea*	GM
Hoya	*H. carnosa*	Wax pl.	*Asclepiada*	*cinerea*	GM
Hyacinthoides	*H.* sp.	Bluebell	*Asparaga*	*cinerea*	GM
	H. non-scripta = Endymion non-scriptus	Wood hyacinth		*cinerea*	GM
Hyacinthus	*H. orientalis*	Common-, garden -, Dutch-hyacinth	*Asparaga*	*cinerea*	GM
				hyacinthi	Fire

Hydrangea	*H.* spp.	Hydrangea, hortensia	*Hydrangea*	*cinerea*	Bl
	H. macrophylla	Bigleaf-, florist's- hy-drangea		*cinerea*	Bl
	H. paniculata	Panicle hydrangea		*cinerea*	Bl
	H. arborescens	Smooth hydrangea		*cinerea*	Bl
	H. quercifolia	Oakleaf hydrangea		*cinerea*	Bl
Hydrastis	*H. canadensis*	Golden seal	*Ranuncula*	*cinerea*	GM
Hylocereus	*H.* spp.	Dragon fr., red pitaya	*Cactaceae*	*cinerea*	GM
	H. megalanthus = *Selenicereus megalanthus*	Pitahaya, yellow pitahya, Pitaya(S)		*cinerea*	GM
	H. undatus	Dragon fr., night blooming cereus		*cinerea*	GM
Hymenocallis	*H.* spp.	Hymenocallis	*Amaryllida*	*cinerea*	GM
	H. amancaes	Amancae, amancay		*cinerea*	GM
	H. caribaea	Caribbean-, variegated- spider-lily		*cinerea*	GM
	H. harrisiana	Spider lily, summer-, Peruvian- daffodil, basket fl.,		*cinerea*	GM
	H. littoralis	Beach spider lily		*cinerea*	GM
	H. narcissiflora	Peruvian daffodil		*cinerea*	GM
	H. speciosa	Green-tinge Spiderlily		*cinerea*	GM
	H.×macrostephana	Day lily		*cinerea*	GM
	H.×festalis	Hybrid		*cinerea*	GM
	H.×spofforthiae	Hybrid		*cinerea*	GM
Hypoestes	*H.* spp.	Mosaic pl.	*Acantha*	*cinerea*	GM
	H. phyllostachya	Polka dot pl.		*cinerea*	GM

(continued)

Table 20.1 (continued)

Genus	Species	Common name[a]	Family[b]	*Botrytis* species	Disease[c]
Hypericum	*H. an-drosaemum*	Tutsan	*Hyperica*	*cinerea*	GM
	H. canariense = *H. canariense* var. *floribundum*	Canary Islands St. John's-wort		*cinerea*	GM
	H. chinense	Chinese St. John's wort fr.		*cinerea*	GM
Hyssopus	*H. officinalis*	Hyssop	*Lamiaceae*	*cinerea*	GM
Iberis	*I.* sp.	Candytuft	*Brassica*	*cinerea*	GM
	I. sempervirens	Evergreen candytuft, perennial candytuft		*cinerea*	GM
Ilex	*I.* spp.	Holly	*Aquifolia*	*cinerea*	GM
	I. aquifolium	Common European holly		*cinerea*	GM
	I. cornuta	Chinese holly		*cinerea*	GM
	I. crenata	Box-leaved holly		*cinerea*	GM
	I. verticillata	Winterberry		*cinerea*	GM
Impatiens	*I.* spp.	Impatiens, jewelweed, touch-me-not, snapweed	*Balsamina*	*cinerea*	Bl
	I. aurantiaca	Impatiens		*cinerea*	Bl
	I. hawkeri	New Guinea impatiens		*cinerea*	Bl
	I. holstii	Busy Lizzy		*cinerea*	Bl
	I. marinae	Balsam		*cinerea*	Bl
	I. oliveri	Olivers touch me not		*cinerea*	Bl
	I. platypetala	–		*cinerea*	Bl
	I. petersiana	Busy Lizzie, balsam		*cinerea*	Bl
	I. wallerana = *I. sultanii*	Busy Lizzie, balsamina, balsam impatiens, miramelindos(S)		*cinerea*	Bl
Ipomoea	*I. batatas*	Sweet potato	*Convolvula*	*cinerea*	GM r

Iris	*I.* spp.	Iris, netted iris, Dutch iris	*Iridaceae*	*convoluta*	Crown r, GM
				narcissicola	Narcissus smoulder
	I. amoena	Pale yellow-, German-, lemon yellow-iris		-"-	-"-
	I. flavescens	Lemon yellow iris		-"-	-"-
	I. junonia	Junonia		-"-	-"-
	I. pumila	Dwarf iris		-"-	-"-
	I. reichenbachii	Yellow-, purple water-flag		-"-	-"-
	I. sambucina	Flags		-"-	-"-
	I. squalens	Flags		-"-	-"-
	I. variegata var. *lurida*	Flags		-"-	-"-
	I. sanibucina×I. variegata	Iris neglecta		-"-	-"-
	I.×Gladiolus	Orchidiola		-"-	-"-
Ischnocarpus	*I. exilis*	NZ native cress	*Brassica*	*cinerea*	GM
Ixia	*I. viridiflora*	Turquoise ixia	*Iridaceae*	*cinerea*	GM
Jacobaea	*J. maritima = Senecio cineraria*	Dusty miller	*Asteraceae*	*cinerea*	GM
Jasminum	*J.* spp.	Jasmine, gelsomino, gelsemium	*Oleaceae*	*cinerea*	GM
	J. officinalis	Common jasmine		*cinerea*	St Bl, GM
	J. polyanthum	Pink-, white- jasmine		*cinerea*	GM
Juglans	*J.* spp.	Walnut	*Juglanda*	*cinerea*	GM
	J. cinerea	Butternut-, white- walnut		*cinerea*	GM
	J. nigra	Eastern black walnut		*cinerea*	GM
	J. regia	English walnut		*cinerea*	GM

(continued)

Table 20.1 (continued)

Genus	Species	Common name[a]	Family[b]	*Botrytis* species	Disease[c]
Juniperus	*J.* sp.	Juniper	*Cupressa*	*cinerea*	GM
Justicia	*J. brandegeeana*	Mexican shrimp pl., water willow, beloperone,	*Acantha*	*cinerea*	GM
Kalanchoe	*K.* spp.	Kalanchoe	*Crassula*	*cinerea*	Bl
	K. blossfeldiana	Flaming Katy, Christmas-, florist- kalanchoe		*cinerea*	Bl
	K. beharensis	Elephant's ear kalanchoe, felt bush, feltbush		*cinerea*	Bl
	K. laciniata	Fir-tree-, Christmas-tree -kalanchoe		*cinerea*	GM
Kalopanax	*K. pictus= K. septemlobus*	Prickly castor oil tree	*Araliaceae*	*cinerea*	GM
Laburnum	*L.* sp.	Golden chain	*Fabaceae*	*cinerea*	GM
Lachenalia	*L. aloides*	Cape cowslip, lachenalia	*Asparaga*	*galanthina*	GM
				paeoniae	Bl
Lactuca	*L. sativa*	Lettuce	*Asteraceae*	*cinerea*	GM
Laelia	*L.* spp.	Laelia	*Orchida*	*cinerea*	Blo Bl
Laeliocattleya	*L.×Cattleya*	Laeliocattleya	*Orchida*	*cinerea*	-"-
Lagenaria	*L. siceraria*	Bottle gourd, calabash	*Cucurbita*	*cinerea*	GM
Lagerstroemia	*L. indica*	Crape-, crepe- myrtle	*Lythraceae*	*cinerea*	GM
Lallemantia	*L. iberica*	Dragon's head	*Lamiaceae*	*cinerea*	GM
Larix	*L. decidua*	European larch	*Pinaceae*	*cinerea*	GM
				douglasii	GM
	L. kaempferi= L. leptolepsis	Japanese larch		-"-	-"-
	L. occidentalis	Western larch		*cinerea*	GM
Lathyrus	*L. odoratus*	Sweet pea	*Fabaceae*	*cinerea*	GM
	L. latifolius	Everlasting-, perennial-pea		*cinerea*	GM
	L. sativus	Grass pea		*cinerea*	GM
Lavandula	*L. angustifolia*	English lavender	*Lamiaceae*	*cinerea*	GM
	L. stoechas	French lavender		*cinerea*	GM

Lens	*L. culinaris*	Lentil	*Fabaceae*	*cinerea*	Bl, GM
				fabae	CS
Lepidium	*L. sativum*	Cress, garden cress	*Brassica*	*cinerea*	GM
Leptospermum	*L.* spp.	Tea-tr.	*Myrtaceae*	*cinerea*	GM
	L. rotundifolium	Round leaf tea tr.		*cinerea*	GM
	L. scoparium	Manuka myrtle, NZ-, broom- tea tr., tea tr.		*cinerea*	GM
Leucadendron	*L.* spp.	Cone-bush	*Proteaceae*	*cinerea*	GM
	L. daphnoides	Leucadendron daphnoides		*cinerea*	GM
	L. discolor	Christmas cones, Inca gold		*cinerea*	GM
	L. floridum	Leucadendron 'Pisa'		*cinerea*	GM
	L. galpinii	Silver balls		*cinerea*	GM
	L. gandogeri	Broadleaf conebush		*cinerea*	GM
	L. laureolum	Grey leaf sugarbush		*cinerea*	GM
	L. salignum	Leucadendron 'Long Tom'		*cinerea*	GM
	L. laureolum×L. salignum	Leucadendron- 'Inca gold', 'safari sunset'		*cinerea*	GM
Leucospermum	*L.* spp.	Protea	*Proteaceae*	*cinerea*	GM
	L. cordifolium	Red pincushion protea		*cinerea*	GM
	L. reflexum	Rocket pincushion, skyrocket leucospermum, pincushion protea		*cinerea*	GM
	L. tottum×L. formosum	Spider pincushion, hybrid		*cinerea*	GM
Levisticum	*L. officinale*	Lovage	*Apiaceae*	*cinerea*	GM
Liatris	*L. spicata*	Liatris	*Asteraceae*	*cinerea*	GM
Ligustrum	*L. vulgare*	Wild privet, common-, European- privet	*Oleaceae*	*cinerea*	GM
	L. ovalifolium	California-, oval-leaved- privet		*cinerea*	GM

(continued)

Table 20.1 (continued)

Genus	Species	Common name[a]	Family[b]	*Botrytis* species	Disease[c]
Lilium	*L.* spp.	Lily	*Liliaceae*	*cinerea*	GM
				elliptica	Fire, lily Bl
				hyacinthi	Fire
				tulipae	Fire
	L. bulbiferum,	Orange-, fire-, tiger- lily		-"-	-"-
	L. candidum	Madonna lily, azucena(S), lis(S)		-"-	-"-
	L. martagon	Martagon, Turk's cap lily		-"-	-"-
	L. regale	King's-, regal-, royal-lily		-"-	-"-
	L. longiflorum	White trumpet lily, easter lily, St. Joseph's lily		-"-	-"-
	L. speciosum	Oriental-, Japanese-lily		-"-	-"-
	L. oriental hybrid	Oriental lily		-"-	-"-
Limnanthes	*L. alba*	Meadowfoam	*Limnantha*	*cinerea*	Bl, GM
Limonium	*L.* spp.	Limonium, statice	*Plumbagina*	*cinerea*	GM
	L. peregrinum	Strandroos		*cinerea*	GM
	L. sinuatum	Wavyleaf sea-lavender		*cinerea*	GM
Linaria	*L.* spp.	Linaria	*Plantagina*	*cinerea*	GM
				viciae	CS
	L. dalmatica	Balkan-, broadleaf-, Dalmatian- toadflax		*cinerea*	GM, Bl
	L. maroccana	Moroccan-, annual-toadflax		*cinerea*	GM
	L. purpurea	Purple toadflax		*cinerea*	GM
	L. vulgaris	Toadflax, yellow toadflax, butter-and-eggs		*cinerea*	GM
Linum	*L. usitatissimum*	Flax, linseed, spring linseed	*Linaceae*	*cinerea*	GM
Liriope	*L.* spp.	Lily grass	*Asparaga*	*cinerea*	GM
Litchi	*L. chinensis = Nephelium litchi*	Lychee	*Sapinda*	*cinerea*	GM

Lithops	*L.* sp.	Pebble pl., living stones	*Aizoaceae*	*cinerea*	GM
Lithospermum	*L.* sp.	Gromwell, stoneseed	*Boragina*	*cinerea*	GM
	L. erythrorhizon	Purple gromwell, red-root gromwell		*cinerea*	GM
Lobelia	*L.* spp.	Lobelia	*Campanula*	*cinerea*	GM
Lobelia	*L. cardinalis = L. fulgens*	Cardinal fl., Indian pink	*Campanula*	*cinerea*	GM
	L. siphilitica	Blue lobelia		*cinerea*	GM
	L. erinus	Edging-, garden-, trailing-lobelia		*cinerea*	GM
Lolium	*L. perenne*	Perennial-, English-, winter-ryegrass	*Poaceae*	*cinerea*	GM
Lonicera	*L.* spp.= *Caprifolium*	Honeysuckle	*Caprifolia*	*cinerea*	GM
	L. similis	Honeysuckle		*cinerea*	GM
Lotononis	*L. bainesii= Amphinomia bainesii*	Lotononis	*Fabaceae*	*cinerea*	GM
Lotus	*L. pedunculatus*	Greater bird's-foot trefoil, Marsh bird's-foot	*Fabaceae*	*cinerea*	GM
Luffa	*L. cylindrica*	Luffa, sponge gourd	*Cucurbita*	*cinerea*	GM
Lunaria	*L.* spp.	Honesty	*Brassica*	*cinerea*	GM
	L. rediviva	Perennial honesty		*cinerea*	GM
	L. annua	Honesty, annual hones		*cinerea*	GM
Lupinus	*L.* spp.	Lupin, lupine	*Fabaceae*	*cinerea*	GM
	L. albus	White lupin		*cinerea*	GM
	L. angustifolius	Australian sweet-, narrow-leafed-lupin		*cinerea*	GM
	L. hirsutus	Blue lupin		*cinerea*	GM
	L. luteus	Yellow lupin		*cinerea*	GM
	L. mutabilis	Andean lupin, tarwi		*cinerea*	GM
	L. polyphyllus	Garden lupin		*cinerea*	GM
	L.×regalis	Rainbow lupin		*cinerea*	GM

(continued)

Table 20.1 (continued)

Genus	Species	Common name[a]	Family[b]	*Botrytis* species	Disease[c]
Lychnis	*L.* sp.	Campion, wall flowers	*Caryophylla*	*cinerea*	GM
	L. chalcedonica = Silene chalcedonica	Maltese-, Jerusalem- cross, burning love, dusky salmon		*cinerea*	GM
	L. tenuis	Narrow-leaved bird's-foot-trefoil		*cinerea*	GM
Macadamia	*M. integrifolia*	Macadamia-, bauple- nut	*Proteaceae*	*cinerea*	GM
	M. tetraphylla	Macadamia-, bauple- nut, prickly macadamia		*cinerea*	GM
Maclura	*M. pomifera*	Osage-orange	*Moraceae*	*cinerea*	GM
Magnolia	*M.* spp.	Magnolia	*Magnolia*	*cinerea*	GM
	M. grandiflora	Southern-, evergreen-magnolia		*cinerea*	GM
Malus	*M. sylvestris* var. *domestica = Pyrus malus*; *M. sylvestris M.×domestica*	Apple, manzana(S)	*Rosaceae*	*cinerea*	GM, dry eye r, blo-end r
				mali	GM
Mammea	*M. americana*	mammee, mammee apple, mamey(S)	*Calophylla*	*cinerea*	GM
Mandevilla =Dipladenia	*M.* spp.	Mandevilla	*Apocyna*	*cinerea*	Bl
	M. laxa	Chilean Jasmine		*cinerea*	Bl
	M. splendens= D. splendens	Mandevilla red riding hood		*cinerea*	Bl
	M. sanderi= D. sanderi	Brazilian jasmine		*cinerea*	Bl
	M.×amoena	Pink allamanda		*cinerea*	Bl
Mangifera	*M. indica*	Mango	*Anacardia*	*cinerea*	Blo Bl
Manilkara	*M. huberi*	Masaranduba, sapotilla, níspero(S)	*Sapotaceae*	*cinerea*	GM
	M. zapota	Sapodilla, zapote(S)		*cinerea*	GM
Maranta	*M.* sp.	Prayer pl.	*Maranta*	*cinerea*	GM
	M. leuconeura	Prayer pl.		*cinerea*	GM
Matricaria	*M. chamomilla = M. recutita*	Chamomile, German-, wild- chamomile, babunag	*Asteraceae*	*cinerea*	GM

Genus	Species	Common name[a]	Family[b]	*Botrytis* species	Disease[c]
Matthiola	*M. incana*	Common stock, hoary stock, ten-week stock	*Brassica*	*cinerea*	GM
Medicago	*M. sativa*	Lucerne, alfalfa	*Fabaceae*	*cinerea*	Blo Bl, GM
Medinilla	*M. magnifica*	Philippine orchid	*Melastomata*	*cinerea*	GM
Melampodium	*M. paldosum*	Melampodium	*Asteraceae*	*cinerea*	GM
Melilotus	*M. alba*	White melilot	*Fabaceae*	*cinerea*	GM
Melissa	*M. officinalis*	Sweet balm, lemon balm, melissa	*Lamiaceae*	*cinerea*	GM
Mentha	*M. longifolia*	Mentha, mint	*Lamiaceae*	*cinerea*	GM
	M. piperita	Peppermint		*cinerea*	GM
Mertensia	*M.* sp.	Tall lungwort, tall-, northern- bluebells	*Boragina*	*cinerea*	GM
Mesembryanthemum	*M. crystallinum*	Common ice-, crystalline ice-, ice- pl.	*Aizoaceae*	*cinerea*	GM
Mespilus	*M. germanica*	Medlar, common medlar	*Rosaceae*	*cinerea*	GM
Metasequoia	*M. glyptostroboides*	Dawn redwood	*Cupressa*	*cinerea*	GM
Mimulus	*M. moschatus*	Monkey flower, muskflower	*Phrymaceae*	*cinerea*	GM
	M.×hybridus	Mimulus, monkey fl.		*cinerea*	Bl
Monarda	*M.* sp.	Bee balm, horsemint, oswego tea, bergamot	*Lamiaceae*	*cinerea*	GM
Monstera	*M. deliciosa*	Monstera, split-leaf philodendron	*Araceae*	*cinerea*	GM
Morus	*M. alba*	White-, common-mulberry	*Moraceae*	*cinerea*	GM
Musa	*M. acuminate*	Plantain	*Musaceae*	*cinerea*	Peduncle r
Muntingia	*M. calabura*	Jamaican-, Singapore-, Sabah- cherry, Panama berry, bajelly-, strawberry- tr.	*Muntingia*	*cinerea*	GM
	M. balbisiana	Plantain		*cinerea*	Peduncle r
	M.×paradisiaca	Plantain		*cinerea*	-"-
	M. sapientum	Banana, cavendish banana		*cinerea*	Tip end r

(continued)

Table 20.1 (continued)

Genus	Species	Common name[a]	Family[b]	*Botrytis* species	Disease[c]
Muscari	*M.* spp.	Grape hyacinth	*Asparaga*	*hyacinthi*	Bl
	M. armeniacum	Grape hyacinth		*hyacinthi*	Bl
	M. azureum	Azure grape hyacinth		*hyacinthi*	Bl
	M. latifolium	Broad-leaved grape hyacinth		*hyacinthi*	Bl
	M. plumosum	Feather hyacinth		*hyacinthi*	Bl
Myosotis	*M.* spp.	Forget-me-not	*Boragina*	*cinerea*	GM
Myrica	*M. faya*	Faya bayberry	*Myricaceae*	*cinerea*	GM
	M. rubra	Yangmei, Chinese-, Japanese-, Red- bayberry		*cinerea*	GM
Myrtillocactus	*M.* sp.	Blueberry cactus	*Cactaceae*	*cinerea*	GM
Narcissus	*N. jonquilla*	Jonquil	*Amaryllida*	*cinerea*	GM
	N. pseudonarcissus	Narcissus, daffodil		*cinerea*	GM
				narcissicola	Yellow lf blotch, smoulder
				polyblastis	Narcissus fire
	N. tazetta	Bunchfl.- daffodil, -narcissus, Chinese sacred lily, Jonqui		-"-	-"-
Nematanthus	*N. gregarious = Hypocyrta nummularia*	Clog-, guppy- Plant, goldfish pl. hardiness	*Gesneria.*	*cinerea*	Bl
Neofinetia	*N. falcata*	feng lan-, pungnan- orchid	*Orchida*	*cinerea*	GM
Neoregelia	*N. carolinae*	Bromeliad, fireball, blushing bromeliad	*Bromelia*	*tulipae*	Bl, GM
Nepeta	*N. cararia*	Catnip	*Lamiaceae*	*cinerea*	GM
Nephrolepsis	*N.* spp.	Ferns	*Lomariopsida*	*cinerea*	GM
	N. exaltata	Sword fern, boston fern		*cinerea*	GM
Nerium	*N. indicum*	Rose bay, dog bane, South sea rose, oleander ceylon tr., adelfa	*Apocyna*	*cinerea*	GM
	N. oleander	Oleander		*cinerea*	GM

Nicotiana	*N. alata*	Ornamental tobacco	*Solanaceae*	*cinerea*	GM
	N. benthamiana	Tobacco	*Solanaceae*	*cinerea*	GM
	N. glutinosa	Sticky tobacco		*cinerea*	GM
	N. plumbaginifolia	Leadwort-leaved tobacco		*cinerea*	GM
	N. rustica	Aztec tobacco, mapacho(S)		*cinerea*	GM
	N. tabacum	Cultivated tobacco		*cinerea*	GM
Nigella	*N.* spp.	Devil-in-a-bush, love in a mist	*Ranuncula*	*cinerea*	GM
Ocimum	*O. americanum*	Limehairy, hoary basil	*Lamiaceae*	*cinerea*	GM
	O. basilicum	Sweet basil, albahaca(S)		*cinerea*	GM
	O. basilicum var. *purpureum*	Red basil		*cinerea*	GM
	O. basilicum var. *thyrsiflora*	Thai basil		*cinerea*	GM
	O. gratissimum	Clove-, African-basil		*cinerea*	GM
	O. kilimandscharicum	Camphor-, African-basil		*cinerea*	GM
	O. tenuiflorum	Holy basil		*cinerea*	GM
	O.×citriodorum	Lemon basil		*cinerea*	GM
Odontoglossum	*O.* sp.	Lace orchid	*Orchida*	*cinerea*	GM
Oenothera	*O. biennis*	Common evening, sundrop primrose, evening star	*Onagraceae*	*cinerea*	GM
	O. macrocarpa	Missouri evening primrose		*cinerea*	GM
Olea	*O. europaea*	Olive	*Oleaceae*	*cinerea*	Bl
Omphalodes	*O. verna*	Navelwort	*Boragina*	*cinerea*	GM
	O. cappadocica	Cappadocian navelwort		*cinerea*	GM
Oncidium	*O.* spp.	Oncidium	*Orchida*	*cinerea*	Blo Bl
Onobrychis	*O. viciifolia* = *O. sativa*	Common sainfoin	*Fabaceae*	*cinerea*	GM, CS
Oplopanax	*O. horridus*	Devil's club	*Araliaceae*	*cinerea*	GM
Opuntia	*O.* sp.	Prickly pear	*Cactaceae*	*cinerea*	Bl
	O. ficusindica	Indian fig opuntia		*cinerea*	Bl

(continued)

Table 20.1 (continued)

Genus	Species	Common name[a]	Family[b]	*Botrytis* species	Disease[c]
Origanum	*O. majorana*	Sweet marjoram	*Lamiaceae*	*cinerea*	GM
	O. syriaca	Syrian marjoram		*cinerea*	GM
	O. vulgare	Oregano, wild marjoram, Orégano(S)		*cinerea*	GM
Ornithogalum	*O.* spp.	Ornithogalum, star of Bethlehem	*Asparaga*	*cinerea*	GM
	O. dubium	Orange star fl., snake fl., sun star		*cinerea*	GM
	O. saundersiae	Giant chincherinchee		*cinerea*	GM
	O. thyrsoides	Wonder Fl., Cape chincherinchee		*cinerea*	GM
Osmanthus	*O. aurantiacus*	Sweet osmanthus, sweet-, tea-, fragrant- olive	*Oleaceae*	*cinerea*	GM
Ostrya	*O. carpinifolia*	European hop-hornbeam	*Betulaceae*	*cinerea*	GM
	O. virginiana	American hop-hornbeam		*cinerea*	GM
Oxalis	*O. latifolia*	Garden pink-, broadleaf woodsorrel	*Oxalidaceae*	*cinerea*	GM
	O. tuberosa	Wood sorrel		*cinerea*	GM
Ozothamnus	*O. diosmifolius= Helichrysum diosmifolius*	Rice fl.	*Asteraceae*	*cinerea*	GM
Pachystachys	*P.* spp.	Pachystachys	*Acantha*	*cinerea*	GM
	P. coccinea	Golden Shrimp		*cinerea*	GM
	P. lutea	Lollipop pl., golden shrimp		*cinerea*	GM
	P. spicata	Cardinal's-guard		*cinerea*	GM
Paeonia	*P. anomala*	Anomala peony	*Paeonia*	*cinerea*	GM
				paeoniae	GM, Bl
	P. lactiflora	Chinese peony, peony, common garden peony		-"-	-"-
	P. peregrina	Balkan peony		-"-	-"-
	P. rockii = Paeonia suffruticosa subsp. *rockii*	Rock's-, Rock's tree- peony		-"-	-"-
	P. suffruticosa	Moutan-, tree- peony		-"-	-"-
	P. tenuifolia	Fern leaf peony		-"-	-"-

Panax	*P. ginseng* = *P. pseudoginseng*	Ginseng (Korean)	*Araliaceae*	*cinerea*	GM
	P. quinquefolis = *Aralia quinquefolia*, *P. americanus*	American ginseng		*cinerea*	GM
Pandanus	*P*. sp.	Pandan, screw pine	*Pandana*	*cinerea*	GM
Papaver	*P*. spp.	Poppy	*Papavera*	*cinerea*	GM
	P. somniferum	Opium poppy		*cinerea*	GM
Parthenium	*P*. sp.	Feverfew	*Asteraceae*	*cinerea*	GM
Parthenium	*P. argentatum*	Guayule	*Asteraceae*	*cinerea*	GM
Paspalum	*P*. spp.	Paspalum	*Poaceae*	*cinerea*	GM
	P. nicorae	Brunswick grass		*cinerea*	GM
	P. notatum	bahiagrass		*cinerea*	GM
	P. scrobiculatum	Koda, varuka, varuku millet		*cinerea*	GM
	P. vaginatum	Water finger-grass		*cinerea*	GM
Passiflora	*P*. spp.	Passion fr., passion fl., passion vine	*Passiflora*	*cinerea*	GM
	P. edulis	Passion fr., maracuyá(S)		*cinerea*	GM
	P. foetida	Wild maracuja		*cinerea*	GM
	P. incarnata	Maypop		*cinerea*	GM
	P. laurifolia	Water lemon		*cinerea*	GM
	P. ligularis	Granadilla(S)		*cinerea*	GM
	P. maliformis	Sweet calabash		*cinerea*	GM
	P. quadrangularis	Giant granadilla, giant tumbo, badea		*cinerea*	GM
	P. tripartita	Banana passionfruit, Curuba(S)		*cinerea*	GM
Pasticana	*P. sativa*	Parsnip	*Apiaceae*	*cinerea*	GM

(continued)

Table 20.1 (continued)

Genus	Species	Common name[a]	Family[b]	*Botrytis* species	Disease[c]
Pelargonium	*P*. spp.	Pelargonium, stork's-bill	*Geraniaceae*	*cinerea*	GM, Bl
				pelargonii	Bl, GM
	P. domesticum	Regal-, Martha Washington- geranium		*cinerea*	Bl, GM
	P. graveolens	Scented geranium		*cinerea*	Bl, GM
	P. inquinans	Florists' geranium		*cinerea*	Bl, GM
	P. peltatum = Geranium peltatum	Ivy-, ivy-leafed- ge-ranium		*cinerea*	Bl, GM
	P. zonale	Zonal geranium		*cinerea*	Bl, GM
	P.×domesticum	Pelargonium hybrid		*cinerea*	Bl, GM
	P.×hortorum	Pelargonium hybrid		*cinerea*	Bl, GM
Penstemon	*P*. sp.	Beardtongue	*Plantagina*	*cinerea*	GM
Pentas	*P. lanceolata*	Pentas	*Rubiaceae*	*cinerea*	GM
Peperomia	*P*. spp. (tens of spp.)	Radiator pl.	*Piperaceae*	*cinerea*	GM
	P. fraseri	Flowering peperomia		*cinerea*	GM
	P. caperata	Emerald ripple peperomia		*cinerea*	GM
	P. argyreia	Watermelon peperomia		*cinerea*	GM
Persea	*P. americana = P. gratissima*	Avocado	*Lauraceae*	*cinerea*	Fr & st end r, fr spots
Pericallis	*P.×hybrida = Senecio×hyb.*	Cineraria	*Asteraceae*	*cinerea*	Bl
Perilla	*P. frutescens = P. nankinensis*	Shiso	*Lamiaceae*	*cinerea*	GM
	P. frutescens var. *crispa*	Shiso, beefsteak pl., perilla mint, Chinese-, wild- basil		*cinerea*	GM
	P. frutescens var. *frutescens*	Egoma		*cinerea*	GM

Petroselinum	*P. crispum* = *P. hortense* var. *tuberosum*	Parsley, flat parsley	*Apiaceae*	*cinerea*	GM
	P. crispum = *P. hortense* var. *crispum*	Curly parsley		*cinerea*	GM
	P. sativum	Parsley root, raiz de perejil(S)		*cinerea*	GM
Phaius	*P. tankervilliae*	Ground orchid	*Orchida*	*cinerea*	Lf Bl
Phalaenopsis	*P.* sp.	Phalaenopsis	*Orchida*	*cinerea*	Blo Bl
Phanerophlebia	*P. caryotidea*	Eared veinfern	*Asplenia*	*cinerea*	GM
Phaseolus	*P. coccineus*	Runner-, scarlet runner-, multiflora- bean	*Fabaceae*	*cinerea*	GM
	P. coccineus subsp. *darwinianus*	Botil bean		*cinerea*	GM
	P. lunatus	Lima bean		*cinerea*	GM
	P. vulgaris	Common bean		*cinerea*	GM
Philadelphus	*P.* spp.	Mock-orange	*Hydrangea*	*cinerea*	GM
	P. coronarius	Sweet mock-orange, English dogwood		*cinerea*	GM
Philodendron	*P.* spp.	Philodendron	*Araceae*	*cinerea*	GM
	P. selloum	Split leaf philodendron		*cinerea*	GM
Phlox	*P. borealis*	Dwarf phlox	*Polemonia*	*cinerea*	Bl, GM
	P. drummondii	Garden phlox		*cinerea*	Bl, GM
	P. pilosa	Woodland phlox		*cinerea*	Bl, GM
	P. subulata	Moss phlox		*cinerea*	Bl, GM
Phoenix	*P. canariensis*	Canary Island date palm	*Arecaceae*	*cinerea*	Rachis lesions
	P. dactylifera	Date palm		*cinerea*	-"-
	P. roebelenii	Pygmy date palm		*cinerea*	LS & Bl
Philadelphus	*P.* spp. and hybrids	Mock-orange	*Hydrangea*	*cinerea*	GM
	P. coronarius	Sweet mock-orange, English dogwood		*cinerea*	GM
	P. lewisii	Lewis's mock-orange		*cinerea*	GM
Phormium	*P. tenax*	Flax, harakeke, NZ flax, NZ hemp	*Xanthorrhoea*	*cinerea*	GM

(continued)

Table 20.1 (continued)

Genus	Species	Common name[a]	Family[b]	*Botrytis* species	Disease[c]
Photinia	*P. fraseri*	Fotinia	*Rosaceae*	*cinerea*	GM
	P. glabra	Red top, Japanese-, red leaf- photinia		*cinerea*	GM
Physallis	*P. peruviana*	Cape gooseberry, uchuva(S)	*Solanaceae*	*cinerea*	GM
Picea	*P*. spp.	Spruce	*Pinaceae*	*cinerea*	GM
	P. abies	Christmas tr., Norway spruce		*cinerea*	GM
	P. engelmannii	Engelmann-, white-, mountain-, silver- spruce		*cinerea*	GM
	P. mariana	Black spruce		*cinerea*	GM
	P. omorika	Serbian spruce		*cinerea*	GM
	P. pungens	Colorado-, blue-, green-, white- spruce		*cinerea*	GM
	P. sitchensis	Sitka spruce		*cinerea*	GM
Pieris	*P. japonica*	Pieris	*Ericaceae*	*cinerea*	GM
Pilea	*P*. spp.	Pilea	*Urticaceae*	*cinerea*	GM
	P. mollis	Moon valley		*cinerea*	GM
	P. microphylla	Artillery pl., artillery fern		*cinerea*	GM
	P. cadierei	Aluminum pl.		*cinerea*	GM
	P. nummulariifolia	Creeping Charlie		*cinerea*	GM
	P. spruceana	Pilea angel wings		*cinerea*	GM
Pinus	*P*. spp.	Pine	*Pinaceae*	*cinerea*	GM, Bl
	P. contorta	Lodgepole-, contorta-, shore-, twisted- pine		*cinerea*	GM, Bl
	P. nigra	Black pine		*cinerea*	GM, Bl
	P. pinaster	Maritime pine		*cinerea*	GM, Bl
	P. ponderosa	Ponderosa-, blackjack-, bull-, western yellow- pine		*cinerea*	GM, Bl
	P. radiata	Monterey pine		*cinerea*	GM, Bl
	P. sylvestris	Scots pine		*cinerea*	GM, Bl
	P. taeda	Loblolly pine		*cinerea*	GM, Bl

Pistacia	*P. atlantica*	Mt. Atlas mastic tr., betoum	*Anacardia*	*cinerea*	Twig Bl
	P. integerrima	Kakar singhi, kakra, Kakring (Hindi)		*cinerea*	Twig Bl
	P. terebinthus	Terebinth, cyprus turpentine		*cinerea*	Twig Bl
	P. vera	Pistachio		*cinerea*	Blo, twig and shoot Bl
	P. atlantica×P. integerrima	-		*cinerea*	Twig Bl
Pisum	*P. sativum*	Field-, dun (grey-brown)-, Kapucijner-pea	*Fabaceae*	*cinerea*	GM
				fabae	CS
	P. sativum var. *arvense*	Winter pea		*cinerea*	GM
Pittosporum	*P. tobira*	Pittosporum, Japanese mock-orange	*Pittospora*	*cinerea*	LS
Platanus	*P. orientalis*	Oriental plane	*Platanaceae*	*cinerea*	GM
	P. occidentalis	American sycamore		*cinerea*	GM
	P.×acerifolia	London plane, London planetree, hybrid plane		*cinerea*	GM
Platycerium	*P*. spp.	Staghorn fern	*Polypodia*	*cinerea*	GM
Platycodon	*P. grandiflorus*	Chinese bellflower		*cinerea*	Bl
Plectranthus	*P*. spp.	Spur flowers	*Lamiaceae*	*cinerea*	GM
	P. amboinicus	Mexican mint		*cinerea*	GM
	P. argentatus	Silver spur fl.		*cinerea*	GM
	P. forsteri	Swedish ivy		*cinerea*	GM
	P. purpuratus	Purple Swedish ivy		*cinerea*	GM
Plumbago	*P*. sp.	Plumbago, leadwort	*Plumbagina*	*cinerea*	GM
	P. auriculata = P. capensis	Blue-, Cape-plumbago, leadwort		*cinerea*	GM
Plumeria	*P*. spp.	Frangipani	*Apocyna*	*cinerea*	GM
	P. rubra	Red paucipan, red-, common-frangipani		*cinerea*	GM
	P. obtusa	Singapore graveyard fl., Lei fl., White plumeria		*cinerea*	GM
Poa	*P*. sp.	Meadow-grass	*Poaceae*	*cinerea*	GM
	P. annua	Annual bluegrass		*cinerea*	GM

(continued)

Table 20.1 (continued)

Genus	Species	Common name[a]	Family[b]	*Botrytis* species	Disease[c]
Podophyllum	*P. peltatum*	Mayapple, May apple	*Berberida*	*cinerea*	GM
Polianthes	*P. tuberosa*	Tuberose	*Asparaga*	*cinerea*	Bl
				elliptica	Spot
Polygonum	*P*. sp.	Knotgrass, bistort	*Polygona*	*cinerea*	GM
Polystichum	*P*. spp.	Polystichum ferns	*Asplenia*	*cinerea*	Bl
	P. tsussimense	Korean rock fern		*cinerea*	Bl
Populus	*P*. spp.	Poplar, aspen, cottonwood	*Salicaceae*	*cinerea*	GM
Potamogeton	*P*. spp.	Pondweed	*Potamogetona*	*cinerea*	GM
Primula	*P*. spp.	Primrose	*Primulaceae*	*cinerea*	GM
	P. malacoides	Fairy primrose		*cinerea*	Bl
	P. obconica	German primrose		*cinerea*	Bl
	P. sinensis	Chinese primrose		*cinerea*	Bl
	P. vulgaris	English primrose		*cinerea*	Bl
	P.×polyantha	Polyanthus primrose		*cinerea*	GM
	P.×Pruhonicensis	Polyanthus primrose		*cinerea*	Bl
Protea	*P*. spp.	Protea, sugarbush	*Proteaceae*	*cinerea*	GM
	P. compacta	Bot River protea		*cinerea*	GM
	P. cynaroides	King protea		*cinerea*	GM
	P. nerifolia	Oleander leaf protea		*cinerea*	GM
	P. obtusifolia	Limestone sugarbush		*cinerea*	GM
	P. repens	Common sugarbush		*cinerea*	GM

Prunus	*P. armeniaca* = *Armeniaca vulgaris*	Armenian plum	*Rosaceae*	*cinerea*	GM
	P. avium	Wild-, sweet-, bird-cherry		*cinerea*	GM
	P. besseyi	Western sandcherry		*cinerea*	GM
	P. domestica	Plum		*cinerea*	Green fr r
	P. dulcis	Almond		*cinerea*	-"-
	P. glandulosa	Korean cherry, dwarf flowering almond		*cinerea*	GM
	P. laurocerasus	Cherry laurel		*cinerea*	GM
	P. pumila	Sand cherry		*cinerea*	GM
	P. salicina	Japanese plum		*cinerea*	Green fr r
	P. spinosa	Blackthorn, sloe		*cinerea*	GM
	P. persica var. *nucipersica*	Nectarine		*cinerea*	Green fr r
	P. persica var. *persica* = *Persica vulgaris*	Peach		*cinerea*	Green fr r
Pseudotsuga	*P. menziesii* var. *menziesii* = *P. taxifolia*	Douglas-fir, Oregon pine, Douglas spruce	*Pinaceae*	*cinerea*	GM
				douglasii	Bl
Psidium	*P. guajava*	Apple-, common- guava, Goiaba, Guayaba(S)	*Myrtaceae*	*cinerea*	GM
Pteridium	*P. aquilinum*	Bracken	*Dennstaedtia*	*cinerea*	GM
Pteris	*P.* spp.	Table fern	*Polypodia*	*cinerea*	GM
Punica	*P. granatum*	Pomegranate	*Lythraceae*	*cinerea*	GM
Pycnanthemum	*P. virginianum*	Mountain mint, Virginia mountain mint	*Lamiaceae*	*cinerea*	GM
Pyracantha	*P.* spp.	Firethorn pyracantha	*Rosaceae*	*cinerea*	GM
Pyrola	*P.* spp.	Wintergreen	*Ericaceae*	*cinerea*	GM
	P. grandiflora	Large-flowered wintergreen		*cinerea*	GM
	P. rotundifolia	Round-leaved wintergreen		*cinerea*	GM

(continued)

Table 20.1 (continued)

Genus	Species	Common name[a]	Family[b]	*Botrytis* species	Disease[c]
Pyrus	*P. betulaefolia = P. betulifolia*	Pear ornamental, Southworth dancer	*Rosaceae*	*cinerea*	GM
	P. calleryana	Callery pear		*cinerea*	GM
	P. communis	Pear		*cinerea*	Spur & blo bl
	P. pyrifolia = P. serotina	Sand-, pineapple-, apple-, Chinese-, Asian-pear		*cinerea*	GM
Quararibea	*Q. cordata = Matisia cordata*	S. American sapote, chupa- chupa (S), zapote costeño (S)	*Malvaceae*	*cinerea*	GM
Quercus	*Q.* spp.	Oak	*Fagaceae*	*cinerea*	GM
	Q. petraea	Cornish-, Irish-, Durmast- oak		*cinerea*	GM
	Q. robur	Pedunculate oak		*cinerea*	GM
	Q. rubra	Northern red oak		*cinerea*	GM
Ranunculus	*R. asiaticus*	Buttercup	*Ranuncula*	*cinerea*	GM
				ranuculi	Bl
	R. lyallii	Mountain-, Mount Cook (MC)- buttercup, MC lily		*cinerea*	GM
Raphanus	*R. sativus = R. raphanistrum* var. *sativus*	Cultivated radish	*Brassica*	*cinerea*	GM
	R. sativus = R. raphanistrum var. *sativus*	Small radish		*cinerea*	GM
	R. sativus var. *oleiferus*	Oleiferous radish		*cinerea*	GM
Ravenea	*R. rivularis*	Majesty palm	*Arecaceae*	*cinerea*	Bl
Rheum	*R. rhabarbarum*	Rhubarb	*Polygona*	*cinerea*	GM
	R.×hybridum	Rhubarb		*cinerea*	GM

Rhinanthus	*R. minor*	Yellow rattle, cockscomb	*Orobancha*	*cinerea*	GM
Rhododendron	*R.* spp.	Rhododendron, azalea	*Ericaceae*	*cinerea*	Lf & pt bl
	R. indicum	Azalea, azalea Indica, macranthum azalea		*cinerea*	Lf & pt bl
Rhoicissus	*R.* sp.	Cape grape	*Vitaceae*	*cinerea*	GM
Rhus	*R.* sp.	Sumac	*Anacardia*	*cinerea*	GM
Ribes	*R. nigrum*	Black currant	*Grossularia*	*cinerea*	GM
	R. rubrum	Red currant		*cinerea*	GM
	R. sativum	White currant		*cinerea*	GM
	R. alpinum	Alpine currant		*cinerea*	GM
	R. uvacrispa = *R. grossularia*	Gooseberry		*cinerea*	GM
	R. hirtellum	N. American gooseberry		*cinerea*	GM
Ricinus	*R. communis*	Castor bean, castor oil pl., higuerilla(S)	*Euphorbia*	*cinerea*	GM
				ricini	GM, Bl
Robinia	*R.* sp.	Robinia, locusts	*Fabaceae*	*cinerea*	GM
Rosa	*R.* spp.	Rose (>100 spp.), Rosa(S)	*Rosaceae*	*cinerea*	Bl, cane canker
	R. chinensis	China rose		*cinerea*	-"-
	R.×damascene	Damask rose, rose of Castile		*cinerea*	-"-
	Rosa×odorata	Chinese bush rose		*cinerea*	-"-
	R.×hybrida	Rose		*cinerea*	-"-
Rosmarinus	*R. officinalis*	Rosemary	*Lamiaceae*	*cinerea*	Bl, GM

(continued)

Table 20.1 (continued)

Genus	Species	Common name[a]	Family[b]	*Botrytis* species	Disease[c]
Rubus	*R.* spp.	Raspberry	*Rosaceae*	*cinerea*	fr r & blo bl, GM
	R. spp.	Blackberry		*cinerea*	-"-
	R. spp.	Caneberries		*cinerea*	-"-
	R. chamaemorus	Cloudberry, bakeapple, knoutberry		*cinerea*	-"-
	R. crataegifolius	Korean raspberry		*cinerea*	-"-
	R. fruticosus	Common blackberry		*cinerea*	GM
				caroliniana	GM
	R. glaucus	Andean (black) raspberry, mora de Castilla(S)		*cinerea*	GM
	R. idaeus	Red raspberry		*cinerea*	GM
	R. idaeus×R. strigosus	Red raspberry		*cinerea*	GM
	R. laciniatus	Evergreen blackberry		*cinerea*	GM
	R. leucodermis	Blue raspberry		*cinerea*	GM
	R. occidentalis	Black raspberry		*cinerea*	GM
	R. ulmifolius	Elmleaf-, thornless- blackberry, zarzamora(S)		*cinerea*	GM
	R. ursinus	California-, Pacific-, trailing- blackberry		*cinerea*	GM
	R. hybrid	Boysenberry		*cinerea*	GM
Ruellia	*R. simplex*	Britton's wild-, Mexica-petunia, Mexican bluebell	*Acantha*	*cinerea*	GM
	R. strigosus×R. occidentalis	Purple raspberry		*cinerea*	GM
	R. formosa	Wild petunia		*cinerea*	GM
Rumex	*R.* sp.	Docks and sorrels	*Polygona*	*cinerea*	GM
Ruscus	*R. hypoglossum*	Ruscus, butcher's broom	*Asparaga*	*cinerea*	GM
Ruta	*R. graveolens*	Rue, common rue, herb-of-grace	*Rutaceae*	*cinerea*	GM
Saccharum	*S. officinarum*	Sugarcane	*Poaceae*	*cinerea*	GM
Saintpaulia	*S. ionantha*	African violet, violeta africana(S)	*Gesneria*	*cinerea*	GM

Salicornia	*S.* spp.	glasswort, pickleweed, and marsh samphire	*Amarantha*	*cinerea*	GM
	S. bigelovii	dwarf saltwort, dwarf glasswort		*cinerea*	GM
	S. europaea	common glasswort, Norfolk samphire		*cinerea*	GM
	S. herbacea	Salicornia herbacea		*cinerea*	GM
Salix	*S.* spp.	Weeping willow	*Salicaceae*	*cinerea*	GM
	S. babylonica	Babylon willow		*cinerea*	GM
Salvia	*S. officinalis*	Salvia, sage	*Lamiaceae*	*cinerea*	GM
	S. officinalis var. *purpurascens*	Purple sage		*cinerea*	GM
	S. splendens	Scarlet sage		*cinerea*	GM
Sambucus	*S.* sp.	Elder, elderberry	*Caprifolia*	*cinerea*	GM
	S. nigra	Black-, European-elder, European black elderberry		*cinerea*	GM
Sandersonia	*S. aurantiaca*	Sandersonia, Chinese lly lantern, Christmas bells	*Liliaceae*	*cinerea*	GM
Sanguinaria	*S. canadensis*	Bloodroot	*Papavera*	*cinerea*	GM
Sansevieria	*S.* spp.	mother-in-law-tongue	*Agavaceae*	*cinerea*	GM
Sarcococca	*S. confusa*	Sweet box	*Buxaceae*	*cinerea*	GM
	S. hookeriana	Christmas box, sweet box		*cinerea*	GM
	S. ruscifolia	Fragrant sarcococca		*cinerea*	GM
Satureja	*S. hortensis*	Summer savory	*Lamiaceae*	*cinerea*	GM
Saxifraga	*S.* spp.	Saxifrages, rockfoils	*Saxifraga*	*cinerea*	GM
	S. stolonifera	strawberry-begonia		*cinerea*	GM
Scabiosa	*S.* spp.	Pincushion fl., scabious	*Caprifolia*	*cinerea*	Bl
	S. atropurpurea	Sweet scabious		*cinerea*	Bl

(continued)

Table 20.1 (continued)

Genus	Species	Common name[a]	Family[b]	*Botrytis* species	Disease[c]
Schefflera	*S.* spp.	Schefflera	*Araliaceae*	*cinerea*	GM
	S. actinophylla	Umbrella tr.		*cinerea*	GM
	S. arboricola	Dwarf umbrella tr.		*cinerea*	GM
Schisandra	*S. chinensis*	Five flavor berry, pinyin	*Schisandra*	*cinerea*	Bl and fr r
Schizanthus	*S.×wisetonensis*	Butterfly fl., poorman's orchid	*Solanaceae*	*cinerea*	Bl
Schlumbergera	*S.* sp.	Christmas-, Thanksgiving-, crab-, holiday- cactus	*Cactaceae*	*cinerea*	Bl
Scorzonera	*S. hispanica*	Black salsify	*Asteraceae*	*cinerea*	GM
Scutellaria	*S.* spp.	Skullcap	*Lamiaceae*	*cinerea*	GM
	S. baicalensis	Chinese skullcap		*cinerea*	GM
	S. lateriflora	American-, Virginia-, mad dog-, blue- skullcap		*cinerea*	GM
Secale	*S. cereale*	Rye	*Poaceae*	*cinerea*	GM
Sedum	*S.* sp.	Goldmoss, stonecrops	*Crassula*	*cinerea*	Bl
Selaginella	*S.* spp.	Selaginella ferns	*Selaginella*	*cinerea*	GM
	S. braunii	Arborvitae fern		*cinerea*	GM
	S. erythropus	Red selaginella, ruby-red spikemoss		*cinerea*	GM
	S. kraussiana	Golden clubmoss		*cinerea*	GM
	S. lepidophylla	Resurrection pl.		*cinerea*	GM
	S. moellendorfii	Gemmiferous spikemoss		*cinerea*	GM
	S. uncinata	Peacock fern		*cinerea*	GM
Sempervivum	*S.* spp.	Houseleek, liveforever, hen and chicks	*Crassula*	*cinerea*	GM
Senecio	*S. cineraria*	Dusty miller	*Asteraceae*	*cinerea*	GM
	S. cruentis	–		*cinerea*	GM
Serruria	*S. florida*	Blushing bride, pride of Franschhoek	*Proteaceae*	*cinerea*	GM
Senecio	*S.* sp.	Ragworts, groundsels	*Asteraceae*	*cinerea*	Bl

Sequoiadendron	*S. giganteum*	Giant sequoia, giant-, Sierra- redwood	*Cupressa*	*cinerea*	GM
				douglasii	Bl
Sequoia	*S. sempervirens*	Coast-, California- redwood, sequoia	*Cupressa*	-"-	-"-
Sesamum	*S. indicum*	Sesame, sesame seed	*Pedaliaceae*	*cinerea*	GM
Sesbania	*S. cannabina*	Sesbania pea	*Fabaceae*	*cinerea*	GM
Silphium	*S.* spp.	Rosinweed	*Asteraceae*	*cinerea*	GM
	S. integrifolium	Whole-leaf-, entire-leaf-, prairie- rosinweed		*cinerea*	GM
	S. laciniatum	Compass pl.		*cinerea*	GM
	S. perfoliatum	Cup plant, carpenter's weed, cup rosinweed		*cinerea*	GM
	S. terebin-thinaceum	Prairie dock		*cinerea*	GM
	S. trifoliatum	Whorled rosinweed		*cinerea*	GM
Sinapis	*S. alba*	white mustard	*Brassica*	*cinerea*	Grey r
Sinningia	*S.* spp.	Gloxinia	*Gesneria*	*cinerea*	GM
	S. tubiflora	Hardy white gloxinia		*cinerea*	GM
Smallanthus	*S. sonchifolius*	Yacon	*Asteraceae*	*cinerea*	GM
Solanum	*S. capsicastrum*	Winter cherry	*Solanaceae*	*cinerea*	GM
	S. lycopersicum= *Lycopersicon esculentum*	Tomato, tomate(S)		*cinerea*	Ghost spots, GM
	S. melongena	Eggplant, berenjena(S)		*cinerea*	GM
	S. muricatum	Sweet pepino, pepino dulce		*cinerea*	GM
	S. pseudocapsicum	Jerusalem cherry, Madeira winter cherry		*cinerea*	Bl
	S. quitoense	Lulo(S), naranjilla(S)		*cinerea*	GM
	S. tuberosum	Potato		*cinerea*	GM
Solenostemon	*S.* spp.	Coleus	*Lamiaceae*	*cinerea*	GM
	S. scutellarioides	Coleus, flame-, painted- nettle		*cinerea*	GM

(continued)

Table 20.1 (continued)

Genus	Species	Common name[a]	Family[b]	*Botrytis* species	Disease[c]
Sorbus	*S. aria* = *Aria nivea*	Common whitebeam	*Rosaceae*	*cinerea*	GM
	S. aucuparia	Rowan, mountain ash, Amur-, European mountain-ash, quick beam		*cinerea*	GM
	S. torminalis = *Torminalis clusii*	Wild service tr., chequer tr., checker tr.		*cinerea*	GM
Sorghum	*S. bicolor* = *S. japonicum*, *S. vulgare*	Sorghum, durra, jowari, milo	*Poaceae*	*cinerea*	Seed r
	S. bicolor×*S. sudanense*	Sudanense sorghum, sorgusudan		*cinerea*	Seed r
Spathiphyllum	*S.* spp.	Spath, peace lily	*Araceae*	*cinerea*	Bl. GM
	S. cannifolium	Spatheflower		*cinerea*	GM
	S. cochlearispathum	Peace lily		*cinerea*	GM
	S. floribundum	Snow fl., peace lily		*cinerea*	GM
Spinacia	*S. oleracea*	Spinach	*Amarantha*	*cinerea*	GM
Staphylea	*S.* sp.	Bladdernut	*Staphylea*	*cinerea*	GM
	S. pinnata	European bladdernut		*cinerea*	GM
Steirodiscus	*S. tagetes*	Kakhi yellow weed	*Asteraceae*	*cinerea*	GM
Stemona	*S. japonica*	–	*Stemona*	*cinerea*	GM
Stenocereus	*S. queretaroensis*	Pitaya(S)	*Cactaceae*	*cinerea*	GM
Stephanotis	*S. Floribunda* = *S. jasminoides*	Madagascar jasmine, wax fl., Hawaiian wed-ding fl., bridal wreath	*Apocyna*	*cinerea*	GM
				elliptica	Bl
Stevia	*S. rebaudiana*	Stevia, sweet leaf, sugarleaf	*Asteraceae*	*cinerea*	GM
Stewartia	*S. pseudocamellia*= *Stuartia pseudocamellia*	Japanese Stewartia, deciduous camellia	*Theaceae*	*cinerea*	GM
Strelitzia	*S. reginae*	Strelitzia, bird of paradise, ave del paraí-so(S)	*Strelitzia*	*cinerea*	GM
Stokesia	*S. laevis*	Stokes' aster, stokesia	*Asteraceae*	*cinerea*	GM

Stylosanthes	*S.* spp.	Stylo	*Fabaceae*	*cinerea*	GM
	S. guianensis	Brazilian-, common- stylo, Brazilian lucerne,		*cinerea*	GM
	S. hamata	Caribbean stylo, pencil fl.		*cinerea*	GM
	S. humilis	Townsville stylo		*cinerea*	GM
	S. scabra	Shrubby stylo, Capitán Juan(S), pata de terecay(S)		*cinerea*	GM
Styphnolobium	*S. japonicum = Sophora japonica*	Japanese pagoda tr., Chinese scholar tr.	*Fabaceae*	*cinerea*	GM
Sutera	*S.* sp.	Bocopa	*Scrophularia*	*cinerea*	GM
Swainsona	*S. formosa*	Sturt's desert pea	*Fabaceae*	*cinerea*	GM
Sylosanthes	*S. gyianensis*	Brazilian-, common- stylo, Brazilian lucerne, sylosanthes	*Fabaceae*	*cinerea*	GM
Symphoricarpos	*S. orbiculatus*	Coral-, snapb-, buckle-, wolf-, wax-berry	*Caprifolia*	*cinerea*	GM
Symphyotrichum	*S. pilosum = Aster pilosus*	Frost aster	*Asteraceae*	*cinerea*	GM
Syngonium	*S.* sp.	Syngonium	*Araceae*	*cinerea*	GM
Syringa	*S.* spp.	Lilac	*Oleaceae*	*cinerea*	Bl
	S. vulgaris	Common lilac		*cinerea*	Bl
	S. cumini	Jambul, jambolan, jamblang, jamun	*Myrtaceae*	*cinerea*	GM, Bl
Tagetes	*T.* spp.	Marigold	*Asteraceae*	*cinerea*	GM
	T. erecta	African marigold, flor del muerto (S)		*cinerea*	GM
	T. minuta	Tagetes oil			
	T. patula	French marigold		*cinerea*	GM
	T. tenuifolia	Signet marigold		*cinerea*	GM
Tapeinochilos	*T. ananassae*	Indonesian wax ginger, Gingers, bastones(S)	*Costaceae*	*cinerea*	GM
Taraxacum	*T.* sp.	Dandelion	*Asteraceae*	*cinerea*	GM
	T. officinale	Common dandelion		*cinerea*	GM
Taxodium	*T. distichum*	Bald-, swamp- cypress	*Cupressa*	*cinerea*	GM

(continued)

Table 20.1 (continued)

Genus	Species	Common name[a]	Family[b]	*Botrytis* species	Disease[c]
Taxus	*T. baccata*	European-, common- yew	*Taxaceae*	*cinerea*	GM
Terminalia	*T. ivorensis*	Idigbo, black afara, framire	*Combreta*	*cinerea*	GM
Thuja	*T.* spp.	Arborvitae thujas redcedars, whitecedars	*Cupressa*	*cinerea*	GM
	T. occidentalis	Northern-, Atlantic-, Eastern-, yellow-, swamp-, false white-cedar		*cinerea*	GM
	T. plicata	Western-, pacific-, giant- redcedar, canoe cedar		*cinerea*	GM
	T. standishii×plicata	Green giant		*cinerea*	GM
Thymus	*T. citriodorus*	Lemon thyme	*Lamiaceae*	*cinerea*	GM
	T. vulgaris	Common thyme		*cinerea*	GM
Tibouchina	*T.* spp.	Glory bush, glory tr.	*Melastomata*	*cinerea*	GM
	T. lepidota	Silverleafed princess fl.		*cinerea*	GM
	T. semidecandra	Princess fl., glory bush, lasiandra		*cinerea*	GM
Tilia	*T.* sp.	Lime tr., linden, basswood	*Malvaceae*	*cinerea*	GM
Trachelium	*T.* sp.	Trachelium, throatwort	*Campanula*	*cinerea*	GM
	T. caeruleum	Blue throatwort		*cinerea*	GM
Tradescantia	*T. virginica*	Virginia spiderwort	*Commelina*	*cinerea*	Bl
Tragopogon	*T. porrifolius*	Purple-, common- salsify, oyster pl., Jerusalem star	*Asteraceae*	*cinerea*	GM
Tricyrtis	*T.* spp.	Toad lily	*Liliaceae*	*elliptica*	Fire
Trifolium	*T. dasyurum*	Eastern star clover	*Fabaceae*	*cinerea*	Lf bl, GM
				Fabae	CS
	T. pratense	Red clover		*anthophila*, *cinerea*	GM
	T. repens	White clover		*cinerea*	GM
	T. subterraneum	Subterranean-, sub- clover		*cinerea*	GM

Trigonella	*T. foenum-graecum*	Fenugreek	*Fabaceae*	*cinerea*	GM
Triodanis	*T.* sp.	Venus' looking-glass	*Campanula*	*cinerea*	GM
Triticum	*T.* spp.	Wheat	*Poaceae*	*cinerea*	GM
	T. aestivum	Common wheat		*cinerea*	GM
Tropaeolum	*T. azureum*	Blue nasturtium	*Tropaeola*	*cinerea*	GM
	T. speciosum	Flame fl., flame nasturtium		*cinerea*	GM
Tsuga	*T. canadensis*	Canada hemlock	*Pinaceae*	*cinerea*	GM
	T. heterophylla	Western hemlock		*cinerea*	GM
	T. mertensiana	Mountain hemlock		*cinerea*	GM
Tulipa	*T.* spp.	Tulip	*Liliaceae*	*cinerea*	Bl, GM
				tulipae	Tulip fire
	T. suaveolens=T. schrenkii			-"-	-"-
	T. gesneriana	Didier's-, garden- tulip		-"-	-"-
	T. clusiana	Lady tulip		-"-	-"-
	T. greigii	Greigii tulip		-"-	-"-
	T. kaufmanniana	Waterlily tulip		-"-	-"-
	T. sprengeri	Sprenger's tulip		-"-	-"-
Tussilago	*T. farfara*	Coltsfoot	*Asteraceae*	*cinerea*	Root r
Ulmus	*U. americana*	American elm	*Ulmaceae*	*cinerea*	GM, LS
Ursinia	*U.* sp.	–	*Asteraceae*	*cinerea*	GM

(continued)

Table 20.1 (continued)

Genus	Species	Common name[a]	Family[b]	*Botrytis* species	Disease[c]
Vaccinium	*V.* spp.	–	*Ericaceae*	*cinerea*	Bl,GM
	V. angustifolium	Lowbush blueberry		*cinerea*	Bl,GM
				pseudocinerea	Bl,GM
	V. ashei	Rabbiteye blueberry		*cinerea*	Bl,GM
	V. caespitosum	Dwarf bilberry		*cinerea*	Bl,GM
	V. corymbosum	Blueberry, highbush blueberry, Northern arándanos(S)		*cinerea*	Bl,GM
	V. deliciosum	Cascade bilberry		*cinerea*	Bl,GM
	V. erythrocarpum	Cranberry		*cinerea*	Bl,GM
	V. macrocarpon	Cranberry, American -cranberry, -bearberry		*cinerea*	Bl,GM
	V. membranaceum	Mountain bilberry, black- twin-leaved- huckleberry		*cinerea*	Bl,GM
	V. meridionale	Agraz(S) camuezo(S), mortiño(S), vichacha(S)		*cinerea*	Bl,GM
	V. microcarpum	Cranberry, macrocarpon		*cinerea*	Bl,GM
	V. myrtillus	Bilberry, whortleberry, blueberry, wild blueberry		*cinerea*	Bl,GM
	V. ovalifolium	Lingonberry, cowberry		*cinerea*	Bl,GM
	V. oxycoccos	Bog-, small-, wild-, swamp- cranberry		*cinerea*	Bl,GM
	V. parvifolium	Huckleberry		*cinerea*	Bl,GM
	V. uliginosum	Bog-bilberry, -blueberry, -whortleberry, -huckleberry		*cinerea*	Bl,GM
	V. vitisidaea	Lingonberry, cowberry		*cinerea*	Bl,GM
Valerianella	*V. locusta* = *V. olitora*	Corn salad, lamb's-, nut-, fetticus- lettuce, mâche	*Caprifolia*	*cinerea*	GM
Vanda	*V.* spp.	Vanda orchids	*Orchida*	*cinerea*	Blo Bl, GM

Verbena	*V*. spp.	Verbena, vervain	*Verbena*	*cinerea*	GM
	V.×hybrida	Verbena, vervain		*cinerea*	GM
	V. officinalis	Common vervain, common verbena		*cinerea*	GM
Veratrum	*V. californicum*	False hellebore	*Melanthia*	*cinerea*	Bl
	V. grandiflorum	False hellebores, corn lilies		*cinerea*	Bl
Veronica	*V*. spp.	Speedwell, bird's eye, gypsyweed	*Plantagina*	*cinerea*	Bl
	V. longifolia	Beach-, garden- speedwell		*cinerea*	GM
Viburnum	*V. dentatum*	Southern -, roughish-arrowwood, arrowwood viburnum,	*Adoxaceae*	*cinerea*	GM, Bl
	V. opulus	Guelder rose		*cinerea*	Bl
	V. trilobum	American cranberrybush viburnum, Kalyna, high bush cranberry		*cinerea*	GM, Bl
Vicia	*V. articulata*	Bard vetch	*Fabaceae*	*cinerea*	GM
	V. benghalensis	Purple verch		*cinerea*	GM
	V. cracca	Tufted-, cow-, bird-, boreal-vetch		*cinerea*	GM
	V. faba var. *major*	Broad bean, haba(S)		*cinerea*	GM
				faba	CS
				fabiopsis	CS
	V. hirsuta	Hairy tare, hairy-, tiny-vetch		*vicia-hirssutae*	CS
	V. narbonensis	Narbon bean		*cinerea*	GM
	V. pannonica	Hungarian vetch		*cinerea*	GM
	V. sativa	Common vetch, tare		*cinerea*	GM
				faba	CS
				viciae	CS
	V. serratifolia	French vetch		*cinerea*	GM
	V. villosa	Fodder-, hairy- vetch		*cinerea*	GM

(continued)

Table 20.1 (continued)

Genus	Species	Common name[a]	Family[b]	*Botrytis* species	Disease[c]
Vigna	*V. unguiculata*	Cowpea, black-eyed pea	*Fabaceae*	*cinerea*	GM
Vinca	*V. minor*	Periwinkle, lesser-, dwarf- periwinkle	*Apocyna*	*cinerea*	GM
Viola	*V.* spp.	Pansy	*Violaceae*	*cinerea*	Bl, GM
	V. cornuta	Horned pansy, horned violet		*cinerea*	Bl, GM
	V. cucullata	Hooded blue-, marsh blue-, purple- violet		*cinerea*	Bl, GM
	V. labradorica	Alpine-, American dog-, Labrador-, dog- violet		*cinerea*	Bl, GM
	V. odorata	Wood-, sweet-, English-, common-, florist's- violet		*cinerea*	Bl, GM
	V. pedata	Birdsfoot-, bird's-foot- violet		*cinerea*	Bl, GM
	V. rotundifolia	Round leaf yellow violet		*cinerea*	Bl, GM
	V. tricolor	Pansy neck plants		*cinerea*	Bl, GM
	V.×wittrockiana	Pansy, heart's ease, ladies delight, stepmother's fl.		*cinerea*	Bl, GM
Vitis	*V.* spp.	Grapevine	*Vitaceae*	*cinerea*	R, Bl, GM
	V. coignetiae	Crimson glory vine		*cinerea*	GM
	V. labrusca	Fox grape		*cinerea*	GM
	V. riparia	Riverbank grapevine		*cinerea*	GM
	V. rotundifolia= Muscadinia rotundifolia	Muscadine grapevine		*cinerea*	GM
	V. rupestris	Rootstock Rupestris St. George		*cinerea*	GM
	V. vinifera	Grape, uva de mesa(S)		*cinerea*	GM
				pseudocinerea	GM
				sinoviticola	GM

Vriesea	*V. duvaliana*	Vriesea	*Bromelia*	*cinerea*	GM
	V. heliconioides	Vriesea		*cinerea*	GM
	V. platynema	Vriesea		*cinerea*	GM
	V. schwackeana	Vriesea		*cinerea*	GM
Xanthium	*X. sibiricum*	Cockleburs	*Asteraceae*	*cinerea*	GM
	X. pennsylvanicum	Common cocklebur		*cinerea*	GM
Xanthoceras	*X. sorbifolium*	Yellowhorn, shiny leaf yellowhorn, goldenhorn,	*Sapinda*	*cinerea*	Bl
Xanthosoma	*X. sagittifolium*	Arrowleaf elephant ear, arrowleaf elephant's ear	*Araceae*	*cinerea*	Cormel r
Xerochrysum	*X. bracteatum= Helichrysum bracteatum*	Golden everlasting, strawflower	*Asteraceae*	*cinerea*	GM
Yucca	*Y. elephantipes*	Yucca	*Agavaceae*	*cinerea*	GM
Zamioculcas	*Z. zamiifolia*	Zanzibar gem	*Araceae*	*cinerea*	GM
Zantedeschia	*Z. aethiopica*	Calla lily, lily of the Nile	*Araceae*	*cinerea*	GM
Zea	*Z. mays*	Corn, maize	*Poaceae*	*cinerea*	Ear r
Zinnia	*Z.* spp.	Zinnia	*Asteraceae*	*cinerea*	St canker
	Z. elegans = Z. violacea	Elegant zinnia		*cinerea*	-"-
	Z. peruviana	Peruvian zinnia		*cinerea*	-"-
Ziziphus	*Z. mauritiana*	Yunnan jujube	*Rhamna*	*cinerea*	GM
	Z. jujube	Lingwu Changzao, jujube		*cinerea*	GM

Note: The authors will appreciate remarks, corrections, additions and updates for the host plants, *Botrytis* spp., and symptoms in the table
Abreviations: *Bl* Blight, *Blo* Blossom, *CS* Chocolate spot, *Fr* Fruit, *GM* Grey mould, *lf* Leaf, *LS* Leaf spot, *Pe* Petal, *Pl* Plant, *R*; *r* Rot, *SR* Storage rot, *St* Stem, *tr* Tree, *fl* Flower, – not known to authors, 'As above' symptoms are similar to the symptoms caused by the same pathogen in a previousely mentioned host; -"- refers to the several species mentioned in the cell above

[a]Only representative names are given

[b]Some family names were abbreviated; the "ceae" is removed so for inctance 'Euphorbiaceae' is abreviated to read 'Euphorbia.'

[c](*S*) Spanish

References

American Type Culture Collection (2015) Fungi and yeast. http://www.atcc.org/en/Products/Cells_and_Microorganisms/Fungi_and_Yeast/Fungi_and_Yeast_Alphanumeric.aspx
Anderson JP (1924) *Botrytis cinerea* in Alaska. Phytopathology 14:152–155
Anonimous (n.d.) Plant pest information network. New Zealand Ministry for Primary Industries Website http://archive.mpi.govt.nz/applications/ppin/search
Baker KF (1946) Observations on some *Botrytis* diseases in California. Plant Dis Report 30:145–155
Beever RE, Weeds PL (2004) Taxonomy and genetic variation of and *Botryotinia*. In: Elad Y, Williamson B, Tudzynski P, Delen N (eds) *Botrytis*: biology, pathology and control. Kluwer Academic Publishers, Dordrecht, pp 30–31
Chase AR (1997) Foliage plant diseases diagnosis and control. APS Press, St. Paul, 169 pp
Coley-Smith JR, Verhoeff K, Jarvis WR (1980) The biology of *Botrytis*. Academic, London
Conners IL (1967) An annotated index of plant diseases in Canada and fungi recorded on plants in Alaska, Canada and Greenland. Can Dep Agric Res Branch Publ 1251:1–381
Dingley JM (1969) Records of plant diseases in New Zealand. NZ Dept Sci Ind Res Bull 192: 1–298
Elad Y, Williamson B, Tudzynski P, Delen N (eds) (2004) *Botrytis*: biology, pathology and control. Kluwer Academic Publishers, Dordrecht
Elad Y et al (2014) Lists of plant pathogens and plant diseases in Israel. http://www.phytopathology.org.il/pws/page!10949
Farr DF, Bills GF, Chamuris GP, Rossman AY (1989) Fungi on plants and plant products in the United State. APS Press, St Paul, 1252 pp
Grigaliunaite B (2001) Fungi of the genus *Botrytis* P. Micheli: Pers. plants in Lithuania. Biologija 3:14–17
Jarvis WR (1977) *Botrytinia* and *Botrytis* species: taxonomy, physiology, and pathogenicity, a guide to the literature. Monograph No. 15, Canada Department of Agriculture, Ottawa, Canada
MacFarlane HH (1968) Plant host – pathogen index to volumes 1–40 (1922–1961), Rev of Applied Mycology. Commonwealth Mycological Institute, Kew. Viii
Moor WC (1959) British parasitic fungi. Cambridge University Press, Cambridge, 435 pp
Palmucci HE, Rivera MC, Wright ER (1997) *Botrytis cinerea* Pers. en plantas ornamentales cultivadas en Argentina. Bol San Veg Plagas 23:295–299
Ponce de León I, Pablo Oliver J, Castro A et al (2007) *Erwinia carotovora* elicitors and *Botrytis cinerea* activate defense responses in *Physcomitrella patens*. BMC Plant Biol 7:52
Smith IM, Dunez J, Phillips DH, et al (eds) (2009) European handbook of plant diseases. Wiley Online Library, UK. Blackwell, Oxford, p 583. doi:10.1002/9781444314199.fmatter
Verhoeff K, Malathrakis NE, Williamson B (1992) Recent advances in *Botrytis* research. Pudoc Scientific Publishers, Wageningen

Web Sites

BFW – Austrian federal forest research- http://bfw.ac.at/ws/sdis.schadenstyp_e?sdis_id_in=1
Chuan Hong Control of ornamental diseases, home ornamentals- pubs.ext.vt.edu/456/456-018/Section_4_Home_Ornamentals.pdf
Das grünen Lexikon Hortipendium- http://www.hortipendium.de/Botrytis
Index of Plant Diseases in South Carolina- http://media.clemson.edu/public/masternaturalist/plant_disease_index.pdf
STIHL – Lexicon of tree/shrub diseases, Affected tree/shrub types- http://fr.stihl.ca/detail-view-of-wood-damage.aspx?idDisease=4
University of Illinois Extension web site- http://urbanext.illinois.edu/focus/index.cfm?problem=gray-mold-ibotrytisi